Edited by
Robert T. Mathers and
Michael A. R. Meier

Green Polymerization Methods

Related Titles

Handbook of Green Chemistry

12 volumes
ISBN: 978-3-527-31404-1

Loos, K. (ed.)

Biocatalysis in Polymer Chemistry

2011
ISBN: 978-3-527-32618-1

Lendlein, Andreas/Schroeter, Michael (eds.)

Handbook of Biodegradable Polymers
Isolation, Synthesis, Characterization and Applications

2011
ISBN: 978-3-527-32441-5

Dubois, P., Coulembier, O., Raquez, J.-M. (eds.)

Handbook of Ring-Opening Polymerization

2009
ISBN: 978-3-527-31953-4

Janssen, Leon/Moscicki, Leszek (eds.)

Thermoplastic Starch
A Green Material for Various Industries

2009
ISBN: 978-3-527-32528-3

Fessner, W.-D., Anthonsen, T. (eds.)

Modern Biocatalysis
Stereoselective and Environmentally Friendly Reactions

2009
ISBN: 978-3-527-32071-4

Lapkin, A., Constable, D. (eds.)

Green Chemistry Metrics

Hardcover
ISBN: 978-1-4051-5968-5

Perosa, A., Zecchini, F.

Methods and Reagents for Green Chemistry
An Introduction

2007
ISBN: 978-0-471-75400-8

Sheldon, R. A., Arends, I., Hanefeld, U.

Green Chemistry and Catalysis

2007
ISBN: 978-3-527-30715-9

Matyjaszewski, K., Gnanou, Y., Leibler, L. (eds.)

Macromolecular Engineering
Precise Synthesis, Materials Properties, Applications

4 volumes
2007
Hardcover
ISBN: 978-3-527-31446-1

Cornils, B., Herrmann, W. A., Muhler, M., Wong, C.-H. (eds.)

Catalysis from A to Z
A Concise Encyclopedia

3 volumes
2007
ISBN: 978-3-527-31438-6

Edited by Robert T. Mathers and Michael A. R. Meier

Green Polymerization Methods

Renewable Starting Materials, Catalysis and Waste Reduction

WILEY-VCH Verlag GmbH & Co. KGaA

The Editors

Prof. Dr. Robert T. Mathers
Pennsylvania State University
Department of Chemistry
3550 Seventh Street Rd.
New Kensington, PA 15068
USA

Prof. Dr. Michael A. R. Meier
Karlsruhe Institute of Technology (KIT)
Institute of Organic Chemistry
Fritz-Haber-Weg 6, Building 30 . 42
76131 Karlsruhe
Germany

All books published by **Wiley-VCH** are carefully produced. Nevertheless, authors, editors, and publisher do not warrant the information contained in these books, including this book, to be free of errors. Readers are advised to keep in mind that statements, data, illustrations, procedural details or other items may inadvertently be inaccurate.

Library of Congress Card No.: applied for

British Library Cataloguing-in-Publication Data
A catalogue record for this book is available from the British Library.

Bibliographic information published by the Deutsche Nationalbibliothek
The Deutsche Nationalbibliothek lists this publication in the Deutsche Nationalbibliografie; detailed bibliographic data are available on the Internet at <http://dnb.d-nb.de>.

© 2011 WILEY-VCH Verlag & Co. KGaA, Boschstr. 12, 69469 Weinheim, Germany

All rights reserved (including those of translation into other languages). No part of this book may be reproduced in any form – by photoprinting, microfilm, or any other means – nor transmitted or translated into a machine language without written permission from the publishers. Registered names, trademarks, etc. used in this book, even when not specifically marked as such, are not to be considered unprotected by law.

Composition   Laserwords Private Ltd., Chennai, India
Printing and Binding   betz-druck GmbH, Darmstadt
Cover Design   Schulz Grafik-Design, Fußgönheim

Printed in the Federal Republic of Germany
Printed on acid-free paper

ISBN: 978-3-527-32625-9

Contents

List of Contributors XIII

Part I Introduction 1

1 Why are Green Polymerization Methods Relevant to Society, Industry, and Academics? 3
Robert T. Mathers and Michael A. R. Meier
1.1 Status and Outlook for Environmentally Benign Processes 3
1.2 Importance of Catalysis 4
1.3 Brief Summaries of Contributions 5
References 6

Part II Integration of Renewable Starting Materials 9

2 Plant Oils as Renewable Feedstock for Polymer Science 11
Michael A. R. Meier
2.1 Introduction 11
2.2 Cross-Linked Materials 12
2.3 Non-Cross-Linked Polymers 15
2.3.1 Monomer Synthesis 15
2.3.2 Polymer Synthesis 18
2.4 Conclusion 24
References 25

3 Furans as Offsprings of Sugars and Polysaccharides and Progenitors of an Emblematic Family of Polymer Siblings 29
Alessandro Gandini
3.1 Introduction 29
3.2 First Generation Furans and their Conversion into Monomers 30
3.2.1 Furfural and Derivatives 30
3.2.2 Monomers from Furfural 31
3.2.3 Hydroxymethylfurfural 35

Green Polymerization Methods: Renewable Starting Materials, Catalysis and Waste Reduction
Edited by Robert T. Mathers and Michael A. R. Meier
Copyright © 2011 WILEY-VCH Verlag GmbH & Co. KGaA, Weinheim
ISBN: 978-3-527-32625-9

3.3	Polymers from Furfuryl Alcohol	35
3.4	Conjugated Polymers and Oligomers	39
3.5	Polyesters	41
3.6	Polyamides	42
3.7	Polyurethanes	43
3.8	Furyl Oxirane	45
3.9	Application of the Diels–Alder Reaction to Furan Polymers	45
3.9.1	Linear Polymerizations	46
3.9.2	Non-Linear Polymerizations	49
3.9.3	Reversible Polymer Cross-Linking	52
3.9.4	Miscellaneous Systems	52
3.10	Conclusions	53
	References	53

4   Selective Conversion of Glycerol into Functional Monomers via Catalytic Processes 57

François Jérôme and Joël Barrault

4.1	Introduction	57
4.2	Conversion of Glycerol into Glycerol Carbonate	58
4.3	Conversion of Glycerol into Acrolein/Acrylic Acid	62
4.4	Conversion of Glycerol into Glycidol	63
4.5	Oxidation of Glycerol to Functional Carboxylic Acid	65
4.5.1	Catalytic Oxidation of Glycerol to Glyceric Acid	65
4.5.2	Oxidative-Assisted Polymerization of Glycerol	68
4.5.2.1	Cationic Polymerization	68
4.5.2.2	Anionic Polymerization	69
4.6	Conversion of Glycerol into Acrylonitrile	71
4.7	Selective Conversion of Glycerol into Propylene Glycol	72
4.7.1	Conversion of Glycerol into Propylene Glycol	72
4.7.1.1	Reaction in the Liquid Phase	73
4.7.1.2	Reaction in the Gas Phase	75
4.7.2	Conversion of Glycerol into 1,3-Propanediol	76
4.8	Selective Coupling of Glycerol with Functional Monomers	78
4.9	Conclusion	84
	References	84

Part III   Sustainable Reaction Conditions 89

5   Monoterpenes as Polymerization Solvents and Monomers in Polymer Chemistry 91

Robert T. Mathers and Stewart P. Lewis

5.1	Introduction	91
5.2	Monoterpenes as Monomers	92
5.2.1	Terpenic Resins Overview	92
5.2.2	Concepts of Cationic Olefin Polymerization	93

5.2.3	Cationic Polymerization of β-Pinene	98
5.2.4	Cationic Polymerization of Dipentene	104
5.2.5	Cationic Polymerization of α-Pinene	106
5.2.6	Characteristics of Terpenic Resins	112
5.2.7	Applications of Terpenic Resins	113
5.2.8	Commercial Production and Markets of Terpenic Resins	113
5.2.9	Environmental Aspects of Terpenic Resin Production	115
5.3	Monoterpenes as Solvents and Chain Transfer Agents	116
5.3.1	Possibilities for Replacing Petroleum Solvents	116
5.3.2	Ring-Opening Polymerizations in Monoterpenes	117
5.3.3	Metallocene Polymerizations in Monoterpenes	121
5.4	Conclusion	124
	Acknowledgments	124
	References	125

6	**Controlled and Living Polymerization in Water: Modern Methods and Application to Bio-Synthetic Hybrid Materials**	**129**
	Debasis Samanta, Katrina Kratz, and Todd Emrick	
6.1	Introduction	129
6.2	Ring-Opening Metathesis Polymerization (ROMP)	130
6.2.1	Water Soluble ROMP Catalysts	133
6.3	Living Free Radical Methods for Bio-Synthetic Hybrid Materials	136
	Acknowledgments	141
	References	141

7	**Towards Sustainable Solution Polymerization: Biodiesel as a Polymerization Solvent**	**143**
	Marc A. Dubé and Somaieh Salehpour	
7.1	Introduction	143
7.2	Solution Polymerization and Green Solvents	144
7.3	Biodiesel as a Polymerization Solvent	144
7.4	Experimental Section	146
7.4.1	Materials	146
7.4.2	Polymerization	147
7.4.3	Characterization	148
7.5	Effect of FAME Solvent on Polymerization Kinetics	148
7.5.1	Chain Transfer to Solvent Constant	149
7.5.2	Rate Constant	151
7.6	Effect of Biodiesel Feedstock	155
7.6.1	Polymerization Kinetics	157
7.6.2	Polymer Composition	158
7.7	Conclusion	160
	References	160

Part IV Catalytic Processes *163*

8 Ring-Opening Polymerization of Renewable Six-Membered Cyclic Carbonates. Monomer Synthesis and Catalysis *165*
Donald J. Darensbourg, Adriana I. Moncada, and Stephanie J. Wilson
8.1 Introduction *165*
8.2 Preparation of 1,3-Propanediol from Renewable Resources *166*
8.3 Preparation of Dimethylcarbonate from Renewable Resources *169*
8.4 Synthesis of Trimethylene Carbonate *171*
8.5 Six-Membered Cyclic Carbonates: Thermodynamic Properties of Ring-Opening Polymerization *171*
8.6 Catalytic Processes Using Green Catalysts Methods *172*
8.6.1 Cationic Ring-Opening Polymerization *173*
8.6.2 Anionic Ring-Opening Polymerization *176*
8.6.3 Enzymatic Ring-Opening Polymerization *178*
8.6.4 Coordination–Insertion Ring-Opening Polymerization *181*
8.6.4.1 Groups 13- and 14 Based Catalysts *182*
8.6.4.2 Groups 4–12 Based Catalysts *186*
8.6.4.3 Lanthanide-Based Catalysts *190*
8.6.4.4 Groups 1 and 2 Based Catalysts *191*
8.6.5 Organocatalytic Ring-Opening Polymerization *193*
8.7 Thermoplastic Elastomers and their Biodegradation Processes *194*
8.8 Concluding Remarks *197*
Acknowledgments *197*
References *197*

9 Poly(lactide)s as Robust Renewable Materials *201*
Jan M. Becker and Andrew P. Dove
9.1 Introduction *201*
9.1.1 The Lactide Cycle *202*
9.2 Ring-Opening Polymerization of Lactide *204*
9.2.1 Coordination–Insertion Polymerization *205*
9.2.2 Organocatalytic Ring-Opening Polymerization *208*
9.3 Poly(lactide) Properties *210*
9.3.1 PLA Properties and Processing Effects *211*
9.3.2 Polymer Blends *213*
9.3.2.1 Poly(Lactide)/Poly(ε-Caprolactone) Blends *213*
9.3.2.2 Other Biodegradable/Renewable Polyesters *214*
9.4 Thermoplastic Elastomers *214*
9.5 Future Developments/Outlook *216*
References *216*

10	**Synthesis of Saccharide-Derived Functional Polymers** 221
	Julian Thimm and Joachim Thiem
10.1	Introduction 221
10.2	Polyethers 223
10.3	Polyamides 226
10.4	Polyurethanes and Polyureas 229
10.5	Glycosilicones 230
	References 234

11	**Degradable and Biodegradable Polymers by Controlled/Living Radical Polymerization: From Synthesis to Application** 235
	Nicolay V. Tsarevsky
11.1	Introduction 235
11.2	(Bio)degradable Polymers by CRP 238
11.2.1	Linear (Bio)degradable Polymers 239
11.2.1.1	Polymers with a Degradable Functional Group 239
11.2.1.2	Polymers with a Degradable Polymeric Segment 242
11.2.1.3	Polymers with Multiple Cleavable Groups or Polymeric Segments 243
11.2.2	Degradable Star Polymers 244
11.2.3	Degradable Graft Polymers (Polymer Brushes) 245
11.2.4	Hyperbranched Degradable Polymers 250
11.2.5	Cross-Linked Degradable Polymers 252
11.3	Conclusions 254
	Abbreviations 254
	References 255

Part V   Biomimetic Methods and Biocatalysis 263

12	**High-Performance Polymers from Phenolic Biomonomers** 265
	Tatsuo Kaneko
12.1	Introduction 265
12.2	Coumarates as Phytomonomers 266
12.3	LC Properties of Homopolymers 267
12.3.1	Syntheses and Structures 267
12.3.2	Solubility 268
12.3.3	Thermotropic Property 269
12.3.4	Ordered Structures 270
12.3.5	Cell Compatibility 273
12.4	LC Copolymers for Biomaterials 274
12.4.1	Lithocholic Acid as Co-monomer 274
12.4.2	Cholic Acid as Co-monomer 276
12.5	LC Copolymers for Photofunctional Polymers 279
12.5.1	Syntheses of P(4HCA-co-DHCA)s 279
12.5.2	Phototunable Hydrolyzes 279

12.5.3	Photoreaction of Nanoparticles	282
12.6	LC Copolymers for High Heat-Resistant Polymers	282
12.6.1	P(4HCA-co-DHCA) Bioplastics	282
12.6.2	Biohybrids	286
12.7	Conclusion	288
	Acknowledgments	289
	References	289

13 Enzymatic Polymer Synthesis in Green Chemistry 291
Andreas Heise and Inge van der Meulen

13.1	Introduction	291
13.2	Polymers	292
13.2.1	Polycondensates	292
13.2.1.1	Polyesters by Ring-Opening Polymerization	293
13.2.1.2	Polyesters by Condensation Polymerization	296
13.2.2	Polyphenols	298
13.2.3	Vinyl Polymers	300
13.2.4	Polyanilines	301
13.3	Green Media for Enzymatic Polymerization	303
13.3.1	Ionic Liquids	303
13.3.2	Supercritical Carbon Dioxide	304
13.4	Conclusions and Outlook	306
	References	307

14 Green Cationic Polymerizations and Polymer Functionalization for Biotechnology 313
Judit E. Puskas, Chengching K. Chiang, and Mustafa Y. Sen

14.1	Introduction	313
14.2	Enzyme Catalysis	313
14.2.1	Lipases	315
14.2.2	*Candida antarctica* Lipase B	321
14.2.3	CALB-Catalyzed Transesterification Reactions	323
14.3	"Green" Cationic Polymerizations and Polymer Functionalization Using Lipases	325
14.3.1	Ring-Opening Polymerization	325
14.3.2	Enzyme-Catalyzed Polymer Functionalization	328
14.4	Natural Rubber Biosynthesis – the Ultimate Green Cationic Polymerization	330
14.4.1	Anatomy of the NR Latex, and Structure of Natural Rubber	331
14.4.1.1	Structure of Natural Rubber	332
14.4.2	Biochemical Pathway of NR Biosynthesis	333
14.4.2.1	Monomer	333
14.4.2.2	Initiators	334
14.4.2.3	Catalyst: Rubber Transferase	335
14.4.3	Chemical Mechanism of Natural Rubber Biosynthesis	337

14.4.4	*In vitro* NR Biosynthesis *339*	
14.5	Green Synthetic Cationic Polymerization and Copolymerization of Isoprene *341*	
	Acknowledgments *342*	
	References *342*	

Index *349*

List of Contributors

Joël Barrault
Université de Poitiers
CNRS
Laboratoire de Catalyse en
Chimie Organique
86022 Poitiers
France

Jan M. Becker
University of Warwick
Department of Chemistry
Coventry CV4 7AL
UK

Chengching K. Chiang
The University of Akron
Department of Polymer Science
170 University Avenue
Akron, OH 44325
USA

Donald J. Darensbourg
Texas A&M University
Department of Chemistry
College Station, TX 77843
USA

Andrew P. Dove
University of Warwick
Department of Chemistry
Coventry CV4 7AL
UK

Marc A. Dubé
University of Ottawa
Department of Chemical and
Biological Engineering
161 Louis Pasteur Pvt.
Ottawa, Ontario K1N 6N5
Canada

Todd Emrick
University of Massachusetts
Conte Center for Polymer
Research
Polymer Science & Engineering
Department
120 Governors Drive
Amherst, MA 01003
USA

Alessandro Gandini
University of Aveiro
CICECO and Chemistry
Department
3810-193 Aveiro
Portugal

List of Contributors

Andreas Heise
Dublin City University
School of Chemical Sciences
Glasnevin
Dublin 9
Ireland

and

Eindhoven University of
Technology
Laboratory of Polymer Chemistry
Den Dolech 2, P.O. Box 513
5600 MB Eindhoven
The Netherlands

François Jérôme
Université de Poitiers
CNRS
Laboratoire de Catalyse en
Chimie Organique
86022 Poitiers
France

Tatsuo Kaneko
Japan Advanced Institute of
Science and Technology
School of Materials Science
1-1 Asahidai
Nomi 923-1292
Japan

Katrina Kratz
University of Massachusetts
Conte Center for Polymer
Research
Polymer Science & Engineering
Department
120 Governors Drive
Amherst, MA 01003
USA

Stewart P. Lewis
Innovative Science, Inc.
3154 State Street
STE 2300
Blacksburg, VA 24060
USA

Robert T. Mathers
Pennsylvania State University
Department of Chemistry
3550 Seventh Street Road
New Kensington, PA 15068
USA

Michael A. R. Meier
Karlsruhe Institute of
Technology (KIT)
Institute of Organic Chemistry
Fritz-Haber-Weg 6
Building 30 . 42
76131 Karlsruhe
Germany

Inge van der Meulen
Eindhoven University of
Technology
Laboratory of Polymer Chemistry
Den Dolech 2, P.O. Box 513
5600 MB Eindhoven
The Netherlands

Adriana I. Moncada
Texas A&M University
Department of Chemistry
College Station, TX 77843
USA

Judit E. Puskas
The University of Akron
Department of Polymer Science
170 University Avenue
Akron, OH 44325
USA

Somaieh Salehpour
University of Ottawa
Department of Chemical and
Biological Engineering
161 Louis Pasteur Pvt.
Ottawa, Ontario K1N 6N5
Canada

Debasis Samanta
University of Massachusetts
Conte Center for Polymer
Research
Polymer Science & Engineering
Department
120 Governors Drive
Amherst, MA 01003
USA

Mustafa Y. Sen
The University of Akron
Department of Polymer Science
170 University Avenue
Akron, OH 44325
USA

Joachim Thiem
University of Hamburg
Faculty of Science
Department of Chemistry
Martin-Luther-King-Platz 6
20146 Hamburg
Germany

Julian Thimm
University of Hamburg
Faculty of Science
Department of Chemistry
Martin-Luther-King-Platz 6
20146 Hamburg
Germany

Nicolay V. Tsarevsky
Southern Methodist University
Department of Chemistry
3215 Daniel Avenue
Dallas, TX 75275
USA

Stephanie J. Wilson
Texas A&M University
Department of Chemistry
College Station, TX 77843
USA

Part I
Introduction

1
Why are Green Polymerization Methods Relevant to Society, Industry, and Academics?

Robert T. Mathers and Michael A. R. Meier

1.1
Status and Outlook for Environmentally Benign Processes

In June 1992, the "Rio Declaration on Environment and Development" (*Rio declaration*) of the United Nations Conference on Environment and Development (UNCED) announced in Principle 1 [1] that human beings are at the center of concerns for sustainable development and that they are entitled to a healthy and productive life in harmony with nature. Since the Rio declaration, the necessity for *sustainable development* has become obvious [2]. Most frequently, sustainable development is defined as development that meets the needs of the present without compromising the ability of future generations to meet their own needs [3]. Much has happened since then and the principles of *green chemistry* [4] are now known and applied by chemists worldwide. Recently, Paul T. Anastas stated in his keynote speech at the 2010 ACS (American Chemical Society) national meeting in San Francisco: "Building a sustainable world is the most taxing intellectual exercise we have ever engaged in. It is also the most important for the future of our world" [5]. Thus, great challenges remain and in the field of green chemistry there are plenty of possibilities in the future for innovation and environmentally friendlier consumer products.

As the use of polymers is becoming increasingly more common for many applications in modern society, *polymer science* is able to make diverse contributions to the rapidly growing field of green chemistry. In particular, polymer science offers manifold possibilities for the sustainable use of renewable raw materials. Even though utilizing renewable resources to meet current needs without creating adverse health or environmental impacts can be challenging, renewable resources offer potentially less toxic products as these resources can be expected to be biodegradable and, more importantly, biocompatible. However, we are fully aware that this is a generalization and a careful case by case evaluation is absolutely necessary! Moreover, nature offers a great synthetic potential to the polymer chemist, and it is up to us to develop new methods to incorporate renewable resources into polymeric materials. This development has to begin now in order to be ready to apply these methods industrially in a few decades, as fossil reserves continue to

---

Green Polymerization Methods: Renewable Starting Materials, Catalysis and Waste Reduction
Edited by Robert T. Mathers and Michael A. R. Meier
Copyright © 2011 WILEY-VCH Verlag GmbH & Co. KGaA, Weinheim
ISBN: 978-3-527-32625-9

deplete and become more expensive. Equally important, we need more sustainable routes toward known polymeric products in order to avoid waste, contamination, high energy consumption, and many other environmental concerns. In the United States, the *National Research Council*, in its report entitled "Sustainability in the Chemical Industry: Grand Challenges and Research Needs," has advocated that all areas of the chemical industry focus on long-term strategies to minimize toxicity and environmental impact while creating sustainable processes [6].

Therefore, we are certain that this edited volume will assist in *training a future generation* of scientists and engineers to consider green chemistry and sustainability within the field of polymer science as the most beneficial long-term strategy. Because these peer-reviewed chapters come from departments of polymer science, chemical engineering, chemistry, and materials science, we anticipate that this volume will build upon previous polymer science [7, 8] and green chemistry [9] books to provide a *state-of-the-art resource for industry and academia*. Moreover, this variety clearly reflects the need for collaboration between these (and other) disciplines to reach our final goal of *sustainability*. Specifically, new catalytic and biomimetic methods, alternative reaction media, and the utilization of renewable resources are described in this edited volume. Additionally, these discussions cover emerging areas in condensation, controlled free radical, anionic, cationic, and metathesis polymerizations. Based on the excellent contributions in this volume, which originate from a number of science and engineering venues, we can only assume that the idea of a *green polymerization method* will continue to be an important part of polymer science for many years.

1.2
Importance of Catalysis

In 1836, Berzelius described his newly coined concept of *"catalysis"* and *"catalytic power"* in an article for *The Edinburgh New Philosophical Journal* entitled "Considerations respecting a New Power which acts in the Formation of Organic Bodies". He described these new idioms as "a power, which is capable of effecting chemical reactions in unorganized substances, as well as organized bodies" [10]. Years later, Karl Zeigler and Giudio Natta received the 1963 Nobel Prize for catalysis research related to polyolefins. More recently, Nobel prizes have been awarded for asymmetric catalysis (2001) and olefin metathesis catalysts (2005).

At the present time, refined ideas regarding catalysis have become very common in science and engineering disciplines, as evidenced by the large number of journal articles devoted to this subject each year. From an industrial standpoint, catalysts have played an integral role in the manufacture of chemical raw materials [11], polyolefins [12], and many other polymeric materials. To gain a perspective on the importance of catalysis in *green polymer chemistry*, it is helpful to mention that during the formulation of the principles of green chemistry [4], catalysis was described as a foundational pillar [13]. Since that time, major advances in organocatalysis and biocatalysis have continued to emerge as complementary methods to traditional metal-based catalysts [14]. As a result, this edited volume contains an emphasis on catalytic processes that includes metal-based catalysts, organocatalysts, and biocatalysts.

Considering the expense of developing or licensing particular catalysts, why are catalysts such an integral part of green chemistry? Certainly, the ability to avoid stoichiometric amounts of reagents or recycle catalysts on heterogeneous supports promotes atom economy and reduces waste [15]. The efficiency is normally measured by *turnover numbers (TONs)*, the number of catalytic cycles or catalyst activity (kg polymer/mol catalyst·h). To accurately understand the amount of waste produced in relation to the amounts of the starting materials, it is helpful to consider the well-known E-factor [16] value for reactions. The *E-factor* concept quantifies the amount of waste produced (in kilograms) during a reaction compared with kilograms of the desired product. The ideal E-factor would be zero, but many reactions do not result in 100% conversion, show side reactions, necessitate protecting groups, or require (toxic) solvents. Several contributions in this volume contain discussions of alternative reaction media. Another compelling reason to incorporate catalytic processes is the ability to utilize *renewable resources* in ways that would not otherwise be possible. For instance, catalytic methods to convert glycerol into acrylic acid or to make cyclic carbonates with CO_2 represent some of the numerous examples of non-petroleum based monomer synthesis in this volume.

1.3
Brief Summaries of Contributions

Creating *sustainable polymers* presents a significant multidisciplinary challenge and this volume represents a diverse effort to utilize a broad range of renewable resources such as lignin, triglycerides, polysaccharides, monoterpenes, furans, lactides, and natural rubber. The methods represented are also diverse. In the first section of the book, some authors report on utilizing renewable resources directly. For example, Michael Meier describes new advances in the use of renewable feedstocks based on plant oils, showing that plant oils are a perfectly suitable renewable resource for the polymer industry. In Chapter 3, Alessandro Gandini summarizes the versatility of furans as monomers in the synthesis of resins, conjugated polymers, and reversible cross-links. Francois Jerome and Joel Barrault examine processes for converting glycerol into functional monomers. Their chapter discusses the integration of biomass with heterogeneous catalysis.

In the second section, the focus involves *sustainable reaction conditions* that reduce waste or eliminate petroleum solvents. In this regard, methods for integrating petroleum-based polymerizations with renewable starting materials are mentioned by Stewart Lewis and Robert Mathers, who provide a current review of monoterpenes in cationic and ring-opening polymerizations. Debasis Samanta, Katrina Kratz, and Todd Emrick discuss controlled and living polymerizations in water as an important method for synthesizing bio-synthetic hybrid materials.

Their chapter highlights possibilities for decreasing volatile organic solvents. Marc Dube and Somaieh Salehpour show that biodiesel is a very useful polymerization solvent for free-radical polymerizations. Fatty acid methyl esters are definitely an alternative to conventional petroleum based solvents.

The third section focuses on *catalytic processes* to synthesize monomers and polymers. Donald Darensbourg, Adriana Moncada, and Stephanie Wilson nicely summarize the ring-opening polymerization of renewable carbonates made from 1,3-propanediol. These six-membered cyclic carbonates, for instance, are suitable monomers for thermoplastic elastomers and biomaterials. Jan Becker and Andrew Dove detail the synthesis and polymerization of lactides using organocatalysts. Julian Thimm and Joachim Thiem assess polysaccharides as a major component of biomass. In Chapter 11, Nicolay Tsarevsky reviews the synthesis of macromolecules with biodegradable linkages using controlled radical polymerization.

In the fourth section, *biomimetic methods* and *biocatalysis* are discussed. Tatsuo Kaneko discusses the use of phenolic biomonomers in developing high-performance liquid crystalline polymers. Andreas Heise and Inge van der Meulen focus on recent advances in enzymatic polymer synthesis related to polyesters, polyphenols, polyanilines, and green media for enzymes. Judit Puskas, Chengching Chiang, and Mustafa Sen summarize green cationic polymerizations using biotechnology. Specifically, they describe the biosynthesis of natural rubber.

We hope that this book will be of value to its readers and promote the concepts of green chemistry and sustainability within polymer science. Last but not least, we would like to express our sincere thanks to all authors for their excellent contributions to this edited volume.

References

1. United Nations Conference on Environment and Development (1992) Report of the United Nations Conference on Environment and Development, Rio de Janeiro. http://www.un.org/esa/sustdev. (accessed on 29 December 2010).
2. Eissen, M., Metzger, J.O., Schmidt, E., and Schneidewind, U. (2002) *Angew. Chem. Int. Ed. Engl.*, **41**, 414–436.
3. Brundtland, G. (1987) *Our Common Future*, Oxford University Press, Oxford.
4. Anastas, P.T. and Warner, J.C. (1998) *Green Chemistry: Theory and Practice*, Oxford University Press, Oxford.
5. Baum, R. (2010) *Chem. Eng. News* **88** (Mar 29), 8.
6. National Research Council Board on Chemical Sciences and Technology (2005) Sustainability in the Chemical Industry: Grand Challenges and Research Needs – A Workshop Report. www.nap.edu. (accessed on 2005)
7. Anastas, P. Bickart, P.H., and Kirchhoff, M.M. (2000) *Designing Safer Polymers*, John Wiley & Sons, Inc., New York.
8. Belgacem, M.N. and Gandini, A. (eds) (2008) *Monomers, Polymers and Composites from Renewable Resources*, Elsevier Ltd., Oxford.
9. Sheldon, R.A., Arends, I., and Hanefeld, U. (2007) *Green Chemistry and Catalysis*, Wiley-VCH Verlag GmbH, Weinheim.
10. Berzelius, M. (1836) *Edinburgh New Philos. J.*, **XXI**, 223–228.
11. Cavani, F., Centi, G., Perathoner, S., and Trifiro, F. (2009) *Sustainable Industrial Chemistry*, Wiley-VCH Verlag GmbH, Weinheim.

12. Hoff, R. and Mathers, R.T. (eds) (2010) *Handbook of Transition Metal Polymerizations Catalysts*, John Wiley & Sons, Inc., Hoboken.
13. Anastas, P.T. and Kirchhoff, M.M. (2002) *Acc. Chem. Res.* **35**, 686–694.
14. List, B. (2007) *Chem. Rev.* **107**, 5413–5415.
15. Trost, B.M. (2002) *Acc. Chem. Res.* **35**, 695–705.
16. Sheldon, R.A. (1992) *Chem. Ind. (London)*, 903–906.

Part II
Integration of Renewable Starting Materials

2
Plant Oils as Renewable Feedstock for Polymer Science
Michael A. R. Meier

2.1
Introduction

Plant oils are mainly composed of triglycerides (triacylglycerols) with varying fatty acid compositions [1]. Commonly, the fatty acids of these triglycerides are unbranched carboxylic acids with 8–24 carbon atoms containing 0–3 Z configured C=C double bonds, depending on the type of *plant oil*. The major *fatty acids* of linseed oil, for instance, are linoleic **2** and linolenic acid **3**, whereas new rapeseed oil is rich in oleic acid **1** and palm kernel oil is rich in lauric acid. However, fatty acids that naturally occur in plant oils can contain conjugated double bonds or functional groups, such as hydroxyls or epoxides. Castor oil is an example of an oil containing such a functionalized fatty acid, namely ricinoleic acid **7**, with a content of up to 90%, thus providing additional natural chemical functionality for modifications, cross-linking, or polymerization. Figure 2.1 shows some of the most interesting plant oil derived fatty acids, which can be used for the synthesis of renewable monomers and polymers. In addition, Figure 2.1 summarizes the composition of some of the industrially important plant oils. It is important to note that the presented values are average values and the actual compositions can vary, as is typical for natural products.

An interesting recent development in terms of fatty acid composition of plant oils is the introduction of high oleic sunflower oil with oleic acid contents of up to 93%. Such plant oils are ideal as renewable resources, as they can be considered as chemicals with technical purity and can be used for chemical transformations without the need for time and energy consuming purifications steps. All fatty acids displayed in Figure 2.1 are directly available from plant oils with the exception of 10-undecenoic acid **8**, which is derived from castor oil via pyrolysis [2]. Nevertheless, **8** is included in Figure 2.1, as it is a renewable platform chemical and offers manifold opportunities for polymers synthesis, as will be discussed within this contribution.

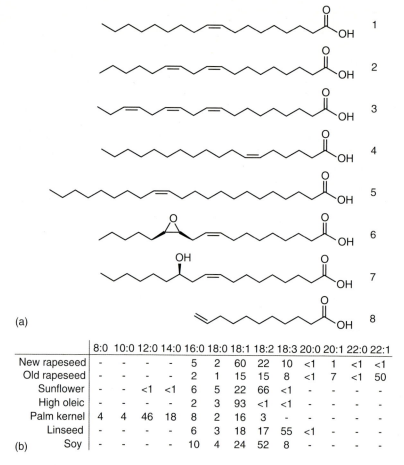

	8:0	10:0	12:0	14:0	16:0	18:0	18:1	18:2	18:3	20:0	20:1	22:0	22:1
New rapeseed	-	-	-	-	5	2	60	22	10	<1	1	<1	<1
Old rapeseed	-	-	-	-	2	1	15	15	8	<1	7	<1	50
Sunflower	-	-	<1	<1	6	5	22	66	<1	-	-	-	-
High oleic	-	-	-	-	2	3	93	<1	<1	-	-	-	-
Palm kernel	4	4	46	18	8	2	16	3	-	-	-	-	-
Linseed	-	-	-	-	6	3	18	17	55	<1	-	-	-
Soy	-	-	-	-	10	4	24	52	8	-	-	-	-

Figure 2.1 (a) Fatty acids as starting materials for the synthesis of monomers and polymers: oleic acid (**1**), linoleic acid (**2**), linolenic acid (**3**), petroselinic acid (**4**), erucic acid (**5**), vernolic acid (**6**), ricinoleic acid (**7**), 10-undecenoic acid (**8**); (b) Typical composition of some important plant oils (in %; numbers a : b = chain length in carbon atoms : number of double bonds).

2.2
Cross-Linked Materials

A large variety of resins can be obtained from plant oils, including, for example, epoxy, polyurethane, or acrylate resins [3, 4]. One of the oldest and best known applications of plant oils in polymer science are probably *alkyd resins*, a term introduced by Kienle [5, 6], for polyester resins with numerous application possibilities as coatings, adhesives, plastics, and textile fibers [6]. Typically, alkyd resins are produced from acid anhydrides (e.g., phthalic anhydride or maleic anhydride), polyols (e.g., glycerol or pentaerythritol), and unsaturated fatty acids to give them drying properties. Generally, oils are classified as *drying oils* if they are highly

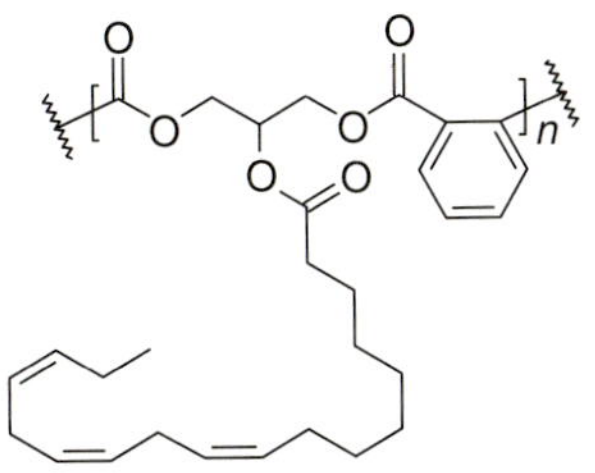

Figure 2.2 Typical structure of a drying alkyd resin before cross-linking.

unsaturated and their iodine values are higher than 170 (e.g., linseed oil). In the case of alkyd resins (see Figure 2.2 for a generalized structure) multiple unsaturated fatty acids are introduced to allow the prepolymers to cross-link in the presence of oxygen and driers (catalysts). The initial reaction step is the abstraction of a bis-allylic hydrogen atom followed by a radical trapping by oxygen, finally leading to the formation of hydroperoxides [7]. The curing then proceeds via multiple reaction steps, which are summarized in the literature [8]. However, Weijnen and coworkers recently summarized some of the possibilities for obtaining alkyd resins from renewable resources [9].

Another very import class of plant oil derived resins are *epoxy resins*. Plant oils can easily be epoxidized through various methods, whereby the epoxidation of fatty acids with peracids is the most commonly applied method in the chemical industry [10]. Moreover, a large variety of metal- and also enzyme-catalyzed processes have been developed [10], leading to epoxidized vegetable oils (or fatty acid derivatives) as shown in Figure 2.3.

Starting from such *epoxidized plant oil derivatives*, rubbers, resins, or coatings can be obtained in a straightforward fashion. For instance, epoxidized oils with high linolenic acid contents were evaluated as environmentally friendly cross-linkers for powder coatings [11], epoxidized castor oil was cationically polymerized with a latent thermal catalyst resulting in materials with relatively high glass-transition temperatures and low coefficients of thermal expansion [12], and flame retardant materials were obtained from epoxidized and phosphorus containing monomers based on **8** and diamino hardeners [13, 14]. In addition, epoxidized linseed oil and 3-glycidylpropylheptaisobutyl-T8-polyhedral oligomeric silsesquioxane (G-POSS) were used as co-monomers to obtain bionanocomposites with nanoscale reinforcement of some of the thermal and mechanical properties [15]. Moreover, other silicon containing materials with application possibilities as flame retardants were

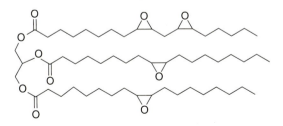

Figure 2.3 Epoxidized plant oil triglyceride.

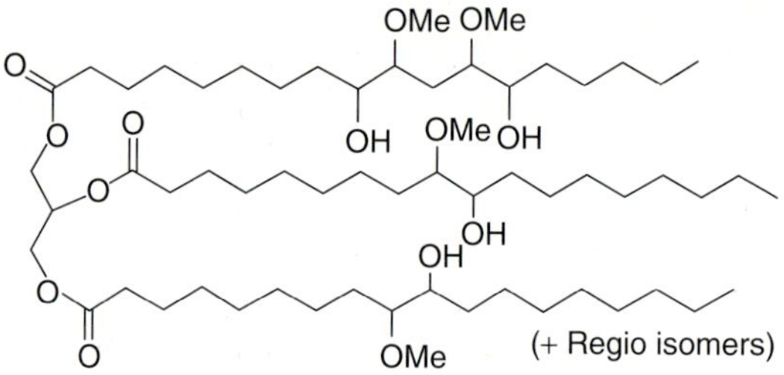

Figure 2.4 Plant oil derived polyol for polyurethane synthesis.

obtained by cationic polymerization of soybean oil, styrene, divinylbenzene, and p-trimethylsilylstyrene [16].

The epoxide functional groups of epoxidized plant oils (as depicted in Figure 2.3) can be ring opened with methanol to yield *polyols for polyurethane synthesis* (consider Figure 2.4) [17]. After polymerization with methylene 4,4'-diphenyldiisocyanate (MDI), these polyols gave polyurethane resins with T_g (glass transition temperature) values varying between 24 and 77 °C and tensile strengths of 15–56 MPa, depending on the oil used and its fatty acid composition. In summary, a higher degree of unsaturation gives polyols with higher hydroxyl numbers, and thus polyurethanes with higher T_g and higher tensile strength [17]. It is also interesting to note that castor oil can be used directly for polyurethane synthesis due to its inherent OH-group functionality [18]. Soybean oil based waterborne polyurethane dispersions were recently prepared successfully from similar polyols [19].

Plant oil derived *acrylic resins* display interesting application possibilities. For instance, soybean oil was first epoxidized and then acrylated [20]. The monomer resulting, if copolymerized with styrene, gave materials exhibiting moduli in the 1–2 GPa range and T_g values ranging from 70 to 120 °C, depending on the resin composition. These renewable resins were also reinforced with glass fibers or natural fibers (flax and hemp) to yield *composite materials* with tensile strengths of 129–463.2 and 20–35 MPa, respectively. Also, hybrid reinforced materials have been prepared that combine the low cost of natural fibers with the high performance of synthetic fibers, resulting in material properties spanning a wide range, as emphasized by these researchers [20]. Interestingly, a new route to acrylated triglycerides via singlet oxygen photooxygenation of high oleic sunflower oil was recently demonstrated [21]. The reduction of the resulting hydroperoxide derivatives led to new hydroxyl-containing triglycerides, which were subsequently functionalized with acrylate moieties to allow radical cross-linking. These authors mentioned that their new and environmentally friendlier route resulted in renewable monomers with similar properties, reactivity, and degree of functionality to the previously reported acrylate triglyceride-based materials and, consequently, are candidates to produce thermosets with similarly good properties [21].

However, plant oils can be bromoacrylated by acrylic acid and N-bromosuccinimide [22]. The subsequent *radical copolymerization* of these functional oils with styrene resulted in rigid thermoset polymers. In addition, epoxidized soybean oil was functionalized with maleate and cinnamate esters, which were then applied as renewable monomers for various resins [23, 24]. Finally, acrylamide functional triglycerides were obtained via Ritter reaction with acrylonitrile and radically copolymerized with styrene to yield semi-rigid polymers with T_g values between 30 and 40 °C [25]. Moreover, resins from plant oils were also prepared by *photoperoxidation* of high oleic sunflower oil and subsequent cross-linking of the resulting α,ω-unsaturated ketone via aza-Michael addition reactions with amines [26].

Recently, Tian and Larock demonstrated several ways to obtain resins from plant oils via *olefin metathesis*. These investigations started with model acyclic diene metathesis (ADMET) polymerizations of ethylene glycol dioleate and glyceryl trioleate and were then transferred to soybean oil to yield polymeric materials ranging from sticky oils to rubbers [27]. Subsequently, norbornene functionalized castor and linseed oils were copolymerized with cyclic monomers by metathesis revealing that the prepared materials were (mostly) stable (5% weight loss) up to temperatures of 300 °C [28–30]. It was also shown that such monomers can be used to prepare glass fiber reinforced renewable resins with significantly improved tensile modulus and toughness [31]. Most recently, this group has also shown that the ring-opening metathesis polymerization (ROMP) of norbornenyl-functionalized fatty alcohols affords materials comparable to petroleum-based plastics, such as HDPE (high-density polyethylene) and poly(norbornene) [32].

2.3 Non-Cross-Linked Polymers

2.3.1 Monomer Synthesis

Industrially the most important synthesis of diacids from plant oils is the *ozonolysis* of oleic acid to yield azelaic (nonanedioic) and pelargonic (nonanoic) acids. As far as can be established, Emery Oleochemicals is at present the only producer to use this route for the production of azelaic acid, which has manifold application possibilities for the synthesis of high-performance copolyesters for use in adhesives, films, and fibers [33]. Recently, the possibilities of oxidative C=C cleavage (and other oxidation reactions of fatty acid derivatives) that can also lead to the desired azelaic acid have been extensively reviewed [34]. In addition to azelaic acid, several companies produce dimer fatty acids from unsaturated fatty acids. Even if these dimer fatty acids are rather undefined and consist of several different isomers, they are valuable starting materials for the synthesis of polyamides, polyesters, and lubricants. For instance, they have been shown to improve the properties of polycarbonates [35]. Therefore, polyester pre-polymers were prepared from dimer fatty acids and 1,6-hexanediol and subsequently converted into polycarbonates with bisphenol A and phosgene. The polymers thus obtained had improved flexibility at low temperatures and high heat deflection temperatures [35]. Moreover, an interesting range of bifunctional monomers is accessible via *biotransformations*. For instance, long chain dicarboxylic acids (cis-9-octadecenedioic and cis-13-docosenedioic acids) were synthesized with high conversion from oleic and erucic acids by fermentation with *Candida tropicalis* ATCC20962 [36]. The subsequent polymerization of these diesters with diols was catalyzed by *Candida antarctica* Lipase B and resulted in unsaturated polyesters with high molecular weights, fairly low melting points (due to the unsaturation, which disrupted crystallization), and good thermal stability.

Within the last few years olefin metathesis of fatty acid derivatives has provided major breakthroughs for the *synthesis of ω-functionalized fatty acids* with possible applications in polyester and polyamide synthesis [37, 38]. Boelhouwer and coworkers reported the first successful metathesis reaction with oleochemicals in 1972 [39]. The *self-metathesis* of methyl oleate (MOA) was performed with WCl_6/Me_4Sn as the catalyst and resulted in the formation of 9-octadecene and dimethyl-9-octadecene-1,18-dioate in an equilibrium reaction [39]. Since then, this reaction has been greatly improved by the development of new catalysts and the lowering of the catalyst amounts, but the main problem of the reaction remains the formation of the products in an equilibrium, which is usual for olefin metathesis reactions [37]. This drawback can be avoided by performing *cross-metathesis* reactions with an excess of one of the reaction partners in order to shift the equilibrium toward the products, as was first demonstrated for the cross-metathesis of fatty acid derivatives with ethylene [37]. More recently, owing to the development of more functional group tolerant metathesis catalysts by Grubbs and others [40–47], it has become possible to introduce functional groups directly into fatty acid derivatives via this approach (Figure 2.5). The first example of a systematic study to use cross-metathesis with functional group containing olefins as a versatile tool for the synthesis of renewable monomers was published in 2007 by Rybak and Meier, who described the synthesis of different chain length α, ω-diesters via cross-metathesis of fatty acid esters with methyl acrylate [48]. Second-generation catalysts, and especially the Hoveyda–Grubbs *second-generation catalyst*, provided full conversions and good selectivity at low catalyst loadings (<0.5%). In addition to the desired diesters, shorter chain monoesters were obtained as second products, which are suitable starting materials for detergent applications. Similarly, the same group, and also Dixneuf and coworkers, described the cross-metathesis of fatty acid methyl esters with allyl chloride and acrylonitrile, respectively [49, 50]. Thus, ω-chloro and ω-nitrile fatty acid esters are also available via this approach (Figure 2.5). Owing to the highly coordinative character of acrylonitrile, this

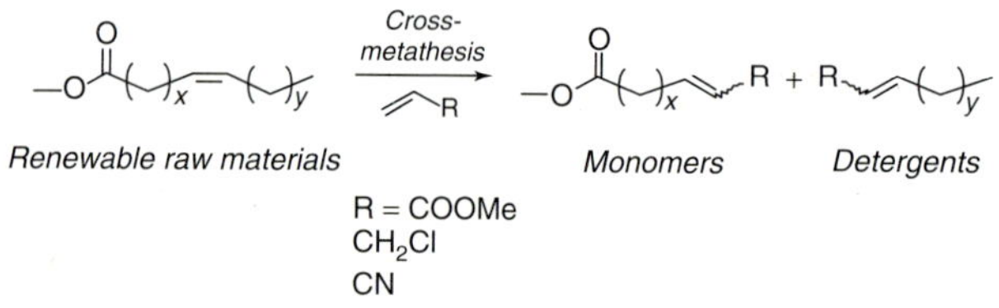

Figure 2.5 Cross-metathesis of fatty acid methyl esters with several functional group containing olefins to yield a variety of monomers for polyesters and polyamides.

2.3 Non-Cross-Linked Polymers

particular cross-metathesis reaction is rather challenging and requires somewhat higher amounts of catalyst. Nevertheless, these examples clearly demonstrate the versatility of this approach and open several new ways toward renewable polyesters and polyamides. Furthermore, Dixneuf and coworkers described the synthesis of bifunctional aldehyde derivatives starting from 10-undecenal (an aldehyde derived from 8) via self- and cross-metathesis reactions (with acrolein, acrylonitrile, acrylic acid, and methylacrylate) [51].

Moreover, Bruneau and coworkers recently introduced 1,3-diene systems to fatty acid methyl esters via a *sequence of a cross-metathesis* with ethene and a subsequent cross-metathesis with an alkyne [52]. This sequence was necessary in order to convert the internal double bond of methyl oleate (methyl ester of 1) into terminal ones prior to the ene–yne cross-metathesis, as the direct cross-metathesis of methyl oleate with alkynes only led to self-metathesis products, even at high catalyst loadings (Figure 2.6). Some of the above mentioned reactions (e.g., cross-metathesis of methyl oleate with methyl acrylate) were also investigated with a second-generation Hoveyda–Grubbs catalyst that was immobilized on the surface of magnetic nanoparticles [53]. Most interestingly, the catalyst could be easily separated by using a magnet and it was possible to reuse it several times with sustained activity.

Furthermore, the *cross-metathesis of fatty alcohols with methyl acrylate* was investigated by Rybak and Meier, revealing that these starting materials were not well tolerated by the catalysts [54], most probably because alcohols can lead to catalyst degradation [55]. Thus, the desired ω-hydroxy fatty acid esters could only be efficiently synthesized via metathesis, if the primary hydroxyl group was protected with

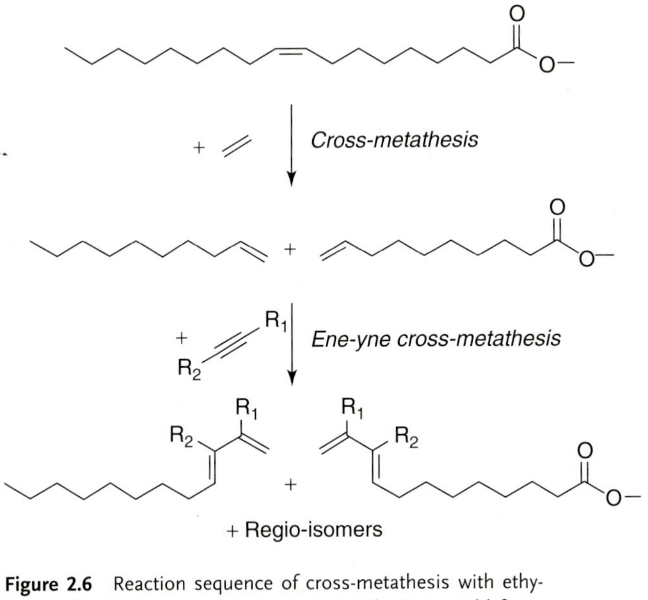

Figure 2.6 Reaction sequence of cross-metathesis with ethylene and subsequent ene–yne cross-metathesis to yield fatty acid derivatives with 1,3-butadiene functional groups.

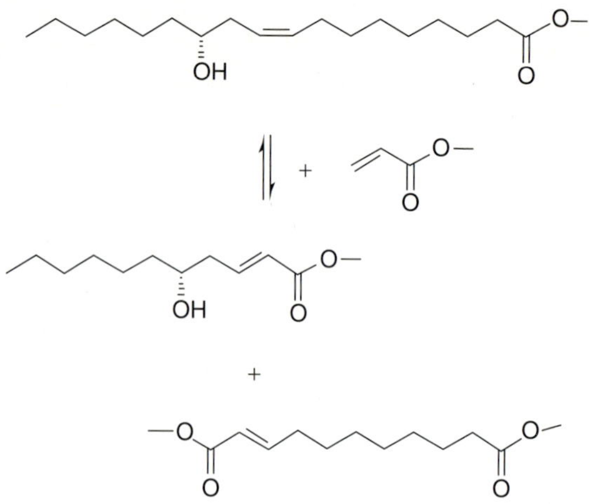

Figure 2.7 Cross-metathesis of methyl ricinoleate with methyl acrylate to yield two monomers for polyesters (one of which is chiral).

an acetate [54]. When cross-metathesis reactions of methyl ricinoleate (methyl ester of **7**) were investigated, this protecting group strategy was not necessary, leading to the conclusion that secondary alcohols are better tolerated by the investigated catalysts (Hoveyda–Grubbs second-generation and Zhan catalysts) [56]. The latter reaction was optimized using a design-of-experiments approach, resulting in quantitative correlations between the investigated parameters (temperature, catalyst loading, excess methyl acrylate, reaction time). The thus obtained optimal reaction conditions were then applied for the synthesis of two renewable monomers in one step (Figure 2.7) [56].

2.3.2
Polymer Synthesis

A large number of polyesters, polyamides, polyacrylates and other polymers based on fatty acid derivatives have been described in the literature. Within the following pages, only a few of the most recent and most promising examples will be discussed. For a more complete overview please consult a recent review article [8]. A very facile and recent example of obtaining renewable polyesters from undec-10-enol (a fatty alcohol derived from **8**) via cobalt-catalyzed step growth polymerization with CO, has been demonstrated by Quinzler and Mecking [57]. This alkoxycarbonylation route led, after optimization, to renewable poly(dodecyloate) with degrees of polymerization exceeding 100 and melting-points of around 65 °C.

Also, for the synthesis of non-cross-linked polymers, olefin metathesis revealed itself as an efficient tool. In this context, ADMET [58] of the castor oil derived undecylenyl undecenoate resulted in unsaturated polyesters with one ester and one C=C double bond every 20 C-atoms (Figure 2.8) [59].

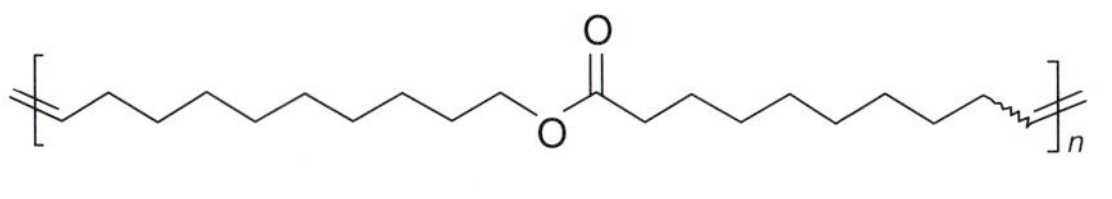

Figure 2.8 Unsaturated polyester derived from castor oil and obtained via ADMET polymerization.

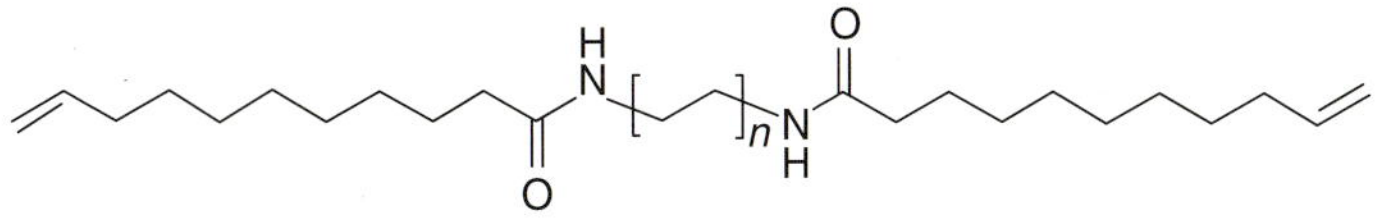

Figure 2.9 Monomers that proved themselves to be unsuitable for obtaining renewable polyamides X,20 via ADMET.

When the ADMET polymerization was performed with chain-stoppers, either telechelics of defined molecular weight or ABA triblock copolymers were obtained, depending on the nature of the chain-stopper utilized [59]. Similar amide containing monomers (Figure 2.9) on the other hand, revealed themselves as almost unpolymerizeable by ADMET, even if the modern functional group tolerant catalysts were used [60].

One of the main reasons for this observation is the high melting-point of these monomers, making it necessary to polymerize in solution because the catalysts studied are not stable above 100 °C. Moreover, these monomers were only soluble in solvents such as N,N'-dimethylformamide (DMF) and dimethylsulfoxide (DMSO), which are not well tolerated by the catalysts. Therefore, 10% catalyst (and more!) was necessary to obtain fairly low molecular weight oligomers [60]. The successful strategy for obtaining the same X,20 polyamides is summarized in Figure 2.10 and relies on the self-metathesis of methyl 10-undecenoate (methyl ester of **8**), followed by a catalytic amidation with the strong organic guanidine base 1,5,7-triazabicyclo[4.4.0]dec-1-ene (TBD). It is interesting to note that this new catalytic amidation polymerization proved to be a straightforward method for the preparation of aliphatic polyamides and resulted in polyamides with molecular weights of up to 15 kDa [60].

These partially renewable (up to 80%) polyamides revealed thermal properties that depended on the structure of the applied monomers. For instance, and as expected, the melting-point of these polyamides increased from 180 to 226 °C with decreasing chain length of the diamine used, due to an increasing number of hydrogen bonding sites. With the exception of polyamide 2,20, all materials showed degradation temperatures of 400 °C and above, making them interesting candidates for several applications [60]. The monomer synthesis as depicted in Figure 2.10 revealed itself to be more difficult than expected, as double bond isomerization side reactions took place concurrently with the self-metathesis, resulting not only in the desired C_{20} monomer but in several α,ω-diesters with different chain lengths [61]. Such side reactions are well known for second-generation ruthenium based metathesis catalysts and are a result of ruthenium hydride formation [62, 63]. On the other hand, Grubbs and coworkers have shown that 1,4-benzoquinone

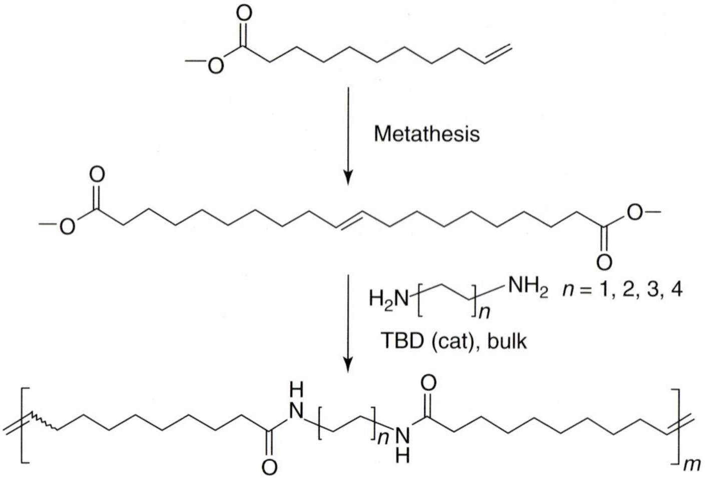

Figure 2.10 Successful strategy for the synthesis of renewable polyamides X,20 via self-metathesis and subsequent catalytic amidation (compare also with Figure 2.9).

can prevent such double-bond isomerizations during the ring-closing metathesis of diallyl ether and other metathesis reactions [64]. If this strategy was applied to the self-metathesis of methyl 10-undecenoate the desired C_{20} monomer was obtained, after extensive optimization, with a conversion of 85% and only 3% isomerization. The optimized reaction conditions were: 50 °C under a continuous vacuum (to remove the ethylene more efficiently) with only 0.05% catalyst and 0.1% benzoquinone to prevent the isomerization [61].

One of the remaining problems was the quantification of such side reactions during ADMET polymerizations, as double-bond isomerization analysis for polymeric materials is much more complicated than for small organic molecules. Thus, a strategy for the quantification was developed based on the preparation of fully renewable polyesters via ADMET and their subsequent transesterification with methanol [65]. The resulting monomer units were then analyzed by gas chromatography–mass spectrometry (GC-MS) and the quantification of the extent of the isomerization side reactions during ADMET polymerizations became possible [65]. These studies revealed, for instance, that the first generation catalyst from Grubbs showed little isomerization (<5%) up to a temperature of 90 °C, whereas it showed as yet unidentified side products at 100 °C and above in high amounts. The second-generation catalyst from Grubbs, on the other hand, showed a temperature dependent isomerization tendency giving up to 76% isomerization, depending on the polymerization conditions [65]. Thus, highly defined polymers could be obtained with Grubbs I, whereas Grubbs II provided rather ill-defined polymeric architectures, as can be clearly visualized from the GC-MS measurements as depicted in Figure 2.11. The influence of the regularity of the repeating units on the polymer properties was also evidenced by a difference in the melting-point of the

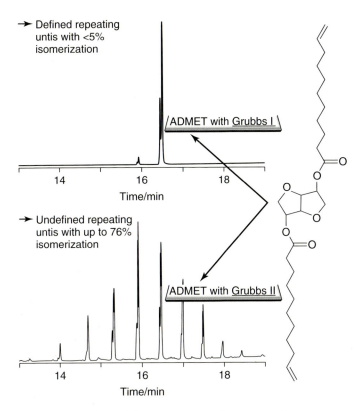

Figure 2.11 Studying olefin isomerization side reactions of ADMET polymers derived from fully renewable monomers by transesterification of the resulting polyesters with methanol and subsequent GC-MS analysis.

defined and undefined polymers of almost 40 °C, clearly showing that a defined polymer architecture is a prerequisite for good material properties [65].

After the method for the quantification of such side reactions had been established, it was applied for the optimization of the ADMET reaction conditions with second-generation metathesis catalysts in order to obtain as little isomerization as possible with these catalysts [66]. As already described above for small organic molecules, in this instance the use of benzoquinone could also largely reduce the amount of olefin isomerization side reactions, especially for the Hoveyda–Grubbs second-generation type catalysts. After optimization it was possible to reduce the extent of the isomerization side reactions from over 90% to less than 10%, thus also making these functional group tolerant catalysts available for precision polymer synthesis [66]. In addition, ADMET polymerizations were used to prepare fatty acid derived flame retardant materials by copolymerization of phosphorus containing monomers [67, 68]. A limiting oxygen index (LOI) of 23.5 was thus obtained for these renewable polyesters with a phosphorus content of only 3.1% [67]. This value was further improved to an LOI of 25.7 by preparing copolymers with hydroxyl functional groups that were functionalized with acrylic acid after polymerization and then radically cross-linked [68].

Interestingly, ADMET can also be used for the polymerization of triglycerides, as shown for soybean oil. This resulted in polymeric materials ranging from sticky oils to rubbers due to cross-linking [27]. By systematically studying the polymerization of a model triglyceride with chain-stoppers, it was however possible to completely avoid cross-linking and to obtain hyperbranched polymers via a one-step procedure [69]. This type of polymerization was termed acyclic triene metathesis (ATMET) with respect to the monomer functionality. Moreover, it was possible to control the molecular weight of these branched polymers by using different amounts of the chain-stopper. As was to be expected, the degree of polymerization was lower, if higher amounts of the chain-stopper were used. If the amount of chain-stopper was too little, cross-linking was observed. This strategy is summarized schematically in Figure 2.12.

The resulting branched polymers were investigated intensively using nuclear magnetic resonance spectroscopy (NMR), gel-permeation chromatography (GPC), and electrospray ionization mass spectrometry (ESI-MS), revealing a molecular architecture, as depicted in Figure 2.12. For instance, the peripheries of these polymers contain functional groups that can be used to further functionalize the system. The investigations also clearly revealed that macrocylces are formed throughout the polymerization [69]. More recently, this strategy could be transferred to the direct polymerization of high oleic sunflower oil [70]. This offers the opportunity to obtain branched polymers directly from a plant oil without any prior chemical modification, thus contributing to further minimizing the environmental impact of such materials. As the condensation product of this polymerization is not ethylene, as it is for the above mentioned model monomer (compare with Figure 2.12), but 9-octadecene, the polymerization of high oleic sunflower oil via ATMET was

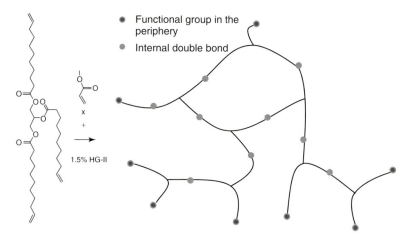

Figure 2.12 Schematic representation of the ATMET polymerization of a model triglyceride and the structure of the resulting branched polymers.

performed under high vacuum, in order to remove 9-octadecene efficiently and to shift the metathesis equilibrium toward polymer formation. During these investigations, oligomers were isolated and identified by ESI-MS/MS, which gave detailed information about the polymer architecture and the polymerization mechanism. All analytic data were thus in accordance with the formation of highly branched high molecular weight polymers using a one-step procedure from high oleic sunflower oil [70].

Apart from metathesis approaches toward fatty acid derived polymers, acrylic fatty acid derivatives are often investigated. For instance, fatty alcohol acrylates or methacrylates are well known in industry and applied as low T_g co-monomers. Fairly recently, such fatty alcohol derived methacrylates were polymerized in a controlled fashion through atom transfer radical polymerization in bulk [71]. It was crucial to use a phase-transfer catalyst for these polymerizations in order to obtain a homogeneous mixture of catalyst and monomer [72]. High monomer conversions, low polydispersities, and predefined molecular weight were thus accessible. All chain lengths of fatty alcohols ranging from C_{10} to C_{18} were investigated, revealing a linear correlation of the observed melting-points of the thus derived polymers with the carbon chain length of the fatty alcohol side chain [71]. In addition, an oxazoline monomer with soybean fatty acid side chains was polymerized via cationic ring-opening polymerization under microwave irradiation [73]. Thus, well-defined polymers with unsaturated side chains, which could be used for subsequent cross-linking, were obtained (Figure 2.13). Subsequently, this monomer was also used to prepare block and also statistical copolymers and core-cross linked micelles could be obtained from such materials [74, 75]. Unfortunately, owing to the limited length of this chapter, previous approaches toward defined fatty acid containing polymers via controlled and living polymerizations cannot be discussed in detail, but these approaches are summarized in the literature [8].

Bromoacrylated methyl oleate (BAMO) (Figure 2.14), however, was homo- and copolymerized with styrene, methyl methacrylate, and vinyl acetate by thermal and by photoinitiated free radical polymerization to obtain polymers with number averaged molecular weights (M_n) of from 20 to 35 kDa [76]. Of this series of copolymers, the methyl methacrylate copolymer showed the highest T_g of −10 °C [76]. Furthermore, acrylated methyl oleate (AMO) (Figure 2.14) resulted in high molecular weight polymers with high conversion in emulsion polymerizations [77]. Long reaction times were required and this was attributed to the monomer structure having a long aliphatic chain. Owing to its amphiphilic nature, AMO was shown to behave as a co-emulsifier [77]. In addition, acryl amide derivatives of methyl undecenoate (MUA) and MOA (Figure 2.14) were prepared via the Ritter reaction of the mentioned fatty acid methyl esters with acrylonitrile in

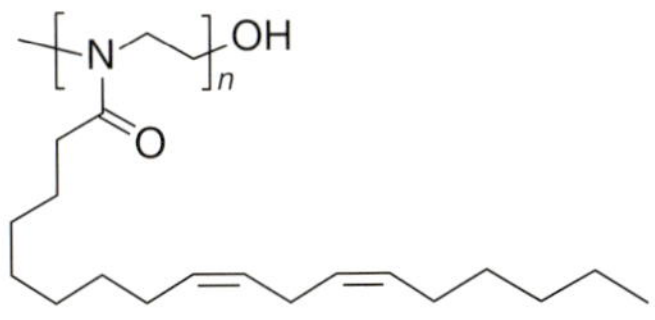

Figure 2.13 Structure of soybean oil derived polyoxazolines.

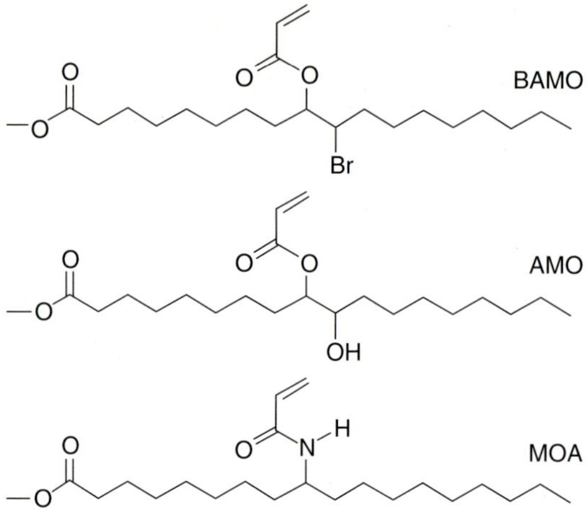

Figure 2.14 Structure of fatty acid derived monomers discussed in the literature.

the presence of SnCl$_4$ [25]. After the behavior of these model monomers had been understood, soybean and sunflower oil triglycerides were used to prepare multifunctional acrylamide monomers, resulting, after copolymerization with styrene, in semi-rigid polymers [25].

The replacement of styrene with methacrylated fatty acid (MFA) monomers in vinyl ester and unsaturated polyester resins has also been discussed [78]. As MFAs are inexpensive, have low volatilities, and free-radically polymerize with vinyl ester, they are ideal candidates for reducing styrene emissions from the above mentioned resins. However, these monomers resulted in a highly increased viscosity and a significantly decreased T_g of the investigated resins. Nevertheless, ternary blends of vinyl ester, styrene (less than 15 wt%, far less than is usually applied), and fatty acid monomers improved the flexural, fracture, and thermo-mechanical properties, and reduced the resin viscosity to acceptable levels [78].

2.4
Conclusion

The examples discussed clearly show why plant oils are already now the most commonly used renewable raw material in the chemical industry. The application possibilities of fatty acids and their derivatives for the synthesis of monomers and polymers are manifold and are almost unlimited, and depend only on the creativity of us scientists. By using fats and oils as renewable feedstock it is possible to efficiently use the synthetic potential of nature, as highlighted by the above described examples, because only a few (efficient and catalytic) synthetic steps for the preparation of monomers and polymers are required. Last but not least, even if the chemical structure of fatty acids seems fairly similar at first sight, a large span of polymer properties can be obtained from this renewable feedstock.

References

1. Gunstone, F.D. (2004) *The Lipid Handbook*, Taylor & Francis, Ltd., Abingdon.
2. Mutlu, H. and Meier, M.A.R. (2010) *Eur. J. Lipid Sci. Technol.*, **112**, 10–30.
3. Galià, M., Montero de Espinosa, L., Ronda, J.C., Lligadas, G., and Cádiz, V. (2010) *Eur. J. Lipid Sci. Technol.*, **112**, 87–96.
4. Günera, F.S., Yagč, Y., and Erciyes, A.T. (2006) *Prog. Polym. Sci.*, **31**, 633–670.
5. Gooch, J.W. (2002) *Lead-Based Paint Handbook*, Kluwer Academic Publishers, New York, Boston, Dordrecht, London, Moscow.
6. Kienle, R.H. (1949) *Ind. Eng. Chem.*, **41**, 726–729.
7. Frankel, E.N. (1980) *Prog. Lipid Res.*, **19**, 1–22.
8. Meier, M.A.R., Metzger, J.O., and Schubert, U.S. (2007) *Chem. Soc. Rev.*, **36**, 1788.
9. van Haveren, J., Oostveen, E.A., Micciche, F., and Weijnen, J.G.J. (2005) *Eur. Coat. J.*, **1–2**, 16–19.
10. Köckritz, A. and Martin, A. (2008) *Eur. J. Lipid Sci. Technol.*, **110**, 812–824.
11. Overeem, A., Buisman, G.J.H., Derksen, J.T.P., Cuperus, F.P., Molhoek, L., Grisnich, W., and Goemans, C. (1999) *Ind. Crop. Prod.*, **10**, 157–165.
12. Park, S.J., Jin, F.L., and Lee, J.R. (2004) *Macromol. Rapid Commun.*, **25**, 724–727.
13. Lligadas, G., Ronda, J.C., Galià, M., and Cádiz, V. (2006) *J. Polym. Sci. Part A: Polym. Chem.*, **44**, 6717–6727.
14. Lligadas, G., Ronda, J.C., Galià, M., and Cádiz, V. (2006) *J. Polym. Sci. Part A: Polym. Chem.*, **44**, 5630–5644.
15. Lligadas, G., Ronda, J.C., Galià, M., and Cádiz, V. (2006) *Biomacromolecules*, **12**, 3521–3526.
16. Sacristán, M., Ronda, J.C., Galià, M., and Cádiz, V. (2009) *Biomacromolecules*, **10**, 2678–2685.
17. Zlatanić, A., Lava, C., Zhang, W., and Petrović, Z.S. (2003) *J. Polym. Sci. Part B: Polym. Phys.*, **42**, 809–819.
18. Petrović, Z.S. and Fajnik, D. (2003) *J. Appl. Polym. Sci.*, **29**, 1031–1040.
19. Lu, Y. and Larock, R.C. (2008) *Biomacromolecules*, **9**, 3332–3340.
20. Khot, S.N., Lascala, J.J., Can, E., Morye, S.S., Williams, G.I., Palmese, G.R., Kusefoglu, S.H., and Wool, R.P. (2001) *J. Polym. Sci. Part A: Polym. Chem.*, **82**, 703–723.
21. Montere de Espinosa, L., Ronda, J.C., Galià, M., and Cádiz, V. (2009) *J. Polym. Sci. Part A: Polym. Chem.*, **47**, 1159–1167.
22. Eren, T. and Kusefoglu, S.H. (2004) *J. Appl. Polym. Sci.*, **91**, 2700–2710.
23. Esen, H., Kusefoglu, S., and Wool, R. (2007) *J. Appl. Polym. Sci.*, **103**, 626–633.
24. Esen, H. and Kusefoglu, S.H. (2003) *J. Appl. Polym. Sci.*, **89**, 3882–3888.
25. Eren, T. and Kusefoglu, S.H. (2005) *J. Appl. Polym. Sci.*, **97**, 2264–2272.
26. Montere de Espinosa, L., Ronda, J.C., Galià, M., and Cádiz, V. (2008) *J. Polym. Sci. Part A: Polym. Chem.*, **46**, 6843–6850.
27. Tian, Q. and Larock, R.C. (2002) *J. Am. Oil Chem. Soc.*, **79**, 479–488.
28. Mauldin, T.C., Haman, K., Sheng, X., Henna, P., Larock, R.C., and Kessler, M.R. (2008) *J. Polym. Sci. Part A: Polym. Chem.*, **46**, 6851–6860.
29. Henna, P.H. and Larock, R.C. (2007) *Macromol. Mater. Eng.*, **292**, 1201–1209.
30. Henna, P. and Larock, R.C. (2009) *J. Appl. Polym. Sci.*, **112**, 1788–1797.
31. Henna, P.H., Kessler, M.R., and Larock, R.C. (2008) *Macromol. Mater. Eng.*, **293**, 979–990.
32. Xia, Y., Lu, Y., and Larock, R.C. (2010) *Polymer*, **51**, 53–61.
33. Emery Oleochemicals http://www.emeryoleo.com/en/ozone-acids/ (accessed 2010).
34. Köckritz, A. and Martin, A. (2008) *Eur. J. Lipid Sci. Technol.*, **110**, 812–824.
35. Dhein, R., Nerger, D., Schreckenberg, M., Morbitzer, L., and Nouvertnt, W. (1986) *Angew. Makromol. Chem.*, **139**, 157–174.
36. Yang, Y., Lu, W., Zhang, X., Xie, W., Cai, M., and Gross, R.A. (2010) *Biomacromolecules*, **11**, 259–268.
37. Rybak, A., Fokou, P.A., and Meier, M.A.R. (2008) *Eur. J. Lipid Sci. Technol.*, **110**, 797–804.

38. Meier, M.A.R. (2009) *Macromol. Chem. Phys.*, **210**, 1073–1079.
39. van Dam, P.B., Mittelmeijer, M.C., and Boelhouwer, C. (1972) *J. Chem. Soc., Chem. Commun.*, 1221–1222.
40. Nguyen, S.T., Johnson, L.K., Grubbs, R.H., and Ziller, J.W. (1992) *J. Am. Chem. Soc.*, **114**, 3974–3975.
41. Schwab, P., Grubbs, R.H., and Ziller, J.W. (1996) *J. Am. Chem. Soc.*, **118**, 100–110.
42. Schwab, P., France, M.B., Ziller, J.W., and Grubbs, R.H. (1995) *Angew. Chem. Int. Ed. Engl.*, **34**, 2039–2041.
43. Scholl, M., Ding, S., Lee, C.W., and Grubbs, R.H. (1999) *Org. Lett.*, **1**, 953–956.
44. Scholl, M., Trnka, T.M., Morgan, J.P., and Grubbs, R.H. (1999) *Tetrahedron Lett.*, **40**, 2247–2250.
45. Garber, S.B., Kingsbury, J.S., Gray, B.L., and Hoveyda, A.H. (2000) *J. Am. Chem. Soc.*, **122**, 8168–8179.
46. Weskamp, T., Schattenmann, W.C., Spiegler, M., and Herrmann, W.A. (1998) *Angew. Chem. Int. Ed.*, **37**, 2490–2492.
47. Huang, J., Stevens, E.D., Nolan, S.P., and Peterson, J.L. (1999) *J. Am. Chem. Soc.*, **121**, 2674–2678.
48. Rybak, A. and Meier, M.A.R. (2007) *Green Chem.*, **9**, 1356–1361.
49. Jacobs, T., Rybak, A., and Meier, M.A.R. (2009) *Appl. Catal., A*, **353**, 32–35.
50. Malacea, R., Fischmeister, C., Bruneau, C., Dubois, J.-L., Couturier, J.-L., and Dixneuf, P.H. (2009) *Green Chem.*, **11**, 152–155.
51. Miao, X., Fischmeister, C., Bruneau, C., and Dixneuf, P.H. (2009) *ChemSusChem*, **2**, 542–545.
52. Le Ravalec, V., Fischmeister, C., and Bruneau, C. (2009) *Adv. Synth. Catal.*, **351**, 1115–1122.
53. Yinghuai, Z., Kuijin, L., Huimin, N., Chuanzhao, L., Stubbs, L.P., Siong, C.F., Muihua, T., and Peng, S.C. (2009) *Adv. Synth. Catal.*, **351**, 2650–2656.
54. Rybak, A. and Meier, M.A.R. (2008) *Green Chem.*, **10**, 1099–1104.
55. Banti, D. and Mol, J.C. (2004) *J. Organomet. Chem.*, **689**, 3113–3116.
56. Ho, T.T.T. and Meier, M.A.R. (2009) *ChemSusChem*, **2**, 749–754.
57. Quinzler, D. and Mecking, S. (2009) *Chem. Commun.*, 5400–5402.
58. Baughman, T.W. and Wagener, K.B. (2005) *Adv. Polym. Sci.*, **176**, 1.
59. Rybak, A. and Meier, M.A.R. (2008) *ChemSusChem*, **1**, 542–547.
60. Mutlu, H. and Meier, M.A.R. (2009) *Macromol. Chem. Phys.*, **210**, 1019–1025.
61. Djigoué, G.B. and Meier, M.A.R. (2009) *Appl. Catal., A*, **368**, 158–162.
62. Schmidt, B. (2004) *Eur. J. Org. Chem.*, 1865–1880.
63. Hong, S.H., Day, M.W., and Grubbs, R.H. (2004) *J. Am. Chem. Soc.*, **126**, 7414–7415.
64. Hong, S.H., Sanders, D.P., Lee, C.W., and Grubbs, R.H. (2005) *J. Am. Chem. Soc.*, **127**, 17160–17161.
65. Fokou, P.A. and Meier, M.A.R. (2009) *J. Am. Chem. Soc.*, **131**, 1664–1665.
66. Fokou, P.A. and Meier, M.A.R. (2010) *Macromol. Rapid Commun.*, **31**, 368–373, doi: 10.1002/marc.200900678.
67. Montero de Espinosa, L., Ronda, J.C., Galià, M., Cádiz, V., and Meier, M.A.R. (2009) *J. Polym. Sci. Part A: Polym. Chem.*, **47**, 5760–5771.
68. Montero de Espinosa, L., Meier, M.A.R., Ronda, J.C., Galià, M., and Cádiz, V. (2010) *J. Polym. Sci. Part A: Polym. Chem.*, **48**, 1649–1660.
69. Fokou, P.A. and Meier, M.A.R. (2008) *Macromol. Rapid Commun.*, **29**, 1620–1625.
70. Biermann, U., Metzger, J.O., and Meier, M.A.R. (2010) *Macromol. Chem. Phys.*, **211**, 854–862.
71. Çayli, G. and Meier, M.A.R. (2008) *Eur. J. Lipid Sci. Technol.*, **110**, 853–859.
72. Chatterjee, D.P. and Mandal, B.M. (2006) *Polymer*, **47**, 1812–1819.
73. Hoogenboom, R. and Schubert, U.S. (2006) *Green Chem.*, **8**, 895–899.
74. Hoogenboom, R., Leenen, M.A.M., Huang, H., Fustin, C.-A., Gohy, J.-F., and Schubert, U.S. (2006) *Colloid Polym. Sci.*, **284**, 1313–1318.
75. Huang, H., Hoogenboom, R., Leenen, M.A.M., Guillet, P., Jonas, A.M., Schubert, U.S., and Gohy, J.-F. (2006) *J. Am. Chem. Soc.*, **128**, 3784–3788.

76. Eren, T. and Kusefoglu, S.H. (2004) *J. Appl. Polym. Sci.*, **94**, 2475–2488.
77. Bunker, S.P. and Wool, R.P. (2002) *J. Polym. Sci., Part A: Polym. Chem.*, **40**, 451–458.
78. La Scala, J.J., Sands, J.M., Orlicki, J.A., Robinette, E.J., and Palmese, G.R. (2004) *Polymer*, **45**, 7729–7737.

3
Furans as Offsprings of Sugars and Polysaccharides and Progenitors of an Emblematic Family of Polymer Siblings

Alessandro Gandini

3.1
Introduction

Within the vast array of monomers and polymers from renewable resources [1, 2], *furan* derivatives and furan chemistry occupy a unique position, because their exploitation for the synthesis of macromolecular materials can be planned strategically in a fashion that simulates closely the approach that has been adopted and progressively developed in petroleum, coal, and natural-gas chemistry for the same purpose.

This is qualitatively different from contexts such as the polymerization of terpenes or the chemical modification of polysaccharides, because in both types of approaches, the scope, however interesting and promising, is intrinsically limited by the structure of the given monomer family (terpenes, sugars, vegetable oils, etc.) or of the natural polymer (cellulose, chitin, starch, etc.). Instead, as will be discussed below, a whole original realm of polymer science can be built from a couple of *first-generation furan compounds*, which are readily prepared from sugars and/or polysaccharides, just as a whole host of monomers are prepared from fossil resources to provide the majority of today's macromolecular materials. That said, it is judicious to tone the comparison down somewhat, considering that the presence of the furan heterocycle, or of one of the structures derived from it, in a given polymer architecture, can in some instances give rise to problems related to lower chemical stability than that of its aliphatic or aromatic counterparts, with the same considerations also applying to some polymerization conditions.

Decades of experience in the laboratory have thus provided a useful guide as to which furan systems are robust and which are prone to difficulties, in terms of both synthetic and materials issues. This experience is, however, far from exhaustive and much more fundamental and technical research is needed to reach a comprehensive culture on furan monomers and polymers.

The purpose of this chapter is to attempt to provide a balanced review of the preparation, properties, and polymerization of a wide spectrum of *furan monomers*, and of the interest and potential applications of the ensuing materials. The most urgent and important gaps that need to be filled in within this wide context are also discussed. This is not the first such monograph [3], but recent developments and the makeup of this book justify a fresh presentation of the subject, considering its highly topical character, with a special emphasis being placed on major contributions in the last few years, and on work in progress.

3.2
First Generation Furans and their Conversion into Monomers

3.2.1
Furfural and Derivatives

Furfural, or 2-formylfuran, (F) was first described about two centuries ago and was first produced industrially about one century ago. Its present annual world output is around 300 kt (at ~$1 per kg), with a geographically widespread distribution of large and small units (and a growing contribution from China). This peculiar feature is readily explained by the fact that in every region of the globe that has some vegetation there is at least one biomass residue bearing sufficient *pentoses* (xylans) to represent a good raw material for the production of F [4]. The list is long, but the most common substrates are sugarcane bagasse, corn cobs, rice hulls, wood chips, and olive husks.

The acid-catalyzed depolymerization of these hemicelluloses yields xylose, which is converted into F by the loss of three water molecules. Other aldehydes such as 5-methyl furfural (MF) from rhamnose are also formed in much smaller proportions and the resulting mixture can be separated by distillation. Figure 3.1 illustrates the latter stage of this mechanism [3].

A substantial proportion of this F is converted by a straightforward industrial process into furfuryl alcohol (FA) through a Cu-catalyzed hydrogenation reaction [4], which is still today the most significant furan compound and monomer, given its important applications, including both the traditional and novel "resins," as discussed below. The other highly relevant F derivatives in the present context are 2-furoic acid and its esters, furfurylamine, and 2-furanacrylic acid, together with their corresponding MF homologs (Figure 3.2), all prepared by standard procedures [3].

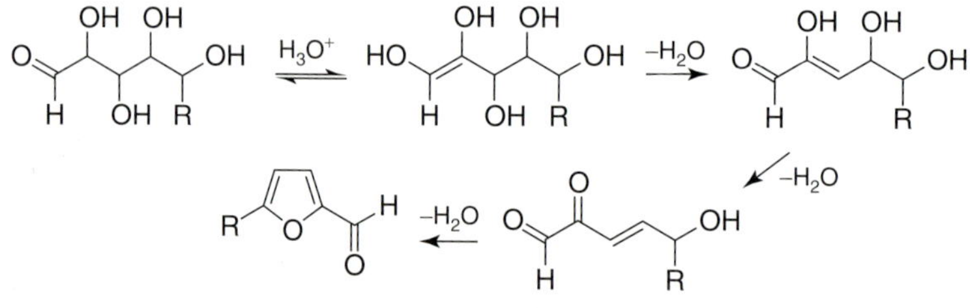

Figure 3.1 The mechanism of formation of furfural (R=H) and 5-methylfurfural (R=CH$_3$) from xylose and rhamnose, respectively.

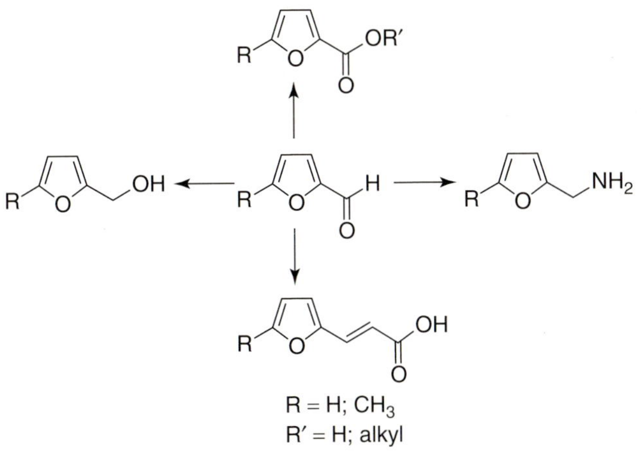

Figure 3.2 The most relevant furfural derivatives in the context of the present chapter.

R = H; CH$_3$
R′ = H; alkyl

3.2.2
Monomers from Furfural

There are two different strategies associated with the *synthesis of furan monomers* from F, namely [3]: (i) the preparation of monofuran structures suitable for chain-growth polymerizations and (ii) the reactions leading to bisfuran molecules suitable for step-growth polymerizations. Only the first general approach can also be applied to monomers derived from MF.

Figure 3.3 illustrates a non-exhaustive series of monomers arising from the appropriate incorporation of polymerizable moieties onto the furan heterocycle starting from F and MF, or one of the derivatives shown in Figure 3.2. It is important to emphasize that all these monomers have already been prepared and characterized [3]. Their *polymerization or copolymerization* leads to macromolecules in which the furan (or tetrahydrofuran) ring appears as a side group appended to them and hence their structure–properties relationships are conditioned by this common architectural feature.

It is also important to note that these monomers owe their originality to the presence of the *heterocycle* in their structure, because the alkenyl, isocyanate, or oxirane functions responsible for their activation are the classical moieties encountered in both aliphatic and aromatic counterparts, that is, monomers such as styrene, acrylates and methacrylates, vinyl ethers, propylene oxide, and so on.

The response of these monomers to the various types of initiations and propagations depends on two factors [3c,e]: (i) the specific aptitude of the polymerizable moiety to be activated by free radical, cationic, and anionic systems and (ii) the possible beneficial or detrimental role of the furan heterocycle in the processes.

The first aspect does not imply any new concept, as it is related to classical reactivity notions, for example, acrylic monomers do not polymerize cationically and vinyl ethers only polymerize through this type of activation and the presence of the ring does not alter them qualitatively.

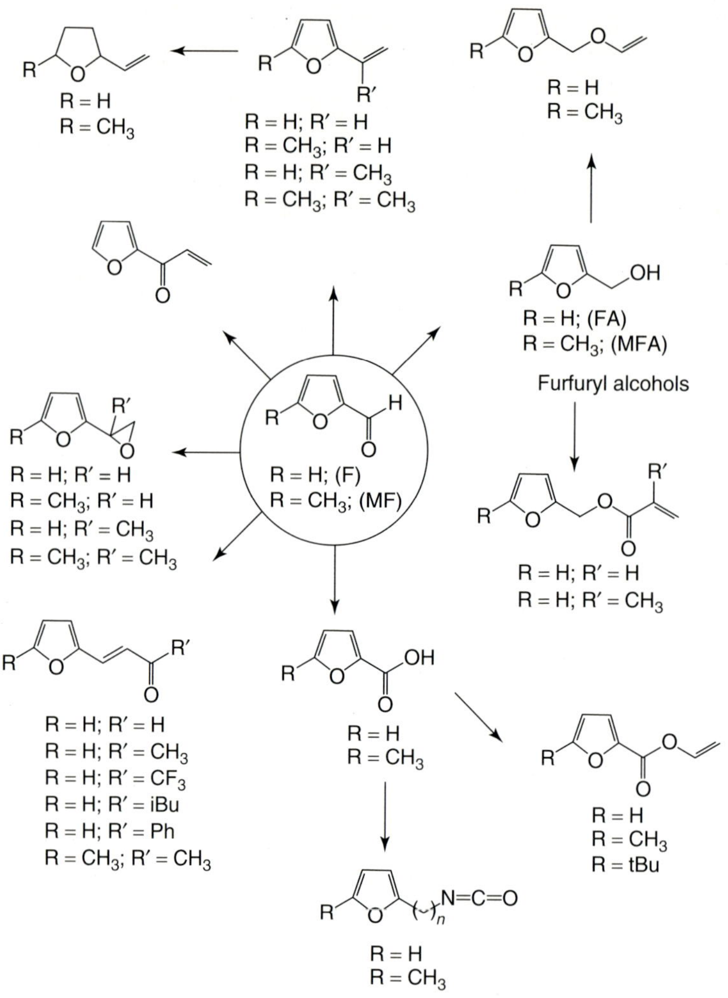

Figure 3.3 Monomers that have been prepared from F or MF.

The second aspect is highly structure-specific in the sense that the intervention of the ring in a given mechanism, which should in principle only involve the appended polymerizable moiety, depends on competitive reactivity factors. In the case of free radicals, for instance, the relative degree of stabilization between the radicals formed with the appended moiety and those arising by addition to the heterocycle will essentially establish two typical extreme situations, that is [3c]:

- if the expected propagating radical is well stabilized, the polymerization will proceed normally, as in the case of furfuryl acrylate;
- if the radical formed with the furan ring is much better stabilized, then little or no polymerization will ensue because the heterocycle plays the role of an internal inhibitor, as in the case of 2-vinyl furoate.

3.2 First Generation Furans and their Conversion into Monomers

Of course intermediate situations have also been observed, in which the polymerization does not reach completion because the furan rings pendant to the polymer chains play a retarding role, as in the case of 2-vinyl furan.

These features, peculiar to the dienic character of the furan heterocycle, have been put to good use [3c] by pushing the free radical stabilization to the extreme through conjugated structures such as **1**, which now function as very powerful *inhibitors*, as the addition of a radical X° generates a highly delocalized entity such as **2**. Of course this inhibiting role is not confined to radical polymerization, but operates in all systems displaying chain reactions conducted by free radicals, such as oxidations, photolyses, and so on.

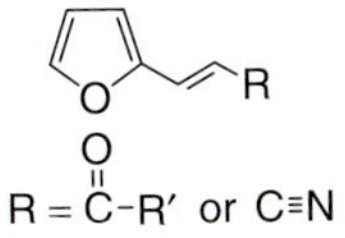

$$R = \overset{O}{\underset{\|}{C}}-R' \text{ or } C \equiv N$$
1

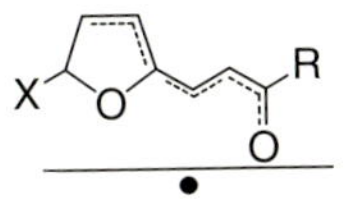

2

Cationic polymerization systems involving furan monomers with an unsubstituted C5 position are liable to electrophilic substitution with the formation of a branched macromolecule [3c], because of the strong tendency of that site to undergo electrophilic substitution with regiospecific control, as shown in Figure 3.4.

This side reaction is of course readily avoided by appending a substitutent at C5, for example, a methyl group, so that monomers derived from MF are immune to it. As in the case of free radical systems, but more so here, the alkylation reaction has been exploited to prepare functionalized oligomers by dominant transfer with a furan agent and block copolymers with difuran bridging molecules [3c].

The second route to the synthesis of monomers from F or its derivatives (but *not* from MF) calls upon a different approach based on the *acid-catalyzed condensation* of a variety of 2-substituted furan compounds with an aldehyde or a ketone following the general scheme shown in Figure 3.5 [3].

These monomers belong to another category of polymerizations, namely *polycondensation reactions* and have been used successfully to prepare a whole variety of materials [3], such as polyesters, polyamides, polyurethanes, and so on. Here, contrary to the monomers shown in Figure 3.2, the ensuing macromolecules

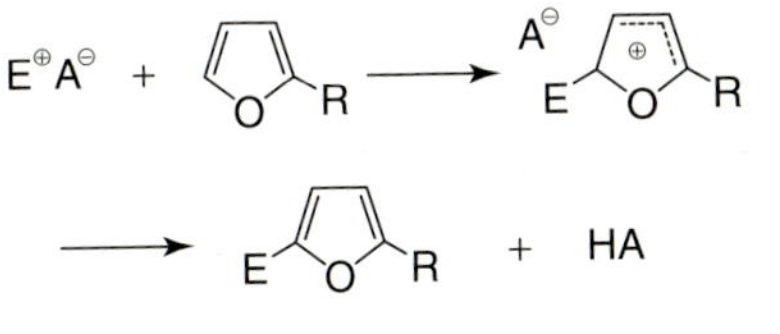

Figure 3.4 Mechanism of electrophilic substitution at the C5 site of a furan ring.

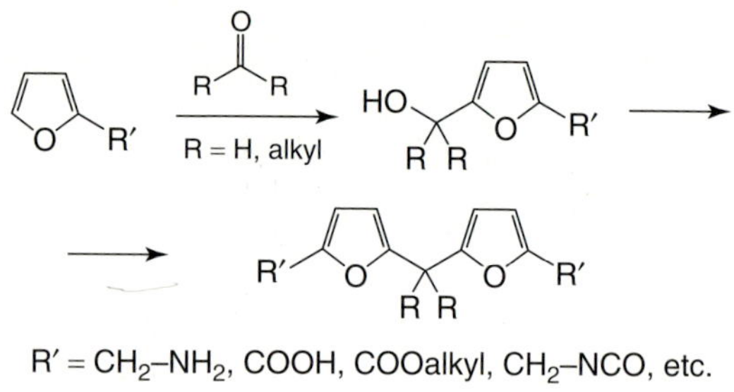

Figure 3.5 Synthesis of difuran monomers from F derivatives.

bear the furan moieties as an integral part of the chain backbone and hence the structure–properties relationships follow different criteria, including the nature of the moieties (R) on the carbon atom bridging the two rings. The possibility of preparing a wide variety of difunctional monomers of this type from largely available monosubstituted furans constitutes a good alternative to the synthesis of similar structures based on a single ring (see next section), and is waiting for hydroxymethylfurfural (HMF) to become commercially available.

Finally, 2,5-bis(hydroxymethyl)furan can be prepared by the hydroxymethylation of FA with formaldehyde [4] and has been used in the synthesis of polyurethanes, polyethers, and polyesters [3] (but see also Figure 3.6 for its synthesis from HMF).

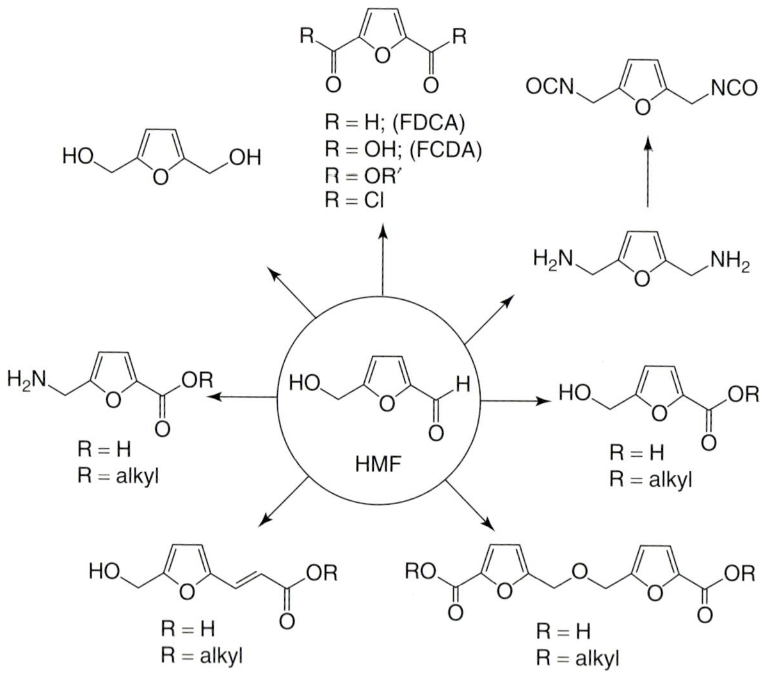

Figure 3.6 A selection of monomers derived from HMF.

3.2.3
Hydroxymethylfurfural

HMF, or 2-hydroxymethyl-5-formylfuran, is the other basic first-generation furan compound, which is prepared from hexoses, in the form of the corresponding mono-, oligo-, and polysaccharides, through a sequential mechanism entirely equivalent to that shown in Figure 3.1. The *industrial production of HMF* has been hindered by difficulties in terms of isolating it in good yields and purity. The interest in optimizing its synthesis has literally exploded in the last few years with an incessant output of publications [5].

The reason for this sudden surge in research is related to the importance of furan derivatives (i.e., molecules from renewable resources) both as alternative energy sources and chemical synthons, mostly in the shape of monomers. Given the promising improvement in yields and selectivity associated with these studies, and also the attention paid to developing green processes, it seems highly probable that HMF will be available as a chemical commodity very soon. Because of its known sensitivity to degrade upon storage, it is in fact one of its stable derivatives that will be produced, for example, by its *in situ* oxidation to either the dialdehyde (FCDA) or the diacid (furan dicarboxylic acid, FDCA) (see Figure 3.6).

Within the specific context of this discussion, HMF must be viewed as a particularly promising precursor to very interesting monomers, many of which have already been studied [3], some of which are shown in Figure 3.6.

Together with those shown in Figure 3.5, these monomers are obvious candidates for polycondensation reactions for polyesters, polyamides, and so on, as discussed below. The key structural feature associated with the monomers prepared from HMF is their close resemblance to *aromatic counterparts* and hence the interest in using them to synthesize polymers that simulate widely used materials such as Kevlar or poly(ethylene terephthalate) (PET).

3.3
Polymers from Furfuryl Alcohol

The exploitation of FA as a monomer represents the oldest approach to *furan-based polymers* and interest in these "resins" is still very much alive, not only in the use of the more traditional materials such as foundry binders, corrosion-resistant cements and mortars, and so on [3, 6], but also in the search for novel formulations and applications, which are briefly outlined here.

FA undergoes complex transformations into equally complex macromolecular structures when treated with an acidic initiator, even under mild conditions of acid strength and concentration at low temperatures. The mechanisms of these reactions, including the basic head-to-tail self-condensation and the chemistry associated with color formation and polymer cross-linking, have been unraveled in a systematic study [7], which led to the formulation of a set of structural changes

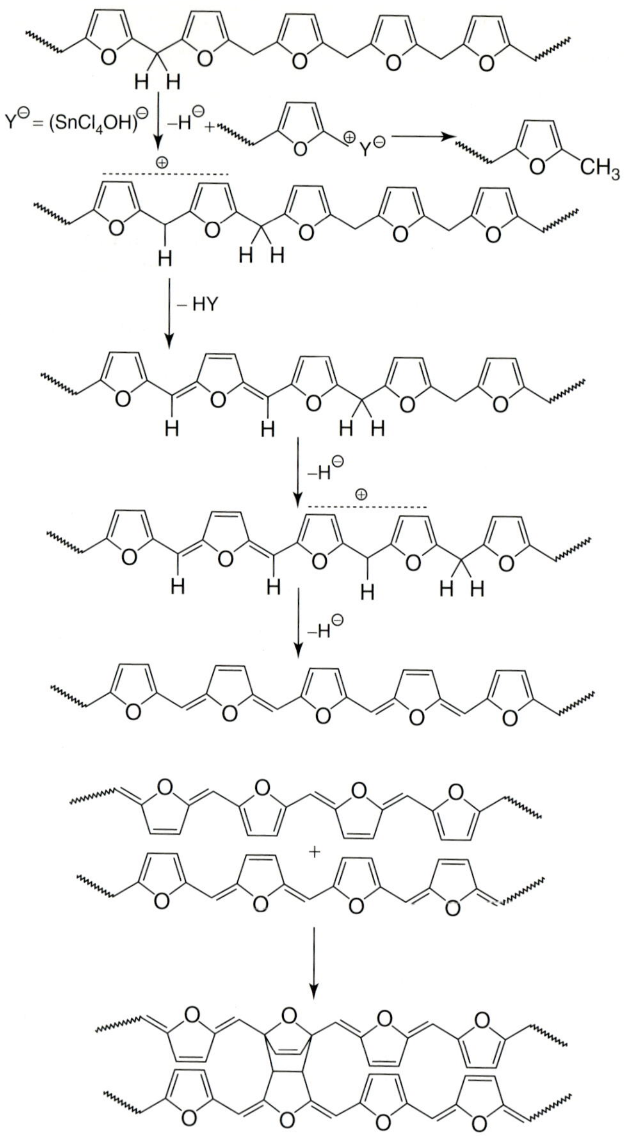

Figure 3.7 The two mechanisms responsible for conjugated sequences (color formation) and subsequent cross-linking in the acid-catalyzed polymerization of FA [7].

whose essence is depicted in Figure 3.7. The understanding of these intricacies also showed that they are intrinsic to the structure of the initial simple *linear oligomers*, that is, that they are not avoidable side reactions. This does not imply necessarily that the ideal poly-FA **3** is unattainable, but it certainly means that, despite considerable efforts in that direction, this simple, but potentially very interesting thermoplastic material, remains a chimera.

$$\left[\begin{array}{c} \\ \\ O \end{array} \right]_n$$

3

The central problem here is how to preserve the *integrity of the methylene moiety* bridging the rings, because its C–H bonds are extremely labile with the loss of a proton, a hydrogen atom or a hydride ion, depending on the medium. This is a result of the ensuing carbanion, free radical, or carbocation, respectively, all being efficiently stabilized by the two adjacent furan rings, which delocalize the charges or the unpaired electron, thanks to their marked dienic character [3].

A recent *chemo-rheological study* of this polymerization [8] confirmed both the clear-cut evidence and the mechanisms previously put forward [7] and provided an interesting kinetic insight into the competition between linear growth and the growing role of interchain Diels–Alder (DA) couplings, responsible for the early attainment of a diffusion-controlled regime. Similar conclusions were drawn following a detailed infrared investigation of this *polycondensation* [9]. Infrared, Raman, and electronic spectroscopies were also employed to follow the FA polymerization in the H-Y confined domains of a protonic Y zeolite [10]. The diminished reaction rate allowed some intermediate structures, including active carbenium ion species, to be identified. This study was extended to the carbonization process undergone by the cross-linked poly-FA, as is mentioned further below in a wider set of investigations.

Within the context of *novel materials*, FA has attracted considerable attention in recent years through studies aimed at preparing and characterizing carbonaceous and other materials, through its polymerization and subsequent pyrolysis. Examples include mesoporous [11] and microporous [12] carbon, silica nanocomposites [13], carbon nanocomposites [14], glass-like carbon [15], ZnO–carbon composites [16], carbon films [17], and foams [18], and also mesoporous crystalline TiO_2 [19].

Another area in which *FA polymerization* is gaining momentum is in the synthesis of *organic–inorganic hybrids* [20], including nanoscopic morphologies [21] and biobased nanomaterials [22]. An original aspect associated with some of these systems is the formation of furfuryl–alkoxide intermediates, for example, from siloxanes incorporating one or several furfuryl moieties through alkoxide exchanges, followed by the acid-promoted polymerization of the furfuryl moieties and/or the sol–gel mechanism applied to the hydrolyzed siloxanes, as shown in Figure 3.8.

Notwithstanding the novel approach, the structural features of the macromolecules generated by the furan moieties within these *hybrid constructs* (denoted as Poly-FA in Figure 3.8) were entirely similar to those obtained in the conventional acid-catalyzed polycondensation of FA, discussed above. The interest in these materials resides therefore primarily in their nanomorphology associating the furan resin and the silica particle.

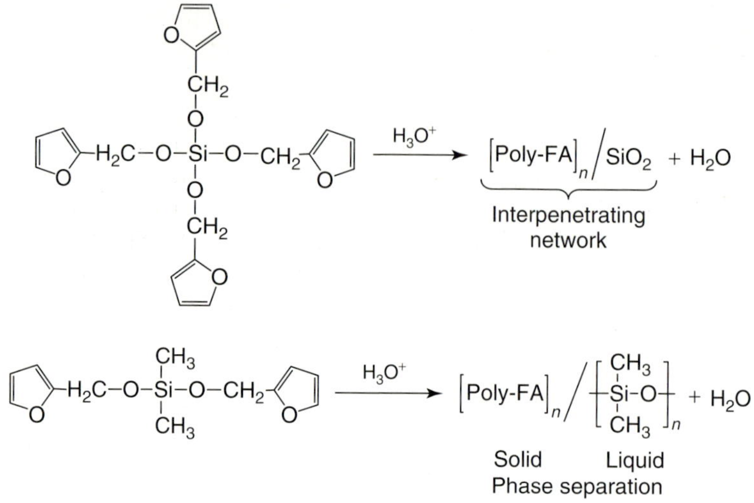

Figure 3.8 Two examples of organic–inorganic hybrids based on FA (adapted from Ref. [21]).

A promising field of application of FA, which has been recently revived with a good measure of success, is *wood preservation* and modification through impregnation with the monomer and its subsequent *in situ* polymerization promoted by acidic catalysts [23]. The process has been optimized to provide remarkable improvements in such properties as dimensional stability, mechanical and chemical strengthening, excellent resistance to microbial decay and insect attack, and also ecological soundness, which have led to its commercialization [23]. Little work on the chemistry associated with this impregnation has accompanied the thorough technical development as yet. Recent attempts to unravel the basic query of whether the polymerizing FA structures react with any of the wood components has thus far been limited to the study of lignin model compounds [24], but even such a specific investigation still only provided unconvincing conclusions. More fundamental work would undoubtedly be welcome here.

This state of affairs is fairly common within the realm of the wide range of studies related to the many applications of poly-FA [3], where empirical approaches dominate to the detriment of deeper mechanistic investigations, the outcome of which would provide the means to improve the performance of the ensuing materials.

Fresh additions to the multifarious applications of poly-FA include its role as a mechanical reinforcement for highly porous polymeric matrices, obtained by the vapor-phase adsorption of FA into their cavities and its subsequent *in situ* polymerization [25] and the elaboration of a novel FA–lignin thermoset [26].

3.4
Conjugated Polymers and Oligomers

Although *polyfuran* has been the subject of a large number of studies, unlike its two homologs polythiophene and polypyrrole, the structural evidence pointing to a regular sequence of 2,5-substituted heterocycles has never been provided, whatever the synthetic procedure adopted [3]. This is still today a critical missing link, not so much because of insufficient scrutiny of the materials, but rather because, during these polymerizations, the pronounced dienic character of the furan ring induces different modes of insertion, hence generating irregular macromolecular structures, which are very difficult to minimize.

As a consequence, the development of materials associated with such a potentially useful conjugated polymer has been hampered and attention has therefore been shifted to the better controlled synthesis of copolymers with other heterocycles such as thiophene [27] and porphyrins [28], or to the polymerization of monomers incorporating furan moieties together with aniline [29] and phenylene motifs [30].

Although these well-defined materials displayed interesting opto-electronic properties, the contribution of the presence of the furan ring in their structure is modest and much less relevant, both scientifically and technologically, than that of the very rich variety of macromolecules incorporating the thiophene and pyrrole counterparts.

A completely different situation characterizes the conjugated furan polymer with a repeat unit made up of the heterocycle and an additional external alkenyl moiety, that is poly(2,5-furylene vinylene) (PFV). This macromolecular structure represents undoubtedly the best achievement in the context of materials incorporating the furan ring within a linear sequence of conjugated elements, without the intervention of other cyclic co-monomers.

The classical approach, usually applied to the synthesis of the aromatic homologs poly(vinylidene arylene)s, was extended to the corresponding furan precursor 2,5-bis(tetrahydrothiopheniomethyl)furan dichloride [31]. However, this procedure proved rather laborious compared with the much simpler base-catalyzed polycondensation of MF, which provided a straightforward route to both PFV and its oligomers and the additional advantage that each macromolecule ended with a formyl moiety [3, 32]. The simplicity of this synthetic approach is complemented by the remarkable fact that the monomer involved is a fairly cheap chemical commodity.

The PFV structure shown in Figure 3.9 bears this *terminal aldehyde group*, which proved particularly useful in a variety of chemical modifications of these oligomers and polymers, including chain extensions, grafting, and preparation of Schiff

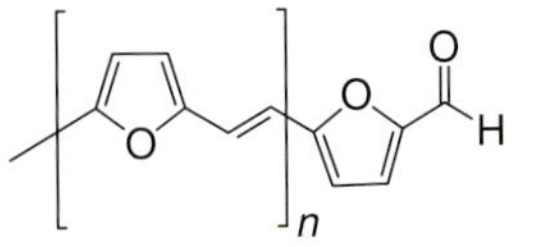

Figure 3.9 The structure of PFV generated by the polycondensation of MF.

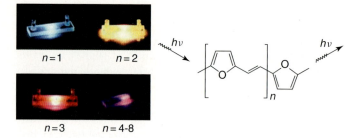

Figure 3.10 Photoluminescence of oligo(furylene vinylene)s of various DPs [33].

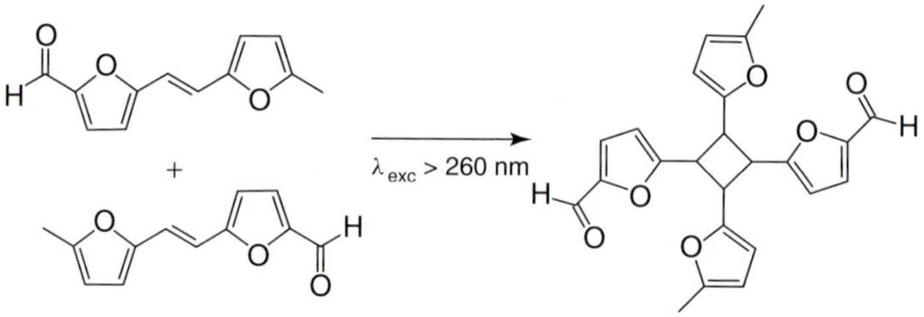

Figure 3.11 Molecular photodimerization of PFV dimer [34].

bases [3, 32]. These straightforward reactions gave rise to materials that combined the classical properties of conjugated polymers with such important additional features as ionic conduction, photoreactivity, and elastomeric behavior, among others.

While this polymer displays good conductivity, solubility in common solvents, and resistance to atmospheric oxidation, and can also be plasticized by attaching polyether chains to its reactive endgroup [3, 32], the individual oligomers, namely dimer to pentamer, display photo-, and electro-luminescence covering essentially the entire visible spectrum. Figure 3.10 shows this behavior in the case of the *luminescence* generated following appropriate laser excitation [33].

The peculiar *photochemical behavior* of the dimer [34], illustrated in Figure 3.11, was, moreover, exploited to prepare photo-cross-linkable materials from poly(vinyl alcohol) (PVA) [35] and chitosan [36]. Again, the aldehyde function was essential to the grafting of these chromophores onto PVA, through acetalization with its OH groups, and onto chitosan, through the formation of Schiff bases with its NH_2 groups. The irradiation of these modified polymers in the near-UV (ultraviolet) produced their cross-linking by intermolecular photo-couplings based on the reaction shown in Figure 3.11.

In conclusion, MF, which is obtained industrially as a secondary product in the manufacture of F, is a very useful precursor to a series of conjugated molecules and macromolecules with high-tech applications.

A very recent contribution to the field of furan-based conjugated polymers describes the use of a *bisfuranyl monomer* containing a 1,6-methano[10]annulene moiety to prepare conducting materials [37].

3.5 Polyesters

Linear *polyesters* incorporating furan heterocycles have been reported since the middle of the last century, but the first thorough study dates from the late 1970s with the pioneering work of Moore and Kelly [3, 38], who approached the issue in a fairly systematic fashion with FDCA (or its dichloride) as the key monomer. The structural and DP (degree of polymerization) features of these polyesters suggested, however, that the adopted synthetic procedures had not been the best and hence gave rise to the feeling that more work was needed to attain materials with optimized properties (see below).

After a slack decade, the interest was shifted to the use of monomers bearing two furan rings [3, 39–41], such as the diacids, diesters, and diacid chlorides shown in Figure 3.5. A comprehensive array of novel *furan polyesters* were prepared and fully characterized in terms of structure, molecular weight, thermal properties, solubility, and so on. Figure 3.12 shows the general formula of these polymers, in which R, R', and R" were varied to include aliphatic and aromatic moieties, in addition to specific groups such as perfluoro substituents, which provided the establishment of a wide-ranging structure–properties relationship.

In recent years, the focus of attention has returned to the study of furan polyesters derived from FDCA, its diesters and dichloride [42, 43]. The first and foremost contribution of this ongoing investigation was the synthesis and characterization of the most obvious, and yet hitherto ignored, homolog, namely poly(2,5-ethylene furandicarboxylate) (PEF), which is the heterocycle homolog of the most important commercial polyester, PET. The best results were achieved through the polytransesterification reaction shown in Figure 3.13, which gave a regular polymer with DPs higher than 200, high crystallinity, and T_g (glass transition temperature) and T_m (melting temperature) values similar to those of PET [42].

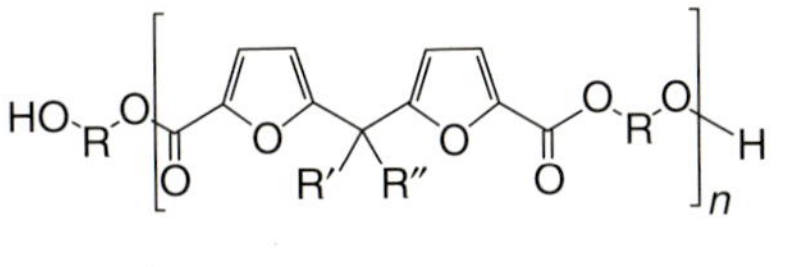

Figure 3.12 The general structure of polyesters incorporating difuran units [3, 39, 40].

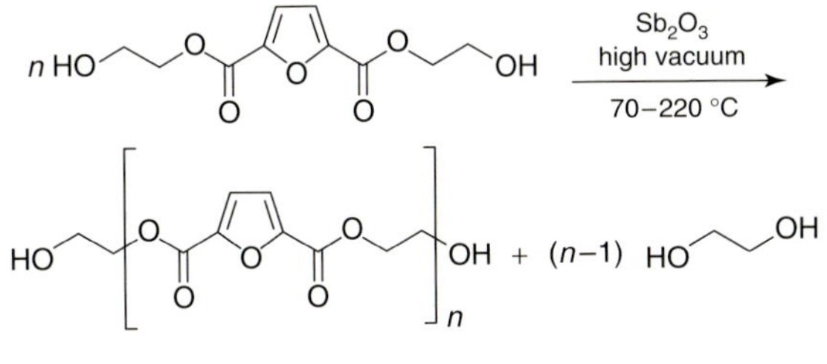

Figure 3.13 Polytransesterification mechanism leading to high-DP PEF.

The study has now been extended to other poly(2,5-furan dicarboxylate)s using aliphatic diols such as propylene glycol, benzylic structures such as 1,4-bishydroxymethyl benzene, bisphenols such as hydroquinone, and sugar-derived diols such as isosorbide [43]. The last polyester **4**, entirely based on renewable resources, gave an amorphous material, because of the non-equivalence of the OH groups of isosorbide, and a high T_g (~ 180 °C), because of the stiffening role of that diol.

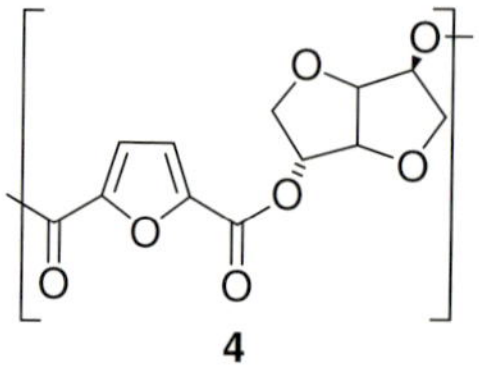

4

This project has set a new stage, in which polyesters with a very wide range of properties, and hence of possible applications, can be readily prepared using one monomer, if not both, from renewable resources. The recent upsurge of interest in improving the process of *HMF production* [5] is particularly beneficial to the future of these polyesters and of other polymers based on FDAC and FDCA. Indeed, very recently, a series of furan–aliphatic polyesters, prepared by the polytranesterification of the dimethyl ester of FDCA and a series of diols with 3–18 methylene groups, was described, including their thermal transitions and dynamic mechanical properties [44].

HMF can be readily converted into the unsaturated hydroxyester **5** via the Wittig reaction, whose polytransesterification gave a stiff material unsuitable for coating [45]. Its copolymerization with an aliphatic homolog produced a soft photosensitive material that readily cross-linked upon near-UV irradiation following an interchain mechanism similar to that shown in Figure 3.11, and was therefore deemed useful as a potential *negative photoresist* [45].

5

3.6
Polyamides

As in the case of polyesters, two families of furan *polyamides* have been reported [3], that is, (i) those based on FDCA and its chloride, incorporating a single heterocycle in each monomer unit and (ii) those prepared with difuran diacid monomers (see Figure 3.5).

Given the fact that amino furans are thermodynamically unstable, the fully heterocyclic equivalents of aramides cannot be prepared. The closest homologs were synthesized using FDCA and aromatic diamines [46] and polyamide **6** was the most thoroughly studied, because of its resemblance to Kevlar, and indeed its properties simulated in all aspects those of the classical aromatic counterpart.

6

Fully furanic polyamides such as **7** were also prepared and characterized, but their thermal stability was hampered by the relative lability associated with the methylene moieties attached to the amine precursor in the polymer unit.

7

This investigation [46], carried out some 20 years ago, deserves to be widened in the light of the renewed interest in FDCA and the progress in the *polycondensation* techniques, as is in fact already happening with a very recent study of furan–aliphatic polyamides based on the dimethylester of FDCA and a series of diamines with 6–12 methylene spacers [44].

A systematic study of polyamides based on bisfuranic diacids covered a wide range of structures [47], as in the case of the homologous polyesters discussed above. Figure 3.14 shows the monomers used and the approach followed for the *interfacial polymerizations*. The ensuing polymers displayed properties, such as glass transition and melting temperatures, which varied considerably, as expected, according to the nature of the diamine used.

3.7
Polyurethanes

A comprehensive approach to the synthesis of *polyurethanes* incorporating furan moieties was conducted in the 1990s [48]. Given the lack of published information in this realm, the project involved a preliminary study of the preparation and reactivity of a selection of model compounds and the characterization of the ensuing urethanes. This was an important introduction to this novel field because it set the stage, with critical information about both electronic and steric factors affecting the kinetics of urethane formation. One of the most surprising results was the exceedingly high reactivity of 2-furyl isocyanate toward alcohols (some 100 times higher than that of the phenyl homolog!), so much so that the corresponding 2,5-diisocyanate could not be reasonably used as a monomer for obvious practical

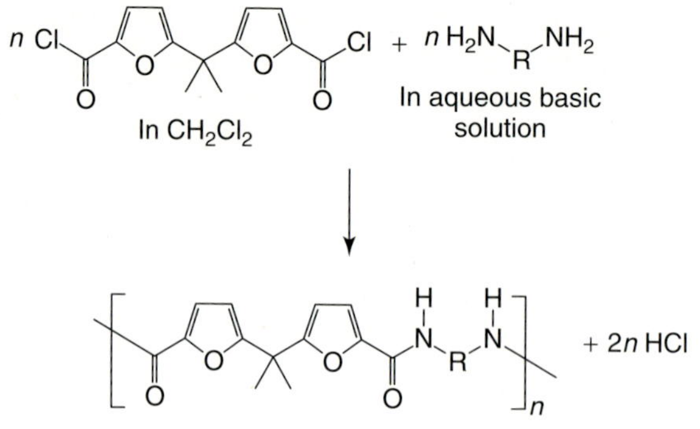

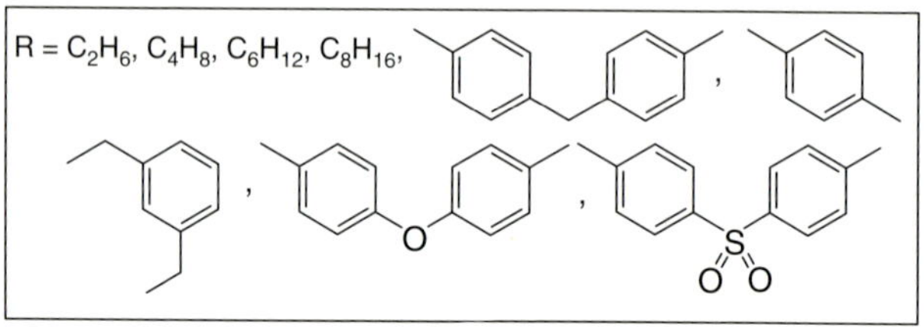

Figure 3.14 Interfacial synthesis of difuran polyamides using a variety of diamines.

reasons. The origin of this extreme reactivity was attributed to the pronounced dienic character of the furan heterocycle that induced this enhancement through an electronic conjugated effect. The actual polymerizations involved a rich choice of furan diols and diisocyanates (mostly novel compounds), which were combined among themselves or with aliphatic and aromatic homologs to give a very wide coverage of macromolecular structures with furan rings both within the chain and as pendant moieties. Their thorough characterization provided sound criteria for the structure–properties relationships, and also clear-cut evidence of the fact that all these systems proceeded without the interference of side reactions [48]. Examples of such materials are given below, including a fully furanic (**8**), a furan–aromatic (**9**), and a furan–aliphatic (**10**) polyurethane.

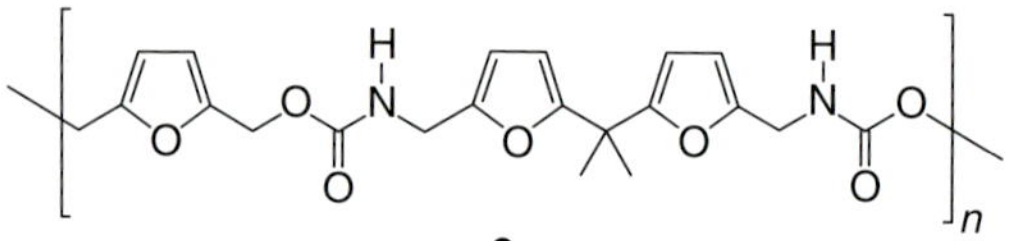

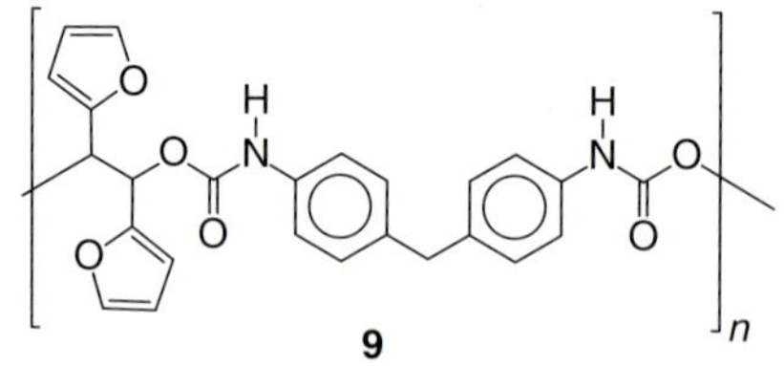

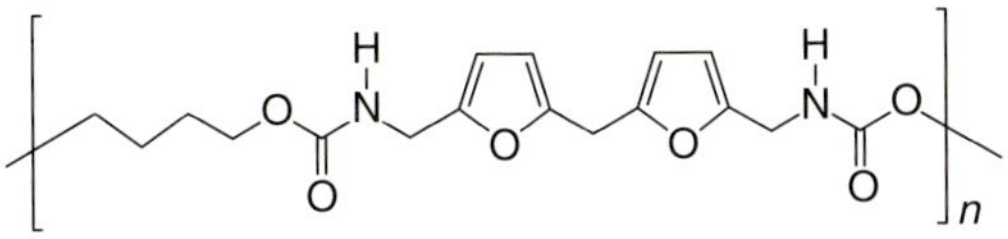

10

3.8
Furyl Oxirane

The remarkable reactivity of furyl oxirane (FO) toward anionic initiation by nucleophiles as weak as alcohols stems from the dienic character of the furan ring [49]. The ensuing polyether **11** has a T_g around room temperature. Interesting materials can be prepared using the OH groups from macrodiols and triols, PVA or cellulose, among other polymer precursors, to initiate FO polymerization in order to synthesize block and graft copolymers. Although this original feature was assessed in preliminary trials, much remains to be done to exploit it in a systematic approach.

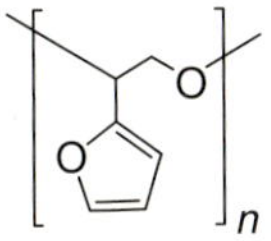

11

3.9
Application of the Diels–Alder Reaction to Furan Polymers

The most important recent contributions to the realm of *furan polymers* is arguably the lively eruption of studies on the use of the DA reaction to synthesize a wide variety of macromolecular architectures possessing, among other original properties, the common and peculiar feature of *thermal reversibility* within a readily accessible domain of temperatures [3]. The more relevant aspects of these contributions to the topic have been reviewed [3, 50], but its salient features and the latest important publications and work in progress deserve appropriate treatment here.

As already mentioned, of the most common five-member ring unsaturated heterocycles, furan displays the most pronounced *dienic character*, when compared with thiophene and pyrrole [3], which makes it highly suitable to intervene as a diene in that beautiful example of *click chemistry* known as the *Diels–Alder reaction*. Organic chemistry textbooks typically call upon furans in conjunction with such dienophiles as maleic anhydride or maleimides to illustrate the mechanism of this reaction. In addition to its clean-cut character with respect to the usual absence of side events perturbing its course, the DA reaction is also remarkable for its temperature sensitivity.

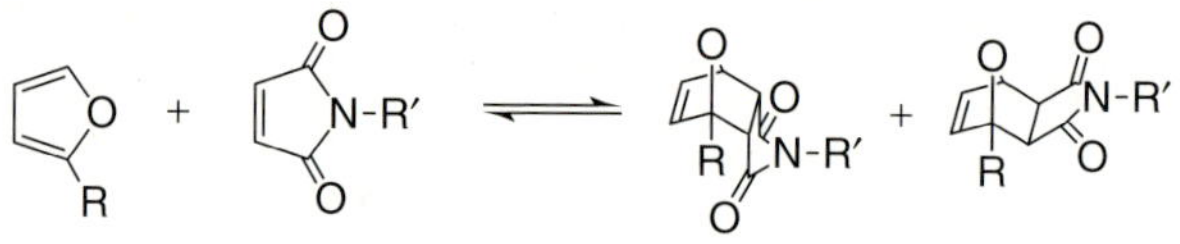

Figure 3.15 The DA reaction between a furan and a maleimide.

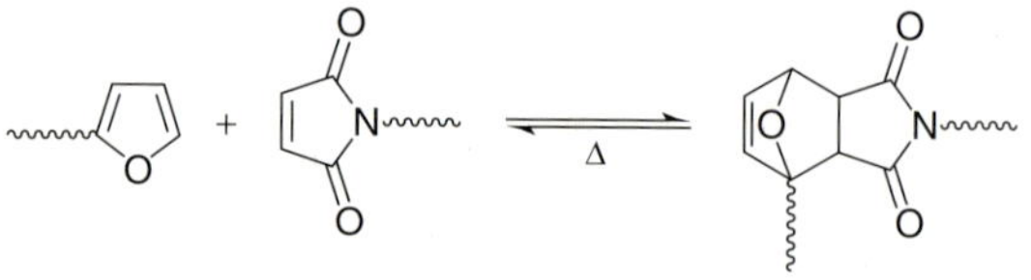

Figure 3.16 The DA equilibrium between growing species bearing, respectively, furan and maleimide endgroups.

The coupling of a furan compound (diene) with a maleimide derivative shown in Figure 3.15 illustrates the DA reaction being applied to the most frequently used combination in *furan macromolecular synthesis*.

Whereas the *stereochemical aspects* related to the relative abundance of the *exo-* and *endo-*forms of the adducts play a significant role in both synthetic and fundamental organic chemistry, the application of this reaction to the construction and deconstruction of macromolecules is not concerned with them, as they do not affect the outcome of either process. The DA equilibrium can therefore be expressed in that context as the more general form of Figure 3.16.

With these two complementary structures, the forward reaction (*polymerization*) can be carried out at temperatures up to about 65 °C while keeping the equilibrium still very heavily shifted to the right, whereas the reverse reaction (*depolymerization*) becomes dominant at about 110 °C, with the equilibrium now inducing the predominant formation of the unreacted moieties. The numerous exploitations of this reversible click reaction in polymer synthesis are briefly illustrated below, following the specific nature of the system involved.

3.9.1
Linear Polymerizations

This approach was the first to be studied some 40 years ago, but the multiplication of reaction combinations and conditions has blossomed in the last few years [3, 50, 51] often with little added originality. Virtually all these studies focused on the polycondensation of a difuran (A-A) compound with a bismaleimide (B-B) and on the application of the retro-Diels–Alder (r-DA) reaction to depolymerize the ensuing polyadduct. On the whole, the focus is on the *qualitative aspects of polymer synthesis* and *thermal degradation* with little or no attention to structures, molecular weights, or kinetic features.

In an attempt to deepen the understanding of these systems, the Aveiro group recently carried out a detailed investigation of some A-A + B-B polymerizations [43, 52] using the four novel monomers **12–15**, which were thoroughly purified and characterized. Two complementary spectroscopic techniques, namely UV and ^{1}H NMR (nuclear magnetic resonance), were adopted to follow both the linear polyadduct growth at 65 °C and its thermal reversion to the starting monomers at 110 °C.

12

13

14

15

Figure 3.17 illustrates one of these systems. Working with initial stoichiometric conditions and using modest monomer concentrations, the progressive decrease in the absorption at 300 nm of the maleimide chromophore, arising from its corresponding loss of conjugation associated with the formation of the adduct (forward reaction, Figure 3.16) provided a quantitative means of following the polymerization reactions and hence their second-order kinetic behavior. Concurrently, the progress of these *polymerizations* was followed by ^{1}H NMR spectroscopy using higher monomer concentrations and these experiments gave a detailed insight into the structural changes occurring with chain growth, that is, as the proportion of adducts increased to the detriment of the terminal furan and maleimide endgroups. Visually, these reactions were accompanied by an increase in viscosity, reflecting the progressive rise in DP of the polymers being built up by the DA couplings. Additionally, this technique was utilized to search for side reactions, but none were detected.

The same tools were adopted to monitor the *depolymerization* reactions, which these polyadducts underwent through the r-DA reaction at 110 °C. The UV time scans showed the progressive return of the absorption due to maleimide moieties following a first-order kinetic pattern until both monomers were fully regenerated. The ^{1}H NMR spectra again provided structural information about the course of these depolymerizations, which were accompanied by a progressive decrease in viscosity.

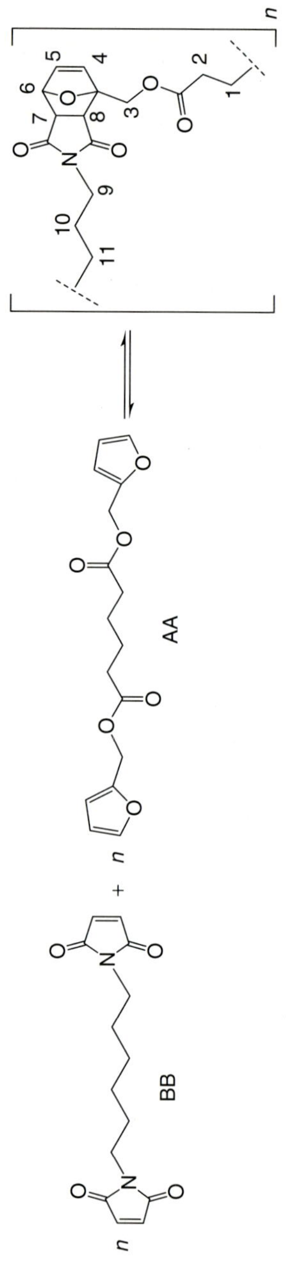

Figure 3.17 Reversible DA polycondensation between complementary bifunctional monomers.

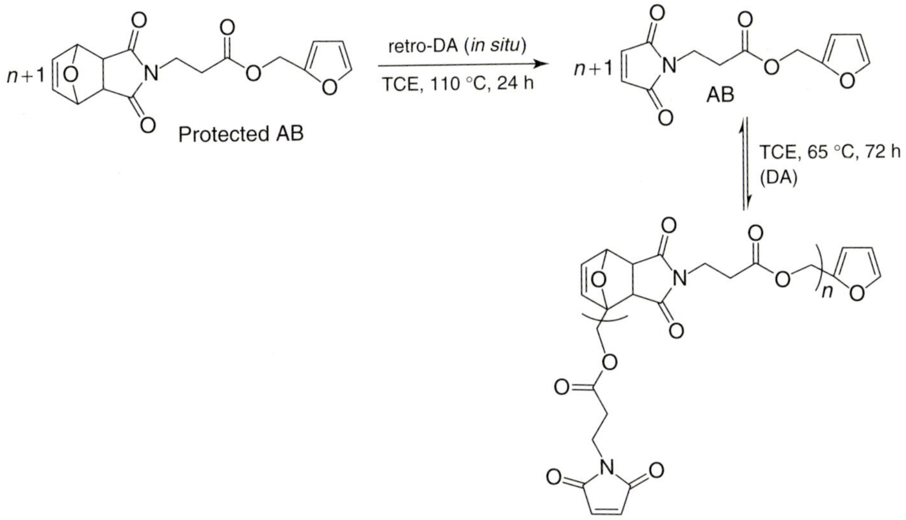

Figure 3.18 Deprotection and reversible DA polymerization of an AB monomer.

In all these systems, the classical problem of ensuring the exact *monomer stoichiometry* inevitably cropped up, as suggested by the incomplete moiety consumption at the end of the polymerizations, and hence molecular weights that were not optimized. In order to circumvent this difficulty, a different strategy was applied, based on the use of an A-B monomer, namely a molecule incorporating both a furan and a maleimide moiety. However, for obvious reasons, only monomers of this type, where one of the reactive group is kept in a masked configuration, can be prepared and characterized before activating them through the appropriate deprotection. Figure 3.18 illustrates this approach through a specific example [43, 53] (A. Gandini, D. Coelho, and A.J.D. Silvestre, unpublished results).

The results of this ongoing study are encouraging and suggest that indeed the use of an A-B monomer provides a better polymerization outcome in terms of degree of conversion and hence molecular weight.

3.9.2
Non-Linear Polymerizations

The study of systems involving A3 + B-B or A-A + B3 monomer combinations was conducted using the same temperature and solvent conditions and called upon the same spectroscopic techniques [43] (A. Gandini, D. Coelho, and A.J.D. Silvestre, unpublished results). The trifunctional monomers used in conjunction with the difunctional counterparts **12–15**, were the furan homologs **16** and **17** and the maleimide **18**.

Figure 3.19 provides the principle of one such *non-linear DA polymerization*. In these systems, the [A]/[B] molar ratio was varied to generate situations either leading to the formation of a network at different degrees of conversions, or ending

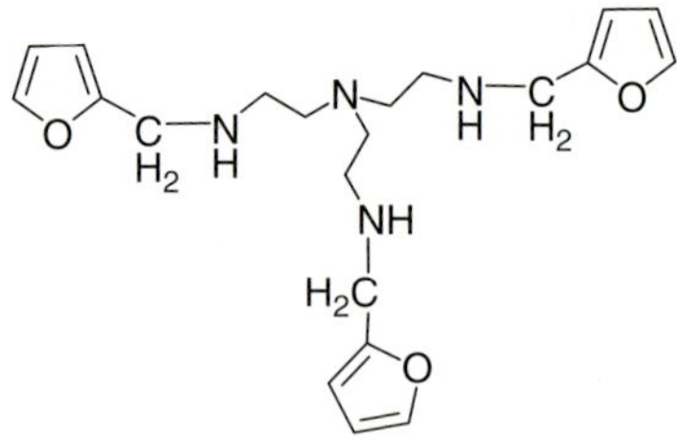

16

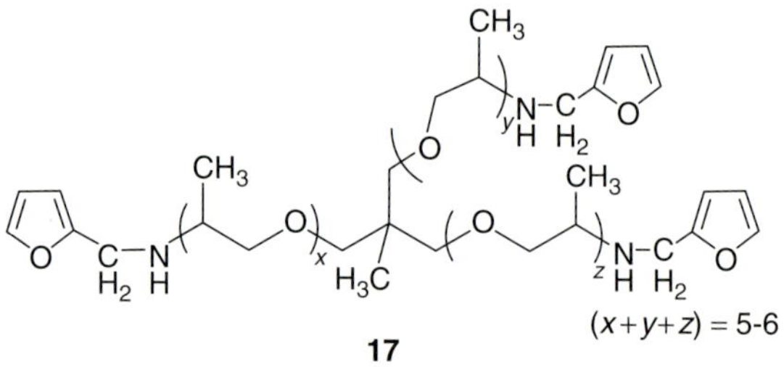

17 $(x+y+z) = 5\text{-}6$

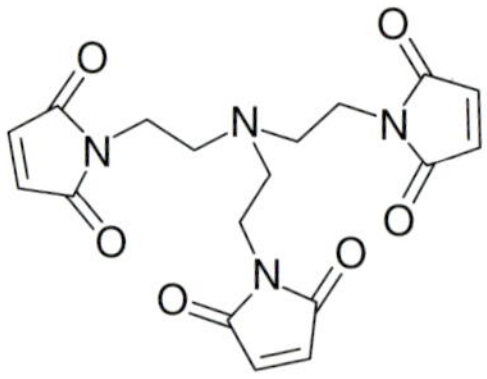

18

before its attainment, as predicted by the *Flory–Stockmayer equation*. In addition to the spectroscopic features already characterized with the linear systems, the increase in viscosity during these polymerizations was now very pronounced and the appearance of a gel phase was clearly detected with expected cross-linking conditions.

The r-DA reactions revealed visually the gradual dissolution of the gel particles (when present), while the UV and ^{1}H NMR spectra gave, once more, the details of the depolymerization processes leading the systems steadily back to the monomers.

This work in progress also includes the synthesis and polymerization of AB_n and A_nB monomers ($n > 1$) and the characterization of the ensuing hyperbranched macromolecular materials, as well as other *macromolecular architectures*, namely graft- and comb-shaped copolymers.

Mendable polymers based on epoxy [54] and poly(ethylene adipate) [55] structures bearing polyfunctional furan and maleimide moieties were recently reported, but these contributions did not add any substantial novelty to this well-characterized topic [50].

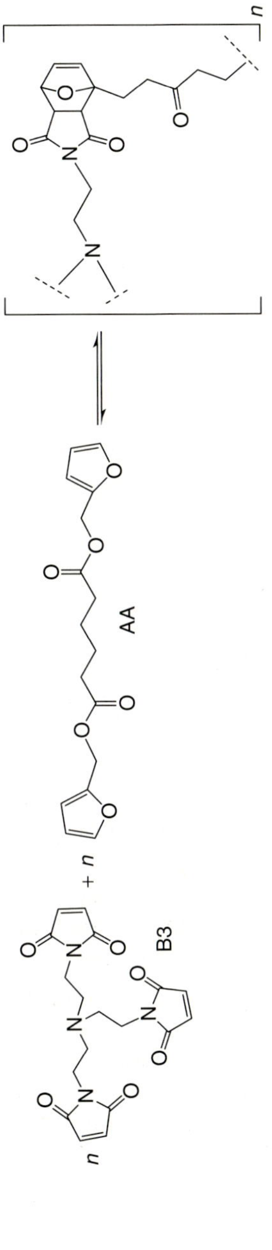

Figure 3.19 Reversible DA non-linear polycondensation between a tris-maleimide and a difuran monomer.

3.9.3
Reversible Polymer Cross-Linking

The idea of *thermoreversible networks* based on the application of the DA reaction to linear polymers bearing furan or maleimide moieties was put forward in the early 1990s, and was investigated more systematically in the first years of the new millennium [3, 50]. The fact that the r-DA reaction can be effectively applied at temperatures around 100 °C is particularly interesting in applications such as mendable polymers or the recycling of rubber-like elastomers and hence these important contributions.

Numerous other studies have been published in the last several years on the *reversible coupling* of furan and maleimide heterocycles aimed at preparing thermally reversible networks. The strategies vary somewhat, but the overall scenario is essentially the same, namely to build a cross-linked material, which can be readily reversed to the starting monomers or polymers. In other words, the original ideas developed for this general purpose [3, 50] are maintained, and only specific issues are in fact modified.

Thus, copolymers bearing furfuryl methacrylate units were cross-linked with a bis-maleimide [56]; in addition shape-memory materials were prepared thanks to this reversible DA reaction [57]; and thermally reversible cross-linked polyamide [58], epoxy [59–61], hydrogels [62], and biobased polymers [63], including self-healing structures [64] were described. As already pointed out in the previous sections for DA polymerization reports prior to the contribution from the Aveiro group, these studies focused essentially on the syntheses of monomers and polymers and on the r-DA reaction applied to the ensuing networks, with little or no emphasis on kinetic aspects and materials properties. An original approach to *strippable imaging* materials was recently described [65] based on the polymerization of a DA-adduct diacrylate cross-linker, which can be thermally broken down to regenerate linear polymers.

Plaisted and Nemat-Nasser provided a substantial contribution to this field by a quantitative study of the phenomena associated with *multiple healing* and *reversible cross-linking* using a tetrafuran–bismaleimide system [66].

3.9.4
Miscellaneous Systems

Other ingenious applications of the *DA reaction* to polymer chemistry have been described, such as the synthesis of reversible dendritic structures and hyperbranched macromolecules [3, 50, 67]. Recent contributions in the same vein include a furan–maleimide star-shaped polymer, which was reversibly dismembered through the DA/r-DA reaction [68], and block dendrimers that were joined/disjoined through the same mechanism [69].

In a different vein, the r-DA reaction of the N-phenylmaleimide–FA adduct was studied in various polymer matrices kept in a viscous state in order to assess the role of diffusion limitations on its decoupling [70]. In addition, multi-walled carbon nanotubes (MWCNTs) were decorated with both furan and maleimide moieties and the subsequent inter-MWCNT DA reactions studied [71]. Also, gold nanoparticles were reversibly coupled with conjugated polymers [72].

Finally, in yet another novel application, *drug-delivery vehicles* constructed through the furan–DA coupling of antibodies to polymer nanoparticles have recently been reported [73].

3.10
Conclusions

Recent research on macromolecular materials based on furan chemistry and furan monomers has confirmed the relevance of a strategy based on the exploitation of renewable resources. The potential availability of a vast array of *furan monomer structures* has set the stage for the development of a novel branch of polymer chemistry capable in principle of providing materials with properties, and hence applications, simulating those of the conventional counterparts derived from fossil resources. In addition to this aspect, which is related to the potential replacement of dwindling sources by the biomass, the peculiarities of furan chemistry also gives access to original polymers, otherwise inaccessible, suitable for utilization in high-tech and added-value domains.

References

1. Belgacem, M.N. and Gandini, A. (eds) (2008) *Monomers, Polymers and Composites from Renewable Resources*, Elsevier, Amsterdam.
2. Gandini, A. (2008) *Macromolecules*, **41**, 9491.
3. (a) Gandini, A. (1977) *Adv. Polym. Sci.*, **25**, 47; (b) Gandini, A. (1990) *ACS Symp. Ser.*, **433**, 195; (c) Belgacem, M.N. and Gandini, A. (1997) *Prog. Polym. Sci.*, **22**, 1203; (d) Moreau, C., Gandini, A., and Belgacem, M.N. (2004) *Top. Catal.*, **27**, 9; (e) Gandini, A. and Belgacem, M.N. (2008) Furan derivatives and furan chemistry at the service of macromolecular materials. In: *Monomers, Polymers and Composites from Renewable Resources* (eds M.N. Belgacem and A. Gandini), Elsevier, Amsterdam, Ch 6; 115; (f) Gandini, A. and Belgacem, M.N. (2010) *Polym. Chem.*, **1**, 245.
4. (a) Zeitsch, K.J. (2000) *The Chemistry and Technology of Furfural and its Many By-Products*, Elsevier, Amsterdam, (b) TransFuran Chemicals bv, http://www.furan.com/tfc.html. (accessed on 2010).
5. (a) Carlini, C., Patrono, P., Galletti, A.M.R., Sbrana, G., and Zima, V. (2004) *Appl. Catal. A-Gen.*, **275**, 111; (b) Bicker, M., Kaiser, D., Ott, L., and Vogel, H. (2005) *J. Supercrit. Fluids*, **36**, 118; (c) Román-Leshkov, Y., Chheda, J.N., and Dumesic, J.A. (2006) *Science*, **312**, 1933; (d) Asghari, F.S. and Yoshida, H. (2006) *Carbohydr. Res.*, **341**, 2379; (e) Asghari, F.S. and Yoshida, H. (2006) *Ind. Eng. Chem. Res.*, **45**, 2163; (f) Zhao, H., Holladay, J.E., Brown, H., and Zhang, C. (2007) *Science*, **316**, 1597; (g) Amarasekara, A.S., Williams, L.D., and Ebede, C.C. (2008) *Carbohydr. Res.*, **343**, 3021; (h) Hu, S., Zhang, Z., Zhou, Y., Han, B., Fan, H., Li, W., Song, J., and Xie, Y. (2008) *Green Chem.*, **10**, 1280; (i) Yong, G., Zhang, Y., and Ying, J.Y. (2008) *Angew. Chem. Int. Ed.*, **47**, 9345; (j) Qi, X., Watanabe,

M., Aida, T.M., and Smith, R.L. Jr. (2008) *Green Chem.*, **10**, 799; (k) Qi, X., Watanabe, M., Aida, T.M., and Smith, R.L. Jr. (2008) *Ind. Eng. Chem. Res.*, **47**, 9234; (l) Hu, S., Zhang, Z., Song, J., Zhou, Y., and Han, B. (2009) *Green Chem.*, **11**, 873; (m) Román-Leshkov, Y. and Dumesic, J.A. (2009) *Top. Catal.*, **52**, 297; (n) Chan, J.Y.G. and Zhang, Y. (2009) *ChemSusChem*, **2**, 731; (o) Li, C., Zhang, Z., Bao, Z., and Zhao, K. (2009) *Tetrahedron Lett.*, **50**, 5403; (p) Qi, X., Watanabe, M., Aida, T.M., and Smith, R.L. Jr. (2009) *Green Chem.*, **11**, 1327; (q) Gorbanev, Y.Y., Klitgaard, S.K., Woodley, J.M., Christensen, C.H., and Riisager, A. (2009) *ChemSusChem*, **2**, 672; (r) Mascal, M. and Nikitin, E.B. (2009) *ChemSusChem*, **2**, 859; (s) Verevkin, S.P., Emel'yanenko, V.N., Stepurko, E.N., Ralys, R.V., and Zaitsau, D.H. (2009) *Ind. Eng. Chem. Res.*, **48**, 10087; (t) Binder, J.B. and Raines, R.T. (2009) *J. Am. Chem. Soc.*, **131**, 1979; (u) Boisen, A., Christensen, T.B., Fu, W., Gorbanev, Y.Y., Hansen, T.S., Jensen, J.S., Klitgaard, S.K., Pedersen, S., Riisager, A., Ståhlberg, T., and Woodley, J.M. (2009) *Chem. Eng. Res. Des.*, **87**, 1318; (v) Qi, X., Watanabe, M., Aida, T.A., and Smith, R.L. Jr. (2009) *ChemSusChem*, **2**, 944; (w) Hu, S., Zhang, Z., Song, J., Zhou, Y., and Han, B. (2009) *Green Chem.*, **11**, 1746; (x) Hansen, T.S., Woodley, J.W., and Riisager, A. (2009) *Carbohydr. Res.*, **344**, 2568; (y) Zhang, Z. and Zhao, Z.K. (2009) *Bioresour. Technol.*, **101**, 1111; (z) Ilgen, F., Ott, D., Kralisch, D., Palmberger, A., and König, B. (2009) *Green Chem.* **11**, 1948; (aa) Casanova, O., Iborra, S., and Corma, A. (2009) *ChemSusChem*, **2**, 1138; (ab) Stalberg, T., Sorensen, M.G., and Riisager, A. (2010) *Green Chem.*, **12**, 321; (ac) Zhang, Z., Zhao, Z.K. (2010) *Biores. Technol.* **101**, 1111.

6. McKillip, W.J. (1989) *ACS Symp. Ser.*, **385**, 408.
7. Choura, M., Belgacem, N.M., and Gandini, A. (1996) *Macromolecules*, **29**, 3839.
8. Guigo, N., Mija, A., Vincent, L., and Sbirrazzuoli, N. (2007) *Phys. Chem. Chem. Phys.*, **9**, 5359.
9. Barsberg, S. and Thygesen, L.G. (2009) *Vib. Spectrosc.*, **49**, 52.
10. Bertarione, S., Bonino, F., Cesano, F., Damin, A., Scarano, D., and Zecchina, A. (2008) *J. Phys. Chem. B*, **112**, 2580.
11. Kawashima, D., Aihara, T., Kobayashi, Y., Kyotani, T., and Tomita, A. (2000) *Chem. Mater.*, **12**, 3397.
12. Yao, J., Wang, H., Liu, J., Chan, K.-Y., Zhang, L., and Xu, N. (2005) *Carbon*, **43**, 1709.
13. Zarbin, A.J.G., Bertholdo, R., and Oliveira, M.A.F.C. (2002) *Carbon*, **40**, 2413.
14. (a) Wang, H. and Yao, J. (2006) *Ind. Eng. Chem. Res.* **45**, 6393; (b) Yo, B., Rajagopalan, R., Foley, H.C., Kim, U.J., Liu, X., and Ecklund, P.C. (2006) *J. Am. Chem. Soc.*, **128**, 11307.
15. Hirasaki, T., Meguro, T., Wakihara, T., Tatami, J., and Komeya, K. (2007) *J. Mater. Sci.*, **42**, 7604.
16. Casano, F., Scarano, D., Bertarione, S., Bonino, F., Damin, A., Bordiga, S., Prestipino, C., Lamberti, C., and Zecchina, A. (2008) *J. Photochem. Photobiol. A Chem.*, **196**, 143.
17. Bertarione, S., Bonino, F., Cesano, F., Jain, S., Zanetti, M., Scarano, D., and Zecchina, A. (2009) *J. Phys. Chem. B.*, **113**, 10571.
18. (a) Tondi, G., Pizzi, A., Pasch, H., Celzard, A., and Rode, K. (2008) *Europ. Polym. J.*, **44**, 2938; (b) Tondi, G., Pizzi, A., Pasch, H., and Celzard, A. (2008) *Polym. Degrad. Stabil.*, **93**, 968; (c) Pizzi, A., Tondi, G., Pasch, H., and Celzard, A. (2008) *J. Appl. Polym. Sci.*, **110**, 1451; (d) Tondi, G., Pizzi, A., Delmonte, L., Parmentier, J., and Gadiou, R. (2010) *Ind. Crops Prod.*, **31**, 327.
19. Yao, J. and Wang, H. (2007) *Ind. Eng. Chem. Res.*, **46**, 6264.
20. Zhai, Y., Tu, B., and Zhao, D. (2009) *J. Mater. Chem.*, **19**, 131.
21. (a) Grund, S., Kempe, P., Baumann, G., Seifert, A., and Spange, S. (2007) *Angew. Chem. Int. Ed.*, **46**, 628; (b) Spange, S. and Grund, S. (2009) *Adv. Mater.*, **21**, 2111.
22. Pranger, L. and Tannenbaum, R. (2008) *Macromolecules*, **41**, 8682.

23. Lande, S., Westin, M., and Schneider, M. (2008) *Mol. Cryst. Liq. Cryst.*, **484**, 367.
24. Nordstierna, L., Lande, S., Westin, M., Karlsson, O., and Furó, I. (2008) *Holzforschung*, **62**, 709.
25. Lépine, O., Birot, M., and Deleuze, H. (2009) *Macromol. Mater. Eng.*, **294**, 599.
26. Guigo, N., Mija, A., Vincent, L., and Sbirrazzuoli, N. (2010) *Eur. Polym. J.*, **46**, 1016.
27. Alakhras, F. and Holze, R. (2007) *Synth. Met.*, **157**, 109.
28. Umeyama, T., Takamatsu, T., Tezuka, N., Matano, Y., Araki, Y., Wada, T., Yoshikawa, O., Sagawa, T., Yoshikawa, S., and Imahori, H. (2009) *J. Phys. Chem. C*, **113**, 10798.
29. Baldwin, L.C., Chafin, A.P., Deschamps, J.R., Hawkins, S.A., Wright, M.E., Witker, D.L., and Prokopuk, N. (2008) *J. Mater. Sci.*, **43**, 4182.
30. Yoneyama, H., Kawabata, K., Tsujimoto, A., and Goto, H. (2008) *Electrochem. Commun.*, **10**, 965.
31. Cho, B.R., Kim, T.H., Son, K.H., Kim, Y.K., Lee, Y.K., and Jeon, S.-J. (2000) *Macromolecules*, **33**, 8167.
32. Gandini, A., Coutterez, C., Goussé, C., Genheim, R., and Waig Fang, S. (2001) *ACS Symp. Ser.*, **784**, 98, and references therein.
33. Coutterez, C. (1998) Synthèse de polymères furaniques conjugués. Doctorate thesis, Grenoble Polytechnic Institute.
34. Baret, V., Gandini, A., and Rousset, E. (1997) *J. Photochem. Photobiol.*, **A103**, 169.
35. Waig Fang, S., Timpe, H.J., and Gandini, A. (2002) *Polymer*, **43**, 3505.
36. Gandini, A., Ariri, S., and Le Nest, J.-F. (2003) *Polymer*, **44**, 7565.
37. Pearl, P.A. and Tovar, J.D. (2009) *Macromolecules*, **42**, 4449.
38. (a) Moore, J.A. and Kelly, J.E. (1978) *Macromolecules*, **11**, 568; (b) Moore, J.A. and Kelly, J.E. (1978) *J. Polym. Sci., Polym. Chem. Ed.*, **16**, 2407; (c) Moore, J.A. and Kelly, J.E. (1979) *Polymer*, **20**, 627; (d) Moore, J.A. and Kelly, J.E. (1984) *J. Polym. Sci., Polym. Chem. Ed.*, **22**, 863.
39. Gandini, A., Khrouf, A., Boufi, S., and El Gharbi, R. (1998) *Macromol. Chem. Phys.*, **199**, 2755.
40. Chaabouni, A., Gharbi, S., Abid, M., Boufi, S., El Gharbi, R., and Gandini, A. (1999) *J. Soc. Chim. Tunisie*, **4**, 547.
41. Abid, M., Kamoun, W., El Gharbi, R., and Fradet, A. (2008) *Macromol. Mater. Eng.*, **293**, 39.
42. Gandini, A., Silvestre, A.J.D., Pascoal Neto, C., Sousa, A.F., and Gomes, M. (2009) *J. Polym. Sci. A Polym. Chem.*, **47**, 295.
43. (a) Gandini, A., Coelho, D., Gomes, M., Reis, B., and Silvestre, A. (2009) *J. Mater. Chem.*, **19**, 8656; (b) Gomes, M. (2009) Síntese de poliésteres a partir do ácido 2,5-furandicarboxílico. MSc thesis, University of Aveiro.
44. Grosshardt, O., Fehrenbacher, U., Kovollik, K., Tübke, B., Dingenouts, N., and Wilhelm, M. (2009) *Chem. Ing. Techn.*, **81**, 1829.
45. Lasseuguette, E., Gandini, A., Belgacem, A.M.N., and Timpe, H.J. (2005) *Polymer*, **46**, 5476.
46. Gandini, A. and Mitiakoudis, A. (1991) *Macromolecules*, **24**, 830.
47. (a) Abid, S., El Gharbi, R., and Gandini, A. (2004) *Polymer*, **45**, 5793; (b) Gharbi, S. and Gandini, A. (2004) *J. Soc. Chim. Tunisie*, **6**, 17.
48. (a) Gandini, A., Quillerou, J., Belgacem, M.N., Rivero, J., and Roux, G. (1989) *Polym. Bull.*, **21**, 555; (b) Gandini, A., Quillerou, J., Belgacem, M.N., Rivero, J., and Roux, G. (1989) *Eur. Polym. J.*, **25**, 1125; (c) Gandini, A., Belgacem, M.N., and Quillerou, J. (1993) *Eur. Polym. J.*, **29**, 1217; (d) Gandini, A., Belgacem, M.N., Boufi, S., and Quillerou, J. (1993) *Macromolecules*, **26**, 6706; (e) Gandini, A., Boufi, S., and Belgacem, M.N. (1995) *Polymer*, **36**, 1689.
49. (a) Gandini, A., Salon, M.C., and Amri, H. (1990) *Polym. Commun.*, **31**, 210; (b) Gandini, A., Amri, H., and Belgacem, M.N. (1996) *Polymer*, **37**, 4815; (c) Gandini, A., Amri, H., Belgacem, M.N., and Signoret, C. (1996) *Polym. Int.*, **41**, 427; (d) Gandini, A., Boufi, S., and Belgacem, M.N. (1997) *Polym. J.*, **29**, 479.

50. (a) Belgacem, M.N. and Gandini, A. (2007) *ACS Symp. Ser.*, **954**, 280; (b) Bergman, S.D. and Wudl, F. (2008) *J. Mater. Chem.*, **18**, 41.
51. (a) Teramoto, N., Arai, Y., and Shibata, M. (2006) *Carbohydr. Polym.*, **64**, 78; (b) Watanabe, M. and Yoshie, N. (2006) *Polymer*, **47**, 4946; (c) Ahmad, J., Ddamba, W.A.A., and Mathokgwane, P.K. (2006) *Asian J. Chem.*, **18**, 1267.
52. Gandini, A., Coelho, D., and Silvestre, A.J.D. (2008) *Eur. Polym. J.*, **44**, 4029.
53. Gandini, A., Silvestre, A.J.D., and Coelho, D. (2010) *J. Polym. Sci. A Polym. Chem.*, **48**, 2053.
54. Tian, Q., Rong, M.Z., Zhang, M.Q., and Yuan, Y.C. (2010) *Polymer*, **51**, 1779.
55. Yoshie, N., Watanabe, M., Araki, H., and Ishida, K. (2010) *Polym. Degrad. Stab.*, **95**, 826.
56. (a) Kavitha, A.A. and Singha, N.K. (2007) *J. Polym. Sci. A Polym. Chem.*, **45**, 4441; (b) Kavitha, A.A. and Singha, N.K. (2007) *Macromol. Chem. Phys.*, **208**, 2569; (c) Kavitha, A.A. and Singha, N.K. (2009) *ACS Appl. Mater. Interfaces*, **1**, 1427; (d) Kavitha, A.A. and Singha, N.K. (2010) *Macromolecules*, **43**, 3193.
57. Yamashiro, M., Inoue, K., and Iji, M. (2008) *Polym. J.*, **40**, 657.
58. (a) Liu, Y.-L., Hsieh, C.-Y., and Chen, Y.-W. (2006) *Polymer*, **47**, 2581; (b) Liu, Y.-L. and Chen, Y.-W. (2007) *Macromol. Chem. Phys.*, **208**, 224.
59. Liu, Y.-L. and Hsieh, C.-Y. (2006) *J. Polym. Sci. A Polym. Chem.*, **44**, 905.
60. Peterson, A.M., Jensen, R.E., and Palmese, G.R. (2009) *ACS Appl. Mater. Interfaces.*, **1**, 992; (2010) *ACS Appl. Mater. Interfaces.* doi: 10.1021/am9009378, **2**, 2170.
61. Tian, Q., Yuan, Y.C., Rong, M.Z., and Zhang, M.Q. (2009) *J. Mater. Chem.*, **19**, 1289.
62. Wei, H.-L., Yang, Z., Zheng, L.-M., and Shen, Y.-M. (2009) *Polymer*, **50**, 2836.
63. Ishida, K. and Yoshie, N. (2008) *Macromol. Biosci.*, **8**, 916.
64. Zhang, Y., Broekhuis, A.A., and Picchioni, F. (2009) *Macromolecules*, **42**, 1906.
65. Heath, W.H., Palmieri, F., Adams, J.R., Long, B.K., Chute, J., Holcombe, T.W., Zieren, S., Truitt, M.T., White, J.L., and Willson, C.G. (2008) *Macromolecules*, **41**, 719.
66. Plaisted, T.A. and Nemat-Nasser, S. (2007) *Acta Mater.*, **55**, 5684.
67. (a) Szalai, M.L., McGrath, D.V., Wheeler, D.R., Zifer, T., and McElhanon, J.R. (2007) *Macromolecules*, **40**, 818; (b) Polaske, N.W., McGrath, D.V., and McElhanon, J.R. (2010) *Macromolecules*, **43**, 1270.
68. Aumsuwan, N. and Urban, M. (2009) *Polymer*, **50**, 33.
69. Kose, M.M., Yesilbag, G., and Sanyal, A. (2008) *Org. Lett.*, **10**, 2353.
70. Jegat, C. and Mignard, N. (2008) *Polym. Bull.*, **60**, 799.
71. Chang, C.-M. and Liu, Y.-L. (2009) *Carbon*, **47**, 3041.
72. (a) Liu, X., Zhu, M., Chen, S., Yuan, M., Guo, Y., Song, Y., Liu, H., and Li, Y. *Langmuir*, **24**, 11967; (b) Liu, X., Liu, H., Zhou, W., Zheng, H., Yin, X., Li, Y., Guo, Y., Zhu, M., Ouyang, C., Zhu, D., and Xia, A. (2010) *Langmuir*, **26**, 3179.
73. (a) Shi, M., Wosnick, J.H., Ho, K., Keating, A., and Shoichet, M.S. (2007) *Angew. Chem. Int. Ed.*, **46**, 6126; (b) Shi, M. and Shoichet, M.S. (2008) *J. Biomater. Sci. Polym. Ed.*, **19**, 1143.

4
Selective Conversion of Glycerol into Functional Monomers via Catalytic Processes

François Jérôme and Joël Barrault

4.1
Introduction

Over the last decade, *glycerol* has received a lot of attention because of the rapid development of the vegetable oil industry, especially for the production of biodiesel, lubricants, and solvents, among others. Glycerol is the main co-product of the *oleochemistry* processes and the growing demand on biodiesel is now creating a significant glut in glycerol production. For instance, since 1992, the production of glycerol has been increasing by almost 3.75% per year. In order to favor a better industrial development of vegetable oils, it seems obvious that there is an urgency to find innovative solutions for the use of glycerol. It should be noted that glycerol can also be produced by enzymatic fermentation or catalytic hydrogenolysis of cellulose, thus providing a non-edible route to the production of glycerol. The growing interest of researchers in cellulose or lignocellulose is also expected to significantly increase glycerol production in the near future.

Glycerol is a natural C_3-triol and can be considered as a star-shaped molecule. Therefore, more and more studies are being devoted to its possible use as a monomer for the synthesis of safer and potentially more biodegradable polymers. From the view point of *green chemistry*, utilization of glycerol as a renewable organic building block for polymer syntheses provides many advantages, stemming from its low price (~0.5 € per kg in 2009), its biodegradability, its non-toxicity, and its ready availability. However, even though numerous polymers exhibit a glyceryl skeleton in their structures, the direct use of glycerol for their synthesis is still rare. To date, glycerol is essentially copolymerized with diacid derivatives to produce polyesters [1]. This tendency comes from the fact that the conversion of glycerol into functional monomers requires chemists to overcome numerous obstacles as a consequence of: (i) its high viscosity, (ii) its high hydrophilicity, and (iii) the presence of three hydroxyl groups with similar pK_a that can potentially lead to the formation of side products. For these reasons, with the aim of introducing a glyceryl unit into a polymer, more active monomers are generally preferred, such as glycidol, epichlorhydrin, or chloropropanediol. However, these molecules are much more toxic and expensive than natural glycerol.

Green Polymerization Methods: Renewable Starting Materials, Catalysis and Waste Reduction
Edited by Robert T. Mathers and Michael A. R. Meier
Copyright © 2011 WILEY-VCH Verlag GmbH & Co. KGaA, Weinheim
ISBN: 978-3-527-32625-9

The rapid developments in *material chemistry* have now given access to a diverse array of sophisticated heterogeneous catalysts that have been found to be particularly efficient for the selective conversion of glycerol [2]. These recent advances in the field of material chemistry have opened up new and elegant strategies for converting glycerol into a wide range of monomers that are usually synthesized from fossil propene.

In this chapter we will present the major results recently reported for the selective conversion of glycerol into functional monomers. In particular, we will make a comparison between the existing C_3-propene platform with that of glycerol.

4.2
Conversion of Glycerol into Glycerol Carbonate

In the field of polymers, *glycerol carbonate* appears to be a key monomer for the synthesis of various polymers such as polyurethanes, polycarbonates, or polyethers. Like glycerol, glycerol carbonate is cheap, biodegradable, and non-toxic. The biggest advantage of glycerol carbonate stems from the presence of four reactive electrophilic centers making possible its utilization as a monomer for the synthesis of a wide range of different polymers [3, 4]. In this context, Rokicki et al. reported the anionic polymerization of glycerol carbonate using partially deprotonated trimethylolpropane as an initiator [5]. This anionic polymerization reaction is accompanied by a spontaneous decarboxylation of the glycerol carbonate leading to the formation of hyperbranched aliphatic polyethers (Scheme 4.1). Glycerol

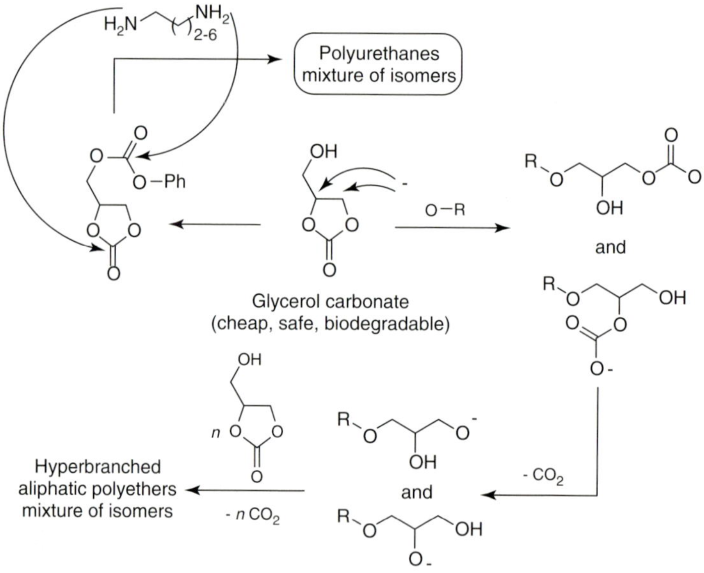

Scheme 4.1 Glycerol carbonate as valuable monomers.

carbonate can also be readily transformed into phenoxycarbonyloxymethylethylene carbonate, which is used not only as a coupler but also as a monomer for the synthesis of polyurethanes [6].

With the increasing demand for the design of safer polymers, the production of glycerol carbonate from glycerol has emerged as a fascinating topic and numerous elegant catalytic routes have been elaborated.

The first report concerning the conversion of glycerol into glycerol carbonate involved phosgene in the presence of basic catalysts such as NaOH, Na_2CO_3, or CaOH [7]. Even if satisfactory yields were obtained (90%), the environmental impact of the process is no longer acceptable. With the aim of by-passing the use of phosgene, alternative routes have been developed.

Transcarbonatation of cyclic organic carbonates (ethylene or propylene carbonate) with glycerol has been investigated over solid bases, such as metal oxides (Al_2O_3, ZnO, TiO_2, etc.), anion exchange resins (A26 HCO_3^- form), or zeolites [8]. Anion exchange resins (A26 HCO_3^- and OH^- form) were reported to be the most efficient solid catalysts, affording the corresponding glycerol carbonate with 85% yield at a lower temperature (110 °C) than the temperatures generally required with common solid catalysts (Scheme 4.2). Although these processes afford the glycerol carbonate with high yield, the side production of diol requires specific post-treatment in order to recover the glycerol carbonate with a high purity. Note that diethylcarbonate can also be employed, and in this instance NaOH is generally used as catalyst. Ethanol is then continuously removed from the reaction medium by distillation, thus offering a more convenient work-up.

Very recently, Corma and coworkers correlated the catalytic activity of various metal oxides (calcined hydrotalcites, regenerated hydrotalcites, MgO, CaO, among

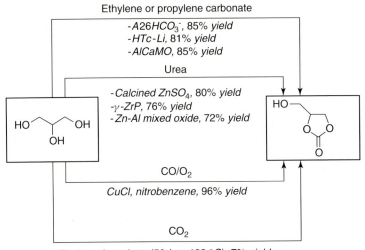

Scheme 4.2 Catalytic conversion of glycerol into glycerol carbonate.

others) with their basic properties [9]. The transcarbonatation of ethylene carbonate with glycerol was chosen as a model reaction because of the environmentally friendly properties of ethylene carbonate. In particular, they showed that metal oxides with strong Lewis basic sites allowed the reaction to proceed under very mild conditions (35 °C) and still with very high selectivity to glycerol carbonate. The best solid catalysts were obtained by: (i) the calcination of a hydrotalcite prepared by the partial substitution of Mg^{2+} with Li^+ (HTc-Li) and (ii) the calcination of a mixture of $CaCO_3$ + $Al(OH)_3$ (AlCaMO). These solid catalysts (0.5 wt%) were able to afford the glycerol carbonate with 81 and 85% yield, respectively, at 35 °C after only 15 min of reaction (Scheme 4.2). Note that HTc-Li and AlCaMO were stable and were successfully recycled at least three times without any significant drop in activity and selectivity.

Glycerol carbonate can also be obtained by reaction of glycerol with urea, which is cheap and also available on a large scale from biomass. This reaction is generally catalyzed by calcined metal oxides. Until very recently, the best result had been obtained using calcined $ZnSO_4$ as catalyst (80% yield, Scheme 4.2) [10]. Even though this reaction involves two natural and cheap organic building blocks, the purity of the final glycerol carbonate is still problematic, due to the formation of side products and to the partial dissolution of the catalyst. In 2009, Aresta *et al.* investigated the catalytic activity of a zirconium phosphate (γ-ZrP) catalyst in the *glycerolysis of urea* [11]. The reaction was conducted at 140 °C and under 2×10^{-4} bar (1 bar = 10^5 Pa) in order to facilitate the removal of the released ammonia. Starting from a glycerol/urea molar ratio of 1, a 76% yield of glycerol carbonate was obtained (Scheme 4.2). Interestingly, as compared with the above-described calcined $ZnSO_4$, the γ-ZrP was easily removed by filtration at the end of the reaction. However, a decrease in the glycerol carbonate was observed cycle after cycle. This drop in the selectivity was attributed by these workers to an increase in the catalyst acidity during the reaction, thus leading to the formation of side products. Moreover, when the γ-ZrP was thermally treated at the end of the reaction, the catalyst activity and the glycerol selectivity were maintained for at least for six catalytic cycles.

Taking into account that acid sites were able to activate the carbonyl group of urea while basic sites activate glycerol, Corma and coworkers investigated the *activity of solid catalysts* with well-balanced acid–basic properties [9]. In this context, they showed that a Zn–Al mixed oxide was highly efficient in the formation of glycerol carbonate from urea. Indeed, over Zn–Al mixed oxide (5 wt%) at 145 °C, with a glycerol/urea molar ratio of 1 and under vacuum (4×10^{-2} bar), the glycerol carbonate was obtained with 72% yield, thus pushing forward the beneficial effect of acid and basic sites on the reaction selectivity (Scheme 4.2). Moreover, the Zn–Al mixed oxide was found to be stable under these conditions and was successfully recycled over at least six consecutive catalytic cycles. It should be noted that the industrial viability of such processes closely relies on the utilization of ammonia, which is released here as a by-product. One of the solutions would require the coupling of these processes to the formation of urea, which could be synthesized directly from the released ammonia and carbon dioxide.

4.2 Conversion of Glycerol into Glycerol Carbonate

Oxycarbonylation of glycerol is also one of the possible routes for the synthesis of glycerol carbonate [12]. In this context, the best result was obtained using CuCl as catalyst in the presence of a gas mixture of $CO-O_2$ at a ratio of 95:5. This reaction takes place at 130 °C in nitrobenzene and affords the corresponding glycerol carbonate with 96% yield (Scheme 4.2).

With respect to the cost and waste minimization, the *direct carbonatation of glycerol* with carbon dioxide is now emerging as a fascinating and challenging task. Aresta et al. reported that tin-based catalysts were able to promote the coupling of glycerol with carbon dioxide without the assistance of any organic solvent [13]. However, the reaction yield was low, because under optimized conditions (50 bar, 180 °C, 8 h) only 7% yield of glycerol carbonate was obtained (Scheme 4.2). More recently, Munshi et al. showed that addition of methanol resulted in a spectacular increase in the glycerol carbonate yield [14]. Indeed, using dibutyltin(IV) oxide as a catalyst and methanol as solvent, glycerol carbonate was obtained with 35% yield (35 bar of CO_2, 80 °C, 4 h), thus opening an attractive avenue in the search for innovative processes for the direct carbonatation of glycerol with CO_2 (Scheme 4.2).

It should be noted that the *six-membered cyclic carbonates* derived from glycerol are also used as valuable monomers in polymer chemistry, for instance for the synthesis of hyperbranched polycarbonate or polycarbonate esters. However, to the best of our knowledge, no catalytic procedure is capable of converting glycerol into the six-membered ring glycerol carbonate with high yield. Indeed, in all cases the five-membered ring glycerol carbonate is obtained. Therefore, synthesis of the six-membered ring glycerol carbonate is still achieved through the conventional route using protective and deprotective agents (Scheme 4.3) [15].

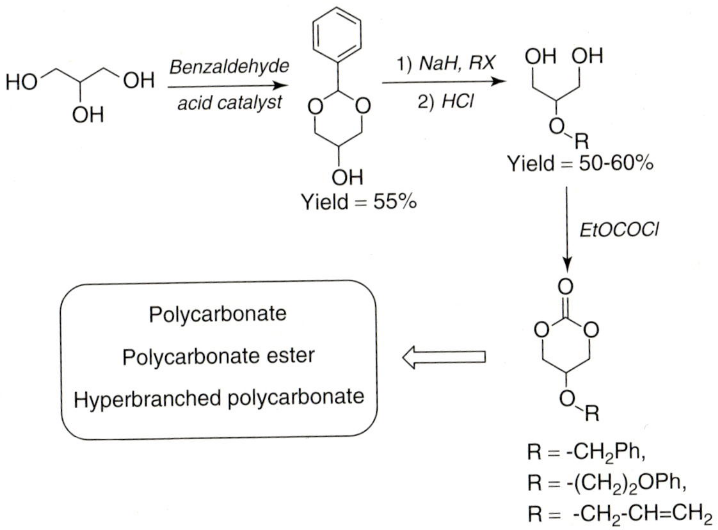

Scheme 4.3 Synthesis of the six-membered ring glycerol carbonate.

4.3
Conversion of Glycerol into Acrolein/Acrylic Acid

Acrolein and acrylic acid represent two key monomers for the synthesis of a wide range of polymers such as polyester resin or polyurethane. In particular, these chemicals are used in the manufacture of various plastics, coatings, adhesives, elastomers, and paints, among others.

Traditionally *acrolein* is prepared industrially by oxidation of propene, a gaseous petrochemical feedstock, using multicomponent mixed oxide catalysts. *Acrylic acid* is then obtained by oxidation of acrolein. Within the framework of green chemistry, the possibility of using glycerol as a renewable raw material for the synthesis of acrolein and acrylic acid represents an attractive prospect (Scheme 4.4).

Dehydration of glycerol to acrolein requires a high reaction temperature (250–340 °C) and/or vacuum in order to remove water from the reaction media and to displace the reaction toward the formation of acrolein. Dehydration of glycerol can be performed in the liquid or gaseous phase. In most instances, mixtures of glycerol and water are used in the presence of an *acid solid catalyst*. The most efficient acid solid catalysts are metal phosphates, zeolites, zirconia, and heteropolyacids [16]. For more information regarding the activity of these catalysts, see a recent review by Dumeignil and coworkers [17]. However, even if these processes are able to convert glycerol into acrolein (best result: 73% yield at total conversion over silicotungstic acid supported on silica with mesopores of 10 nm), the economical viability is still problematic. Indeed, these processes lead to the side production of hydroxypropanone, propanaldehyde, acetaldehyde, acetone, and polyglycerol derivatives, which have to be removed and disposed off. Moreover, a deactivation of these solid catalysts also occurs due to the formation of coke. The possibility of performing this reaction in sub- and supercritical water has also been explored [18]. Despite interesting results, the conversion of glycerol (50%), the selectivity to acrolein (75%), and the reaction conditions (250 bar, 360 °C) are not yet satisfactory for an industrial application. In 2005, the company Arkema patented an efficient process for the catalytic dehydration of glycerol to produce acrolein [19]. This process involves a WO_3/ZrO_2 solid catalyst, which displayed a

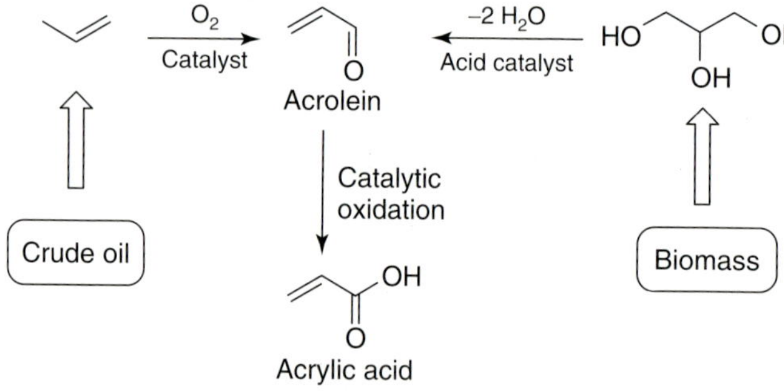

Scheme 4.4 Synthesis of acrolein and acrylic acid from glycerol.

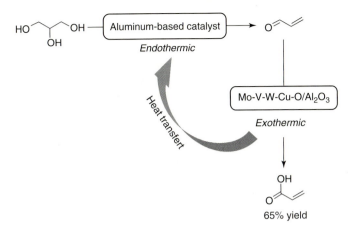

Scheme 4.5 Catalytic conversion of glycerol into acrylic acid.

very high activity and selectivity to acrolein (80% of acrolein has been obtained at total conversion).

As a continuation of their efforts, Arkema reported a new way for directly converting glycerol into acrylic acid without isolating acrolein [20]. This process is divided into two steps involving two consecutive catalytic reactors (Scheme 4.5).

The dehydration of glycerol to acrolein takes place over an *aluminum-based catalyst* impregnated with phosphoric acid and silica. Once produced, acrolein is immediately transferred into a second reactor and oxidized to acrylic acid over a Mo–V–W–Cu–O catalyst supported over alumina. The reaction was performed between 250 and 350 °C under a diluted oxygen flow (1–5 bar), which favors not only the oxidation of acrolein but also limits the formation of secondary products (propanaldehyde, acetone, hydroxypropanone) and the deactivation of the catalyst. Under these conditions, up to 65% yield of acrylic acid was obtained directly from glycerol. Interestingly, the exothermicity of the oxidation reaction was combined with the endothermicity of the dehydration reaction, thus considerably limiting the energy consumption of the process. Note that the secondary products formed were also burned to provide energy to the process. Even if the actual tonnage of glycerol is not high enough to totally replace propene in the synthesis of acrylic acid, Arkema directly combined the actual propene-based process with that of glycerol, thus limiting the dependency of this process on fossil carbon (acrolein being independently produced from propene and glycerol before conversion into acrylic acid).

4.4
Conversion of Glycerol into Glycidol

Glycidol is a widely used monomer in polymer chemistry and is produced industrially from the oxidation of allylic alcohol using hydrogen peroxide and a tungsten-based catalyst. It is not only used for the introduction of a glycerol moiety on a polymer but can also be polymerized to polyethers or *polyglycerol* either by cationic or anionic polymerization [21]. Polyglycerols and their modified derivatives represent an important class of polymers that display numerous applications, such as surfactants, vectors for pharmaceutical ingredients, among others. Direct synthesis of polyglycerol from glycerol has become a challenging task, which is now the subject of numerous investigations. Glycerol can be polymerized either in the presence of an acid or basic catalyst [22]. In particular, a theoretical study on the role of surface basicity and Lewis acidity on the etherification of glycerol has recently been reported [22c]. However, compared with glycidol, the resulting polyglycerol was obtained with a very low degree of polymerization [22]. Therefore, some groups are now exploring the possibility of converting glycerol into glycidol, which would represent a direct biomass-based route for the synthesis of polyglycerol with a high degree of polymerization.

The first studies concerning the *production of glycidol from glycerol* were reported by Bruson, Riener, and coworkers in 1952–1953 [23]. In these reactions, glycidol was produced from a mixture of glycerol and a cyclic carbonate such as ethylene carbonate. After heating the reaction at a temperature of between 125 and 260 °C, 60% yield of glycidol was obtained (glycidol was isolated by continuous vacuum distillation). Even if the mechanism was not clearly understood, these workers initially suspected a transcarbonatation of ethylene carbonate with glycerol, followed by the thermal decomposition of the glycerol carbonate, produced *in situ*, to glycidol (Scheme 4.6). Later, Malkemus and Currier reported that addition of 0.001–1 wt% of a metal salt (for example, LiCl, NaBr, Ca$_2$CO$_3$, among others) led to a significant increase in the glycidol yield [24]. Indeed, at a temperature range of 175–225 °C, a pH range of 6–7, and under a pressure of from 1.3×10^{-3} to 0.13 bar, 80–90% yield of glycidol was obtained.

More recently, Moulougui and coworkers optimized this process [25]. Glycerol was reacted with glycerol carbonate either in the presence of a zeolite (type A) or γ-alumina. The reaction was performed in a temperature range of 170–210 °C and

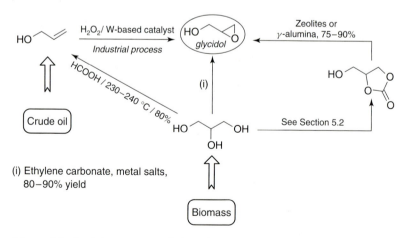

Scheme 4.6 Catalytic conversion of glycerol into glycidol.

under a pressure of from 3.3×10^{-2} to 0.1 bar in order to continuously produce the distillate glycidol. Under these conditions glycidol was obtained with 75–90% yield (Scheme 4.6). Compared with the above-described reaction, the main advantage of this process stems from the utilization of two reactants directly issued or produced from biomass: glycerol and glycerol carbonate.

In 2009, Bergman and co-workers reported an exciting new way of converting glycerol into allylic alcohol, thus opening a new route for accessing *glycidol directly from biomass* [26]. In particular, on heating glycerol with formic acid at a temperature of around 230–240 °C, these workers observed the formation of allylic alcohol. Proceeding under a stream of nitrogen, they completely eliminated charring, which was mainly responsible for the formation of undesired products, and allylic alcohol was isolated with 80% yield. Note that during the reaction, allylic alcohol is continuously removed from the reaction medium by distillation. Possible formation of allylic alcohol from glycerol is noteworthy and opens up access to a very attractive and competitive route to glycidol (Scheme 4.6).

4.5
Oxidation of Glycerol to Functional Carboxylic Acid

Oxidation of glycerol has emerged as a very powerful reaction because the products resulting from this reaction already have, and could have various other, applications in polymer chemistry. Indeed, oxidation of glycerol affords a wide range of functional carboxylic acids. These functional carboxylic acids can be divided into two families: (i) the hydroxycarboxylic acids and (ii) the dicarboxylic acids. Scheme 4.7 summarizes the main functionalized carboxylic acids that can be obtained from glycerol.

Because of the presence of three hydroxyl groups on the glycerol moiety, the control of the reaction selectivity is rather complex and, in most instances a mixture of oxygenated derivatives is usually obtained after oxidation. Among the possible oxygenated adducts, *glyceric acid* appears as a platform chemical (Scheme 4.8). Indeed, through the over oxidation of glyceric acid, various functional monomers can be produced, such as tartronic, mesoxalic, oxalic, and hydropyruvic acids. Therefore, most of the reported oxidative catalytic processes focused on the selective conversion of glycerol into glyceric acid and a number of elegant strategies have recently been reported. Note that we will not discuss here the oxidation of glycerol to dihydroxyacetone because the market of this derivative is not dedicated to polymer chemistry. Comprehensive articles concerning the oxidation of glycerol to dihydroxyacetone, which generally requires assistance of promoters such as Bi, can be found in the literature [27].

4.5.1
Catalytic Oxidation of Glycerol to Glyceric Acid

Using air as oxidant, Besson and coworkers found that 5 wt% Pd supported over activated carbon was able to selectively catalyze the conversion of glycerol into

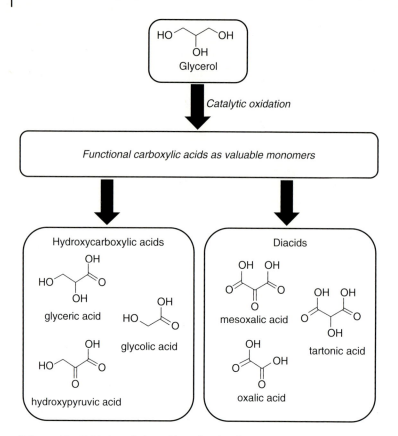

Scheme 4.7 Oxidation of glycerol into functional monomers.

glyceric acid in water [27c]. The pH of the solution was found to be crucial either for the catalyst activity or the selectivity to glyceric acid. Under optimized conditions (pH = 11, 60 °C), 70% selectivity of glyceric acid was obtained at total conversion. 5 wt% Pt deposed over activated carbon was found to be much more active than Pd/C [TOF (turnover frequency) = 375 h^{-1} versus 100 h^{-1} for Pd/C] but far less selective due to the over oxidation of glyceric acid [27]. Indeed, under the same conditions, glyceric acid was obtained with only 55% selectivity at 90% conversion over Pt/C.

Later, Hutchings and coworkers investigated the *catalytic activity* of 1 wt% of gold dispersed over activated carbon or graphite [28]. Compared with platinum, gold is more resistant to the oxygen poisoning and, therefore, catalytic processes can be performed under pressure of dioxygen. In particular, they found that after optimization (6 bar of O_2, 0.5 mol% of Au dispersed over graphite, 2 equiv. of NaOH), glyceric acid can be produced in water with 92% selectivity at 91% conversion after only 3 h of reaction. These workers pointed out that the particles size dramatically influenced the catalyst activity and the selectivity to glyceric acid. In particular, whereas gold nanoparticles of 2–4 nm are generally required for

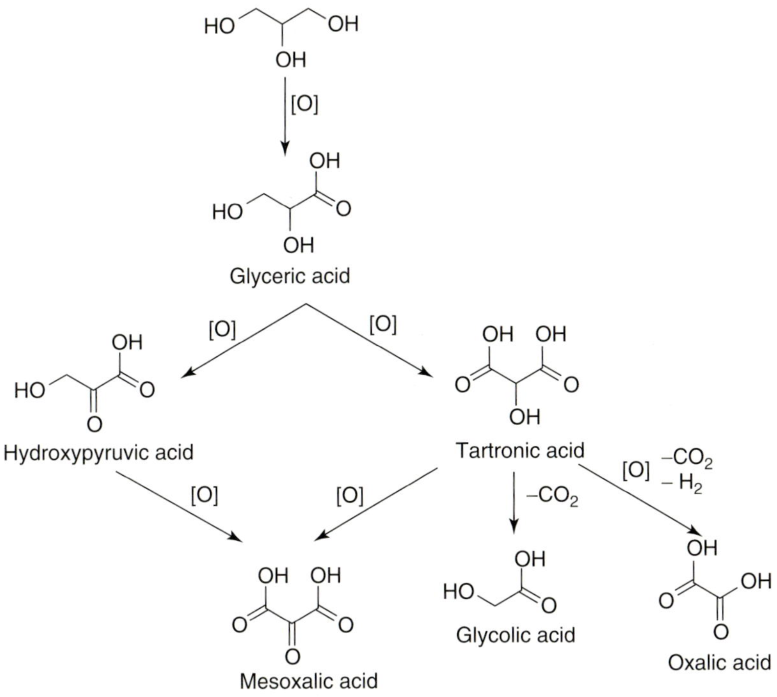

Scheme 4.8 Glyceric acid as a platform chemical.

the low temperature oxidation of carbon monoxide, they found that an optimal gold nanoparticle size of 15–30 nm was the best for the selective oxidation of glycerol to glyceric acid. As a continuation of this work, Porta *et al.* showed that using well-dispersed gold nanoparticles with an average diameter of 6 nm, the selectivity dramatically dropped from the initial to the final stage of the reaction [29]. Conversely, using larger particles size (>20 nm) a constant selectivity of glyceric acid was achieved from the beginning to the end of the reaction, thus confirming the results previously reported by Hutchings and coworkers. Under optimized conditions, (30 °C, 4 equiv. of NaOH, glycerol/Au molar ratio = 500), 92% selectivity of glyceric acid was obtained at 90% conversion using 1 wt% Au/C prepared by the citrate-protected gold-sol method.

Whereas most of the gold-based processes were performed under oxygen pressure (3–6 bar), Claus and coworkers recently reported that *nanosized gold catalysts deposited over carbon were active under atmospheric pressure and at constant pH* [30]. Activity of these gold-based catalysts closely depended on the nature of the carbonaceous solid support. The best results were obtained using activated carbon NSX1G. Under these conditions, glyceric acid was successfully produced with 53% selectivity at 50% conversion. The superior activity of activated carbon NSX1G as compared with other activated carbons that were tested was attributed to a lower fraction of micropores, thus making the Au nanoparticles more accessible to the glycerol.

As suggested by previous studies, this work confirms that the catalyst activity and the reaction selectivity are not strictly governed by the metal particles size, but that the nature of the carbonaceous support also plays a significant role on the reaction mechanism. However, as observed in the chemistry of glucose, it should be noted that the size of the gold particles plays the major role in the catalyst activity [31]. As observed previously, using platinum as a *promoter*, the catalytic activity of the resulting Pt-Au/C catalyst was significantly boosted (40% conversion after 1 h versus 18% conversion for the unpromoted Au/C). Indeed, whereas Pt/C is rapidly deactivated during the reaction, the bimetallic Pt-Au/C was found to be much more stable. However, over Pt-Au/C, the reaction selectivity is now shifted toward the main formation of dihydroxyacetone, in accordance with what was reported above on the work by Besson and coworkers with Pt/C. Indeed, in this instance, the selectivity to glyceric acid drops from 44 to 34% (at 50% conversion) when the Pt/Au molar ratio increases from 0 to 1. Although XRD (X-ray diffraction) measurements detected the formation of Pt-Au alloys, the plot of the glycerol conversion versus time does not clearly highlight a synergistic effect between Pt and Au, and the collected results seem to be an average between what is obtained from Au/C and from Pt/C.

4.5.2
Oxidative-Assisted Polymerization of Glycerol

4.5.2.1 Cationic Polymerization
Searching further improvements to the *selective oxidation of glycerol*, Kimura successfully achieved the direct synthesis of a polyketomalonate by catalytic oxidation of an aqueous solution of glycerol, thus showing the potential of the above-described functional monomers for the synthesis of safer polymers [32].

In earlier reports, Kimura and coworkers investigated the *oxidative-assisted polymerization of glycerol* using a pair of fixed-bed reactors filled with two different solid catalysts: CeBiPtPd/C for the oxidation of glycerol to tartronic acid and then BiPt/C for the oxidation of tartronic acid to ketomalonic acid. Under acidic conditions, they successfully observed the polymerization of ketomalonic acid. However, as mentioned by these workers, this approach proved to be technically difficult. With the aim of simplifying the process, Kimura and coworkers then prepared a new type of solid catalyst, a 0.45% Ce–0.6% Bi–3% Pt/C, which was expected to oxidize simultaneously both the primary and secondary hydroxyl groups of glycerol, thus allowing the utilization of a single catalyst. When the concentration of glycerol in water was about 7.5%, a mixture of 32% mesoxalic acid and 30% oxalic acid (resulting from the decarboxylation of mesoxalic acid) was obtained. Remarkably, when the aqueous concentration of glycerol was then raised from 7.5 to 30%, polymeric materials (amounting to about 55%) were produced. By means of infrared and ^{13}C-NMR studies, this polymeric material was identified as a polyketomalonate, a biodegradable polyether with pendent carboxyl groups attached to the carbon backbone. During the reaction, the polymeric mixture was gradually decarboxylated to form a random copolymer of mesoxalic and glyoxylic acids (Scheme 4.9).

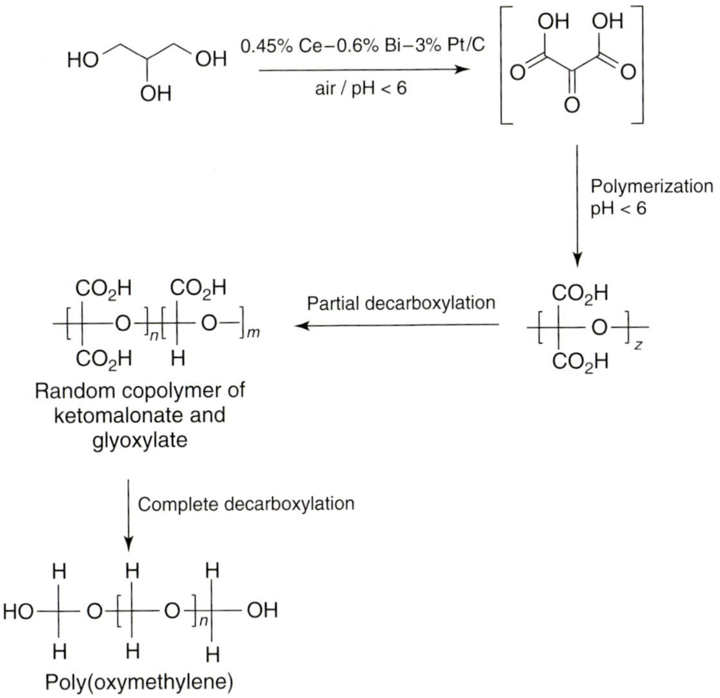

4.5 Oxidation of Glycerol to Functional Carboxylic Acid

Scheme 4.9 Oxidative/cationic polymerization of glycerol.

Through GPC (gel permeation chromatography), the average molecular weight of this polymer was estimated to be 5000–30 000 g mol^{-1}. It should be noted that complete decarboxylation of this polymer leads to the formation of poly(oxymethylene), which is used largely in the plastics industry [33].

4.5.2.2 Anionic Polymerization

As a continuation of their efforts, Kimura then investigated the *oxidation/anionic polymerization of glycerol* [34]. To this end, three different types of solid catalysts were prepared. Catalyst compositions and the results are listed in Table 4.1.

According to the solid catalyst composition, various polymers (M_w 7300–1 000 000) with different polydispersities can be obtained at a pH higher than 10 directly from glycerol. Even if the amount of polymeric material was much lower than in the case of the oxidative/cationic polymerization of glycerol, we can clearly note that the substitution of bismuth by tellurium significantly increases this amount from 0.2 to 7.2% (Table 4.1).

By means of NMR studies, Kimura showed that, under basic conditions, the tartronic acid sodium salt, the intermediate produced *in situ* during the glycerol oxidation, plays the role of an initiator for the polymerization of the mesoxalic acid sodium salt (Scheme 4.10).

Interestingly, compared with the oxidation/cationic polymerization of glycerol, the resulting *polycarboxylate* was found to be much more stable and, in alkaline

4 Selective Conversion of Glycerol into Functional Monomers via Catalytic Processes

Table 4.1 Oxidative/anionic polymerization of glycerol over CeBiPd/C catalysts.

Catalyst composition				Polymer content (%)[a]	M_w[b]	M_n[b]	M_w/M_n[b]
Ce	Te	Bi	Pd				
0.9	–	1.2	3.0	0.10	1 000 000	344 000	2.91
				0.12	81 000	70 000	1.16
0.9	–	3.0	3.0	0.68	456 000	278 000	1.64
				0.60	80 000	70 000	1.14
				0.28	17 600	16 000	1.10
0.9	1.5	–	3.0	0.14	345 000	284 000	1.21
				3.05	45 000	38 000	1.18
				3.98	7 300	4 300	1.70

[a] 100 g of catalyst were packed into the reactor, glycerol/NaOH molar ratio = 1 : 2 (continuous flow rate of 16.0 g h^{-1}), oxygen rate = 2.3 1 h^{-1}, 50 °C.
[b] Determined by GPC.

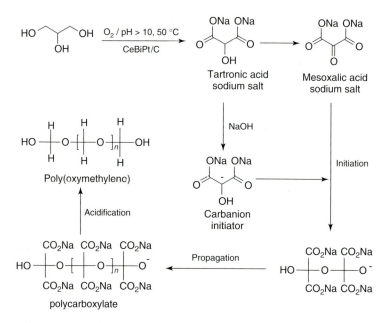

Scheme 4.10 Oxidative/anionic polymerization of glycerol.

solution, no decarboxylation takes place. Note that this biodegradable polycarboxylate was found to be stable up to 250 °C. However, as observed above, acidification of the reaction media leads to an important decarboxylation of the polymer yielding the corresponding poly(oxymethylene).

High molecular weight polycarboxylates are used in the manufacture of detergents. Whatever the considered mechanism (cationic or anionic), this work opens an aqueous and bio-based alternative to the process patented by Monsanto, consisting in the anionic polymerization of dimethylester of ketomalonate in dichloromethane [35].

4.6
Conversion of Glycerol into Acrylonitrile

Acrylonitrile is a highly valuable monomer involved in the manufacture of various polymers, such as polyacrylonitrile (for acrylic fibers), synthetic rubber, among others. Its dimerization affords adiponitrile, which is widely used for the synthesis of Nylons. Acrylonitrile is nowadays produced industrially on a large scale (6 million tons in 2005) through the Sohio process by ammoxidation of propylene over a molybdenum-based catalyst.

As acrolein is an intermediate in the *ammoxidation* of propene, Guerrero-Perez and Bañaress recently investigated the possibility of producing acrylonitrile directly from glycerol (Scheme 4.11). In this study, they investigated the catalytic activity and selectivity of a VSb-based catalyst supported over γ-alumina, which is a solid catalyst similar to that used in the ammoxidation of propene [36]. The reaction was performed in the gas phase at 400 °C from glycerol using a mixture of $NH_3 + O_2$ (25% + 8.6%) diluted in helium. Interestingly, VSb/Al_2O_3 was found to be particularly efficient yielding 56% selectivity to acrylonitrile at 72% conversion of glycerol. Introduction of niobium afforded an even more selective catalyst. Indeed, using $VSbNb/Al_2O_3$, 58% selectivity to acrylonitrile was achieved at 83% conversion, thus opening a new bio-based route for the synthesis of acrylonitrile.

Further inspections of the catalyst surface by Raman spectroscopy clearly showed the presence of V–Nb–O mixed phases, which played a key role in the reaction mechanism. By means of counter experiments, workers showed that

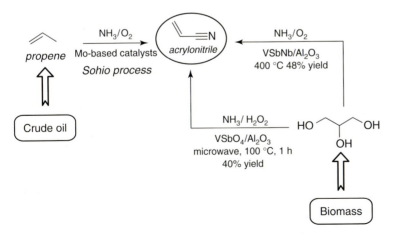

Scheme 4.11 Catalytic conversion of glycerol to acrylonitrile.

the vanadium sites are necessary for the glycerol activation, whereas antimony sites are responsible for the formation of nitrile. Addition of niobium to the VSb/Al_2O_3 catalyst increased its acidity, which facilitated the formation of acrolein and enhanced the activation of ammonia. It should be noted that, at 400 °C, a deactivation of the VSbNb/Al_2O_3 catalyst occurred due to the formation of coke and polyacrylonitrile-derived deposits.

More recently, Calvino-Casilda et al. investigated the production of acrylonitrile from glycerol in the liquid phase and in the presence of microwave irradiation (0–100 W with a heating ramp = 10 °C min^{-1}) (Scheme 4.11) [37]. The same catalysts as those discussed above were used. The reaction was performed at 100 °C in neat glycerol (0.5 mmol) and in the presence of ammonia (NH_4OH, 57 mmol) and hydrogen peroxide (15 mmol). Among all the solid catalysts tested, the rutile $VSbO_4$/Al_2O_3 yielded the best results. Indeed, the $VSbO_4$/Al_2O_3 catalyst afforded, after only 1 h of reaction, 84% selectivity to acrylonitrile at 47% conversion of glycerol. Note that without microwave activation, the rutile $VSbO_4$/Al_2O_3 catalyst led to only 14% conversion of glycerol (63% selectivity to acrylonitrile) after 20 h of reaction, thus promoting the effectiveness of the microwave irradiation in this reaction. Note that if this microwave assisted route offers an attractive alternative to the conversion of glycerol into acrylonitrile at 400 °C, the large excess of ammonia, hydrogen peroxide, and water remains a huge problem for industrial applications, and optimizations are still needed.

4.7
Selective Conversion of Glycerol into Propylene Glycol

4.7.1
Conversion of Glycerol into Propylene Glycol

Propylene glycol is used as monomer in the manufacture of unsaturated polyester resins. This monomer is produced industrially through the hydration of propylene oxide, which is itself obtained from propylene either by the chlorohydrin or the hydroperoxide process (Scheme 4.12). Possible formation of propylene glycol directly from glycerol as renewable feedstock has emerged as a feasible approach. In this context, a great deal of work has been dedicated to this reaction. Conversion of glycerol into propylene glycol is achieved under hydrogen pressure and in the presence of a catalyst. In most instances, these works required a dilute aqueous solution of glycerol (30 wt%). Therefore, energy consuming processes are necessary in order to remove water from the propylene glycol. Moreover, owing to the harsh conditions (high temperature and high hydrogen pressure), the selectivity of these processes is still unsatisfactory and numerous side products, such as ethylene glycol, lactic acid, acetol, acrolein, among others, are formed. In this section, we will only discuss the recent significant results obtained in this field, and provide an overview of the most suitable processes that allow the conversion of glycerol into propylene glycol with high yields.

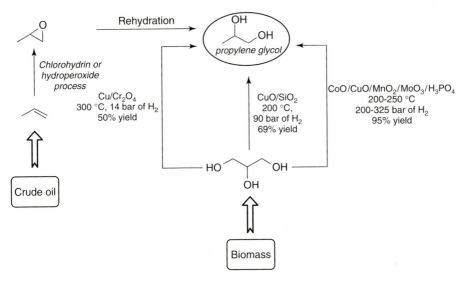

Scheme 4.12 Catalytic conversion of glycerol into propylene glycol in liquid phase.

4.7.1.1 Reaction in the Liquid Phase

Schuster and Eggersdorfer reported an efficient catalytic route to the production of propylene glycol directly from glycerol [38]. All catalysts employed do not contain support and consist of from 40 to 70 wt% cobalt (as CoO), 13–17 wt% copper (as CuO), 3–8 wt% manganese (as MnO_2), 0.1–5 wt% phosphorus (as H_3PO_4), and 0.5–5 wt% molybdenum (as MoO_3). Prior to reaction, the catalyst was activated by reduction in a hydrogen stream at from 200 to 400 °C. Remarkably, compared with previous studies, this catalytic process was found to be particularly efficient when starting from a concentrated solution of glycerol (86–100 wt%). At 200–250 °C and 200–325 bar of hydrogen, these workers claimed a propylene glycol yield as high as 95%, thus opening an alternative route to the traditional use of propylene for the synthesis of propylene glycol (Scheme 4.12).

With the aim of limiting the *hydrogen dependency* of the process, in 2005 Suppes and coworkers investigated the conversion of glycerol into propylene glycol over nickel, palladium, platinum, copper, and copper–chromite catalysts [39]. In particular, they found that copper–chromite catalyst was able to convert a concentrated solution of glycerol (80–90 wt%) into propylene glycol at much lower hydrogen pressure than that usually required (14 bar), thus significantly decreasing the energy consumption of the process. After optimization (reaction temperature of 300 °C, catalyst reduction temperature of 300 °C, 10–20 wt% of water), 50% yield of propylene glycol was obtained at 70% conversion (Scheme 4.12).

Interestingly, by means of counter experiments, a plausible reaction pathway was proposed involving (i) a dehydration of glycerol to acetol followed by (ii) a reduction to propylene glycol (Scheme 4.13).

Considering that copper was an excellent candidate for use in the conversion of glycerol into propylene glycol, Xia and coworkers recently showed that CuO/SiO_2

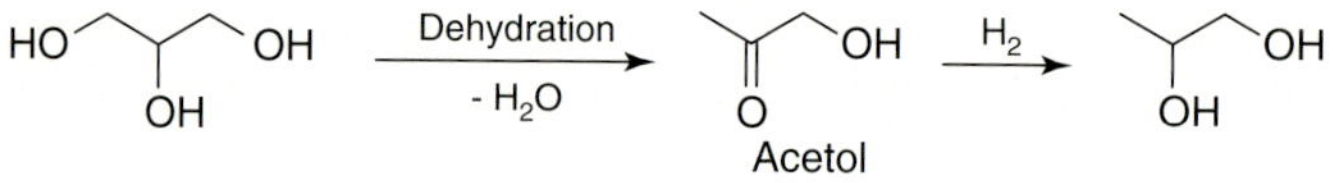

Scheme 4.13 Plausible reaction mechanism for the formation of propylene glycol from glycerol.

prepared by a precipitation–gel technique was a very efficient catalyst [40]. In particular, starting from a concentrated solution of glycerol (30 wt% in water), propylene glycol can be produced with 94% selectivity at 73% conversion (i.e., yield = 69%) at 200 °C and 90 bar of hydrogen. Compared with other tested catalysts, the remarkable activity of the CuO/SiO_2 catalyst relies on: (i) the presence of highly dispersed copper nanoparticles with an average diameter of 6 nm, (ii) strong metal–support interaction, and (iii) high resistance to sintering due to the presence of Cu^+ species on the catalyst surface formed during the prereduction treatment of the catalyst. Therefore, compared with most of the copper-based catalysts, the CuO/SiO_2 was found to be much more stable and no deactivation takes place.

In 2008, Jacobs and coworkers reported what is probably one of the most ingenious processes for the production of propylene glycol from glycerol [41]. Instead of using expensive bio- or petrochemical derived hydrogen (which is the main limitation of the actual glycerol-based processes), hydrogen was produced *in situ* by reforming of glycerol (230 °C, 20 wt% of glycerol in water).

Using a basic Pt/NaY *zeolite as catalyst*, they found that mainly gaseous compounds (81% H_2, 10% CO_2, 5% CO) were formed at the initial stage of the reaction. Then, in the presence of water, CO_2 reacted with the water to form carbonic acid, which then interacted with Pt/NaY to generate a partially exchanged Pt/H-NaY zeolite. This enhancement of the acidity of the Pt/NaY zeolite promoted the dehydration of glycerol to acetol, which was then reduced over Pt in the presence of the H_2 produced *in situ* (Scheme 4.14). Under these conditions, propylene glycol has been obtained with 64% selectivity at 85% conversion. Considering that intrazeolitic Pt favors gas formation, it has been suggested that large extra-framework Pt particles were probably responsible for the hydrogenation of the acetol. It should be noted that acidity of the catalyst plays also a major role on the reaction selectivity and a good balance between the acid strength and the metal particle have to be found.

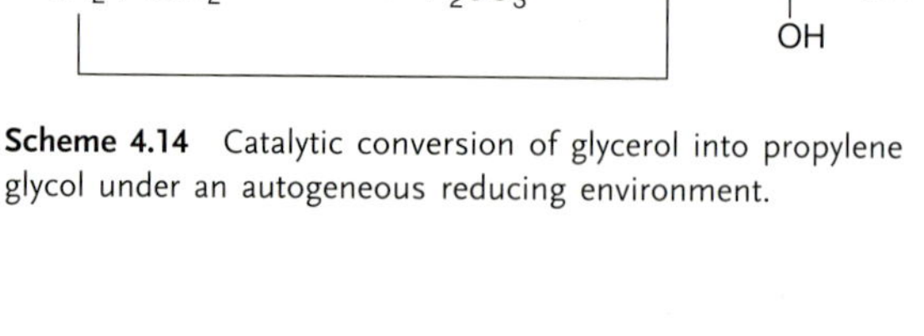

Scheme 4.14 Catalytic conversion of glycerol into propylene glycol under an autogeneous reducing environment.

4.7.1.2 Reaction in the Gas Phase

Continuing the work of Suppes, Zhu and coworkers recently investigated the conversion of a concentrated solution of glycerol (60 wt% glycerol in water) into propylene glycol in the gas phase using a *fixed-bed reactor* [42]. Among a large library of solid catalysts that were tested, the $Cu/ZnO/Al_2O_3$ solid catalyst was found to display the best activity. Indeed, at 50 bar of hydrogen and 200 °C, 80% selectivity of propylene glycol at 20% conversion has been obtained with a weight hourly space velocity of $0.08\,h^{-1}$. After optimization of the reaction conditions ($T = 190\,°C$; $H_2 = 64$ bar; H_2/glycerol molar ratio $= 140$), 92% selectivity of propylene glycol was obtained at 96% conversion (Scheme 4.15). By means of Gibbs energy calculations, these workers supported the above-described mechanism proposed by Suppes and coworkers based on the synthesis of acetol as an intermediate. The remarkable activity of the $Cu/ZnO/Al_2O_3$ catalyst was ascribed by them to the concomitant presence of acid and reductive sites on the catalyst surface. It has been proposed that the acid sites of ZnO enhance the dehydration of glycerol to acetol, while the reduction of acetol to propylene glycol would occur over copper.

In a similar way, in 2009 Sato et al. investigated, the conversion of glycerol into propylene glycol over Cu/Al_2O_3 in the gas phase (30 wt% of glycerol in water) and at atmospheric pressure of hydrogen [43]. In particular, they showed that the dehydration of glycerol to acetol and the hydrogenation of acetol to propylene glycol were closely dependent on the reaction temperature. Dehydration of glycerol to acetol is favored at high temperature (>190 °C), whereas the hydrogenation of

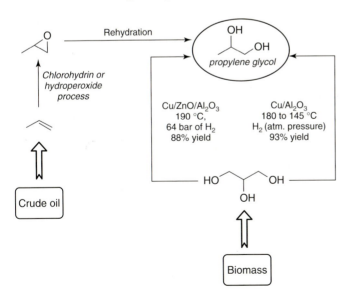

Scheme 4.15 Catalytic conversion of glycerol into propylene glycol in the gas phase.

acetol to propylene glycol requires a lower temperature (<190 °C). Indeed, at high temperature (>190 °C), workers found that propylene glycol can be dehydrogenated over copper to regenerate the acetol, resulting in the formation of by-products. Under atmospheric pressure of H_2 (H_2 flow = 360 cm^3 min^{-1}) and using a gradient of temperatures between 180 and 145 °C, they successfully obtained a 93% yield of propylene glycol (Scheme 4.15).

4.7.2
Conversion of Glycerol into 1,3-Propanediol

In the field of polymer chemistry, the possibility of converting glycerol into 1,3-*propanediol* is much more valuable than the production of propylene glycol. Indeed, 1,3-propanediol is a very important monomer involved in the synthesis of various polyesters, which find various applications today, such as composites, adhesives, laminates, coatings, moldings, fibers, among others. 1,3-Propanediol is produced industrially either through the Degussa–Dupont process via hydration of acrolein, or through the Shell process via hydroformylation of ethylene oxide (Scheme 4.16). With the depletion of the fossil oil reserves, the direct conversion of glycerol into 1,3-propanediol has become of greater interest. In this context, Dupont's Sorona® [poly(trimethyleneterephthalate)] is among the first industrial

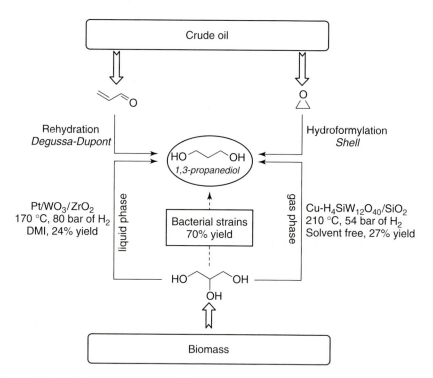

Scheme 4.16 Catalytic conversion of glycerol into 1,3-propanediol.

materials in which 1,3-propanediol, one of the two starting monomers, is made from glycerol. The resulting material contains 37% renewably sourced ingredients by weight. In this process, 1,3-propanediol is produced through a fermentation route from corn as a renewable raw material. This enzymatic process involves the conversion of glucose into glycerol and then the conversion of glycerol into 1,3-propanediol.

The direct catalytic conversion of glycerol into 1,3-propanediol is rather complex and, in most instances, the 1,2-propanediol is obtained as the main product (see Section 4.7.1). In 2005, Chaminand et al. investigated the conversion of glycerol into 1,3-propanediol over an Rh/C solid catalyst [44]. In particular, they showed that on heating at 180 °C and under 80 bar of H_2, a solution of glycerol in sulfolane with 0.3 mol% of tungstic acid, (230 g of glycerol l^{-1}), 1,3-propanediol can be produced with a higher selectivity than propylene glycol. However, despite these interesting results, the yield of 1,3-propanediol remains very low (4%) making the scale-up of this process difficult.

Considering that the selectivity to 1,3-propanediol can be improved in the presence of tungstic acid, noble metal, and polar aprotic solvent, Sasaki and coworkers later investigated the catalytic activity of noble metal dispersed over WO_3/ZrO_2 [45]. All reactions were performed in an autoclave at 170 °C and 80 bar of H_2. Instead of sulfolane, they used the 1,3-dimethyl-2-imidazolidinone (DMI) as the polar aprotic solvent (glycerol concentration = 15 mol l^{-1}), which is known to be more stable. Among all the noble metals tested (Pt, Pd, Rh, Ru, Ir), the $Pt/WO_3/ZrO_2$ afforded the best results. Indeed, under these conditions, 1,3-propanediol was obtained with an unprecedented yield of 24%, which corresponds to a selectivity twice that of propylene glycol (Scheme 4.16). It should be noted that the catalyst preparation plays a very important role here and the best method consists of a sequential impregnation of WO_3 and then Pt on ZrO_2.

In 2009, Huang et al. investigated the conversion of glycerol into 1,3-propanediol in the gas phase [46]. Considering that (i) $H_4SiW_{12}O_{40}/SiO_2$ promotes the dehydration of glycerol to acrolein and (ii) copper favors the reduction of aldehyde, they studied the catalytic activity of $Cu-H_4SiW_{12}O_{40}/SiO_2$. Reactions were carried out in a fixed-bed reactor at 210 °C and 54 bar of H_2. Under these conditions, 1,3-propanediol was produced with 32% selectivity at 83% conversion (i.e., 27% yield). Even if no significant improvement of yield was obtained as compared with the work of Sasaki and coworkers, the main advantages of this work stem from the absence of organic solvent and precious metal, thus affording an environmentally friendlier process (Scheme 4.16).

Compared with the conversion of glycerol into propylene glycol, direct synthesis of 1,3-propanediol from glycerol is rather complex and yields are still not yet competitive with the *fermentation process*. Indeed, using bacterial strains, 70% yield of 1,3-propanediol has been obtained from aqueous glycerol (Scheme 4.16). However, the fermentation reaction rate is rather low, and it shows that there is no doubt that the progress in the knowledge of the active sites of enzymes will soon provide new opportunities for the design of suitable solid catalysts for the conversion of glycerol into 1,3-propanediol.

4.8
Selective Coupling of Glycerol with Functional Monomers

Selective coupling of glycerol with functional monomers is an interesting issue, as it allows the direct introduction of hydrophilic pendent arms in a polymeric structure. In this context, glycerol acrylate or methacrylate and allyl ether glycerol have been widely reported in the literature for the synthesis of a wide range of safer polymers, such as polyesters, polycarbonate, epoxidized polymers, among others. The selective coupling of glycerol with functional monomers is fairly complex, mainly due to the presence of three hydroxyl groups that unfortunately lead to the formation of side products and waste, which necessitate treatment and disposal. Therefore, the assistance of protective agents or more reactive reagents are often used in order to obtain higher yields of the desired glycerol-based monomers (Scheme 4.17) [47].

Although these routes give access to numerous safe and potentially biodegradable glycerol-based monomers, the economical and environmental impact of these processes strongly limit the advantage of using glycerol. In order to circumvent this issue, solid catalysts have emerged as a fascinating tool, allowing the direct and selective coupling of glycerol with functional monomers, according to an environmentally friendlier route.

In this context, Jérôme and coworkers investigated the direct and selective esterification of glycerol with *hydroxylated fatty acids* [48]. The resulting *amphiphilic monomers* exhibit a broad polymorphism, which is of great importance

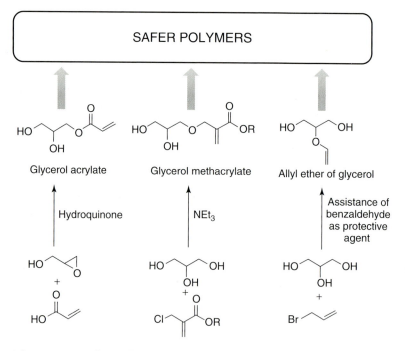

Scheme 4.17 Synthesis of glycerol-based functional monomers.

in supramolecular chemistry. In particular, these monoglycerides of hydroxylated fatty acids are used for the synthesis of safer and biodegradable amphiphilic polyesters. In a first set of experiments, workers investigated the selective and solvent-free esterification of glycerol with juniperic acid. Because of the presence of a hydroxyl group on the fatty chain, side polymerization reactions took place, making the isolation of the corresponding monoglyceride with high yields difficult. For instance, when *p*-toluenesulfonic acid (PTSA), a cation exchange resin (A119), a sulfonated carbon or an HFAU zeolite were used as an acid catalyst, the corresponding monoglycerides were always obtained with low yields (12–35%) at total conversion (Scheme 4.18).

This poor selectivity of the reaction stems from the side polymerization reaction of both the juniperic acid and the functional monoglycerides produced under acidic conditions.

In order to better control the reaction selectivity, sulfonic acid sites were grafted over a hydrophilic solid support with the aim of: (i) limiting the adsorption of the juniperic acid on the catalyst surface and (ii) rapidly desorbing the monoglycerides of hydroxylated fatty acids from the catalyst surface, in order to avoid their subsequent polymerization. As solid acid catalysts, sulfonic acid sites were anchored on a siliceous surface using the co-condensation procedure. By means of polarity measurements, Macquarrie *et al.* have shown that these silica-supported sulfonic acid sites were highly hydrophilic [49]. As was expected, this consideration of the catalyst surface hydrophilicity led to a spectacular improvement in the reaction selectivity and the monoglycerides were now obtained with 91% yield, thus presenting a new route for the green synthesis of amphiphilic monomers from glycerol and hydroxylated fatty acids. Other naturally available hydroxylated fatty acids, such as aleuritic, thapsic, and 12-hydroxystearic acids, were found to be eligible for this process. In

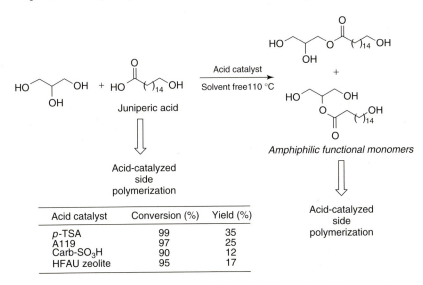

Scheme 4.18 Acid-catalyzed esterification of glycerol with juniperic acid.

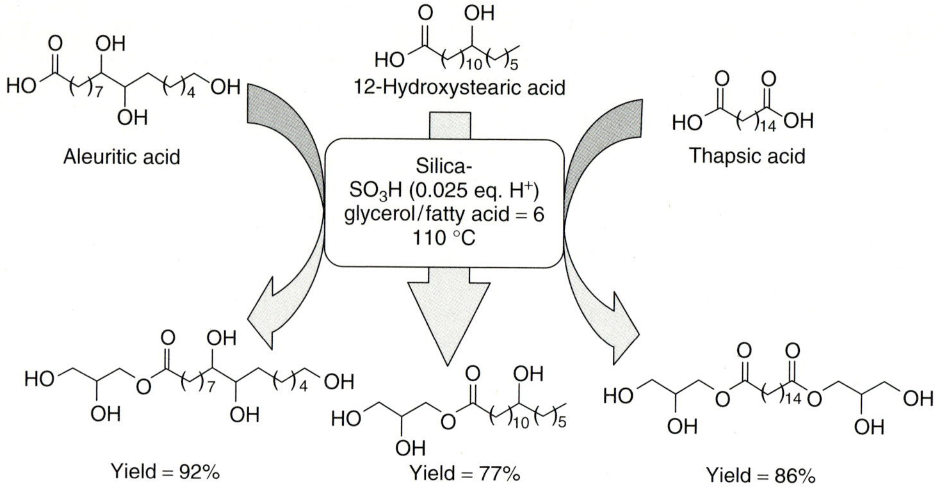

80 | 4 Selective Conversion of Glycerol into Functional Monomers via Catalytic Processes

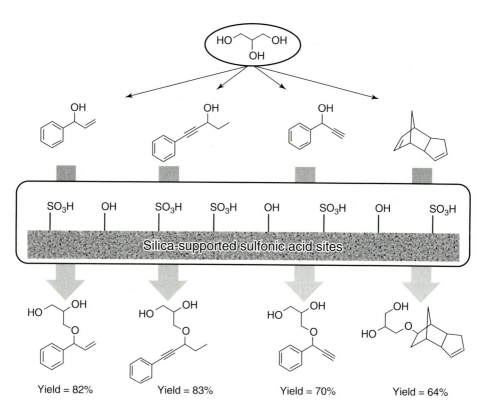

Scheme 4.19 Selective esterification of glycerol with hydroxylated fatty acids.

Scheme 4.20 Coupling of glycerol with readily polymerizable derivatives.

4.8 Selective Coupling of Glycerol with Functional Monomers

all instances, whereas all other solid acid catalysts tested mainly drove the reaction toward the polymerization of the corresponding monoglyceride, the siliceous solid support functionalized with sulfonic sites afforded the monoglyceride, with a yield range of 77–92% (Scheme 4.19).

Proceeding on the same lines, Gu *et al.* then generalized this process to the coupling of glycerol with other readily polymerizable derivatives such as allylic alcohols, propargylic alcohols, and olefins [50].

As summarized in Scheme 4.20, over silica-supported sulfonic acid sites, glycerol was successfully converted into a diverse array of functional monomers in a one-step process and without addition of any organic solvent. Interestingly, the silica-supported sulfonic acid sites were found to be more active and more selective than the conventionally used acid catalysts, such as PTSA, methane sulfonic acid, Carb-SO_3H, Amberlyst 16, and HFAU zeolite.

As suggested above, the spectacular selectivity obtained over the silica-supported sulfonic acid sites was ascribed to the strong hydrophilicity allowing a rapid desorption of the glycerol-based monomer from the catalyst surface. It should be noted that in all instances the corresponding glycerol-based monomer was produced as a mixture of two regioisomers (alkylation on the primary or secondary hydroxyl group) in a molar ratio of 9:1 in favor of the alkylation on the primary hydroxyl group.

Finally, functionalization of glycerol with a readily polymerizable olefin can be achieved through the *telomerization of glycerol* with butadiene. This reaction is catalyzed by palladium and generally affords a mixture of mono-, di-, and triethers of glycerol (Scheme 4.21). In this reaction monoethers of glycerol are the most valuable derivatives, as they can be used not only as amphiphilic monomers but also their saturated analogs may be used as plasticizer for polymers, such as PVC [poly(vinyl chloride)].

In 2003, Behr and Urschey reported the first example of telomerization of pure glycerol with butadiene [51]. The reaction was performed in water, at 80 °C, and in the presence of 0.06 mol% of Pd(acac)$_2$ and a TPPTS [3,3′,3″-Phosphinidynetris(benzenesulfonic acid)trisodium salt] ligand (P/Pd = 5) (Scheme 4.21). Interestingly, in water, monotelomers were not miscible in the reaction medium and were readily separated from the glycerol/catalyst phase by simple phase decantation. Therefore, the process was highly selective affording the monoethers of glycerol with more than 95% selectivity [i.e., 58% yield, the amount of di-ethers remained much lower (<1%)]. As the palladium catalyst was soluble in water, catalyst recycling was easily undertaken and four catalytic cycles were performed. However, a drop in the catalyst activity was unfortunately observed due to (i) the partial oxidation of the TPPS ligand and (ii) the formation of inactive palladium black.

Continuing with their efforts, Behr *et al.* investigated the *leaching of the palladium* from the water phase into the organic phase [52]. Using water as the solvent, they found that 73 ppm of palladium leached into the organic phase. Interestingly, upon addition of 2-methyl-2-butanol as the organic solvent, the leaching was significantly decreased from 73 to only 8 ppm, while keeping a similar yield for glycerol ethers under neat conditions (34% after 3 h of reaction). Remarkably, addition of

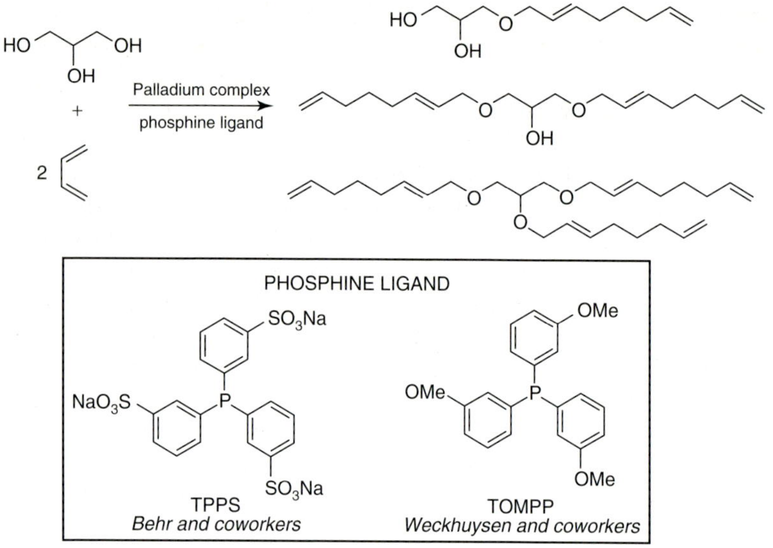

Scheme 4.21 Telomerization of glycerol with butadiene over a palladium-based catalyst.

P-octa-2,7-dienyl-*P,P,P*-[tri(3-sulfonatophenyl)-phosphonium hydrogencarbonate] trisodium salt stabilized the palladium catalyst and, under these conditions, the catalyst activity remained constant for at least 230 h (60 kg$_{prod}$ kg$_{Pd}^{-1}$ h^{-1}). When methylated β-cyclodextrin (Me-β-CD) was added in water, the leaching of palladium was also limited (46 ppm, no contamination of the reaction product with Me-β-CD) while keeping a high yield of glycerol ethers (65% after 18 h). Finally, these workers also found that addition of 0.01 mol% of 4-*tert*-butylbenzene-1,2-diol suppressed the side polymerization of butadiene, generally produced as waste in most of the telomerization processes.

Recently, Weckhuysen and coworkers studied the effect of different types of *phosphine ligands* on the activity of palladium [53]. In particular, whereas the above-described Pd/TPPS exhibited a TOF of 248 h^{-1}, this group showed that, under similar conditions [0.06 mol% of Pd(acac)$_2$, phosphine ligand with P/Pd = 5, 80 °C] but using a methoxy-functionalized PPh$_3$ ligand [TOMPP, tris-(*o*-methoxyphenyl)phosphine], the catalyst activity was significantly boosted and TOF values of 3418 h^{-1} were obtained (Scheme 4.21). This greater activity of the Pd/TOMPP complex was attributed to a higher electron density of the palladium caused by the presence of electron donating methoxy groups on the arylphosphine ligand. However, although the Pd/TOMPP catalyst was far more active than the Pd/TPPS complex, Pd/TOMPP was less selective affording, under similar conditions, a mixture of mono-, di-, and triethers with a selectivity of 58, 29, and 14%, respectively (collected for a yield of glycerol ethers of 67%). Under biphasic conditions, this drop in selectivity observed with the Pd/TOMPP catalyst was attributed to its partial dissolution in the monoethers of the glycerol phase, thus enhancing the polyetherification of glycerol. Note that the Pd/TOMPP catalyst

4.8 Selective Coupling of Glycerol with Functional Monomers

was active for crude glycerol, because starting from a butadiene/crude glycerol molar ratio of 4, a 73% yield of the glycerol ethers was obtained (20% selectivity to monoethers).

Weckhuysen and coworkers continued with their work on optimizing the reaction conditions [54]. In particular, they investigated the addition of a base (triethylamine) to the reaction medium. Interestingly, although *addition of a base* has a significant influence on the catalytic activity of numerous palladium-based catalysts, it was shown that addition of triethylamine to Pd/TOMPP only facilitated the formation of the active species, but had no effect on the type of catalytically active species formed and the resulting catalytic cycle. With respect to the minimization of waste, the possibility of catalyzing the telomerization of butadiene with glycerol over Pd/TOMPP without assistance of a base is a very attractive route. They also showed that a minimum phosphine/palladium molar ratio of 2 had to be used, as below this value *inactive black palladium* was produced. Interestingly, using a butadiene/glycerol molar ratio of 4, a 92% yield of glycerol ethers along with a selectivity to monoethers of 40% was obtained with a Pd(acac)$_2$/TOMPP catalyst (TOF = 5252 h^{-1}).

In 2009, Rothenberg and coworkers reported for the first time the telomerization of glycerol with isoprene, which is a much more lucrative diene than butadiene (Scheme 4.22) [55]. This reaction yields terpene derivatives that have already found numerous applications, including in polymer chemistry.

The reaction was conducted using a mixture of polyethylene glycol (PEG-200) and dioxane as solvent in order to ensure a better contact between glycerol and isoprene. In this instance, both PEG-200 and glycerol were telomerized. The reaction was catalyzed by a palladium carbene catalyst generated *in situ* by reaction of 1,3-dimesitylimidazolium mesylate (IMesHCl) with Na-*t*-OBu [0.05 mol% of Pd(acac)$_2$ + 0.075 mol% of IMesHCl + 10 mol% of Na-*t*-OBu]. At 90 °C, 20 bar of helium, with an isoprene/glycerol molar ratio = 5, a PEG-200/dioxane molar ratio of 1, and a PEG-200/glycerol molar ratio of 2.5, monoethers of glycerol were obtained with 99% selectivity (70% yield), and no diethers were detected. Note that the major products were always the tail-to-head product (tail-to-head/head-to-head molar ratio = 7/3) (Scheme 4.22).

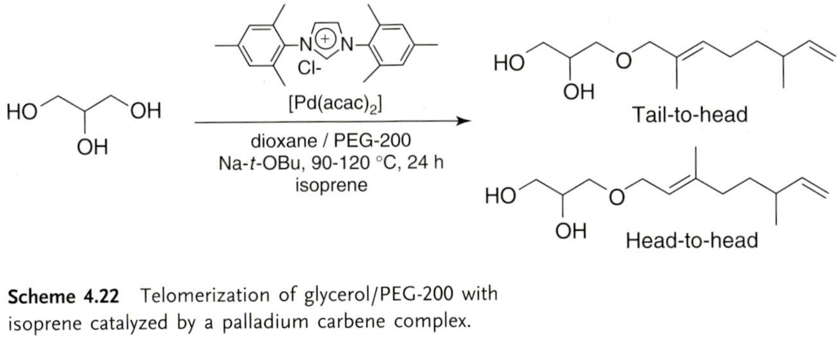

Scheme 4.22 Telomerization of glycerol/PEG-200 with isoprene catalyzed by a palladium carbene complex.

4.9
Conclusion

The recent advances made in the field of heterogeneous catalysis present very efficient ways to selectively convert glycerol into a wide range of C_3-monomers such as acrolein, acrylic and glyceric acids, glycidol, allylic alcohol, acrylonitrile, 1,3-propanediol, and propylene glycol. These biomass-based routes now offer suitable alternatives for preparing a diverse array of polymers with a high content of renewably sourced ingredients. As for propene-based processes, the conversion of glycerol into C_3-monomers requires harsh conditions, such as a high temperature and, in some instances, a high pressure. Moreover, owing to the presence of three hydroxyl groups on the glycerol, many secondary products can be formed under such reaction conditions. The control of the reaction selectivity determines the viability of these glycerol-based processes. Ideally, the C_3-monomers produced from glycerol have to be synthesized with at least the same purity as those traditionally obtained from propene. Compared with propene, the presence of water in glycerol is also a serious problem that has to be addressed. Indeed, according to the industrial process, industrial grade glycerol that is available contains 5–20% of water, which may cause serious damage to solid catalysts (residual salts might be also present). Owing to their high thermal stability, metal oxides appear as promising solid catalysts. Interaction of the metal–support and the dispersion of the catalytic sites on the oxide surface represent the two key factors governing the activity and selectivity of the catalyst.

Even though it is clear that glycerol will not replace propene in the industrial production of C_3-monomers, one of the most attractive solutions consists of the direct addition of a reasonable amount of glycerol to propene-based processes in order to reduce the dependency of such processes on fossil reserves.

References

1. As selected articles see (a) Himmelsbach, D. and Holser, R.A. (2009) *Vib. Spectrosc.*, **51** (1), 142–145; (b) de Brioude, M., Guimarães, D.H., da Fiúza, R.P., de Prado, L.A.S.A., and José, N.M. (2007) *Mater. Res.*, **10** (4), 335–339; (c) Wyatt, V.T., Nuñez, A., Foglia, T.A., and Marmer, W.M. (2006) *J. Am. Oil Chem. Soc.*, **83** (12), 1033–1039; (d) Kiyotsukuri, T., Kanaboshi, M., and Tsutsumi, N. (1994) *Polymer Int.*, **33** (1), 1–8; (e) Nagata, M., Kiyotsukuri, T., Ibuki, H., Tsutsumi, N., and Sakai, W. (1996) *React. Funct. Polym.*, **30** (1–3), 165–171.
2. (a) Jérôme, F., Pouilloux, Y., and Barrault, J. (2008) *ChemSusChem*, **1**, 586–613; (b) Zhou, C.-H., Beltramini, J.N., Fan, Y.X., and Lu, G.Q. (2008) *Chem Soc. Rev.*, **37** (3), 527; (c) Pagliaro, M., Ciriminna, R., Kimura, H., Rossi, M., and Della Pina, C. (2007) *Angew. Chem. Int. Ed.*, **46**, 4434; (d) Behr, A., Eilting, J., Irawadi, K., Leschinski, J., and Lindner, F. (2008) *Green Chem.*, **1**, 13; (e) Zheng, Y., Chen, X., and Shen, Y. (2008) *Chem. Rev.*, **108** (12), 5253.
3. Simao, A.-C., Lynikaite-Pukleviciene, B., Rousseau, C., Tatibouët, A., Cassel, S., Sackus, A., Rauter, A.P., and Rollin, P. (2006) *Lett. Org. Chem.*, **3**, 744–748.
4. Clements, J.H. (2003) *Ind. Eng. Chem. Res.*, **42**, 663–674.

5. Rokicki, G., Rakoczy, P., Parzuchowski, P., and Sobiecki, M. (2005) *Green Chem.*, **7**, 529–539.
6. (a) Fricke, N., Keul, H., and Möller, M. (2009) *Macromol. Chem. Phys.*, **210**, 242–255; (b) Pasquier, N., Keul, H., Heine, E., and Moeller, M. (2007) *Biomacromolecules*, **8**, 2874–2882; (c) Ubaghs, L., Fricke, N., Keul, H., and Höcker, H. (2004) *Macromol. Rapid. Commun.*, **25**, 517–521.
7. Strain, F. (1948) US Patent 2, 446,145.
8. (a) Akira, S. and Yoshio, S. (1994) JP 6, 329,663; (b) Mouloungui, Z., Yoo, J-W., Gachen, C.-A., Gaset, A., and Vermeersch, G. (1996) EP 0, 739,888.
9. Climent, M.J., Corma, A., De Frutos, P., Iborra, S., Noy, M., Velty, A., and Conception, P. (2010) *J. Catal.*, **269** (1), 140–149.
10. (a) Claude, S., Moulougui, Z., Yoo, J.-W., and Gaset, A. (2000) US Patent 6, 025,504; (b) Yoo, J.-W. and Moulougui, Z. (2003) *Stud. Surf. Sci. Catal.*, **146**, 757; (c) Okutsu, M. and Kitsuki, T. (2002) US Patent 6, 495,703.
11. Aresta, M., Dibenedetto, A., Nocito, F., and Ferragina, C. (2009) *J. Catal.*, doi: 10.1016/j.jcat.2009.09.008.
12. Teles, J.H., Rieber, N., and Harder, W. (1994) US Patent 5, 359,094.
13. Aresta, M., Dibenedetto, A., Nocito, F., and Pastore, C. (2006) *J. Mol. Catal. A. Chem.*, **257**, 149–153.
14. George, J., Patel, Y., Muthukumaru Pillai, S., and Munshi, P. (2009) *J. Mol. Catal. A. Chem.*, **304**, 1–7.
15. As selected works see (a) Parzuchowski, P.G., Jaroch, M., Tryznowski, M., and Rokicki, G. (2008) *Macromolecules*, **41**, 3859–3865; (b) Mei, H., Zhong, Z., Long, F., and Zhuo, R. (2006) *Macromol. Rapid Commun.*, **27**, 1894–1899; (c) Wolinsky, J.B., Ray, W.C., Colson, Y.L., and Grinstaff, M.W. (2007) *Macromolecules*, **40**, 7065–7068; (d) Ray, W.C. and Grinstaff, M.W. (2003) *Macromolecules*, **36**, 3557–3562.
16. As selected examples see (a) Matsunami, E., Takahashi, T., Kasuga, H., and Arita, Y. (2007) Intl. Publ. No. WO119,528; (b) Li, X.Z. (2007) CN 1, 01,070,276A; (c) Tsukuda, E., Sato, S., Tajahashi, R., and Sodesewa, T. (2007) *Catal. Commun.*, **8**, 1349–1353; (d) Chai, S.-H., Whang, H.-P., Liang, Y., and Xu, B.Q. (2007) *Green Chem.*, **9**, 1130–1136; (e) Atia, H., Armbruster, U., and Martin, A. (2008) *J. Catal.*, **258**, 71–82; (f) Chai, S.-H., Whang, H.-P., Liang, Y., and Xu, B.Q. (2007) *J. Catal.*, **250**, 342–349; (g) Chai, S.-H., Whang, H.-P., Liang, Y., and Xu, B.Q. (2009) *Appl. Catal.*, **353**, 213–222; (h) Chai, S.-H., Wang, H.-P., Liang, Y., and Xu, B.-Q. (2008) *Green Chem.*, **10**, 1087–1093.
17. Katryniok, B., Paul, S., Capron, M., and Dumeignil, F. (2009) *ChemSusChem*, **2**, 719–730.
18. As selected examples see (a) Ramayya, S., Brittain, A., De Almeida, C., Mok, W., and Antal, M.J. (1987) *Fuel*, **66**, 1364–1371; (b) Bühler, W., Dinjus, E., Ederer, H.J., Kruse, A., and Mas, C. (2002) *J. Supercrit Fluids*, **22**, 37–53; (c) Lehr, V., Sarlea, M., Ott, L., and Vogel, H. (2007) *Catal. Today*, **121**, 121–129; (d) Watanabe, M., Lida, T., Aizawa, Y., Aida, T.M., and Inomata, H. (2007) *Bioresour. Technol.*, **98**, 1285–1290; (e) Suzuki, N. and Takahashi, M. (2006) JP 290,815; (f) Ott, L., Bicker, M., and Vogel, H. (2006) *Green Chem.*, **2**, 214–220.
19. (a) Dubois, J.L., Duquenne, C., Hoelderich, and W.F. (2005) FR 2, 882,052; (b) Dubois, J.L., Duquenne, C., Hoelderich, W.F., and Kervennal, J., (2005) FR 2, 882,053; (c) Dubois, J.L., Duquenne, C., Hoelderich, and W.F. (2005) FR 2, 884,817; (d) Dubois, J.L., Duquenne, C., Hoelderich, and W.F. (2006) FR 2, 884,818; (e) Ulgen, A. and Hoelderich, W.F. (2009) *Catal. Lett.*, **131**, 122–128.
20. Dubois, J.L., Duquenne, C., and Hoelderich, W.F. (2006) Intl. Publ. No. WO 114,506.
21. As a recent review see Wilms, D., Stiriba, S.-E., and Frey, H. (2009) *Acc. Chem. Res.*, doi: 10.1021/ar900158p.
22. (a) Barrault, J. and Jérôme, F. (2008) *Eur. J. Lipid. Sci. Technol.*, **110** (9), 825–830; (b) Ruppert, A.M., Meeldijk, J.D., Kuipers, B.W.M., Erné, B.H., and Weckhuysen, B.M. (2008) *Chem. Eur. J.*, **14** (7), 2016–2024; (c) Calatayud,

M., Ruppert, A.M., and Weckhuysen, B.M. (2009) *Chem. Eur. J.*, **15** (41), 10864–10870.

23. (a) Bruson, H.A. and Reiner, T.W. (1952) *J. Am. Chem. Soc.*, **74** (8), 2100–2101; (b) Bruson, H.A., Heights, S., Reiner, and T.W. (1953) US Patent 2, 636,040.

24. Malkemus, J.D. and Currier, V.A. (1958) US Patent 2, 856,413.

25. Yoo, J.W., Mouloungui, Z., and Gaset, A. (1998) Intl. Publ. No. WO/1998/040371.

26. Arceo, E., Marsden, P., Bergman, R.G., and Ellman, J.A. (2009) *Chem. Commun.*, **23**, 3357–3359.

27. As selected example see (a) Kimura, H. (1993) *Appl. Catal. A.*, **105**, 147–158; (b) Kimura, H. and Tsuto, K. (1993) *Appl. Catal. A.*, **96**, 217–228; (c) Garcia, R., Besson, M., and Gallezot, P. (1995) *Appl. Catal. A.*, **127**, 165–176.

28. (a) Carrettin, S., McMorn, P., Johnston, P., Griffin, K., and Hutchings, G.J. (2002) *Chem. Commun.*, 696–697; (b) Carrettin, S., McMorn, P., Johnston, P., Griffin, K., Kiely, C.J., and Hutchings, G.J. (2003) *Phys. Chem. Chem. Phys.*, **5**, 1329–1336.

29. (a) Porta, F. and Prati, L. (2004) *J. Catal.*, **224**, 397–403; (b) Dimitratos, N., Messi, C., Porta, F., Prati, L., and Villa, A. (2006) *J. Mol. Catal. A.*, **256**, 21–28.

30. (a) Demirel-Güllen, S., Lucas, M., and Claus, P. (2005) *Catal. Today*, **102–103**, 166–172; (b) Demirel, S., Lehnert, K., Lucas, M., and Claus, P. (2007) *Appl. Catal. B.*, **70**, 637–643.

31. Comotti, M., Della Pina, C., Matarrese, R., Rossi, M., and Siani, A. (2005) *Appl Catal. A.*, **291**, 204–209.

32. Kimura, H. (1996) *J. Polym. Sci. A.*, **34**, 3607–3614.

33. Kimura, H. (2001) *Polym. Adv. Technol.*, **12**, 697.

34. (a) Kimura, H. (1998) *J. Polym. Sci. A.*, **36**, 195–205; (b) Kimura, H. (1998) *J. Polym. Sci. A.*, **36**, 189–193.

35. Papanu, V.D. (1979) US Patent 4, 169,934.

36. Olga Guerrero-Perez, M. and Bañares, M.A. (2008) *ChemSusChem*, **1**, 511–513.

37. Calvino-Casilda, V., Olga Guerrero-Perez, M., and Bañares, M.A. (2009) *Green Chem.*, **11**, 939–941.

38. Schuster, L. and Eggersdorfer, M. (1997) US Patent 5, 616,817.

39. Dasari, M.A., Kiatsimkul, P.-P., Sutterlin, W.R., and Suppes, G.J. (2005) *Appl. Catal. A.*, **281**, 225–231.

40. (a) Huang, Z., Cui, F., Kang, H., Chen, J., Zhang, X., and Xia, C. (2008) *Chem. Mater.*, **20**, 5090–5099; (b) Huang, Z., Cui, F., Kang, H., Chen, J., and Xia, C. (2009) *Appl. Catal. A.*, **366**, 288–298.

41. D'Hondt, E., Van de Vyver, S., Sels, B.F., and Jacobs, P. (2008) *Chem. Commun.*, 6011–6012.

42. Huang, L., Zhu, Y.-L., Zheng, H.-Z., Li, Y.-W., and Zeng, Z.-Y. (2008) *J. Chem. Technol. Biotechnol.*, **83**, 1670–1675.

43. Sato, S., Akiyama, M., Inui, K., and Yokota, M. (2009) *Chem. Lett.*, **38** (6), 560–561.

44. Chaminand, J., Djakovitch, L., Gallezot, P., Marion, P., Pinel, C., and Rosier, C. (2004) *Green Chem.*, **6**, 359–361.

45. Kurosaka, T., Maruyama, H., Naribayashi, I., and Sasaki, Y. (2008) *Catal. Commun.*, **9**, 1360–1363.

46. Huang, L., Zhu, Y., Zheng, H., Ding, G., and Li, Y. (2009) *Catal. Lett.*, **131**, 312–320.

47. As recent examples see (a) Topham, P.D., Sandom, N., Read, E.S., Madsen, J., Ryan, A.J., and Armes, S.P. (2008) *Macromolecules*, **41**, 9542–9547; (b) He, F., Wang, Y.-P., Liu, G., Jia, H.-L., Feng, J., and Zhuo, R.-X. (2008) *Polymer*, **49**, 1185–1190; (c) Tallon, C., Moreno, R., and Nieto, M.I. (2007) *J. Am. Ceram. Soc.*, **90** (5), 1386–1393; (d) Avci, D. and Mathias, L.J. (2004) *Polymer*, **45**, 1763–1769.

48. Karam, A., Gu, Y., Jérôme, F., Douliez, J.-P., and Barrault, J. (2007) *Chem. Commun.*, **22**, 2222–2224.

49. Macquarrie, D.J., Jackson, D.B., Mdoe, J.E.G., and Clark, J.H. (1999) *New J. Chem.*, **23**, 539–544.

50. Gu, Y., Azzouzi, A., Jérôme, F., Pouilloux, Y., and Barrault, J. (2008) *Green. Chem.*, **10** (2), 164–167.

51. Behr, A. and Urschey, M. (2003) *Adv. Synth. Catal.*, **345**, 1242–1246.

52. Behr, A., Leschinski, J., Awungacha, C., Simic, S., and Knoth, T. (2009) *ChemSusChem*, **2**, 71–76.
53. Palkovits, R., Nieddu, I., Klein Gebbink, R.J.M., and Weckhuysen, B. (2008) *ChemSusChem*, **1**, 193–196.
54. (a) Palkovits, R., Nieddu, I., Kruithof, C.A., Klein Gebbink, R.J.M., and Weckhuysen, B. (2008) *Chem. Eur. J.*, **14**, 8995–9005; (b) Palkovits, R., Parvulescu, A.N., Hausoul, P.J.C., Kruithof, C.A., Klein Gebbink, R.J.M., and Weckhuysen, B. (2009) *Green Chem.*, **11**, 1155–1160.
55. Gordillo, A., Duran Pachon, L., De Jesus, E., and Rothenberg, G. (2009) *Adv. Synth. Catal.*, **351**, 325–330.

Part III
Sustainable Reaction Conditions

5
Monoterpenes as Polymerization Solvents and Monomers in Polymer Chemistry

Robert T. Mathers and Stewart P. Lewis

5.1
Introduction

Terpenes are a broad class of molecules found in trees, flowers, fruits, and spices that are based on C_5 isoprene variations. Approximately 30 000 of these molecules have been reported in the literature [1, 2]. The structure may be cyclic or acyclic and contain alkenes or other functional groups such as alcohols, aldehydes, ketones, esters, and carboxylic acids. As shown in Figure 5.1, monoterpenes (MTs) are a C_{10} subset of terpenes. Many MTs possess familiar *odors* associated with oranges (d-limonene), lemons (l-limonene), turpentine (α-pinene and β-pinene), peppermint [(−)-menthol], spearmint [(R)-carvone], and caraway seeds [(S)-carvone]. The insect repellent (R)-(+)-citronellal and mandarin peel oil [(−)-perillaaldehyde] both contain aldehyde functional groups. The pleasant smell of many aldehyde and alcohol containing MTs has been used for fragrances. Even aromatic MTs are found in the oils of eucalyptus (p-cymene) and thyme (p-cymene and thymol). Owing the plentiful nature of certain terpenes, this chapter will focus on MTs as monomers, solvents, and chain transfer agents to develop environmentally benign polymerization methods. A wide variety of characteristics and benefits can be envisioned for a *green polymerization system* and may include:

1) A process that relies on chemical feedstocks readily derived from renewable resources in high yields will aid in sustainability. In this regard, large amounts of d-limonene are produced as a by-product of the citrus industry (110–165 million lbs per year) [3].
2) Derivation of starting materials from native and non-invasive renewable resources will reduce dependency on petroleum without threatening the ecosystem.
3) Choosing non-toxic and biodegradable chemical feedstocks will reduce potential environmental consequences of hazardous starting materials.
4) Preferably, renewable resources will facilitate commercially viable process conditions with fewer synthetic steps.

Green Polymerization Methods: Renewable Starting Materials, Catalysis and Waste Reduction
Edited by Robert T. Mathers and Michael A. R. Meier
Copyright © 2011 WILEY-VCH Verlag GmbH & Co. KGaA, Weinheim
ISBN: 978-3-527-32625-9

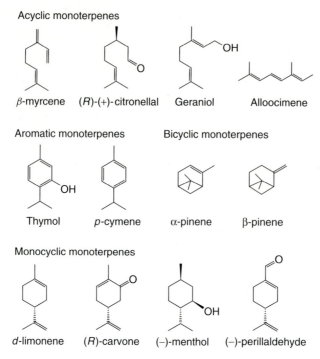

Figure 5.1 Select examples of polar and non-polar monoterpenes with acyclic, aromatic, bicyclic, and cyclic structures.

Utilizing MTs allows numerous possibilities for polymers of commercial significance to meet some of these criteria.

5.2
Monoterpenes as Monomers

5.2.1
Terpenic Resins Overview

Polymers based on MTs have a long history. Interestingly, the oldest reported polymerization is that of turpentine [4]. Commercial preparation became possible with patents in 1909 for sulfuric acid initiated polymerization [5] and 1933 for aluminum chloride coinitiated polymerization [6]. MTs of industrial significance [β-pinene, α-pinene, and dipentene (d,l-limonene)] are readily derived from pine trees as mixtures known as *turpentines*. There are three main types: *gum turpentine* results from the fractional distillation of tappings from live trees, *wood turpentine* is produced from steam distillation of pine stumps, and *sulfate turpentine* is the by-product of the sulfate pulp process (Table 5.1) [7]. d-Limonene is also readily available from by-product rinds of the citrus industry. Although free radical polymerizations with MTs have been accomplished, integrating MTs with cationic

Table 5.1 Main components of US turpentines [7].

Type	α-Pinene (%)	β-Pinene (%)	Dipentene %
Gum	65.6	28.1	3.2
Sulfate	65.5	20.4	8.2
Wood	81.3	2.1	0.9

and ring-opening metathesis polymerization (ROMP) mechanisms offers many benefits, which are the focus of this chapter.

5.2.2
Concepts of Cationic Olefin Polymerization

Cationic olefin polymerization is a chain growth process where the active chain-end is a carbenium ion [8–10]. Monomers that polymerize by this mechanism (Figure 5.2) are electron rich and devoid of basic functionalities. Some exceptions include vinyl ethers, *p*-methoxystyrene, and *N*-vinyl carbazole. During the polymerization, the olefinic functionality of the monomer should be the most nucleophilic species present. As monomer reactivity mirrors carbenium ion stability, monomers bearing substituents that stabilize positive charge inductively, by resonance, or via hyperconjugation, exhibit enhanced reactivity toward polymerization (Figure 5.2). Monomers that give rise to overly stable and/or sterically encumbered carbocations have reduced reactivity and are not likely to polymerize (Figure 5.3) [9]. Given these considerations, MTs that contain conjugated diene and/or alkyl substituents are ideal green monomers for cationic olefin polymerization.

The process of cationic olefin polymerization consists of four distinct reactions: initiation, propagation, chain transfer, and termination. *Initiation* refers to the generation of a carbenium ion capable of undergoing propagation (Scheme 5.1). The initial carbenium ion can be generated (chemically) by ionization of a covalent carbocation source (Scheme 5.1, Reaction 1), addition of a proton to a monomer from a proton source (Scheme 5.1, Reactions 2–3), addition of a stable carbenium or acylium ion to a monomer (Scheme 5.1, Reaction 4), and addition of a metal cation to a monomer (Scheme 5.1, Reaction 5) [8]. Most of these processes are aided by Lewis acids. In this case, the carbenium/acylium ion, proton, or metal

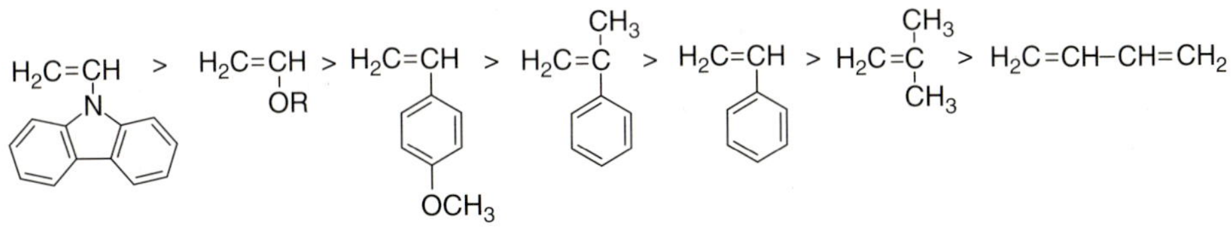

Figure 5.2 Monomer reactivity (most to least reactive) in cationic polymerization.

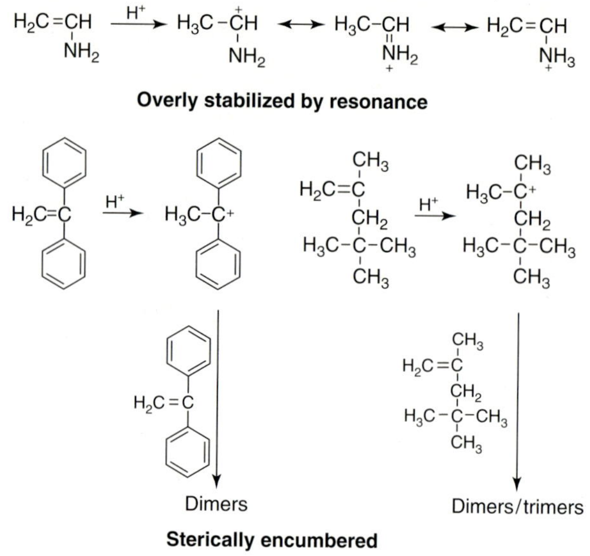

Figure 5.3 Examples of monomers not polymerizable by a cationic mechanism.

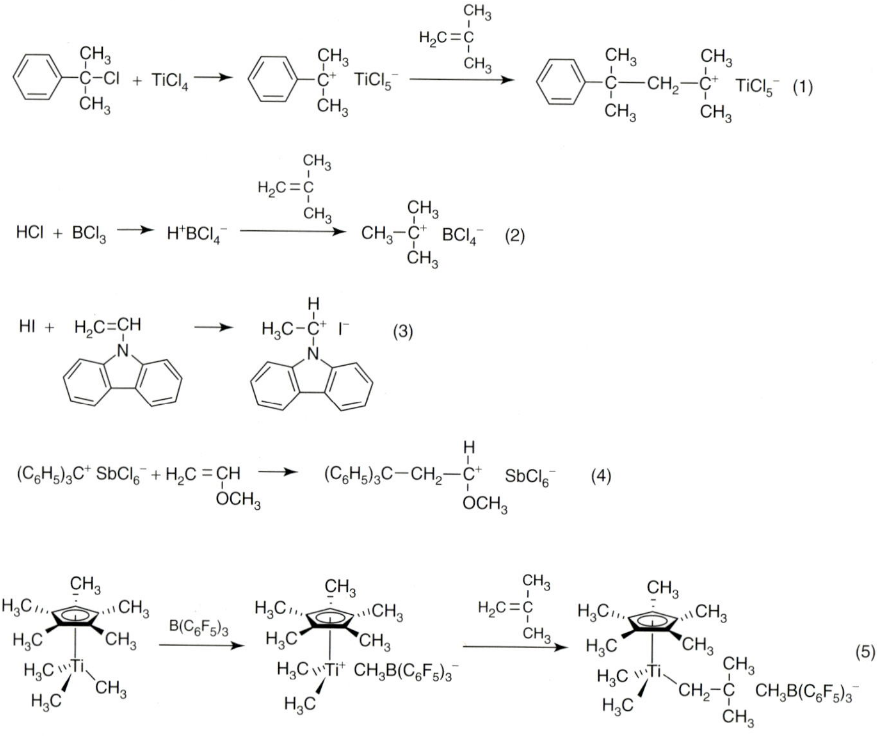

Scheme 5.1 Different forms of initiation for cationic polymerization.

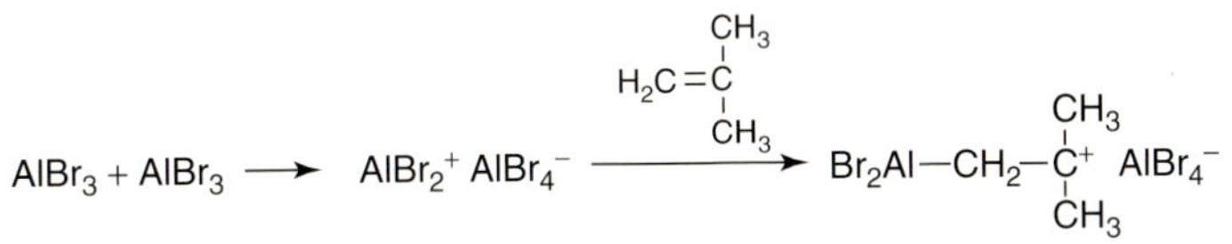

$$\text{AlBr}_3 + \text{AlBr}_3 \longrightarrow \text{AlBr}_2^+ \ \text{AlBr}_4^- \xrightarrow{\text{H}_2\text{C}=\overset{\overset{\text{CH}_3}{|}}{\underset{\underset{\text{CH}_3}{|}}{\text{C}}}} \text{Br}_2\text{Al}-\text{CH}_2-\overset{\overset{\text{CH}_3}{|}}{\underset{\underset{\text{CH}_3}{|}}{\text{C}^+}} \ \text{AlBr}_4^-$$

Scheme 5.2 Initiation of cationic polymerization by self-ionization.

ion are termed *initiators*, whereas the Lewis acid is denoted as the *coinitiator*. With highly reactive monomers, a Brönsted acid or stable carbenium ion salt may initiate polymerization directly (Scheme 5.1, Reactions 3, 4) [11–15]. In the case of self-ionization the Lewis acid functions as both the initiator and coinitiator (Scheme 5.2) [16, 17]. These materials are consumed during initiation and cannot be classified as cocatalysts or catalysts.

One of the major drawbacks of chemical initiator systems used in cationic polymerizations is that the *Lewis acid coinitiator* typically becomes solubilized in the end product. Owing to their reactive nature, most Lewis acid coinitiators undergo destructive hydrolysis during post-polymerization purification of the polymer precluding their reuse while simultaneously generating additional waste streams. This problem can be circumvented by application of Lewis acid coinitiators that are insoluble in the polymerization reaction products. Many chemical initiator systems also only operate efficiently in polar media and necessitate the use of chlorinated solvents that are detrimental to the environment. Depending on the identity of the Lewis acid coinitiator, trace impurities, such as adventitious moisture, can function as initiators. Although water is an environmentally benign initiator, this form of initiation is difficult to control as it involves careful manipulation of minute amounts of impurities and is undesirable in cases when complete control over polymer microstructure is required. To further complicate matters, once the concentration of such impurities exceeds a certain threshold they become deleterious, as they can deactivate/destroy both the Lewis acid coinitiator and any carbocationic species.

Initiation through physical methods, such as γ-radiation, has received much attention. It is assumed that irradiation of monomer results in the ejection of an electron with concomitant formation of a radical cation (Scheme 5.3) that can either form dications [18] or react with monomer to yield tertiary cations and allylic radicals [19]. The counteranion may be a solvated electron. Physical methods of initiation do not contaminate the polymer and can be more environmentally benign than chemical initiator systems. However, at the current time, they are overly sensitive to impurities and impractical for a commercial scale. Some physical methods also require costly handling techniques as they present special hazards such as high energy radiation and associated radioactive wastes, further negating their commercial utility.

Propagation (Scheme 5.4) is repetitive electrophilic addition of the carbenium ion to monomer resulting in the formation of σ-bonds, growth of the polymer, and simultaneous regeneration of the carbenium ion. Depending on the initiating system, propagation can occur in a living or non-living manner with the majority

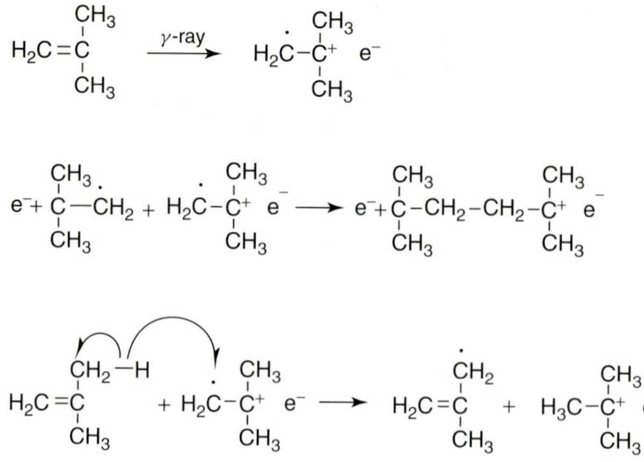

5 Monoterpenes as Polymerization Solvents and Monomers in Polymer Chemistry

$$H_2C=C(CH_3)(CH_3) \xrightarrow{\gamma\text{-ray}} H_2\dot{C}-\overset{+}{C}(CH_3)(CH_3) \; e^-$$

$$e^- + \dot{C}(CH_3)(CH_3)-CH_2 + H_2\dot{C}-\overset{+}{C}(CH_3)(CH_3) \; e^- \longrightarrow e^- + \overset{CH_3}{\underset{CH_3}{\dot{C}}}-CH_2-CH_2-\overset{CH_3}{\underset{CH_3}{\overset{+}{C}}} \; e^-$$

$$H_2C=C(CH_3)(CH_2-H) + H_2\dot{C}-\overset{+}{C}(CH_3)(CH_3) \; e^- \longrightarrow H_2C=C(CH_2^{\bullet})(CH_3) + H_3C-\overset{+}{C}(CH_3)(CH_3) \; e^-$$

Scheme 5.3 Initiation of cationic polymerization by γ-radiation.

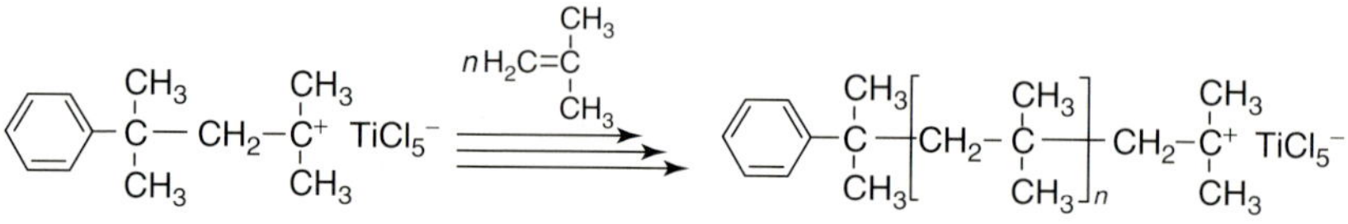

$$\text{Ph}-\overset{CH_3}{\underset{CH_3}{C}}-CH_2-\overset{CH_3}{\underset{CH_3}{\overset{+}{C}}} \; TiCl_5^- \;\; \overset{nH_2C=C(CH_3)(CH_3)}{\rightleftharpoons} \;\; \text{Ph}-\overset{CH_3}{\underset{CH_3}{C}}\left[CH_2-\overset{CH_3}{\underset{CH_3}{C}}\right]_n CH_2-\overset{CH_3}{\underset{CH_3}{\overset{+}{C}}} \; TiCl_5^-$$

Scheme 5.4 Propagation in cationic polymerization of isobutylene.

of polymerizations falling into the latter category. Owing to the greater number of additives and higher concentration of Lewis acid coinitiator required, most living polymerization systems generate larger amounts of waste than non-living systems. Regardless of whether a polymerization occurs in a living fashion, *cationic polymerizations* are generally conducted in solution to safely moderate exothermicity. Additionally, most of the polymeric products have sufficiently low glass transition temperatures (T_g) as to preclude their synthesis by solventless techniques such as gas phase polymerization. In theory, the requisite use of volatile organic solvents can be avoided by carrying out polymerization to limited conversion under bulk conditions, but in practice the rapidity of cationic polymerization is problematic.

Chain transfer will cause deactivation of a propagating polymer chain-end with concomitant production of a species capable of reinitiating polymerization (Scheme 5.5) [8–10]. Living polymerizations do not undergo detectable amounts of chain transfer on a laboratory timescale. Owing to hyperconjugation, β-proton elimination is the predominant method of chain transfer. Even weak bases (solvent, monomer, polymer, counteranion, impurities) will abstract β-protons from the carbenium ion yielding an unsaturated endgroup and a species capable of reinitiation (Scheme 5.5, Reaction 1). *Friedel–Crafts alkylation* is another common form of chain transfer (Scheme 5.5, Reaction 2). Covalent compounds can transfer groups such as halide and alkoxide to a carbenium ion with simultaneous production of a new active species (Scheme 5.5, Reaction 3). As the activation energy for

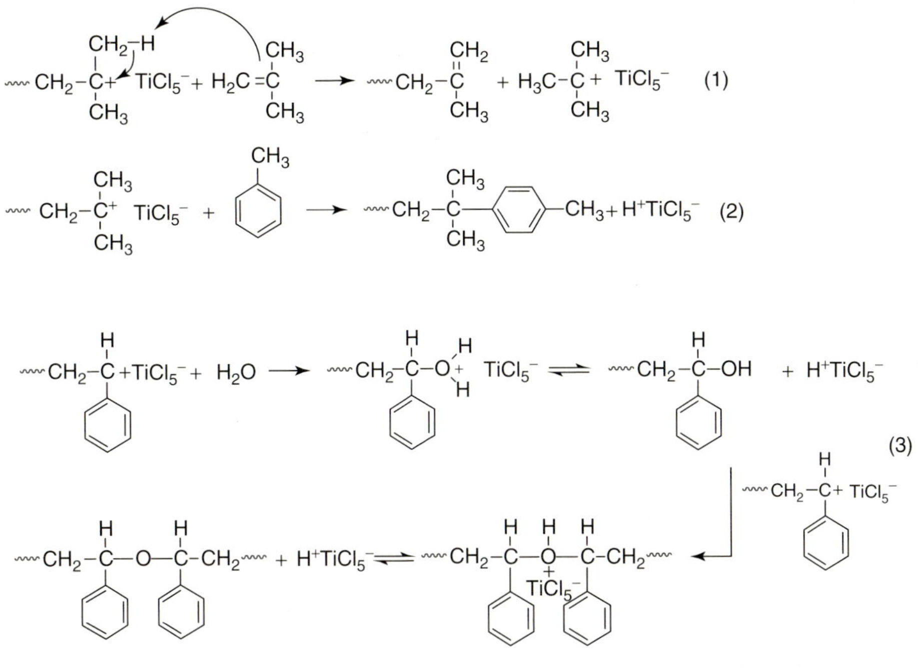

5.2 Monoterpenes as Monomers

Scheme 5.5 Different forms of chain transfer in cationic polymerization.

chain transfer is higher than propagation, it can be suppressed by running the polymerization at reduced temperature [20, 21]. Other methods involve sterically hindered pyridines (SHPs) [22–24] that stop reinitiation by trapping the H$^+$ generated by β-proton elimination or by stabilizing the carbenium ion through use of nucleophilic counteranions [25], nucleophilic additives [26], or salts [14] to prevent loss of β-protons (Figure 5.4).

Chain transfer is the main molecular weight determining event at a given polymerization temperature. When precise control over *polymer molecular weight* is needed, chain transfer becomes an unnecessary side reaction. However, from an energy standpoint, avoiding chain transfer to synthesize high molecular weight polymer is costly and requires energy intensive cryogenic cooling. A potential solution for limiting chain transfer involves initiator systems with weakly coordinating anions [27–43]. Because chain transfer results in initiator efficiencies well in excess

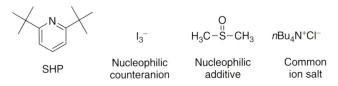

SHP	I_3^-	$H_3C-\overset{O}{\underset{\|}{S}}-CH_3$	$nBu_4N^+Cl^-$
	Nucleophilic counteranion	Nucleophilic additive	Common ion salt

Figure 5.4 Additives for limiting chain transfer in cationic polymerization.

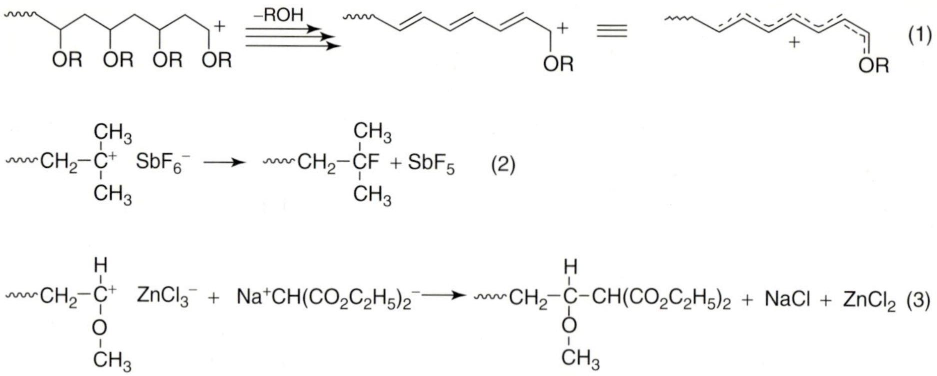

Scheme 5.6 Various termination reactions in cationic polymerization.

of the theoretical value, it can be useful for the synthesis of large amounts of low molecular weight polymer.

Termination (Scheme 5.6) is transformation of the carbenium ion into a species incapable of propagation. This type of permanent deactivation differs from reversible deactivation found in living polymerization systems. Termination may involve rearrangement of the carbenium ion to a more stable and less reactive species (Scheme 5.6, Reaction 1), collapse of the ion pair (Scheme 5.6, Reaction 2), and reaction of the carbenium ion with nucleophiles (Scheme 5.6, Reaction 3). Termination is undesirable as it leads to excess consumption of initiator and loss of control over polymer microstructure. In order to prevent termination, high purity starting materials are needed and may lead to increased operating costs and the generation of unwanted waste.

5.2.3
Cationic Polymerization of β-Pinene

Of the terpenic monomers, β-pinene has the highest reactivity and is the most important in terms of resin production. This is surprising as initiation generates a sterically encumbered carbenium ion (**1**) that should have little propensity toward propagation (Scheme 5.7). For example, *exo*-methylene cyclohexane forms a similar cyclohexyl carbenium ion (**3**, Scheme 5.8) but only affords low molecular

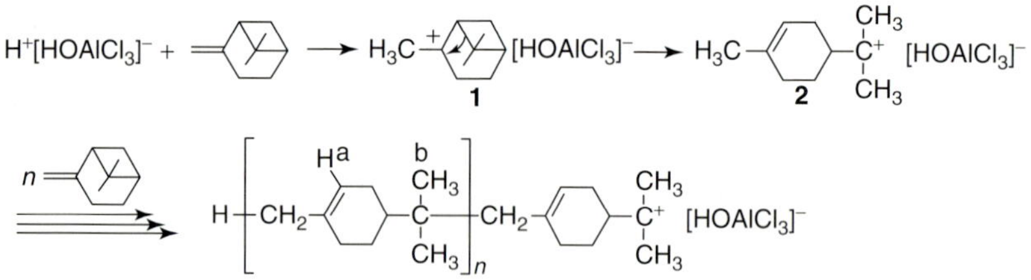

Scheme 5.7 Cationic polymerization of β-pinene.

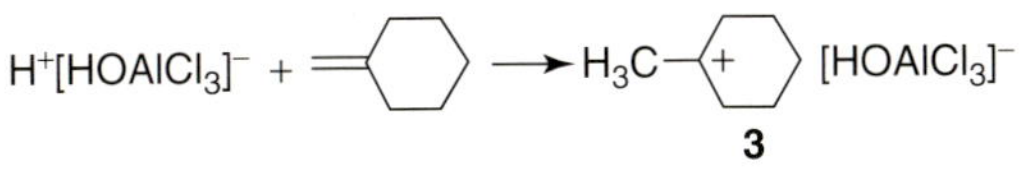

$H^+[HOAlCl_3]^-$ + [cyclohexene with exo-methylene] ⟶ H_3C–[cyclohexyl cation] **3** $[HOAlCl_3]^-$

Scheme 5.8 Cationic polymerization of *exo*-methylene cyclohexane.

weight polymers at extremely low polymerization temperatures (-150 to $-180\,°C$) [44–46]. The reactivity of β-pinene has been attributed to the ability of the initial cyclic carbenium ion to undergo isomerization. The resulting β-scission of the *gem*-dimethyl bridge of the cyclobutane ring yields a less sterically encumbered acyclic carbenium ion (**2**) that can undergo propagation (Scheme 5.7). The net result is *isomerization–polymerization* and yields a polymer with a *p*-menthene repeat unit that can be viewed as a perfectly alternating copolymer of isobutylene and cyclohexene [47–52]. Peracid oxidation and ozonolysis indicate that there is one olefinic group per repeat unit and therefore support this structure [44, 47–52].

Recent studies have shown the actual structure of poly(β-pinene) is more complex and depends on the *method of initiation* [53]. ^1H and ^{13}C NMR (nuclear magnetic resonance) spectroscopy show that the percentage of structural defects (Table 5.2) increases as follows: γ-radiation (most ordered) < free radical < Ziegler–Natta ≈ Friedel–Crafts acids (least ordered). Reported defects are *o*-menthene repeat units and cross-linking. Most likely, *m*-menthene repeat units are the actual structural defect due to rearrangement of a camphenic carbenium ion (Scheme 5.9). This assumption agrees with the postulation that poly(β-pinene) molecular weights are limited by chain transfer involving rearrangement of the propagating cyclic carbenium ion (**1**, Scheme 5.7) to a sterically hindered camphenic carbenium ion (Scheme 5.10) [47–52]. Post-polymerization rearrangement of chiral poly(β-pinene)

Table 5.2 Structural defects for poly(β-pinene)s [53].

Sample	*p*-Menthene (%)	Cross-linking (%)
γ-ray 300 kGy	100	4
γ-ray 600 kGy	100	9
γ-ray 1000 kGy	100	13
γ-ray 2000 kGy	100	12
Free radical	90	23
Friedel–Crafts	70	30
Ziegler–Natta	65	15

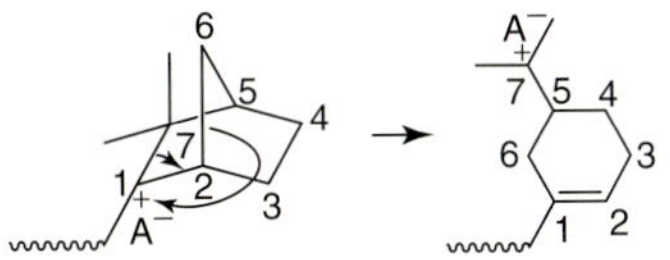

Scheme 5.9 Isomerization of camphenic carbenium ion to an *m*-menthene repeat unit.

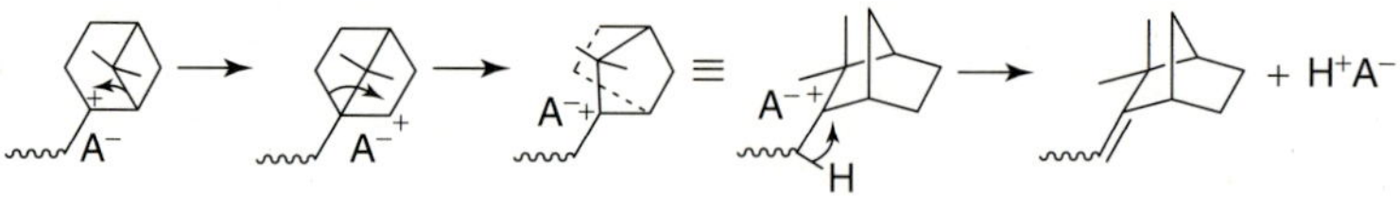

Scheme 5.10 Chain transfer via isomerization to a camphenic carbenium ion.

can also occur as acids catalyze racemization [54]. Cross-linking is common in all systems and appears to be directly related to the activity of the initiating system. It is possible that intramolecular backbiting of a propagating acyclic carbenium ion with a cyclohexene moiety within the polymer chain may actually occur as cross-linking does not appear to affect molecular weights or solubility appreciably. This would be difficult to distinguish by integration of the ^{1}H NMR signals for the methine (position a, Scheme 5.7) and *gem*-dimethyl (position b, Scheme 5.7) substituents as used in the determination of the percentage cross-linking.

Most polymerization systems for β-pinene make use of *Lewis acid coinitiators* in conjunction with adventitious moisture as an initiator. Activity mirrors the strength of the Lewis acid used; in terms of decreasing effectiveness: $AlBr_3 >$ $AlCl_3 > ZrCl_4 > AlCl_3 \cdot Et_2O > BF_3 > BF_3 \cdot Et_2O > SnCl_4 > BiCl_3 > SbCl_3 >$ $ZnCl_2$ [44]. Yields and molecular weights are directly dependent on coinitiator activity (Table 5.3). Subsequent research corroborates this assertion and coinitiator activity was found to decrease as: $TiCl_4 > Et_2AlCl$, $BF_3 \cdot Et_2O > SnCl_4$ [55]. It is of interest to note that Et_2AlCl is an effective coinitiator with water as this initiator combination is reported to be ineffective for the polymerization of isobutylene [56, 57]. This could be due to the higher reactivity of β-pinene or the presence of trace amounts of HCl impurities as Et_2AlCl–HCl is an effective initiator system for isobutylene. Other alkyl aluminum halides ($EtAlCl_2$ and $Et_3Al_2Cl_3$) have been

Table 5.3 β- and α-pinene resins yielded by various Lewis acid coinitiators [44].

Coinitiator[b]	β-Pinene[a]		α-Pinene[a]	
	Yield (%)	SP (°C)[c]	Yield (%)	SP (°C)[c]
$AlBr_3$	93.2	136	35.3	85.0
$AlCl_3$	94.2	134	35.2	84.0
$ZrCl_4$	96.0	132	20.0	91.3
$AlCl_3 \cdot Et_2O$	76.5	102	18.0	67.3
BF_3	54.0	104	14.2	67.0
$BF_3 \cdot Et_2O$	43.0	68.0	11.8	Semi-solid
$SnCl_4$	21.3	–	4.7	–
$BiCl_3$	5.6	–	0.1	–
$SbCl_3$	0.7	–	0.2	–
$ZnCl_2$	0.5	–	0.0	–

[a] 50 wt% pinene in toluene at 40–45 °C.
[b] 5 wt% based on pinene.
[c] Softening point as determined by ring-and-ball method.

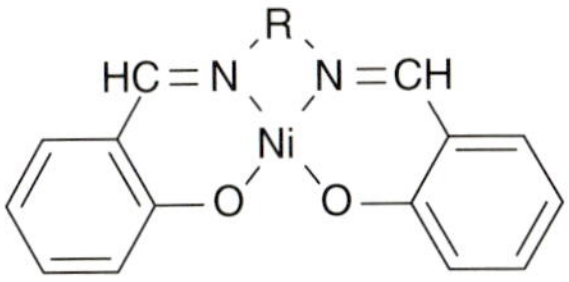

Figure 5.5 Schiff-base nickel complexes used for polymerization of β-pinene.

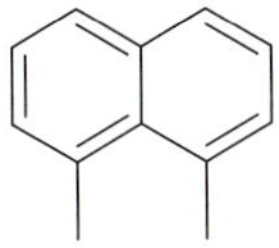

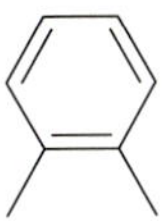

explored as coinitiators [58–60]. In particular, EtAlCl$_2$ produces poly(β-pinene) with the highest reported molecular weight ($\overline{M}_n \approx 40$ kg·mol^{-1}), when conducted in a 50/50% vol/vol mixture of methyl chloride and methylcyclohexane in the presence of an SHP [59]. Molecular weights decrease in the absence of an SHP or when solvent polarity is increased due to greater chain transfer.

Polymerization of β-pinene has been initiated with Ziegler–Natta catalysts including Al(i-Bu)$_3$–TiCl$_4$, ClAl(i-Bu)$_2$–TiCl$_4$, BrAlEt$_2$–TiCl$_4$, Al(i-Bu)$_3$–VOCl$_3$, and ClAl(i-Bu)$_2$–VOCl$_3$ [61, 62]. Polymerizations were conducted in n-heptane at temperatures ranging from 25 to −80 °C. The individual components of each Ziegler–Natta catalyst possessed little to no activity when used alone. Al(i-Bu)$_3$–TiCl$_4$ gave the highest yields (91%) and highest molecular weights, based on softening point (SP), when run at 25 °C [62]. Most likely, initiation involves electrophilic addition of a metal cation derived from the Ziegler–Natta catalyst and is not catalytic in nature. Under such circumstances polymerization would perpetuate by chain transfer.

Schiff-base nickel complexes (Figure 5.5) in conjunction with methylaluminoxane (MAO) have been explored for polymerization of β-pinene [63]. Neither the nickel complex nor MAO alone was active for polymerization. All nickel–MAO complexes produced high molecular weight polymer and exhibited high catalyst activities (Table 5.4). Complexes containing the most flexible substituents had the highest activities. This is thought to result from the production of a more sterically accessible nickel cation. Initiation may involve direct addition of a nickel cation to monomer in a manner similar to that proposed by Baird in the polymerization of isobutylene with Cp*TiMe$_3$–B(C$_6$F$_5$)$_3$ (Scheme 5.1, Reaction 5) [29]. A cationic mechanism is supported by the fact that termination with methanol gives rise to polymer

Table 5.4 Polymerization of β-pinene with Schiff base Ni complexes–MAO [63].

Initiator–coinitiator[a]	Initiator ($\times 10^5$ M)	Yield (%)	Initiator productivity[b]	$\overline{M}_n$[c] (g·mol^{-1})	$\overline{M}_w/\overline{M}_n$[c]
4–MAO	1.7	17.8	2.24×10^6	4610	1.47
5–MAO	1.7	75.5	9.50×10^6	6920	1.66
4–MAO	3.4	100	6.25×10^6	6730	1.81
5–MAO	3.4	100	6.25×10^6	7710	1.76
6–MAO	1.7	100	1.25×10^7	8960	1.92
7–MAO	1.7	100	1.25×10^7	10930	1.70
4–7	1.7	–	–	–	–
MAO	850	–	–	–	–

[a] [β-Pinene] = 1.6 M; [MAO]/[Ni] = 500; 40 °C in toluene for 24 h.
[b] g of poly(β-pinene)/mol of Ni.
[c] GPC, polystyrene calibration.

bearing a methoxy endgroup. It was determined, via NMR spectroscopy, that chain transfer arising from β-proton transfer from the cyclohexyl ring (position a, Scheme 5.11) leads to high catalyst activities. No *exo*-olefinic end groups were detected. High molecular weights were speculated to result from a long lived carbenium ion. As molecular weight increases with conversion and the polymerization is not living, it is more likely that chain transfer to polymer accounts for the high values obtained. This could explain the noticeable absence of *exo*-olefinic chain-ends, whose formation is more favorable from a statistical standpoint (six β-methyl protons compared with one β-cyclohexyl) and are of higher reactivity than *endo*-olefinic groups.

Recently, a system for the *living polymerization* of β-pinene has been developed [64]. The initiating system involves the HCl adduct of 2-chloroethyl vinyl ether as an initiator in combination with TiCl$_3$(O*i*Pr) as a coinitiator and is conducted in methylene chloride at −40 to −78 °C in the presence of n-Bu$_4$NCl. This system affords poly(β-pinene)s with $\overline{M}_n \leq 5 \times 10^3$ and $\overline{M}_w/\overline{M}_n \sim 1.3$ and allows for the production of block and graft copolymers of β-pinene [64].

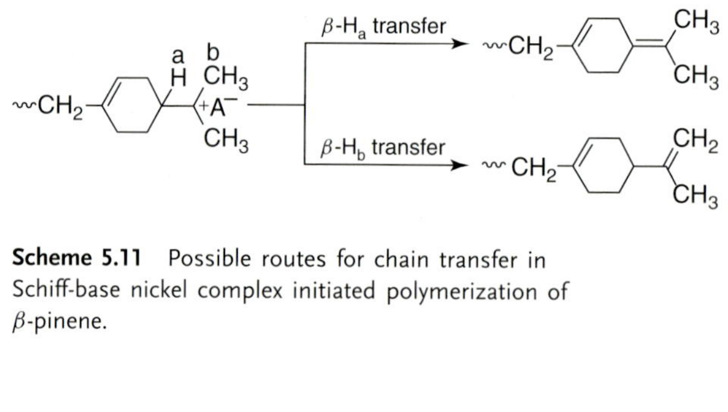

Scheme 5.11 Possible routes for chain transfer in Schiff-base nickel complex initiated polymerization of β-pinene.

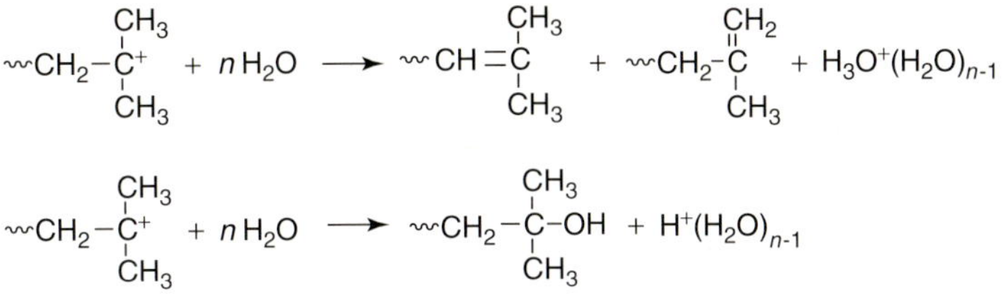

$$\text{\textasciitilde\textasciitilde}CH_2-\underset{CH_3}{\overset{CH_3}{\underset{|}{\overset{|}{C^+}}}} + nH_2O \longrightarrow \text{\textasciitilde\textasciitilde}CH=\underset{CH_3}{\overset{CH_3}{\underset{|}{\overset{|}{C}}}} + \text{\textasciitilde\textasciitilde}CH_2-\underset{CH_3}{\overset{CH_2}{\underset{|}{\overset{\|}{C}}}} + H_3O^+(H_2O)_{n-1}$$

$$\text{\textasciitilde\textasciitilde}CH_2-\underset{CH_3}{\overset{CH_3}{\underset{|}{\overset{|}{C^+}}}} + nH_2O \longrightarrow \text{\textasciitilde\textasciitilde}CH_2-\underset{CH_3}{\overset{CH_3}{\underset{|}{\overset{|}{C}}}}-OH + H^+(H_2O)_{n-1}$$

Scheme 5.12 Termination reactions for γ-ray initiated polymerization of β-pinene.

γ-Radiation has been used to initiate polymerization of β-pinene [65–67]. Minute traces of water have a deleterious effect on yield (Scheme 5.12). It was speculated that water could react with the carbenium ion either by β-proton abstraction or by nucleophilic addition with concomitant production of a hydrated proton of such low acidity that it is incapable of reinitiation. Such reactions would be accelerated by the electrostatic attraction between the carbenium ion and water, estimated to occur in less than 10^{-6} s [68]. In light of recent work conducted on the *aqueous polymerization* of isobutylene, β-proton abstraction by water leading to termination is fairly likely [69–71]. Under extremely dry conditions, 10% yields could be obtained per Mrad dose of γ-radiation. Of the polymer formed, only one half could be dissolved in halogenated aromatic solvents above 120 °C, signifying that it could be a highly cross-linked fraction. A cationic mechanism was inferred by the addition of tripropylamine, which had an inhibiting effect on polymerization.

Numerous *co- and terpolymers of β-pinene* have been prepared. Co- and terpolymers based on phenol, β-pinene, and other monomers such as aldehydes are useful *tackifiers* that possess antioxidant properties or thermosets, respectively [72–84]. They are made by bulk or solution polymerization at 25–150 °C in the presence of a Brönsted acid initiator or Lewis acid coinitiator in conjunction with adventitious moisture or organohalides as initiators. β-Pinene has also been copolymerized with isobutylene by adventitious moisture with either BF_3 or $AlCl_3$ coinitiators in ethyl chloride at −70 to −50 °C [85]. Copolymers with minimal isobutylene content perform as low melting thermoplastics, whereas high isobutylene levels provide vulcanizable rubbery solids. Azeotropic copolymerization of these two monomers with $EtAlCl_2$–H_2O below −130 °C in ethyl chloride has been reported [86]. Additional data from this work indicate that copolymerization with $AlCl_3$ in methyl chloride is also azeotropic in nature at much higher reaction temperatures. β-Pinene has also been copolymerized with styrene [87–89]. Copolymerization was initiated by adventitious moisture with either BF_3 or $AlCl_3$ in a mixture of toluene and petroleum naphtha at 40 °C. These copolymers are excellent tackifiers for a broad range of rubbers, especially styrene–butadiene rubber (SBR). Terpolymers of styrene, isobutylene, and β-pinene are also useful tackifiers [90]. Copolymerization involved the use of the $EtAlCl_2$–H_2O in hexanes at 25–30 °C. These terpolymers have improved wax compatibility due to the incorporation of isobutylene and are better suited for hot melt applications than copolymers of β-pinene and styrene. Copolymers of β-pinene with alkylidene norbornenes are also useful tackifiers [91].

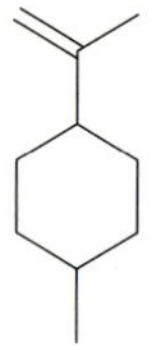

8-*p*-Menthene

Figure 5.6 Structure of 8-*p*-menthene.

Although the polymerization of β-pinene proceeds readily, yielding useful amounts of high softening point resin, a truly *green production process* has yet to be devised. The main flaw with contemporary initiator systems used for this monomer is that the coinitiator becomes trapped in the polymer and cannot be readily recovered and recycled. This leads to needless consumption of Lewis acid and the generation of undesirable waste streams. Initiator systems based on heterogeneous Lewis acids hold promise for the economical production of β-pinene resins in an ecologically friendly manner.

5.2.4
Cationic Polymerization of Dipentene

Of the terpenic monomers, *dipentene* is of intermediate reactivity and significance. NMR spectroscopic, ozonolysis, and perbenzoic acid oxidation analyzes indicate that only one-half of the repeat units have an unsaturation that would be expected for propagation solely through the exocyclic olefinic group [44, 47–52]. Polymerization studies of 8-*p*-menthene (Figure 5.6), an analog of dipentene that lacks the endocyclic double bond, only generate dimer. Control experiments with dipentene support the assumption that polymerization requires involvement of the endocyclic double bond in addition to the exocyclic olefinic group [49, 50]. Several hypotheses have been put forward to explain these findings. One involves initiation of the endocyclic olefinic group followed by subsequent reaction of the resultant carbenium ion with the exocyclic olefinic group of the same monomer molecule, leading to intramolecular cyclization and generation of a new cyclic carbenium ion that can undergo further propagation (Scheme 5.13) [49, 50]. This process is likened to cyclopolymerization of methylene-4-vinylcyclohexane (Scheme 5.14) [92]. A more likely scenario is a repetitive process involving attack of the pendant isopropyl carbenium ion on the endocyclic double bond of the penultimate repeat unit, resulting in ring formation and concomitant production of a cyclic carbenium

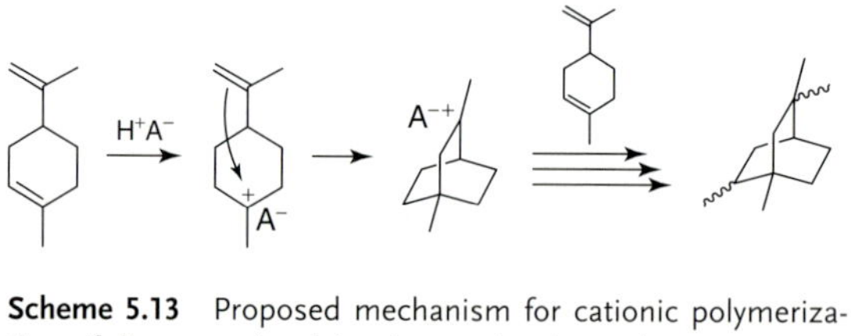

Scheme 5.13 Proposed mechanism for cationic polymerization of dipentene involving intramolecular cyclization.

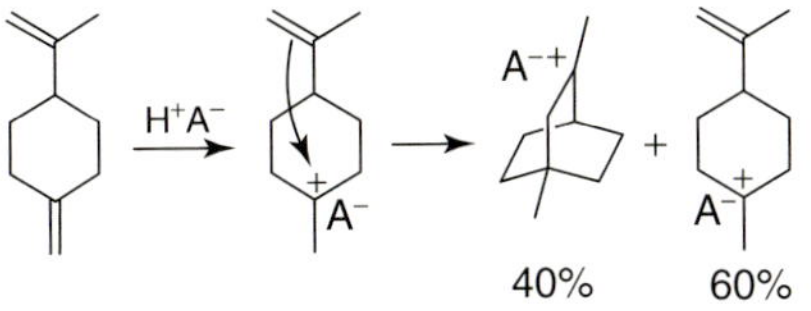

Scheme 5.14 Cationic polymerization of methylene-4-vinylcyclohexane.

ion that undergoes propagation with the *exo*-olefinic group of incoming monomer (Scheme 5.15) [49, 50].

Most of the *initiator systems* developed for α- and β-pinene are used for dipentene and its polymerization behavior is intermediate compared with the former. For example, Ziegler–Natta catalysts [Al(*i*-Bu)$_3$–TiCl$_4$, Al(*i*-Bu)$_3$–TiI$_4$, ClAl(*i*-Bu)$_2$–TiCl$_4$, ClAl(*i*-Bu)$_2$–TiI$_4$, Al(*i*-Bu)$_3$–VOCl$_3$, and ClAl(*i*-Bu)$_2$–VOCl$_3$] are effective for polymerization of dipentene [62]. The individual catalyst components possessed little to no activity for polymerization. The best conversions were obtained with an [alkyl aluminum]/[metal halide] ratio of one. It is most likely that polymerization occurs by a cationic mechanism where initiation involves addition of a metal cation to monomer. Dipentene has been copolymerized with numerous co-monomers including isobutylene [85], styrene [89], phenol [72–84, 93, 94], and other monomers [91, 95].

Similar to β-pinene, *polymerization systems for dipentene* are not ideal green processes. Loss of Lewis acid coinitiator and the generation of unwanted waste streams during purification of the final product limit the green aspects of this chemistry. Use of heterogeneous initiator systems or the development of physical initiator systems that are not overly sensitive to impurities will impart green characteristics to the production of dipentene resins.

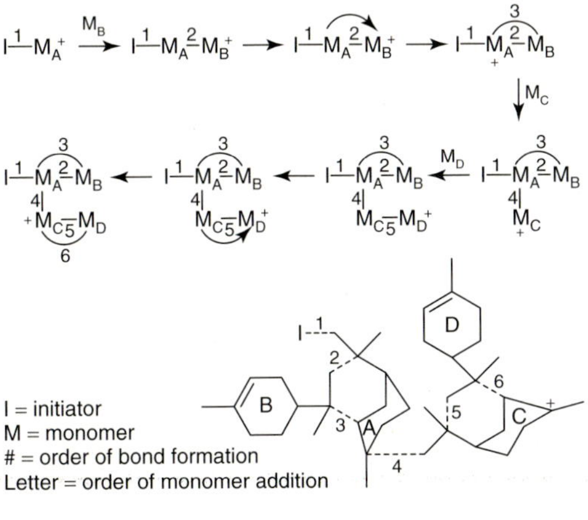

I = initiator
M = monomer
# = order of bond formation
Letter = order of monomer addition

Scheme 5.15 Proposed mechanism for cationic polymerization of dipentene involving alternating reaction of *exo*-olefinic and endocyclic double bonds (anion omitted for clarity).

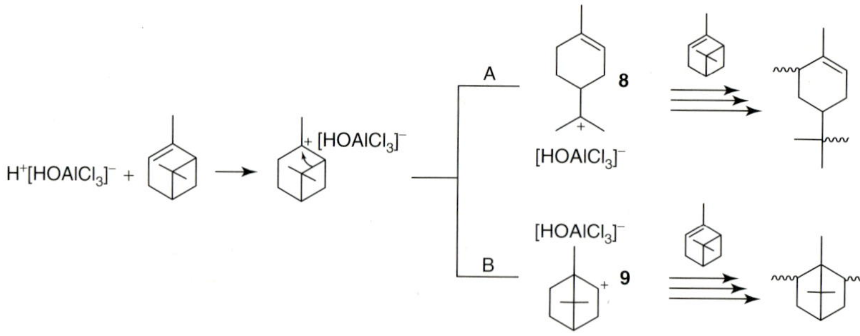

Scheme 5.16 Cationic polymerization of α-pinene initiated by AlCl₃–H₂O.

5.2.5
Cationic Polymerization of α-Pinene

Although α-pinene is the most abundant of the terpenic monomers, the low reactivity makes it the least valuable. It could serve as an ideal green monomer if the polymerizations had acceptable yields and produced a high softening point resin. Initiation generates a carbenium ion identical to that initially formed from β-pinene. This sterically hindered carbenium ion can undergo two basic forms of *isomerization* (Scheme 5.16) [47–52]. Isomerization by β-scission of the *gem*-dimethyl bridge generates a less sterically encumbered acyclic carbenium ion (**8**) of similar stability (Scheme 5.16, path A) that can react with additional α-pinene. Rearrangement can also involve expansion of the cyclobutane ring to form a less hindered (less stable) secondary cyclic carbenium ion (**9**) that can react with additional α-pinene (Scheme 5.16, path B). Propagation is difficult due to steric interactions involving the *gem*-dimethyl group and ring methyl substituents resulting in the formation of large amounts of dimer. Steric compression is avoided in β-pinene due to the presence of an exocyclic-methylene group leading to the large reactivity difference between these two otherwise identical terpenes. ^{1}H NMR analysis of poly(α-pinene)s support structures **10** and **11** [96, 97].

Most polymerization systems for α-pinene make use of Lewis acid coinitiators in conjunction with *adventitious moisture* as an initiator. Their activity is dependent on the strength of the Lewis acid coinitiator (in terms of decreasing effectiveness): $AlBr_3 > AlCl_3 > ZrCl_4 > AlCl_3 \cdot Et_2O > BF_3 > BF_3 \cdot Et_2O > SnCl_4 > SbCl_3 > BiCl_3 > ZnCl_2$ [44]. For a given initiator system, both yields and molecular weights are lower for α-pinene in comparison with β-pinene (Table 5.3).

Subsequent research has focused on the use of $AlCl_3$ with various additives, including organotin halides, antimony(III) halides, organosilicon halides, organogermanium halides, amines, esters, ethers, and ammonium salts, in an attempt to improve the yield of *high softening point resin*. The exact mode of initiation for these systems has yet to be determined but adventitious moisture is most likely the initiator in the majority. All of these systems provide improved yields of higher molecular weight poly(α-pinene) than can be obtained from the $AlCl_3-H_2O$ system; however, none are ideal from a green chemistry standpoint.

Table 5.5 Polymerization of α-pinene with AlCl$_3$: tin Lewis acid systems [98].

System[a]	AlCl$_3$: R$_m$SnX$_n$(wt : wt)	Resin (% yield)	Oil (% yield)	SP (°C)[b]
AlCl$_3$	–	29.1	35.2	92
AlCl$_3$: SnCl$_4$	5.0	33.6	31.9	79
AlCl$_3$: (C$_4$H$_9$)$_2$SnCl$_2$	5.0	48.9	34.5	84
AlCl$_3$: (C$_4$H$_9$)$_2$SnCl$_2$	2.5	85.0	10.0	130
AlCl$_3$: (C$_3$H$_7$)$_2$SnCl$_2$	10.0	65.3	23.6	130
AlCl$_3$: (C$_5$H$_9$)$_2$SnCl$_2$	10.0	60.1	23.1	128

[a] 5 wt% of each coinitiator system based on weight of monomer was used.
[b] Softening point as determined by ring-and-ball method.

Organotin halides were the first *additives* used with AlCl$_3$ (Table 5.5); one part by weight dibutyl tin dichloride to five parts by weight AlCl$_3$ gave the best results [98]. Later it was discovered that an antimony (III or V) halide in conjunction with AlCl$_3$ (in the optional presence of an alkyl, alkenyl, or aryl halide) is an effective initiator system [99]. Resin yields were much higher for systems employing *antimony halides* in place of alkyl tin halides (Table 5.6). A study of this system showed a significant enhancement in polymerization rate upon the addition of SbCl$_3$ with concomitant decrease in dimer formation [96]. This study showed dimer formation is minimized at a molar ratio [SbCl$_3$]/[AlCl$_3$] $\geq$ 0.5 and the resultant polymer possesses structure **10**. These investigators [97] found that AlBr$_3$ and EtAlCl$_2$ are also effective when used in conjunction with SbCl$_3$ (Table 5.7). Polymerization rate increases with the strength of the aluminum containing Lewis acid and decreases as: AlBr$_3$ > AlCl$_3$ > EtAlCl$_2$. Other Lewis acids (BF$_3$·Et$_2$O, SnCl$_4$, TiCl$_4$, and WCl$_6$) produce large amounts of dimer in conjunction with SbCl$_3$. In the absence of SbCl$_3$, the AlCl$_3$–H$_2$O system gives rise to polymers and dimers with a significant number of repeat units of structure **11**.

Table 5.6 Polymerization of α-pinene with AlCl$_3$: antimony Lewis acid systems [99].

AlCl$_3$ (wt%)	SbX$_n$/(wt%)	Halide/(wt%)	Resin(% yield)[a]
3	SbCl$_3$/0.47	–	55
3	SbCl$_3$/0.47	t-BuCl/0.57	85
2	SbCl$_3$/0.93	–	95
3	SbCl$_3$/0.93	C$_6$H$_5$CH$_2$Cl/0.19	98
3	SbCl$_3$/1.40	CH$_2$=CH-CH$_2$Cl/0.23	94
3	SbBr$_3$/2.1	t-BuCl/0.28	86
3	SbCl$_5$/1.2	–	92

[a] 115 °C softening point as determined by ring-and-ball method.

Table 5.7 α-Pinene resins produced by Lewis acid : SbCl₃ systems [97].

System[a]	Conversion (%)	$\overline{M}_n$ (g·mol^{-1})[b]	Dimer content (wt%)
AlCl₃	6	500	27
AlCl₃–SbCl₃	90	1460	5
AlBr₃	22	490	31
AlBr₃–SbCl₃	92	960	17
AlEtCl₂	14	530	34
AlEtCl₂–SbCl₃	29	700	16
BF₃·Et₂O	5	500	40
BF₃·Et₂O–SbCl₃	7	490	37
SnCl₄	3	500	35
SnCl₄–SbCl₃	3	510	24
TiCl₄	8	650	16
TiCl₄–SbCl₃	8	690	16
WCl₆	4	630	21
WCl₆–SbCl₃	6	620	19

[a] [M] = 3.6 M, [MtX$_n$] = 42.5 mM, [SbCl₃] = 21.3 mM, in toluene at −15 °C for 2 h.
[b] Measured by gel permeation chromatography (GPC) excluding dimers.

The improvements seen upon the inclusion of organotin and antimony halides may result from the formation of a *bulkier counteranion*. SbCl₃ could coordinate to the hypothetical Brönsted acid complex formed from H₂O and AlCl₃ (Scheme 5.17, Reaction 1). This species might also be formed by reaction of water with the SbCl₃ adduct of AlCl₃ wherein SbCl₃ acts as a weak Lewis base (Scheme 5.17, Reaction 2). Antimony oxychloride formed from the hydrolysis of SbCl₃ (Scheme 5.17, Reaction 3) might also coordinate with the complex Brönsted acid formed from H₂O (or HCl) with AlCl₃ (Scheme 5.17, Reaction 4). Correspondingly, the same

$$H^+[HOAlCl_3]^- + SbCl_3 \longrightarrow H^+[HOAlCl_3 \cdot SbCl_3]^- \quad (1)$$

$$H_2O + AlCl_3 \cdot SbCl_3 \longrightarrow H^+[HOAlCl_3 \cdot SbCl_3]^- \quad (2)$$

$$H_2O + SbCl_3 \longrightarrow OSbCl + 2HCl \quad (3)$$

$$H^+[XAlCl_3]^- + OSbCl \longrightarrow H^+[XAlCl_3 \cdot OSbCl]^- \quad (4)$$

$$HX + AlCl_3 \cdot OSbCl \longrightarrow H^+[XAlCl_3 \cdot OSbCl]^- \quad (5)$$

$$X = OH, Cl$$

Scheme 5.17 Possible reactions for the AlCl₃–SbCl₃–H₂O initiating system.

Table 5.8 α-Pinene resins yielded by AlCl$_3$: organosilicon systems [103].

Silane[a]	Resin (% yield)	SP (°C)[b]
[n-C$_6$H$_{13}$(CH$_3$)$_2$Si]$_2$O	70	96
n-C$_{18}$H$_{37}$(CH$_3$)$_2$SiCl	69	100
[(CH$_3$)$_2$HSi]$_2$O	38	100
C$_6$H$_5$(CH$_3$)$_2$SiCl	70	122
[C$_6$H$_5$CH(CH$_3$)CH$_2$(CH$_3$)$_2$Si]$_2$O	56	83
[C$_2$H$_5$(CH$_3$)$_2$Si]$_2$O	70	103
(CH$_3$)$_3$SiCl	70	124
[(CH$_3$)$_3$Si]$_2$O	70	118

[a] 5 wt% of each coinitiator system based on weight of monomer was used. For each coinitiator system AlCl$_3$: Silane (wt : wt) = 5 : 1.
[b] Softening point as determined by ring-and-ball method.

species could be produced from reaction of the OSbCl adduct of AlCl$_3$ with H$_2$O or HCl (Scheme 5.17, Reaction 5). In all cases, the resulting counteranion would be larger than HOAlCl$_3^-$ and when bound to the initially formed cyclic carbenium ion might prevent cyclobutane ring expansion and formation of the secondary cyclic carbenium ion **9**; cyclobutane ring opening and formation of the isopropenyl carbenium ion **8** would therefore be favored. In cases where the antimony compound acts as a weak base (Scheme 5.17, Reactions 2–5), the resultant anion would also have increased nucleophilicity resulting in stronger coordination to the initially formed cyclic carbenium ion. As SbCl$_3$ readily forms complexes with aromatic solvents [100–102], SbCl$_3$–AlCl$_3$ adducts may have enhanced solubility in comparison with AlCl$_3$ and this could explain the increased polymerization rates.

Several systems based on *Group 14 congeners* other than organotin halides have been developed. One uses organosilicon halides and disiloxanes (Table 5.8) [103]. Initiator systems based on one part by weight organosilicon compound to five parts by weight AlCl$_3$ gave rise to high yields of higher molecular weight poly(α-pinene)s. Another uses organogermanium halides or alkoxides (Table 5.9) [104]. The best

Table 5.9 α-Pinene resins yielded by AlCl$_3$: organogermanium systems [104].

R$_m$GeX$_n$ (mM)[a]	Yield (%)	SP (°C)[c]
None[b]	28.0	90
(C$_2$H$_5$)$_3$GeCl 15.1	94.5	115
(C$_2$H$_5$)$_3$GeCl 5.87	88.0	113
(CH$_3$)$_3$GeCl 19.1	93.4	115
(C$_2$H$_5$)$_3$GeOCH$_3$ 15.4	89	114

[a] [M] = 3.7 M, [AlCl$_3$] = 133 mM, in xylenes at −20 °C.
[b] [M] = 3.7 M, [AlCl$_3$] = 133 mM, in toluene at −20 °C.
[c] Softening point as determined by ring-and-ball method.

$R_3MX + H_2O \longrightarrow R_3MOH + HX$ (1)

$R_3MOH + AlCl_3 \longrightarrow H^+[R_3MOAlCl_3]^-$ (2)

$R_3MX + H^+[HOAlCl_3]^- \longrightarrow H^+[R_3MOAlCl_3]^- + HX$ (3)

M = Si, Ge, Sn
X = halogen, OR
R = organic substituent or X

Scheme 5.18 Possible reactions in the modification of the $AlCl_3–H_2O$ initiating system with Group 14 compounds.

results were obtained using 3–5 wt% $AlCl_3$ with 0.6 wt% organogermanium compound, based on monomer weight.

Organotin, silicon, and germanium additives might improve the activity of $AlCl_3$ through the formation of a bulkier, more *nucleophilic counteranion*. This anion could be produced by reaction of the hydrolysis products of these additives with $AlCl_3$ (Scheme 5.18, Reactions 1 and 2) or by reaction of these additives with the complex Brönsted acids formed from $AlCl_3$ (Scheme 5.18, Reaction 3). Such an anion might hinder cyclobutane ring expansion that would generate carbenium ion **9** and favor cyclobutane ring opening and formation of carbenium ion **8**. $AlCl_3$ could also abstract the halogen substituent from these additives to form the corresponding cation (Scheme 5.19, Reaction 1). These species could initiate polymerization by direct addition to α-pinene (Scheme 5.19, Reaction 2). They might also react with adventitious moisture to form a reactive proton and produce a bulkier, more nucleophilic counteranion (Scheme 5.19, Reactions 3 and 4). Direct addition would stabilize the initially formed carbenium ion via interactions analogous to a β-silicon effect and may favor formation of carbenium ion **8**. Studies conducted on polymerizations containing weakly coordinating anions (WCAs) in

$R_3MX + AlCl_3 \longrightarrow R_3M^+[XAlCl_3]^-$ (1)

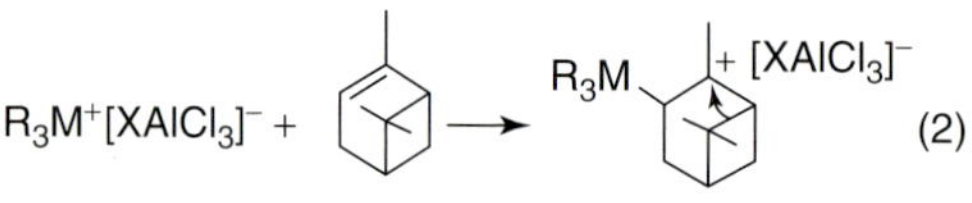

$R_3M^+[XAlCl_3]^- + \text{(α-pinene)} \longrightarrow R_3M\text{-(α-pinene)}^+ [XAlCl_3]^-$ (2)

$R_3M^+[XAlCl_3]^- + H_2O \longrightarrow R_3MOH + H^+[XAlCl_3]^-$ (3)

$R_3MOH + AlCl_3 \longrightarrow H^+[R_3MOAlCl_3]^-$ (4)

M = Si, Ge, Sn
X = halogen, OR
R = organic substituent or X

Scheme 5.19 Direct and indirect initiation of α-pinene polymerization by Group 14 cations.

Table 5.10 Polymerization of α-pinene with amine modified AlCl$_3$ [109].

AlCl$_3$[a] (×10^{-2} mol)	R$_3$N (×10^{-3} mol)	R$_3$N·HCl (×10^{-3} mol)	Yield (%)	SP (°C)[b]
9.00	3.59	8.72	69	115
6.00	3.59–14.3	–	67	105–108

[a] In xylenes at 25–28 °C.
[b] Softening point as determined by ring-and-ball method.

the presence of SHPs indicate that silylium ions might initiate polymerization indirectly via reaction with water to generate a reactive proton [35, 36, 42]. Such findings are flawed in light of more recent research that has shown that SHPs are capable of reacting with metallocenium ions [105] and carbocations [106] when paired with WCAs. The formation of silylium ions is further supported by recent work conducted on the polymerization of isobutylene using novel initiator systems based on aluminoxanes and silicon halides [107]. Studies of this system using the model monomer 2,4,4-trimethyl-1-pentene in the absence of an SHP appear to provide evidence that silylium ions can add directly to alkenes to initiate *carbocationic polymerization* [108].

Initiator systems based on AlCl$_3$ or AlCl$_3$–SbCl$_3$ in conjunction with various base and salt additives have been explored for the *polymerization of α-pinene*. The use of tertiary amines in combination with AlCl$_3$ results in improved yields of poly(α-pinene)s with higher softening points and inclusion of an amine hydrochloride further increases the yields for this system (Table 5.10) [109]. Esters, ethers, and ammonium halides have also been used as additives with the AlCl$_3$–SbCl$_3$ system [110]. Compared with the AlCl$_3$–SbCl$_3$ system, polymerizations in the presence of these additives were found to have reduced rates and result in greater yields of higher molecular weight polymer at a given reaction temperature (Table 5.11). This may result from an increase in both the size and nucleophilicity of the counteranion

Table 5.11 Polymerization of α-pinene with base modified AlCl$_3$–SbCl$_3$ [110].

Base additive[a]	$\overline{M}_n$ (g·mol^{-1})[b]	$\overline{M}_w$ (g·mol^{-1})[b]	Dimer content (wt%)
None	770	1790	11
C$_6$H$_5$CO$_2$C$_2$H$_5$	1030	2320	6.6
CH$_3$CO$_2$C$_2$H$_5$	990	2260	6.9
CH$_2$ClCO$_2$C$_2$H$_5$	1100	2450	6.4
(C$_2$H$_5$)$_2$O	1160	2710	6.1
1,4-Dioxane	1080	2460	6.2
nBu$_4$NCl	1070	2540	6.3

[a] [M] = 3.6 M, [AlCl$_3$] = 42.5 mM, [SbCl$_3$] = 21.3 mM, [base] = 10.6 mM, in toluene at 0 °C.
[b] GPC.

$$H_2O + AlCl_3 \cdot Et_2O \longrightarrow H^+[HOAlCl_3 \cdot Et_2O]^-$$

Scheme 5.20 Example of an $AlCl_3$-base–H_2O initiator system.

due to complex formation (Scheme 5.20). A larger, more nucleophilic counteranion will more strongly associate with the initially formed cyclic carbenium ion and could hinder expansion of the cyclobutane ring, in turn reducing the formation of polymers with structure **11**.

As with other terpenic monomers, numerous *copolymers of α-pinene* have been synthesized. Copolymers based on α-pinene with phenol [72–84, 93], styrene [111], and other monomers [91] have been prepared. Copolymerization is typically conducted in an aromatic solvent using $AlCl_3$ or BF_3 coinitiators in conjunction with adventitious moisture as an initiator.

Significant improvements have been made in the production of *resins from α-pinene*. Even further refinements can be made to the production of these materials that focus on the use of recyclable Lewis acid coinitiators or initiator systems that do not contaminate the product. Recent developments in the field of heterogeneous Lewis acid catalysts hold promise for the commercial production of these resins in a green manner.

5.2.6
Characteristics of Terpenic Resins

Molecular weights and softening points for commercial *terpenic resins* are provided in Table 5.12. Dipentene and α-pinene resins have a lower molecular weight for a given softening point in comparison with β-pinene resins, due to the higher chain flexibility of the latter. The hydrodynamic volumes for dipentene and α-pinene resins are smaller than for β-pinene resins and the former give rise to solutions of lower viscosity at a given temperature [49, 50]. Dipentene resins possess higher color and thermal stability than either α-pinene or β-pinene resins. This might be

Table 5.12 Softening point for terpenic resins as a function of $\overline{M}_n$ [48].

SP (°C)[a]	β-Pinene $\overline{M}_n$ (g·mol^{-1})	Dipentene $\overline{M}_n$ (g·mol^{-1})	α-Pinene $\overline{M}_n$ (g·mol^{-1})
85	815	570	725
100	870	675	775
115	1030[b]	720[c]	815[d]
125	1110	760	830
135	1230	810	870

[a] Softening point as determined by the ring-and-ball method.
[b] $\overline{M}_w/\overline{M}_n = 1.9$.
[c] $\overline{M}_w/\overline{M}_n = 1.4$.
[d] $\overline{M}_w/\overline{M}_n = 1.4$.

due to a multiple stranded chain structure of dipentene resin and the fact that they possess fewer olefinic sites that are susceptible to oxidation [49, 50]. The *cloud point* describes the temperature where a cloud-free melt is obtained. In a 10 : 20 : 20 by weight blend of resin : wax : polyethylene–vinyl acetate, the cloud point is 90 °C for dipentene resins, 115 °C for α-pinene resins, and 175 °C for β-pinene resins [47–52]. The predominant use of dipentene resins is in hot melt adhesives due to their low cloud point, reduced viscosity, and higher thermal stability, whereas β-pinene and α-pinene resins are primarily used in pressure-sensitive adhesives [49, 50].

5.2.7
Applications of Terpenic Resins

Owing to their high softening points, low molecular weights, and compatibility with a broad range of polymers, terpenic resins are excellent *tackifying agents*. They enable a high molecular weight elastomer, which by itself is not capable of adherence, to wet a substrate with little or no pressure. Terpenic resins obtain tack-bestowing properties at relatively low molecular weights, but once above a threshold molecular weight they fail to function effectively as tackifiers as they become more glass-like [49, 50]. They can be viewed as solid solvents that function to bring out smaller, tack-bestowing molecules from their dispersion in the base elastomer to the surface [112]. Optimal tackiness is obtained at a specific loading of tackifying resin, which provides maximum submicroscopic heterogeneity and is dependent on the identity of both the tackifier and the base elastomer [112].

The three main categories of *terpene based adhesives* are pressure-sensitive adhesives, hot-melt adhesives/coatings, and elastomeric sealants. In *pressure-sensitive tapes* the adhesive consists of an elastomer that has been tackified with a terpenic resin [113–117]. This adhesive mixture may be applied to the tape backing either as a solution, an aqueous emulsion, or as a hot-melt system. Of these application methods, aqueous emulsion and hot-melt systems are the most acceptable from an environmental standpoint due to omission of volatile organic compounds (VOCs). *Hot-melt adhesives and coatings* are typically blends of terpenic resin, elastomer, and wax [115, 118–120]. Common *elastomers* are polyethylene–vinyl acetate and thermoplastic elastomers (TPEs). Such hot-melt adhesives have green characteristics in that the tackifying resin is derived from renewable resources and due to the fact that they are 100% solids. Terpenic resins also find use in caulks and other sealants as these materials must be resistant to ultraviolet light, moisture, and temperature extremes encountered in external applications [47–52].

5.2.8
Commercial Production and Markets of Terpenic Resins

Most terpenic resins are made by *solution polymerization* where monomer is gradually added with stirring to the initiator system [47–52]. The prototypical solvent is xylene and $AlCl_3-H_2O$ is the most common initiator system. For β-pinene 2–5 wt% (based on monomer) of $AlCl_3$ is common whereas 4–8 wt% is used for dipentene and α-pinene. Trace moisture is the initiator. Equipment must be corrosion resistant and is either glass lined or stainless steel in construction. Cooling is accomplished through the use of jacketed reactors and/or internal cooling coils. Temperature control is paramount for safety concerns, due to the exothermicity of reaction, and for manipulation of polymer yields/molecular weights. For the $AlCl_3-H_2O$ initiator system, the highest yields are obtained at $-30\,°C$ for β-pinene and $40\,°C$ for α-pinene [44]. The monomer addition rate is of utmost importance for maintaining temperature control and obtaining maximum yields and molecular weights [58]. Polymerizations are conducted under an inert gas to preclude contaminates that adversely affect polymerization and to prevent oxidation. Following monomer addition, reaction may be continued for an additional period of up to 2–3 h to obtain maximum conversion. The coinitiator is then deactivated by the addition of large amounts of water and removed by repeated washings. Polymer is recovered by distillation of volatiles.

Table 5.13 summarizes *production data of terpenic resins* for the period of 1979–2007. Global production peaked in 2000 at 117 700 metric tons and by 2007 had dropped to 68 600 metric tons [121]. According to the latest figures, β-pinene resins make up the majority (45 wt%), followed by dipentene/limonene resins (32 wt%), and then terpene phenolics (23 wt%). Global consumption of terpenic resins has declined due to increasing competition from hydrogenated petroleum resins driven by the higher cost of β-pinene monomer compared with petroleum counterparts [121]. Terpene phenolic resin production has remained steady as no petroleum counterparts exist. The previous and projected market demand in the US for terpenic resins on a usage basis is provided in Table 5.14 [122]. Consumption in the US is projected to grow 2.5% yearly due to an increased demand for products derived from natural materials. The largest producers of terpenic resins are Arizona Chemical (39%), DRT (20%), Eastman (10%), and Yashura (9%) [121].

Table 5.13 Terpenic resin production data.

Year	Region	Total ($\times 10^6$ kg)	β-Pinene (%)	Dipentene (%)	Phenolics (%)	α-Pinene (%)
1979 [48]	US	18	42	35	15	8
1985 [121]	US	23	42	35	15	8
1994 [51]	Global	30	–	–	–	–
2000 [121]	Global	118	44	17	37	2
2004 [121]	Global	72	46	32	22	–
2007 [121]	Global	68.6	45	32	23	–

Table 5.14 US terpenic resin demand [122].

Polyterpene demand (millions of dollars)[a]

Year	1997	2002	2007	2012	2017
Total	40	45	53	60	70
Adhesives	17	20	23	25	29
Food and beverages	9	10	12	14	16
Packaging	5	6	7	7	8
Cosmetics and toiletries	5	5	7	8	9
Other	4	4	4	6	8

[a] Excluding terpene phenolics.

5.2.9
Environmental Aspects of Terpenic Resin Production

The *commercial production of terpenic resins* has progressed in terms of efficiency. The main drawback to contemporary systems is that the coinitiator becomes solubilized in the reaction medium making recycling difficult and resulting in increased material consumption and waste generation. For example, $AlCl_3$ forms adducts with aromatic and olfenic compounds and cannot be easily separated from the product, necessitating de-ashing steps that generate $Al(OH)_3$ and HCl [123]. Physical methods of initiation such as γ-radiation do not contaminate the product but their sensitivity to impurities is so great they cannot be used on a commercial scale.

For a chemical system to be *green and commercially viable*, the coinitiator should have minimal solubility in the product, be of moderate to low cost, possess high activity under a broad range of reaction conditions, and lend itself to recycling. Few initiator systems with these characteristics have been developed. Heterogeneous initiator systems based on strong Brönsted acids in conjunction with Lewis acids or those consisting of strong Lewis acids in combination with weak Lewis acids may allow for green production of terpenic resins [124, 125]. These materials function as solid superacids where a proton derived from either the Brönsted acid or from adventitious moisture initiates polymerization. They do not contaminate the product with Lewis acid and are compatible with continual processes. Owing to non-uniformity of the active sites, these initiator systems give rise to products with broad molecular weight distributions, limiting their utility. The recyclability of these materials is also questionable. Oxidized carbon and carbon supported transition metal oxides have been explored as solid catalysts for the polymerization of and α- and β-pinene [126]. In all cases catalyst activity was low giving rise to resin yields <50% after 50 h of reaction at 90 °C, precluding their application in a commercial setting.

A novel system for the *economical green production* of polyisobutylene and butyl rubber may also be promising for the synthesis of terpenic resins [107, 108]. This

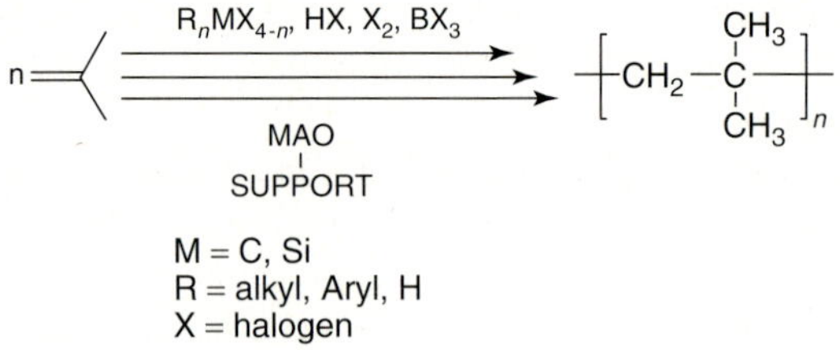

Scheme 5.21 Polymerization of isobutylene using supported alkylaluminoxane coinitiators.

system makes use of heterogeneous coinitiators based on MAO (Scheme 5.21). The coinitiators are insoluble in the reaction medium, can be supported on a variety of substrates, and are compatible with a broad range of initiators. Polymerizations can be conducted in non-polar solvents or in neat monomer and yield polymer with high molecular weights at elevated reaction temperatures. The heterogeneous nature of the coinitiator lends itself to continual production methods and supported versions are readily reactivated with little to no waste generation.

5.3
Monoterpenes as Solvents and Chain Transfer Agents

5.3.1
Possibilities for Replacing Petroleum Solvents

Monoterpenes share several similarities to petroleum solvents, but also several differences. As shown in Table 5.15, the available data for dielectric constants (ε) and densities suggest MTs are more similar to aromatic solvents, such as toluene

Table 5.15 Literature values and extrapolated data for the physical properties of monoterpenes compared with petroleum solvents.

Solvent	Dielectric constant (ε, at 25 °C)	Density (g ml^{-1})	Viscosity (cP)	Boiling point (°C)	References
d-Limonene	2.375	0.838	0.897	175–177	[114, 115]
Myrcene	2.300	0.789	–	167	[114]
α-Pinene	2.179	0.854	1.303	155–156	[114, 115]
β-Pinene	2.497	0.867	1.593	165–167	[114, 116]
Benzene	2.273	0.874	0.605	80	[114, 117, 118]
Toluene	2.379	0.865	0.560	110	[117, 119]
p-Xylene	2.250	0.86	0.603	138	[117, 119]
Cumene	2.374	0.859	0.737	152–154	[117, 119]
Heptane	1.890 (20 °C)	0.680	0.387	98	[118]

and xylene, than saturated hydrocarbons, such as heptane. Although MTs are less volatile and more viscous than petroleum based hydrocarbon solvents, they provide solubility for a wide variety of monomers and catalysts.

The MTs have great potential to replace petroleum solvents in polymerization processes. When designing a *green polymerization system* without petroleum based solvents, the compatibility of initiators and catalysts with MTs must be taken into account. Consider the polymerization mechanism from a standpoint of initiation, propagation, and termination. Depending on the polymerization mechanism, the unsaturation or functional groups in MTs could function as monomers, solvents, chain transfer agents, or inhibitors. For example, the preceding discussion of cationic polymerizations has shown that MTs are often better suited as monomers rather than solvents. Free radical initiators will copolymerize *d*-limonene with styrene and methyl methacrylate to produce alternating copolymers [127, 128]. As certain free radical and cationic initiators consume MTs as monomers, this section will focus on the use of *MTs as solvents* in metallocene (coordination) and ROMPs.

5.3.2
Ring-Opening Polymerizations in Monoterpenes

Ring-opening polymerizations are excellent methods to prepare a wide variety of polyesters, polyethers, polyamides, and unsaturated polyolefins [129–131]. In particular, advances in catalyst efficiency have made ROMP increasingly attractive and allowed the design of experiments with reduced catalyst loading. For example, the high turnover numbers (TONs) for Grubbs ruthenium metathesis complexes allow extremely high monomer to catalyst ratios (100 000 : 1) in the polymerization of cyclooctene and 1,5-cyclooctadiene [132]. Although a large majority of ROMPs are often conducted in methylene chloride or toluene, MTs can replace petroleum solvents [133].

Generally, the molecular weight from polymerizations in MTs was lower than comparable polymerizations in toluene. As shown in Figure 5.7, a variety of *cyclic alkenes* are soluble in *d*-limonene and pinenes. Because the yields in Table 5.16 for polymerizations with cycloheptene and norbornene are similar, the molecular weight differences are not attributed to catalyst inactivity. Instead, vinylidene or vinyl alkene substituents of MTs may participate in ROMP by reacting with the metathesis catalysts. The resulting chain transfer

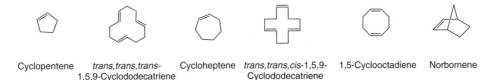

Cyclopentene   trans,trans,trans-   Cycloheptene   trans,trans,cis-1,5,9-   1,5-Cyclooctadiene   Norbornene
                1,5,9-Cyclododecatriene   Cyclododecatriene

Figure 5.7 Sample of cyclic monomers that are soluble in *d*-limonene and pinenes.

Table 5.16 Ring opening metathesis polymerization of alkene monomers[a].

Entry	Monomer	Solvent	Yield (%)	$\overline{M}_w$ (g·mol^{-1})	$\overline{M}_w/\overline{M}_n$ (g·mol^{-1})
1		d-Limonene	16	8 260	2.1
2		d-Limonene	29	2 400	1.6
3		d-Limonene	61	6 320	2.0
4		Toluene	50	50 500	2.6
5		d-Limonene	83	11 100	2.4
6		β-Pinene	74	3 200	1.5
7		Hydrogenated d-limonene	66	56 100	2.0
8		Toluene	68	57 900	2.6
9		d-Limonene	90	26 300	1.3
10		Toluene	88	80 300	1.7

[a] The polymerizations were run for 1 h at 1.3 M and ambient temperature using [monomer]/[catalyst] = 800.

with these accessible alkenes lowers the average molecular weight values. If *molecular weight control* is not necessary, then hydrogenated d-limonene (entry 7, Table 5.16) is a viable option to obtain $\overline{M}_w$ values that are closer to polymerizations in toluene (entry 8). As a function of time, the $\overline{M}_w$ values in Figure 5.8 for ROMP in toluene decrease due to secondary intramolecular and intermolecular metathesis reactions. A comparison between polymerizations in toluene and MTs (Figure 5.8) shows the molecular weight differences due to

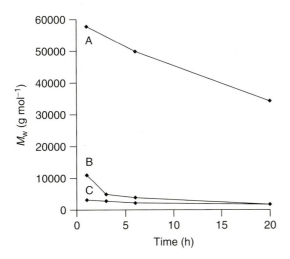

Figure 5.8 Ring opening metathesis polymerization of 1,5-cyclooctadiene (1.3 M) with a second generation ruthenium catalyst ([COD]/[catalyst] = 800) at ambient temperature in: A, toluene; B, d-limonene; and C, β-pinene.

chain transfer. Overall, the $\overline{M}_w$ values for ROMP in different solvents decreased in the following order: toluene ~ hydrogenated d-limonene > d-limonene > β-pinene.

A wide variety of methods have emerged to create *hyperbranched polymers* [134–136]. This section will examine recent methods to integrate *renewable resources* with hyperbranched polymers to develop one-pot procedures that provide an economic alternative to multi-step dendrimer synthesis [137]. In Figure 5.9, the polymerization of dicyclopentadiene (DCPD) with ruthenium metathesis catalysts ring opens both the norbornene and cyclopentene alkenes. The resulting polymerization with neat DCPD becomes cross-linked to yield an insoluble heterogeneous network. However, in the presence of MTs, chain transfer limits the formation of the growing insoluble network. The resulting molecular weight, intrinsic viscosity, and glass transition temperature depend on the type of MT and decrease in the following order: d-limonene > limonene oxide > β-pinene (Table 5.17).

The Mark–Houwink–Sakarada (MHS) equation relates *intrinsic viscosity* ($[\eta]$) to molecular weight [138]. Equation 5.1 is widely used to quantify the influence of solvent, temperature, polymer structure, and branching on $[\eta]$. Essentially, the MHS exponent indicates decreases in the hydrodynamic volume and $[\eta]$ that result from branching. As polymers become more branched, a decrease in the MHS exponent (~0.3–0.6) in a good solvent is observed compared with linear polymers in a good solvent (~0.6–0.8).

$$[\eta] = KM^a \qquad (5.1)$$

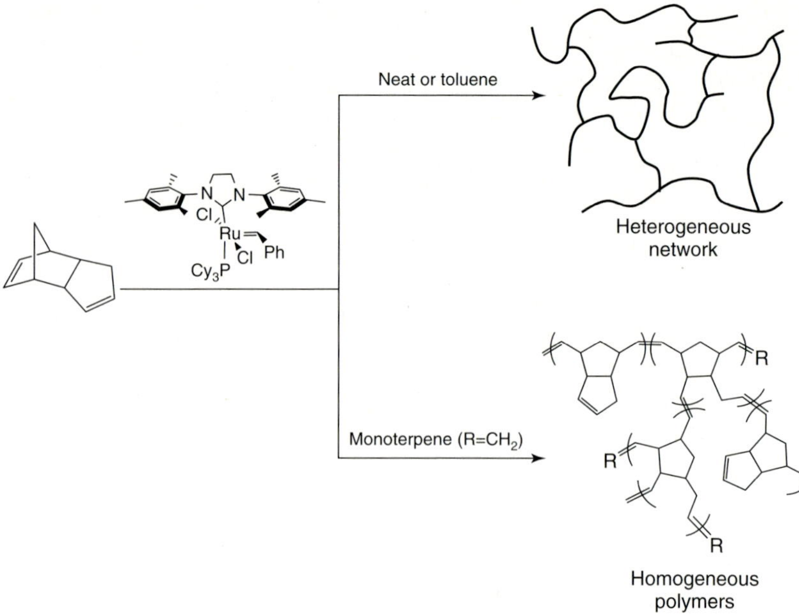

Figure 5.9 Polymerization of dicyclopentadiene with a ruthenium metathesis catalyst.

Table 5.17 Polymerization data for dicyclopentadiene in the presence of monoterpenes[a].

Entry	MT	Toluene	[DCPD]/ [catalyst]	[DCPD]/ [MT]	Yield (%)	$\overline{M}_w$ (g mol^{-1})	$\overline{M}_w/\overline{M}_n$	$[\eta]$ (ml g^{-1})
1	d-Limonene	No	200	0.054	90	6 260	2.0	6.9
2	d-Limonene	No	1000	0.054	93	7 570	2.1	7.9
3	d-Limonene	Yes	1000	5	98	54 400	2.4	24.4
4	Limonene oxide	No	200	0.054	98	4 580	1.3	5.2
5	Limonene oxide	Yes	1000	5	70	26 600	1.5	15.5
6	β-Pinene	No	200	0.054	99	2 120	1.3	3.3
7	β-Pinene	No	1000	0.054	99	2 440	1.2	3.5
8	β-Pinene	Yes	1000	5	78	14 700	1.4	10.5

[a] The polymerizations were run with neat monoterpenes ([DCPD]/[MT] = 0.054) or stoichiometric amounts of monoterpenes in toluene ([DCPD]/[MT] = 5).

where

M = molecular weight
K and a = constants dependent on the polymer, solvent, and temperature

The ability to develop experimental parameters to control the MHS exponent is not always trivial [139, 140]. From the standpoint of waste reduction, slow monomer

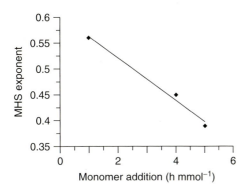

Figure 5.10 Plot of MHS exponent values versus time for slow monomer addition of DCPD. Polymerizations were run in d-limonene at 50 °C with [DCPD]/[catalyst] = 1000.

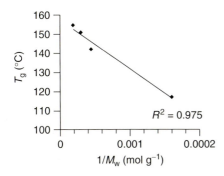

Figure 5.11 Correlation between the glass transition temperature (T_g) and molecular weight for the polymerization of DCPD (0.3 M) in d-limonene at 50 °C after 1 h.

addition avoids the need for high dilution and controls the rate of chain transfer relative to the rate of cross-linking. As shown in Figure 5.10, the MHS exponent for ROMP of DCPD can be varied with the rate of monomer addition.

To evaluate the *physical properties of a homogeneous poly(DCPD) polymer*, the glass transition temperature can be plotted versus the molecular weight. Figure 5.11 shows the data for the polymerization of DCPD ([DCPC]/[initiator] = 1000) in d-limonene after 1 h. The differential scanning calorimetry (DSC) data suggest that at approximately 34 000 g·mol^{-1} the T_g value for homogeneous poly(DCPD) approaches the T_g value that is typical for an insoluble network. A comparison of a literature value with Figure 5.11 suggests the extrapolated value (T_g = 158 °C) for infinite molecular weight is close to an insoluble poly(DCPD) network (maximum T_g = 160 °C) [141]. In addition to good physical properties, these prepolymers have much better processability, could undergo future cross-linking, and do not exhibit the bad odor associated with DCPD.

5.3.3
Metallocene Polymerizations in Monoterpenes

Although the disubstituted and trisubstituted alkenes in MTs may coordinate to the active site in metallocene–MAO catalyst systems, they are generally unable to insert into the active site and homopolymerize. As a result, MTs have potential as benign chain transfer agents and polymerization solvents that can replace petroleum based solvents used for *metallocene polymerizations*. The success of metallocene polymerizations in MTs depends on two main considerations [142]. Firstly, understanding the solubility of monomer, catalyst, and cocatalyst in MTs is important when designing a polymerization system. Secondly, the interaction between catalyst and MT should be evaluated for chain transfer, catalyst inactivity, or polymerization of MT. In this regard, MTs with Lewis basic atoms would be detrimental and standard solvent purification procedures will be needed to remove adventitious water and avoid deactivation of early-metal transition metal catalysts.

The use of MTs with *metallocene catalysts* is complicated by the potential for cationic-type oligomerizations either during the polymerization or during the acidic workup and recovery of polyolefins. To examine the suitability of α-pinene and d-limonene with metallocene catalyst systems, control experiments were conducted with $Et(Ind)_2ZrCl_2$–MAO, Cp_2ZrCl_2–MAO, and Cp_2TiCl_2–MAO. After 4 h at 25 °C, the solution containing catalyst, MAO, and MT was precipitated in methanol containing stoichiometric amounts of hydrochloric acid. For control experiments in d-limonene, oligomers were not detected by GPC (gel permeation chromatography). However, α-pinene was more susceptible to the acidic workup and catalyst system and resulted in 0.8% conversion.

The *polymerization of ethylene* in MTs depended on the catalyst geometry, type of solvent, and ethylene pressure. For the $Et(Ind)_2ZrCl_2$–MAO and Cp_2TiCl_2–MAO catalyst systems, the polymerization activity decreased as follows: toluene > d-limonene > α-pinene. Thermal analysis of the resulting polymers in Table 5.18 suggest that highly crystalline polyethylene results from polymerizations in MTs. Decreases in $\overline{M}_w$ values are attributed to chain transfer to monomer or MT.

Although measuring the concentration of ethylene is not trivial, an accurate determination will allow further calculation of *chain transfer constants* (C_s) for d-limonene. The solubility of gaseous monomers in a solvent can be evaluated based on enthalpy of solution values. In the case of ethylene, the concentration in solution is inversely proportional to the solvent polarity. For example, ethylene concentration at a given pressure is expected to decrease as follows: dodecane >

Table 5.18 Polymerization of ethylene with zirconium and titanium catalysts in monoterpene solvents[a].

Entry	Solvent	Catalyst	Activity (kg·mol^{-1}·h^{-1})	$\overline{M}_w$ (g·mol^{-1})	$\overline{M}_w/\overline{M}_n$	T_m (°C)
1	Toluene	$Et(Ind)_2ZrCl_2$	468	109 000	2.2	130.7
2	d-Limonene	$Et(Ind)_2ZrCl_2$	312	65 000	2.2	126.3
3	α-Pinene	$Et(Ind)_2ZrCl_2$	276	67 000	2.0	128.2
4	Toluene	Cp_2TiCl_2	390	219 000	1.9	136.2
5	d-Limonene	Cp_2TiCl_2	249	180 000	2.1	133.8

[a] The polymerizations were run with monomer (20 psi), solvent (25 ml) and PMAO ([Al]/[M] = 2000) as a cocatalyst.

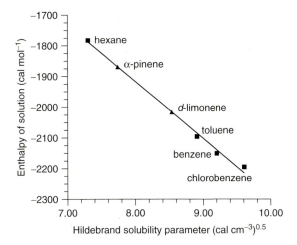

Figure 5.12 Plot of enthalpy of solution values (■) for hexane, benzene, chlorobenzene, and toluene versus the Hildebrand solubility parameters using reported literature values [144]. The line represents the linear regression analysis ($R^2 = 0.994$) of the data. The enthalpy of solution values (▲) for α-pinene and d-limonene were calculated with the linear regression equation and known Hildebrand solubility parameters [145].

hexane > toluene > benzene > acetone [143]. As the *enthalpy of solution* values were not known, they were estimated using Hildebrand solubility parameters. In Figure 5.12, a linear regression ($R^2 = 0.994$) of known Hildebrand solubility parameters versus known enthalpy of solution values was created. Then, the Hildebrand solubility parameters for d-limonene and α-pinene were used to estimate the unknown enthalpy values. Based on this approach, the enthalpy of solution values for d-limonene or α-pinene predict that d-limonene is comparable to toluene while α-pinene is more similar to hexane. This observation and the dielectric constants in Table 5.15 suggest d-limonene and α-pinene are suitable solvents for ethylene polymerizations.

The *Henry relationship* in Equation (5.2) allows the concentration of ethylene to be related to the pressure and enthalpy of solution values [146]. The Henry constants (K_H) for hexane (0.0075 mol·l^{-1}·atm^{-1}), toluene (0.0040 mol·l^{-1}·atm^{-1}), benzene (0.0035 mol·l^{-1}·atm^{-1}), and chlorobenzene (0.0028 mol·l^{-1}·atm^{-1}) were calculated by adapting literature data to Equation 5.2 [143, 144, 147].

$$[\text{Ethylene}] = (P_{\text{ethylene}})(K_H)\, e^{-\Delta H^\circ / RT} \qquad (5.2)$$

The K_H value for ethylene in d-limonene (0.0049 mol·l^{-1}·atm^{-1}) was calculated with the linear regression analysis ($R^2 = 0.997$) of the *Hildebrand solubility parameters* (δ) versus the K_H values for toluene, hexane, chlorobenzene, and benzene (Table 5.19). Using the Mayo equation, the *chain transfer constants* ($C_s \times 10^4$) for ethylene polymerizations using Et(Ind)$_2$ZrCl$_2$–MAO and Cp$_2$TiCl$_2$–MAO with d-limonene are 0.06 and 0.009, respectively. These values are much smaller than the chain transfer

Table 5.19 Calculated enthalpy of solution ($\Delta H°$) values and Henry constants (K_H) for 20 psi of ethylene in α-pinene and d-limonene at 25 °C.

Entry	Solvent	δ (cal·cm^{-3})$^{0.5}$	$\Delta H°$ (cal·mol^{-1})	K_H(mol·l^{-1}·atm^{-1})	[Ethylene] (M)
1	d-Limonene	8.53	−2015	0.0049	0.21
2	α-Pinene	7.74	−1869	0.0065	0.22

constant ($C_s \times 10^4 = 3$) for 1-hexene polymerizations with Et(Ind)$_2$ZrCl$_2$–MAO in d-limonene.

Developing *renewable chain transfer agents* is important from a standpoint of toxicity and waste reduction. Some chain transfer agents, such as vinyl chloride, are much more reactive but also much more toxic [148]. Renewable chain transfer agents also have potential to avoid synthesis procedures that are associated with the preparation of some chain transfer agents. However, more research is needed to fully understand the mechanism of chain transfer and determine if residual amounts of myrcene (~1–2%) in d-limonene and α-pinene interact with the metallocene catalyst system and contribute to the chain transfer process.

5.4
Conclusion

Utilizing MTs as monomers has a long history in polymer science. Given the fact that approximately 1500 terpenes are found in nature [2], many future possibilities for MTs exist. Recent work with limonene [133], carvone [149], and myrcene [150] highlight the ability of MTs to function as chain transfer agents and monomers. With regard to cationic polymerizations, designing systems without volatile and/or toxic solvents, soluble coinitiators that are not recyclable, and the need for very low temperatures, while still obtaining the desired polymer properties, can be a challenge. However, modifying these facets of traditional cationic polymerizations may offer benefits such as lower energy consumption and reduction of waste. In this regard, heterogeneous initiator systems hold promise for the cationic polymerization of terpenes in an economical and environmentally friendly manner [107, 108].

Acknowledgments

RTM thanks the American Chemical Society Petroleum Research Fund and the Army Research Office for financial support.

References

1. Breitmaier, E. (2006) *Terpenes: Flavors, Fragrances, Pharmaca, Pheromones*, Wiley-VCH Verlag GmbH, Weinheim.
2. Connolly, J.D. and Hill, R.A. (1991) *Dictionary of Terpenoids*, vol. 1, Chapman & Hall, London.
3. Braddock, R.J. (1999) *Handbook of Citrus By-Products and Processing Technology*, John Wiley & Sons, Inc., New York.
4. Bishop, W. (1789) *Chem. Essays*, **3** (3), 5–6.
5. Leopolde, E.A. (1909) US Patent 919,248.
6. Cooper, S.M. (1931) US Patent 1,939,320.
7. Derfer, J.M. and Traynor, S.G. (1989), In: *Naval Stores: Production, Chemistry, Utilization* (eds D.F. Zinkel and J. Russell), Pulp Chemicals Association, New York, pp. 225–260.
8. Gandini, A. and Cheradame, H. (1980) *Advances in Polymer Science*, vol. 34/35, Springer-Verlag, New York.
9. Kennedy, J.P. (1975) *Cationic Polymerization of Olefins: A Critical Inventory*, John Wiley & Sons, Inc., New York.
10. Matyjaszewski, K. (ed.) (1996) *Cationic Polymerizations*, Marcel Dekker, New York.
11. Cotrel, R., Sauvet, G., Vairon, J.P., and Sigwalt, P. (1976) *Macromolecules*, **9** (6), 931–936.
12. Pielichowski, J. (1973) *J. Polym. Sci., Polym. Symp.*, **42** (1), 451–456.
13. Sauvet, G., Vairon, J.P., and Sigwalt, P. (1974) *Eur. Polym. J.*, **10**, 501–509.
14. Sawamoto, M., Fukimori, J., and Higashimura, T. (1987) *Macromolecules*, **20** (5), 916–920.
15. Subira, F., Vairon, J.P., and Sigwalt, P. (1988) *Macromolecules*, **21** (8), 2339–2346.
16. Balogh, L., Wang, L., and Faust, R. (1994) *Macromolecules*, **27** (13), 3453–3458.
17. Chmelir, M., Marek, M., and Wichterle, O. (1967) *J. Polym. Sci. C: Polym. Symp.*, **16** (2), 833–839.
18. Stannett, V.T. and Silverman, J. (1983) *ACS Symp. Ser.*, **212**, 435.
19. Lampe, F.W. (1959) *J. Phys. Chem.*, **63**, 1986.
20. Flory, P.J. (1953) *Principles of Polymer Chemistry*, Cornell University Press, Ithaca, p. 218.
21. Thomas, R.M., Sparks, W.J., Frolich, P.K., Otto, M., and Mueller-Cunradi, M. (1940) *J. Am. Chem. Soc.*, **62**, 276–280.
22. Balogh, L. and Faust, R. (1992) *Polym. Bull.*, **28**, 367–374.
23. Gyor, M., Wang, H.-C., and Faust, R. (1992) *J. Macromol. Sci. Pure Appl. Chem.*, **A29** (8), 639–653.
24. Kennedy, J.P. and Chou, R.T. (1979) *Polymer. Prepr.*, **20**, 306.
25. Higashimura, T., Miayamoto, M., and Sawamoto, M. (1985) *Macromolecules*, **18** (4), 611–616.
26. Aoshima, S. and Higashimura, T. (1986) *Polym. Bull.*, **15**, 417–423.
27. Baird, M.C. (1995) US Patent 5,448,001.
28. Baird, M.C. (2000) *Chem. Rev.*, **100**, 1471–1478.
29. Barsan, F., Karan, A.R., Parent, M.A., and Baird, M.C. (1998) *Macromolecules*, **31** (24), 8439–8447.
30. Bochmann, M. and Dawson, D.M. (1996) *Angew. Chem., Int. Ed. Engl.*, **35** (19), 2226–2228.
31. Carr, A.G., Dawson, D.M., and Bochmann, M. (1998) *Macromolecules*, **31**, 2035–2040.
32. Garratt, S., Carr, A.G., Langstein, G., and Bochmann, M. (2003) *Macromolecules*, **36**, 4276–4287.
33. Guerrero, A., Kulbaba, K., and Bochmann, M. (2007) *Macromolecules*, **40** (12), 4124–4126.
34. Jacob, S., Pi, Z., and Kennedy, J.P. (1998) *Polymer. Bull.*, **41**, 503–510.
35. Jacob, S., Pi, Z., and Kennedy, J.P. (1999), In: *Ionic Polymerizations and Related Processes, Nato Science Series, Series E*, vol. 359 (ed. J.E. Puskas), Kluwer, Dordrecht, pp. 1–12.
36. Jacob, S., Pi, Z., and Kennedy, J.P. (1999) *Polym. Mater. Sci. Eng.*, **80**, 495.
37. Kumar, K.R., Hall, C., Penciu, A., Drewitt, M.J., Mcinenly, P.J., and

Baird, M.C. (2002) *J. Polym. Sci. A: Polym. Chem.*, **40**, 3302–3311.

38. Lewis, S.P., Piers, W.E., Taylor, N., and Collins, S. (2003) *J. Am. Chem. Soc.*, **125** (48), 14686–14687.
39. Shaffer, T.D. (1994) US Patent Application 234,782.
40. Shaffer, T.D. (1997), In: *Cationic Polymerization* (eds R. Faust and T.D. Shaffer), American Chemical Society, Washington, DC, pp. 96–105.
41. Shaffer, T.D. and Ashbaugh, J.R. (1996) *Polym. Prepr., Am. Chem. Soc. Div. Polym. Chem.*, **37** (1), 339–340.
42. Shaffer, T.D. and Ashbaugh, J.R. (1997) *J. Polym. Sci. A: Polym. Chem.*, **35** (2), 329–344.
43. Vierle, M., Schön, D., Bohnenpoll, M., Kühn, F.E., and Nuyken, O. (2003) CA Patent 2,421,688.
44. Roberts, W.J. and Day, A.R. (1950) *J. Am. Chem. Soc.*, **72**, 1226–1230.
45. Zlamal, Z. and Ambroz, L. (1958) *J. Polym. Sci.*, **29**, 595–604.
46. Zlamal, Z., Zazda, A., and Ambroz, L. (1966) *J. Polym. Sci., A1*, **4** (7), 367–375.
47. Gonzenbach, C.T., Jordan, M.A., and Yunick, R.P. (1970), In: *Encyclopedia of Polymer Science and Technology*, vol. 13 (eds H.F. Mark, N.M. Bikales, J. Conrad, G.O. Schetly, J. Perlman, K.N. Brown, M.D. Fernandez, and J.M. Ricciardi), John Wiley & Sons, Inc., New York, pp. 575–596.
48. Ruckel, E.R. (1982), In: *Carbocationic Polymerization* (eds J.P. Kennedy and E. Maréchal) John Wiley & Sons, Inc., New York, pp. 491–499.
49. Ruckel, E.R. and Arlt, H.G. Jr. (1989), In: *Naval Stores: Production, Chemistry, Utilization* (eds D.F. Zinkel and J. Russell), Pulp Chemicals Association, New York, pp. 511–530.
50. Ruckel, E.R., Arlt, H.G. Jr., and Wojcik, R.T. (1975), In: *Adhesion Science and Technology Polymer Science and Technology*, vol. 9A, (ed. L.H. Lee), Plenum Press, New York, pp. 395–412.
51. Vairon, J.-P. and Spassky, N. (1996), In: *Cationic Polymerizations* (ed. K. Matyjaszewski), Marcel Dekker, New York, pp. 707–711.
52. Vredenburgh, W., Foley, K.F., and Scarlatti, A.N. (1987), In: *Encyclopedia of Polymer Science and Engineering*, 2nd edn, vol. 7 (eds H.F. Mark, N.M. Bikales, C.G. Overberger, G. Menges, and J.I. Kroschwitz), John Wiley & Sons, Inc., New York, pp. 769–782.
53. Cataldo, F., Angelini, G., Capitani, D., Gobbino, M., Ursini, O., and Forlini, F. (2008) *J. Macromol. Sci. A: Pure Appl. Chem.*, **45** (10), 839–849.
54. Cataldo, F., Gobbino, M., Ursini, O., and Angelini, G. (2007) *J. Macromol. Sci. A: Pure Appl. Chem.*, **44** (11), 1225–1234.
55. Martinez, F. (1984) *J. Polym. Sci. Polym. Chem. Ed.*, **22** (3), 673–677.
56. Kennedy, J.P. and Maréchal, E. (1981) *J. Polym. Sci. Macromol. Rev.*, **16**, 123–198.
57. Kennedy, J.P. and Milliman, G.E. (1969) *Adv. Chem. Ser.*, **91**, 287–305.
58. Guiné, R.P.F. and Castro, J.A.A.M. (2001) *J. Appl. Polym. Sci.*, **82**, 2558–2565.
59. Keszler, B. and Kennedy, J.P. (1992) *Adv. Polym. Sci.*, **100**, 1–9.
60. Satoh, K., Sugiyama, H., and Kamigaito, M. (2006) *Green Chem.*, **8**, 878–882.
61. Marvel, C.S., Hanley, J.R., and Longone, D.T. (1959) *J. Polym. Sci.*, **40**, 551–555.
62. Modena, M., Bates, R.B., and Marvel, C.S. (1965) *J. Polym. Sci. A*, **3**, 949–960.
63. Yu, P., Li, A.-L., Liang, H., and Lu, J. (2007) *J. Polym. Sci. A: Polym. Chem.*, **45** (16), 3739–3746.
64. Lu, J., Kamigaito, M., Sawamoto, M., Higashimura, T., and Deng, Y.-X. (1997) *Macromolecules*, **30**, 22–26.
65. Adur, A.M. and Williams, F. (1981) *J. Polym. Sci. Polym. Chem. Ed.*, **19** (3), 669–678.
66. Bates, T.H., Best, J.V.F., and Williams, T.F. (1960) *Nature*, **188**, 469–470.
67. Bates, T.H., Best, J.V.F., and Williams, T.F. (1962) *J. Chem. Soc.*, 1531–1540.
68. Ingold, C.K. (1957) *Proc. Chem. Soc.*, 279–287.
69. Kennedy, J.P., Collins, S., and Lewis, S.P. (2007) US Patent 7,202,317.

70. Lewis, S.P. (2004) Project 1. Synthesis of PIB-Silsequioxane Stars via the Sol-Gel Process Project 2. Solution and Aqueous Suspension/Emulsion Polymerization of Isobutylene Coinitiated by 1,2-$C_6F_4[B(C_6F_5)_2]_2$ PhD dissertation. The University of Akron.
71. Lewis, S.P., Henderson, L., Parvez, M.R., Piers, W.E., and Collins, S. (2005) *J. Am. Chem. Soc.*, **127**, 46–47.
72. Beck, Koller & Company (1935) GB Patent 467,816.
73. Tenneco Chemicals Inc. (1962) GB Patent 1,012,045.
74. Farbweke Hoechst Aktiengessellschaft (1963) GB Patent 1,043,159.
75. Gobran, R. (1976) US Patent 3,976,606.
76. Gonzenbach, C.T. (1968) US Patent 3,383,362.
77. Hönel, H. and Zinke, A. (1938) US Patent 2,123,898.
78. Kaupp, J. and Blaettner, K. (1967) US Patent 3,347,935.
79. Powers, P.O. (1944) US Patent 2,343,845.
80. Rummelsburg, A.L. (1945) US Patent 2,378,436.
81. Rummelsburg, A.L. (1949) US Patent 2,471,454.
82. Tarassoff, K. (1914) GB Patent 7,560.
83. Tarassoff, K. (1923) GB Patent 223,636.
84. Wuyts, H. (1923) GB Patent 204,754.
85. Ott, E. (1941) US Patent 2,373,706.
86. Kennedy, J.P. and Chou, T.M. (1975) US Patent 3,923,759.
87. Pietila, H., Sivola, A., and Sheffer, H. (1970) *J. Polym. Sci. A1*, **8** (3), 727–737.
88. Snyder, C., McIver, W., and Sheffer, H. (1977) *J. App. Polym. Sci.*, **21** (1), 131–139.
89. Weymann, H.P. and Jen, Y. (1968) US Patent 3,413,246.
90. Hokama, T. and Scardiglia, F. (1976) US Patent 3,959,238.
91. Davis, J.B. (1970) US Patent 3,510,461.
92. Butler, G.B., Miles, M.L., and Bray, W.S. (1965) *J. Polym. Sci. A*, **3** (2), 723–733.
93. Gobran, R. (1975) US Patent 3,927,239.
94. White, R.H. and Hill, F.M. (1975) US Patent 3,929,938.
95. Kralevich, M.L.J., Blok, E.J., Wideman, L.G. and Sandstrom, P.H. (2001) US Patent 6,265,478.
96. Higashimura, T., Lu, J., Kamigaito, M., and Sawamoto, M. (1992) *Makromol. Chem.*, **193**, 2311–2321.
97. Higashimura, T., Lu, J., Kamigaito, M., and Sawamoto, M. (1993) *Makromol. Chem.*, **194**, 3441–3453.
98. Sproat, A.D. (1967) US Patent 3,354,132.
99. Wojcik, R.T. and Ruckel, E.R. (1977) US Patent 4,016,346.
100. Auel, T. and Amma, E. (1968) *J. Am. Chem. Soc.*, **90** (21), 5941–5942.
101. Corriu, R., Coste, C., and Sournia, A. (1970) *Bull. Soc. Chim. Fr*, 2998.
102. Gerbier, J. (1966) *C.R. Hebd. Seances Acad. Sci. Ser. B*, **262**, 685.
103. Barkley, L.B. and Patellis, A.P. (1969) US Patent 3,478,007.
104. Wang, L.S. and Ruckel, E.R. (1977) US Patent 4,011,385.
105. Barsan, F., Karam, A.R., Parent, M.A., and Baird, M.C. (1998) *Macromolecules*, **31** (24), 8439–8447.
106. Jianfang, C., Lewis, S.P., Kennedy, J.P., and Collins, S. (2007) *Macromolecules*, **40** (21), 7421–7424.
107. Lewis, S.P. (2009) Heterogeneous Lewis acid catalysts for cationic polymerization. US Patent Application 2010-0273964A1.
108. Mathers, R.T., Fu, Y., Qian, L., and Lewis, S.P. (2010) *Macromolecules*, submitted for publication.
109. Patellis, A.P. and Nufer, H.L. (1969) US Patent 3,466,271.
110. Higashimura, T., Lu, J., Kamigaito, M., and Sawamoto, M. (1993) *Makromol. Chem.*, **194**, 3455–3465.
111. Khan, A.R., Yousufzai, A.H.K., Jeelani, H.A., and Akhter, T. (1985) *J. Macromol. Sci. Chem.*, **A22** (12), 1673–1678.
112. DeWalt, C. (1970) *Adhes. Age*, **13** (3), 38–45.
113. Antonsen, S.A. and Boaz, D.P. (1979) US Patent 4,163,077.
114. Korpman, R. (1966) US Patent 3,242,110.
115. Orth, G.O.J. (1979) DE Patent 2,282,524.

116. Scholl, W.M. and Kemp, M.H. (1965) DE Patent 1,191,926.
117. Temin, S.C. (1980) DE Patent 3,004,377.
118. Flanagan, T.P. (1971) US Patent 3,573,240.
119. Tse, M.F., Hughes, V., Mehta, A.K., and Milks, R.R. (1992) WO Patent 9,212,212 (A1).
120. Vanhaeren, G. (1988) Eur. Patent 271,254 (A2).
121. Stauffer, D. (2009) *Study of International Rosin Markets*, International Development Associates, Mendenhall.
122. Weizer, W. (2009) *Natural Polymers*, The Freedonia Group, Cleveland.
123. Olah, G.A. (1973) *Friedel-Crafts Chemistry*, John Wiley & Sons, Inc., New York, p. 258.
124. Chen, F.J., Deore, C.L., Spitz, R., and Guyot, A. (1997) US Patent 5,648,580.
125. Chen, F.J., Guyot, A., Hamaide, T., and Deore, C.L. (1997) US Patent 5,607,890.
126. Ramos, A.M., Silva, I.F., Vital, J., and McKee, D.W. (1997) *Carbon*, **35** (8), 1187–1189.
127. Sharma, S. and Srivastava, A.K. (2003) *J. Macromol. Sci., Pure Appl. Chem.*, **A40** (6), 593–603.
128. Sharma, S. and Srivastava, A.K. (2004) *Eur. Polym. J.*, **40** (9), 2235–2240.
129. McGrath, J.E. (ed.) (1985) *Ring Opening Polymerization*, ACS Symposium Series, vol. 286, American Chemical Society, Washington, DC.
130. Grubbs, R.H. (ed.) (2003) *Handbook of Metathesis. Applications in Polymer Synthesis*, 1st edn, vol. 3, Wiley-VCH Verlag GmbH, Weinheim.
131. Hoff, R. and Mathers, R.T. (eds) (2009) *Handbook of Transition Metal Polymerization Catalysts*, John Wiley & Sons, Inc., Hoboken.
132. Bielawski, C.W. and Grubbs, R.H. (2000) *Angew. Chem. Int. Ed.*, **39** (16), 2903–2906.
133. Mathers, R.T., McMahon, K.C., Damodaran, K., Retarides, C.J., and Kelley, D.J. (2006) *Macromolecules*, **39**, 8982–8986.
134. Tomalia, D.A. and Frechet, J.M. (2002) *J. Polym. Sci. A Polym. Chem.*, **40** (16), 2719–2728.
135. Sunder, A., Heinemann, J., and Frey, H. (2000) *Chem. Eur. J.*, **6** (14), 2499–2506.
136. Voit, B. (2005) *J. Polym. Sci. A Polym. Chem.*, **43** (13), 2679–2699.
137. Mathers, R.T., Damodaran, K., Rendos, M.G., and Lavrich, M.S. (2009) *Macromolecules*, **42**, 1512–1518.
138. Patterson, G. (2007) *Physical Chemistry of Macromolecules*, CRC Press, Boca Raton.
139. Hanselmann, R., Holter, D., and Frey, H. (1998) *Macromolecules*, **31**, 3790–3801.
140. Mock, A., Burgath, A., Hanselmann, R., and Frey, H. (2001) *Macromolecules*, **34**, 7692–7698.
141. Lee, J.K., Liu, X., Yoon, S.H., and Kessler, M.R. (2007) *J. Polym. Sci. B Polym. Phys.*, **45**, 1771–1780.
142. Mathers, R.T. and Damodaran, K. (2007) *J. Polym. Sci. A Polym. Chem.*, **45**, 3150–3165.
143. Sahgal, H.M., Hayduk, W. (1978) *Can. J. Chem. Eng.* **56**, 354–357.
144. Wilhelm, E. and Battino, R. (1973) *Chem. Rev.*, **73** (1), 1–9.
145. Paul, P.K.C. (2005) *Org. Biomol. Chem.*, **3**, 1176–1179.
146. Kissin, Y.V. (1985) *Isospecific Polymerization of Olefins*, Springer-Verlag, New York.
147. Waters, J.A., Mortimer, G.A., and Clements, H.E. (1970) *J. Chem. Eng. Data*, **15** (1), 174–176.
148. Gaynor, S.G. (2003) *Macromolecules*, **36**, 4692–4698.
149. Lowe, J.R., Tolman, W.B., and Hillmyer, M.A. (2009) *Biomacromolecules*, **10**, 2003–2008.
150. Kobayahi, S., Lu, C., Hoye, T.R., and Hillmyer, M.A. (2009) *J. Am. Chem. Soc.*, **131**, 7960–7961.

6
Controlled and Living Polymerization in Water: Modern Methods and Application to Bio-Synthetic Hybrid Materials

Debasis Samanta, Katrina Kratz, and Todd Emrick

6.1
Introduction

Living polymerizations are characterized by the successful propagation of monomers into well-defined polymer chains, enabled by continuous propagation events that proceed cleanly, in the absence of substantial interference from competing termination and chain transfer reactions [1]. Classic *living polymerizations* are enabled by the kinetic situation in which the rate of initiation is significantly faster than the rate of propagation. This provides assurance that all of the polymer chains have an approximately identical starting point timeframe, leading to a nearly equivalent degree of polymerization of all the chains. As a result, the polymer products of living polymerization possess a narrow molecular weight distribution (i.e., low polydispersity index or PDI) and predictable molecular weights achieved by varying the monomer-to-initiator ratio.

The practice of living polymerization was dominated traditionally by anionic techniques, especially by the generation of *carbanion initiators* (usually organolithium or organopotassium reagents) at low temperature, for polymerization of monomers such as styrene and methyl methacrylate [2]. Living anionic polymerization requires strict exclusion of water and other deprotonatable reagents, and is best done at low temperatures to avoid competing termination reactions. Living polymerization conducted in this manner is a useful but decidedly energy and solvent intensive process. In the 1980s, researchers at DuPont developed a new technique, termed group transfer polymerization (GTP). GTP permits living polymerization at or near room temperature, improving its amenability to scale-up, and permitting polymerization of a wider range of monomers [e.g., poly(vinyl alcohol) precursors] [3, 4] than traditional living anionic methods.

In recent years, advances in polymer chemistry have enabled the extension of living polymerization into *non-anionic mechanisms*, thus opening avenues to prepare well-defined polymers from monomers that are too sensitive for anionic conditions, and the use of polar solvents, and even water, as polymerization media. Free radical polymerization is prominent among these advances, including nitroxide-mediated polymerization (NMP) [5], atom transfer radical polymerization (ATRP) [6], and

Green Polymerization Methods: Renewable Starting Materials, Catalysis and Waste Reduction
Edited by Robert T. Mathers and Michael A. R. Meier
Copyright © 2011 WILEY-VCH Verlag GmbH & Co. KGaA, Weinheim
ISBN: 978-3-527-32625-9

reversible addition–fragmentation chain transfer (RAFT) [7] polymerization. While these three living free radical techniques use distinctly different chemistries, each with its own advantages, their common principle is to decrease the overall free radical concentration by establishing equilibrium between active and dormant (protected) chains. This active/dormant equilibrium decreases the frequency of termination by biomolecular coupling and disproportionation, while still allowing propagation to proceed at a reasonable rate.

Recent advances in *cyclic olefin polymerization*, specifically ring-opening metathesis polymerization (ROMP), have enabled for the first time its inclusion in the living polymerization category, including that in water. The discovery and development of metathesis catalysts have since led to "living ROMP," with specific advances including: (i) polymerization of cyclic olefins with appropriate initiation/propagation kinetics; (ii) preparation and use of catalysts that maintain activity at the polymer chain-end for sequential monomer addition (i.e., to prepare block copolymers); and (iii) catalysts that function effectively in water.

This chapter examines two distinct topics pertinent to modern living polymerization methods in aqueous environments, specifically: (i) state-of-the-art catalysts for performing ROMP in water and (ii) recent advances in aqueous living free radical polymerization, focusing on the area of biohybrid materials prepared by protein–polymer conjugation.

6.2
Ring-Opening Metathesis Polymerization (ROMP)

Olefin metathesis in organometallic chemistry describes a double bond substituent switching process (a transalkylidenation), catalyzed by transition metals such as ruthenium, osmium, molybdenum, and tungsten, long recognized as useful in small molecule and polymer synthesis. Prominent variants of the metathesis process are shown in Figure 6.1, including acyclic diene metathesis (ADMET) polymerization, ring-closing metathesis (RCM), and ROMP. Advances associated with environmentally conscious or green chemistry in all aspects of metathesis have emerged in recent years, including RCM as an atom-economical approach to small molecule synthesis (i.e., proceeding in fewer steps than conventional synthetic methodology would allow).

ROMP converts cyclic olefin monomers into *unsaturated polyolefins*, giving, for example, polynorbornene, polycyclooctene, polycyclopentene, and polycyclooctatetraene (polyacetylene) from their respective cyclic olefins. ROMP has contributed significantly to the polymer industry, such as in the cases of Norsorex (polynorbornene), Zeonor (a hydrogenated norbornene-based copolymer), Vestenamer (a polycyclooctene used in processing asphalt concrete blends), and a variety of plastics produced by Materia, Inc. for sporting goods [8]. The general concept and practice of *metathesis chemistry* and catalysis was recognized with the awarding of the 2005 Nobel Prize in Chemistry to Robert Grubbs, Richard Schrock, and Yves Chauvin [9]. The green chemistry aspects evolving from metathesis chemistry, exemplified

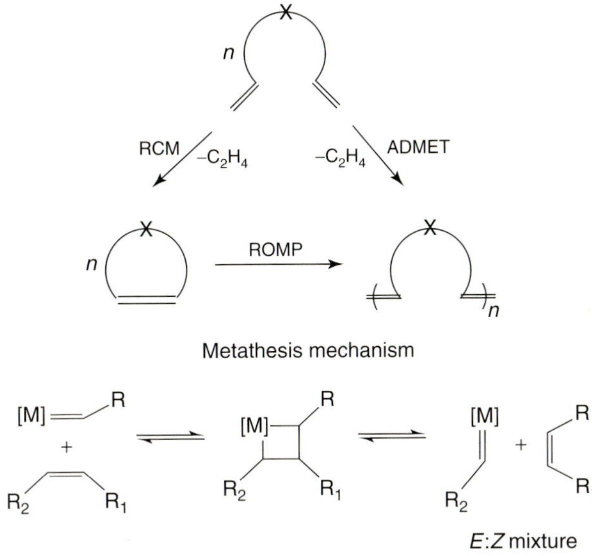

Figure 6.1 Olefin metathesis reactions (RCM, ADMET, and ROMP), and the fundamental metathesis mechanism.

in both ROMP and RCM, highlight salient features of this methodology and explain its rapidly growing popularity and usefulness in academic and industrial settings.

The mechanism of ROMP, elucidated through detailed studies of several groups [10–12], begins with the [2 + 2] *cycloaddition* of a metal carbene and cyclic olefin to give a metallocyclobutane (Figure 6.1). Relief of ring strain energy drives propagation, and strained norbornenes polymerize readily to high monomer conversion. Less-strained monomers such as cyclooctene (COE) and cyclooctadiene (COD) polymerize at slower rates, with more competition from side reactions (e.g., cross-metathesis). A variety of metal complexes catalyze ROMP, some having complex ill-defined structures, and others having discrete and well-characterized structures. Various metal halide complexes with cocatalysts, such as $WOCl_4$–$SnMe_4$ and $MoCl_4$–$AlEt_3$, were found to catalyze ROMP of norbornene and other cyclic olefins. Metal halides of ruthenium, iridium, and osmium hydrate are also effective ROMP catalysts [13]. However, these catalyst systems rely on the *in situ* generation of the active metal-carbene species, and as such proceed without fine control over molecular weight and PDI. In contrast, transition metal-based metallocyclobutanes [14], and molybdenum or tungsten-containing complexes [15, 16], are well-defined and effective in catalyzing ROMP to afford high molecular weight polymers with good control over PDI. However, these catalysts require an inert atmosphere and strict exclusion of water and other protic solvents, and cannot handle monomers with polar groups (e.g., amines, carboxylic acids, aldehydes, etc.), which competitively coordinate to the metal center. As such, these catalysts are not readily amenable to aqueous-based green methods desired in organic and polymer synthesis.

Ruthenium-based alkylidene and benzylidene metathesis catalysts, developed initially in the 1990s and still being exploited and optimized today, offer opportunities for adapting metathesis to green chemistry. In general, these *ruthenium-based structures* are highly active in ROMP, with various coordinating ligands employed to control relative initiation and propagation kinetics. In some instances, ROMP can be performed in a "living" fashion, giving polymers with low PDI, and well-defined block copolymers through sequential monomer addition. The success of these catalysts hinges on their exceptional air, water, and functional group tolerance, making efforts to improve their inherent water solubility a natural and important outgrowth of metathesis research.

Early ROMP catalysts were based on Group 8 *transition metals*, including osmium and ruthenium hydrates, such as $Ru(H_2O)_6(tosylate)_2$ [13]. While these organometallic compounds are active in water, controlled polymerization was not achieved. For example, $RuCl_3$ catalyzes the polymerization of *exo,exo*-2,3-bis(methoxymethyl)-7-oxanorbornene in ethanol–water mixtures, giving relatively high molecular weights (100–300 kDa) with PDI values ~2. $RuCl_3$, $IrCl_3$, and $OsCl_3$ hydrates are active catalysts for ROMP of norbornene [17], and $Ru(H_2O)_6(tosylate)_2$ effects the polymerization of *exo*-5,6-bis(methoxymethyl)-7-oxanorbornene **1** in water, generating very high molecular weight polymer [18]. Irradiation of $(C_6H_6)_2Ru(tosylate)_2$ gives a solvated Ru(II) species that initiates ROMP of norbornene in alcohol–water mixtures, resulting in polynorbornene with PDI values above 2 [19]. Kiessling and coworkers used aqueous solutions of $RuCl_3$ to prepare neoglycopolyolefins by ROMP [20]. These polymers, containing biologically derived units strung pendent to the polymer backbone, were obtained with ~1 : 1 cis- to trans-olefins, and exhibited high molecular mass (~10^6 g mol^{-1}) by gel electrophoresis relative to dextran standards.

Figure 6.2 shows three examples of ruthenium benzylidene catalysts developed by Grubbs and coworkers. These catalysts have enjoyed widespread use in *organic/polymer synthesis* for researchers utilizing ROMP and other metathesis chemistries. As suggested by their ligand structures, the catalysts are hydrophobic and soluble in organic solvents only. The *N*-heterocyclic carbenes shown as **2** and **3** are the more active of the catalysts, but only the pyridine-substituted version enables the preparation of low PDI ROMP polymers [21]. Pyridine-substituted **3** can also

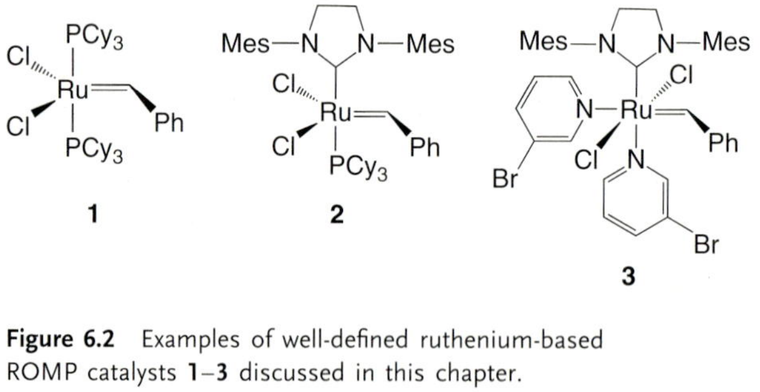

Figure 6.2 Examples of well-defined ruthenium-based ROMP catalysts **1–3** discussed in this chapter.

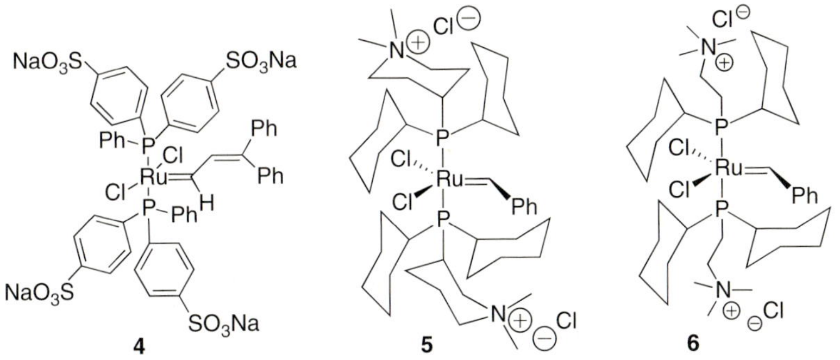

Figure 6.3 Water soluble metathesis catalysts with charged ligands.

handle sterically-demanding monomers, as seen for dendronized polynorbornene prepared by Fréchet and coworkers [22].

The *chemical stability* toward water of the catalysts shown in Figure 6.2 suggest their potential utility in aqueous-based polymerizations. Grubbs and coworkers have reported the polymerization of sugar-substituted norbornenes using ruthenium benzylidene catalysts in an aqueous emulsion system [23]. For this the catalyst is dissolved in a minimal amount of organic solvent, and the emulsion is stabilized with ammonium halide salts. Gnanou and coworkers prepared norbornene latexes by ROMP of norbornene in aqueous mini-emulsions, using both lipophilic and hydrophilic Ru catalysts [24]. This gave polynorbornene latex particles in the 200–500 nm diameter range. When a *lipophilic catalyst* such as **1** was used, the particles coagulate, but when $RuCl_3$ catalyst was used, the particles were more stable. Claverie *et al.* have also reported ROMP using water-soluble catalysts and water-insoluble monomer in emulsion [25], obtaining submicron polynorbornene latex particles using catalyst **5** (Figure 6.3), in which the particles are protected against coagulation by a layer of absorbed surfactant.

6.2.1
Water Soluble ROMP Catalysts

Fully realizing the benefits of well-defined ruthenium benzylidene catalysts in water requires *solubilizing the catalysts* in water with appropriate hydrophilic ligands. Catalysts **4–6** in Figure 6.3, reported by Grubbs and coworkers, are made water soluble by anchoring charged substituents (sulfonate and ammonium ions) to the coordinating phosphine ligands [26]. Interestingly, catalyst **4** did not initiate ROMP in aqueous solution, attributed to the small cone angle and weak electron-donating character of the coordinated triarylphosphine ligands. However, catalysts **5** and **6** led to productive ROMP when used with water-soluble norbornene and oxanorbornene monomers. Although initiation occurred in a well-defined manner, rapid decomposition of the propagating species led to high PDIs and lower molecular weights than anticipated from the monomer-to-catalyst ratios employed. Moreover, clean living polymerization was observed when catalysts **5** and **6** were used in

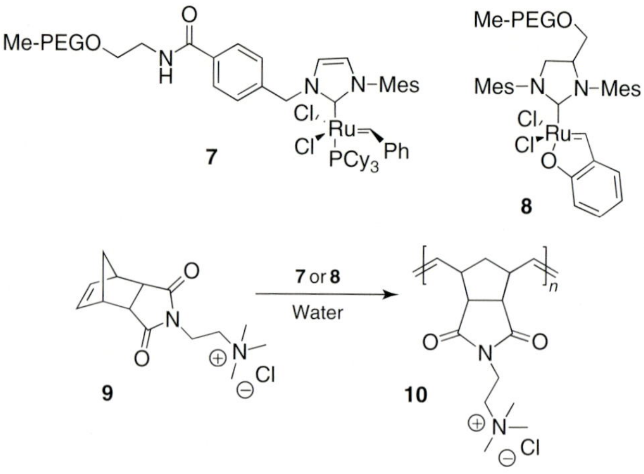

Figure 6.4 Water soluble ruthenium-based metathesis catalysts containing PEG in the ligand structure.

acidic aqueous media. Nuclear magnetic resonance (NMR) spectroscopy revealed that water-soluble monomers were polymerized quantitatively with DCl (0.3–1.0 equiv. relative to monomer) present in solution. Low PDI values (typically ~1.2) suggested a living polymerization, confirmed by the successful extension of the polymer chain upon a second monomer addition. In polymerizations using alkylidenes **5** and **6**, hydronium ions function as phosphine scavengers, thus increasing the rate of metathesis without accelerating catalyst decomposition. Thus, the difference in propagation and termination rates with added acid enables rapid and controlled monomer conversion, and the presence of acid precludes any potential catalyst degradation stemming from hydroxide ion.

Recent efforts to impart water solubility to *ruthenium benzylidene catalysts* have utilized neutral hydrophilic substituents as components of the metal-bound ligands, such as poly(ethylene glycol) (PEG). For example, Grubbs and coworkers reported the preparation of the PEG-based water soluble catalyst **7** shown in Figure 6.4 [27], utilizing a PEGylated imidazolium salt prepared from PEG-amine in two steps. Grubbs and coworkers later reported the synthesis of water-soluble catalyst **8**, which showed improved stability and activity in water [28]. In this case, the PEGylated imidazolium salt was first prepared from a PEGylated diamine, then reacted with the Hoveyda–Grubbs catalyst to obtain water-soluble catalyst **8**. Catalysts **6–8** were all found to perform ROMP successfully on sterically challenging monomers such as *endo*-norbornene **9** containing a quaternary ammonium group. With regards to ROMP of norbornene **9** in water, catalyst **8** provided faster initiation and higher monomer conversion relative to catalysts **6** and **7** [28]. Interestingly, catalysts **6** and **7** showed no activity for cross-metathesis in aqueous medium, but **8** showed excellent activity in the homodimerization of allyl alcohol, and the self-metathesis of *cis*-2-butene-1,4-diol in water.

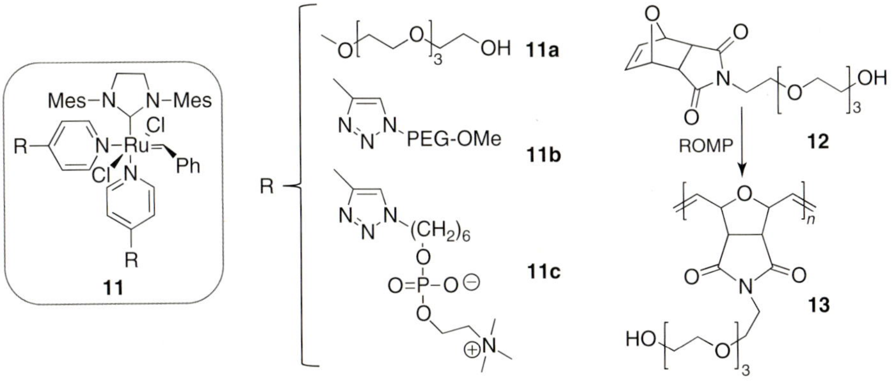

Figure 6.5 Water soluble ruthenium-based metathesis catalysts using hydrophilic pyridine structures as labile ligands.

Emrick and coworkers approached water solubilization of ruthenium benzylidene catalysts from the standpoint of the *labile pyridine ligands* associated with the *Generation 3 catalyst* **3**, resulting in the preparation of novel catalysts **11a–c** shown in Figure 6.5 [29, 30]. For example, catalyst **11a** was synthesized by coordinating PEGylated pyridines to the ruthenium metal. Conveniently, **11a** provided access to homogeneous ROMP in both aqueous and organic media, due to the inherent solubility properties of PEG. In organic solvents, **11a** was very active in ROMP, giving well-defined polynorborenes in high yield and with low PDI (1.1–1.2). Catalyst **11a** proved inactive in neutral water, failing to perform ROMP on PEG-oxanorbornene **12**, but active in aqueous acidic media. Addition of a Brönstead acid activates the catalyst by protonation of the coordinating pyridine, opening a pathway for metallocyclobutane formation. The molecular weights of polyolefins produced under these conditions could be controlled by adjusting the monomer-to-initiator ratio, but the observed molecular weights were typically higher than targeted, attributed in part to slow initiation kinetics. Nonetheless, such PEGylated catalysts are broadly applicable due to their amphiphilicity, such as in interfacial chemistries for producing encapsulant systems and nanostructured materials that contain cyclic olefins [31].

More efficient routes to PEGylated pyridines enabled further research and improved hydrophilic ROMP catalysts. In particular, *click cycloaddition* of alkynyl pyridines with PEG-azide was found to give good yields of PEGylated pyridines, in which the ruthenium-coordinating pyridine and water-solvating PEG group are linked by a triazole group. This concept was applied further to phosphorylcholine azides to produce a second generation of hydrophilic neutral ROMP catalyst systems. For example, PEGylated **11b** and phosphorylcholine (PC)-substituted **11c** were prepared by coordination of the ruthenium center with the corresponding "clicked" pyridine ligands [30]. As with catalyst **11a**, PEGylated **11b** showed only low activity in neutral aqueous media for ROMP of PEG-oxanorbornene **12**. However, 70% conversion of monomer **12** was observed in neutral water containing

CuSO$_4$ as a pyridine scavenger, and quantitative conversion was observed in acidic water. Interestingly, PC catalyst **11c** showed good activity even in the absence of added copper, giving quantitative ROMP of PEG-oxanorbornene **12** in neutral water.

6.3
Living Free Radical Methods for Bio-Synthetic Hybrid Materials

The *functionalization of proteins* with synthetic polymers is an intriguing research area at the intersection of synthetic polymer chemistry and biology, highly relevant to aqueous polymers and green chemistry, and of immense practical importance for advancing the state-of-the-art in medicine [32]. Polymer-functionalized therapeutic proteins, especially PEGylated versions, are now important for treating chronic hepatitis C and anemia, and for use in conjunction with chemotherapeutic drugs, as, for example, in the cases of PEGasys [33] and PEGylated granulocyte colony stimulating factor (GCSF) [34]. The polymer shields the therapeutic protein from the body's natural immune response, increasing its *in vivo* circulatory half-life, and improving its therapeutic effect. Classic chemistries for preparing protein–polymer hybrid materials involve (i) the synthesis of polymers with functional chain-ends, such as *N*-hydroxysuccinimidyl (NHS) esters or aldehydes, and (ii) the reaction of the polymer chain-end with the protein, using, for example, the amines from lysine residues. Such *"grafting-to" methodology* is effective but requires a large excess of functional polymer, and often non-trivial separation of the polymer–protein conjugate from excess polymer and/or unreacted protein, a time-consuming and yield-compromising process.

Recent advances in green polymerization methods, such as aqueous living free radical polymerization, promise to simplify and improve *protein functionalization*. Converting the proteins into macroinitiators enables a *"grafting-from" process* that effectively fixes the polymer chain-end from the outset of the polymerization, leaving little-to-no unreacted polymer for subsequent separations. RAFT and ATRP have proven especially effective in this regard, due to their amenability to aqueous conditions. For example, Davis and coworkers recently reported the synthesis of polymer–protein conjugates using RAFT as the controlled polymerization technique (Figure 6.6) [35]. In this case, a PEG-containing water-soluble pyridyl disulfide macroRAFT agent was prepared. This RAFT agent, designed to be non-interacting (i.e., by electrostatic or hydrophobic interactions) with the protein structure, was conjugated to bovine serum albumin (BSA) at the cysteine-34 residue. Excess RAFT agent was removed by passing the reaction mixture through a size-exclusion centrifuge filter, and mass spectroscopic (matrix assisted laser desorption/ionization time-of-flight, MALDI-TOF) analysis showed a shift in mass (from 66 400 g mol^{-1} of native BSA to 67 600 g mol^{-1}) reflecting the addition of one RAFT agent to the protein. This BSA macroinitiator was then used for RAFT polymerization of *N*-isopropylacrylamide (NIPAM) in PBS (phosphate bufferd saline) at 25 °C in the presence of trace external free radical initiator. The resulting

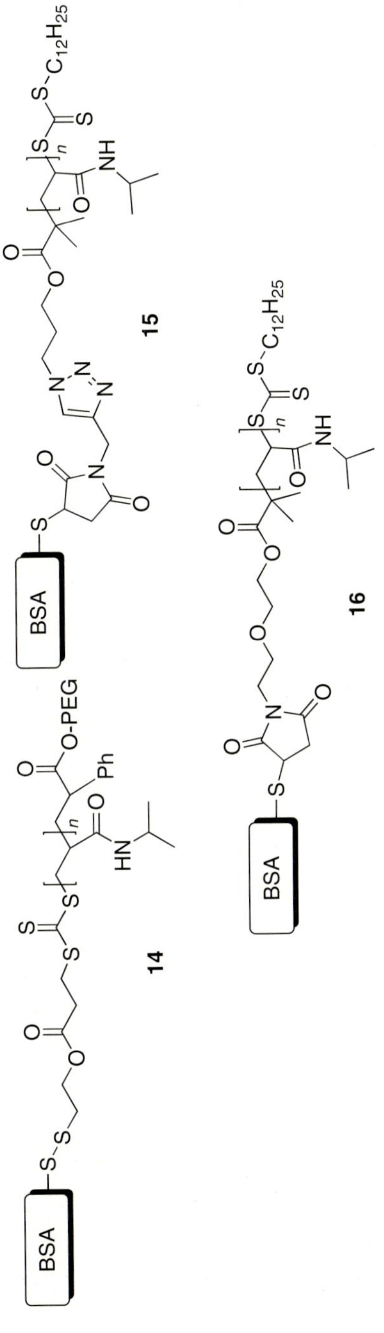

Figure 6.6 Examples of protein–polymer conjugates prepared by RAFT polymerization.

BSA-polyNIPAM conjugate **14** was soluble in water below the lower critical solution temperature (LCST) of polyNIPAM, and gave BSA-polyNIPAM nanoparticles above the LCST. Importantly, ~95% of esterase-like activity of native BSA was retained in the conjugate, with no variance in activity as a function of grafted polyNIPAM molecular weight.

Sumerlin and coworkers also reported RAFT as a useful technique to prepare *polymer–protein conjugates*, combining the concept with copper-catalyzed azide–alkyne ("click") cycloaddition [36]. In one example, BSA was functionalized with an alkyne moiety, by reacting its available cysteine residue with propargyl maleimide. Azide-terminated polyNIPAM was prepared separately by RAFT, and the click conjugation to produce BSA-polyNIPAM conjugate **15** was accomplished in aqueous solution (Figure 6.6). The high yields and mild reaction conditions associated with both RAFT polymerization and *click cycloaddition* are critically important to the success of this strategy. Sumerlin and coworkers also used RAFT to synthesize BSA-polyNIPAM conjugates by the *grafting-from technique*. In this case, a maleimide-functionalized chain transfer agent was applied to afford a BSA-macroinitiator, and subsequently the desired BSA-polyNIPAM conjugate **16** [37]. In this grafting-from case, the stability of the polymer-to-protein linkage requires protein degradation chemistry to determine the molecular weight of the grafted polymer. Isolation of polyNIPAM following protein degradation, at different time-points of the polymerization, revealed a steady increase in molecular weight with monomer conversion, relatively low PDI values (<1.4), and the ability to attain very high molecular weight polymer grafts (>200 000 g mol^{-1}). An additional report by Stayton and coworkers described the preparation of polyNIPAM–streptavidin complexes, in which the dithiocarbamate chain-end on a RAFT-derived polyNIPAM was converted into a free thiol that was then used for biotin conjugation and subsequent steptavidin functionalization. The temperature responsiveness of polyNIPAM enabled the formation of *"smart" particles*, 1–2 μm in diameter, by a reversible complexation and decomplexation of the hybrid structure around the LCST [38].

ATRP as a controlled free radical technique is amenable to aqueous-based (green) conditions, and is very well suited for the *preparation of polymer–protein conjugates* (Figure 6.7). For example, Haddleton and coworkers reported the synthesis of NHS-ester terminated poly [poly(ethylene glycol) methacrylate] [poly(PEGMA] using copper mediated ATRP [39]. ATRP capably produces poly(PEGMA) with low PDI (<1.2), providing well-defined structures for reaction with available amines of the protein. Using lysozyme as a model protein, poly(PEGMA) functionalization to form conjugate **17** was achieved in high yield from the NHS-terminated polymer structures, with size exclusion HPLC (high-performance liquid chromatography) and SDS-PAGE (sodium dodecyl sulfate polyacrylamide gel electrophoresis) showing complete protein functionalization after 6 h reaction time. Haddleton and coworkers also described the synthesis of aldehyde-terminated polyPEGMA by ATRP, and the use of these polymers for lysozyme conjugation [40]. In both the NHS and aldehyde examples, *living free radical polymerization* using functional initiators gives pure, monofunctionalized polymer products, to the exclusion of

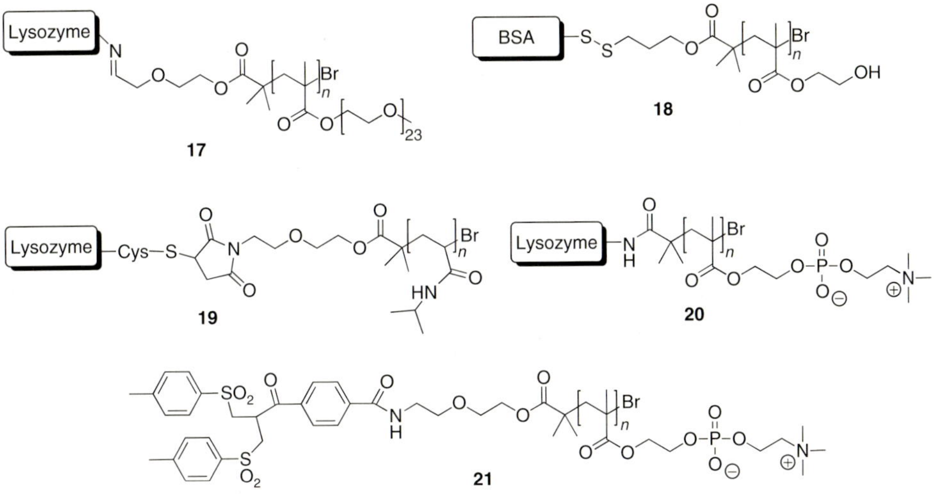

Figure 6.7 Examples of protein–polymer conjugates prepared by ATRP chemistry.

difunctional impurities, which would otherwise lead to unwanted cross-linking during attempted conjugation. Additional reports by Haddleton and coworkers described ATRP methodology to produce maleimide-terminated polymers with molecular weights ranging from 4 to 35 kDa with well-controlled PDI ($\sim$1.1–1.2) [41]. These maleimide-terminated materials are highly water soluble, making them ideal for the synthesis of novel polymer–protein, or polymer–polypeptide, bio-synthetic hybrid materials. They have been employed successfully in conjugation with thiol-containing model substrates, such as reduced glutathione (ç-Glu-Cys-Gly) and BSA.

Maynard and coworkers have also demonstrated the effective use of ATRP to prepare *bio-synthetic hybrid materials* [42], including, for example, poly(hydroxyethyl methacrylate) (polyHEMA) with activated disulfides, such as pyridyl disulfide, at the chain-end. These polymers are well suited for BSA conjugation (**18**), as verified by gel electrophoresis and Ellman's assay, following conjugation. Maynard and colleagues also reported the site-specific conjugation of polyNIPAM to lysozyme (**19**) [43], achieved by first inserting a cysteine residue into the structure, then reacting the modified lysozyme with a maleimide-containing ATRP initiator. Grafting polyNIPAM from the lysozyme macroinitiator gave bio-synthetic hybrid conjugates that retained their native activity despite the polymer modification.

Grafting-to techniques for preparing *polymer–protein conjugates*, with the aim of improvement over PEGylated therapeutics, have recently been reported for the zwitterionic poly(methacryloyloxyethyl phosphorylcholine) (polyMPC). While the high biocompatibility of polyMPC [44] makes it interesting for comparison with PEGylated therapeutics, only recently have aqueous-based controlled free radical polymerization techniques been developed to the point of capably polymerizing the MPC monomer. In bioconjugation, Emrick and coworkers showed that NHS and aldehyde-terminated polyMPC could be prepared by ATRP, and subsequently conjugated to lysozyme (**20**) as a model enzyme, and also erthyropoetin and GCSF as therapeutic examples [45]. In accord with the polymer–protein biohybrids described previously, gel electrophoresis, FPLC (fast protein liquid chromatography), and HPLC provide key characterization tools that reveal an effective conjugation with very little unreacted protein. Shortly following this report, the work by Godwin and coworkers [46] on the synthesis of a sulfide-containing polyMPC derivative by ATRP (**21**) appeared. Conjugation of this structure to interferon-α2a (IFN) gave polyMPC-IFN conjugates, for example, with a 20 kDa polyMPC, that were shown in a mouse model to have an elimination half-life ($t_{1/2}$) of 24 ± 2 h, almost twice that of a PEGylated IFN with a 20 kDa PEG chain.

Finally, it is interesting to note the reports by Nolte and coworkers describing *protein–polymer biohybrids* as surfactants in solution, or *giant amphiphiles* that combine in one structure the properties of classic small molecule amphiphiles, polymeric amphiphilies, and biological function inherent to the protein used [47]. For example, conjugation of polystyrene to the chemically modified protein *Candida antartica lipase B*, using thiol-maleimide chemistry, gave the corresponding hybrid structure. Nolte and coworkers also used click cycloaddition for the preparation of biohybrid amphiphiles, by preparing alkyne-functional proteins (using propargyl maleimide at the CYS 34 residue of BSA) for cycloaddition with the complementary azide-terminated polystyrene. Click cycloaddition was also used as a convenient probe for monitoring polymer–protein conjugation, by the reaction of "profluorescent" 3-azidocoumarin-terminated polymers (such as PEG) with alkyne-substituted proteins (such as BSA) [48]. Such methodology had been shown to be effective for fluorescent labeling, by click, of modified virus particle *"nanocages"* [49]. The triazole cycloadducts are strongly fluorescent, while no fluorescence is seen in the starting materials, enabling a tracking of the conjugation reaction.

In summary, it is becoming clear that new monomers, catalysts, and synthetic methodology in organic and polymer synthesis are in place, and are developing further, enabling the use of *aqueous media to produce polymer materials* with a much higher level of structural and molecular weight precision than was previously possible. This is advantageous for decreasing the use of volatile organic solvents in chemical processes, and increasing our understanding of organic and polymerization reactions outside the framework of conventional organic solvents. Such aqueous-based techniques are amenable to advancing green approaches that produce known and novel polymer materials, and are also adaptable to biologically relevant systems, including the preparation of protein–polymer hybrid materials of both fundamental interest and biomedical relevance. Combining the metathesis and bio-hybrid synthetic techniques with other emerging synthetic methodologies, such as click cycloaddition chemistry, serves to further advance environmentally appropriate chemical and materials syntheses.

Acknowledgments

The authors acknowledge support of their research on these topics from the National Science Foundation (CBET-0932781) and the NSF-supported Materials Research Science & Engineering Center on Polymers at UMass Amherst (DMR-0820506) and the Nanoscale Science and Engineering Center at UMass Amherst (DMI-0531171).

References

1. Webster, O.W. (1991) *Science*, **251**, 887–893.
2. De Gunzbourg, A., Favier, J.-C., and Hamery, P. (1994) *Polym. Int.*, **35**, 179–188.
3. De Gunzbourg, A., Maisonnier, S., Favier, J.-C., Maitre, C., Masure, M., and Hemery, P. (1998) *Macromol. Symp.*, **132**, 359–370.
4. Satoh, K., Kamigaito, M., and Sawamoto, M. (1999) *Macromolecules*, **32**, 3827–3832.
5. Hawker, C.J., Bosman, A.W., and Harth, E. (2001) *Chem. Rev.*, **101**, 3661–3688.
6. Wang, J.S. and Matyjaszewski, K. (1995) *Macromolecules*, **28**, 7901–7910.
7. Chiefari, J., Chong, Y.K., Ercole, F., Krstina, J., Jeffery, J., Le, T.P.T., Mayadunne, R.T.A., Meijs, G.F., Moad, C.L., Moad, G., Rizzardo, E., and Thang, S.H. (1998) *Macromolecules*, **31**, 5559–5562.
8. Mol, J.C. (2004) *J. Mol. Catal. A. Chem.*, **213**, 39–45.
9. The Swedish Academy of Sciences (2005) The Nobel Prize in Chemistry 2005. Press release. www.nobelprize.org (accessed 5 October 2005).
10. Herison, J.L. and Chauvin, Y. (1971) *Makromol. Chem.*, **141**, 161–176.
11. McGinnis, J., Katz, T.J., and Hurtwitz, S. (1976) *J. Am. Chem. Soc.*, **98**, 605–606.
12. Grubbs, R.H. (1978) *Prog. Inorg. Chem.*, **24**, 1–50.
13. Trnka, T.M. and Grubbs, R.H. (2001) *Acc. Chem. Res.*, **34**, 18–29.
14. Gilliom, L. and Grubbs, R.H. (1988) *J. Mol. Catal.*, **46**, 255–256.
15. Schrock, R.R., Clark, D.N., Sancho, J., Wengrovius, J.H., Rocklage, S.M., and Pederson, S.F. (1982) *Organometallics*, **1**, 1645–1651.
16. Schaverien, C.J., Dewan, J.C., and Schrock, R.R. (1986) *J. Am. Chem. Soc.*, **108**, 2771–2773.
17. Ivin, K.J. (1983) *Olefin Metathesis*, Academic Press, London.
18. Novak, B.M. and Grubbs, R.H. (1988) *J. Am. Chem. Soc.*, **110**, 7542–7543.
19. Hafner, A., Van der Schaff, P.A., and Muehlebach, A. (1996) *Chimia*, **50**, 131–134.
20. Manning, D.D., Strong, L.E., Hu, X., Beck, P.J., and Kiessling, L.L. (1997) *Tetrahedron*, **53**, 11937–11952.
21. Trnka, T.M., Morgan, J.P., Sanford, M.S., Wilhelm, T.E., Scholl, M., Choi, T.L., Ding, S., Day, M.W., and Grubbs, R.H. (2003) *J. Am. Chem. Soc.*, **125**, 2546–2558.
22. Boydston, A.J., Holcombe, T.W., Unruh, D.A., Frechet, J.M.J., and Grubbs, R.H. (2009) *J. Am. Chem. Soc.*, **131**, 5388.
23. Fraser, C. and Grubbs, R.H. (1995) *Macromolecules*, **28**, 7248–7255.
24. Quemener, D., Heroguez, V., and Gnanou, Y. (2005) *Macromolecules*, **38**, 7977–7982.
25. Claverie, J.P., Viala, S., Maurel, V., and Novat, C. (2001) *Macromolecules*, **34**, 382–388.
26. Lynn, D.M., Mohr, B., Grubbs, R.H., Henling, L.M., and Day, M.W. (2000) *J. Am. Chem. Soc.*, **122**, 6601–6609.
27. Gallivan, J.P., Jordan, J.P., and Grubbs, R.H. (2005) *Tetrahedron Lett.*, **46**, 2577–2580.
28. Hong, S.H. and Grubbs, R.H. (2006) *J. Am. Chem. Soc.*, **128**, 3508–3509.
29. Breitenkamp, K. and Emrick, T. (2005) *J. Polym. Sci. A. Polym. Chem.*, **43**, 5715–5721.
30. Samanta, D., Kratz, K., Zhang, X., and Emrick, T. (2008) *Macromolecules*, **41**, 530–532.

31. Rangirala, R., Hu, Y.X., Joralemon, M., Zhang, Q.L., He, J.B., Russell, T.P., and Emrick, T. (2009) *Soft Matter*, **5**, 1048–1054.
32. Hinds, K.D. and Kim, S.W. (2002) *Adv. Drug Deliv. Rev.*, **54**, 505–530.
33. Zeuzem, S., Pawlotsky, J.M., Lukasiewicz, E., von Wagner, M., Goulis, I., Lurie, Y., Gianfranco, E., Brolijk, J.M., Esteban, J.L., Hezode, C., Lagging, M., Negro, F., Soulier, A., Verheij-Hart, E., Hansen, B., Tl, R., Ferrari, C., Schalm, S.W., and Neumann, A.U. (2005) *J. Hepatol.*, **43**, 250–257.
34. Rajan, R.S., Tiansheng, L., Aras, M., Sloey, C., Sutherland, W., Arai, H., Briddell, R., Kinstler, O., Leuras, A.M.K., Zhang, Y., Yeghnazar, H., Treuheit, M., and Brems, D.N. (2006) *Protein Sci.*, **15**, 1063–1075.
35. Boyer, C., Bulmus, V., Liu, J., Davis, T.P., Stenzel, M.H., and Barner-Kowolik, C. (2007) *J. Am. Chem. Soc.*, **129**, 7145–7154.
36. Li, M., De, P., Gondi, S.R., and Sumerlin, B.S. (2008) *Macromol. Rapid Commun.*, **29**, 1172–1176.
37. De, P., Li, M., Gondi, S.R., and Sumerlin, B.S. (2008) *J. Am. Chem. Soc.*, **130**, 11288–11289.
38. Kulkarni, S., Schilli, C., Muller, A.H.E., Hoffman, A.S., and Stayton, P.S. (2004) *Bioconjug. Chem.*, **15**, 747–753.
39. Lecolley, F., Tao, L., Mantovani, G., Durkin, I., Lautru, S., and Haddleton, D.M. (2004) *Chem. Commun.*, 2026–2027.
40. Tao, L., Mantovani, G., Lecolley, F., and Haddleton, D.M. (2004) *J. Am. Chem. Soc.*, **126**, 13220–13221.
41. Mantovani, G., Lecolley, F., Tao, L., Haddleton, D.M., Clerx, J., Cornelissen, J.J.L.M., and Velonia, K. (2005) *J. Am. Chem. Soc.*, **127**, 2966–2973.
42. Bontempo, D., Heredia, K.L., Fish, B.A., and Maynard, H.D. (2004) *J. Am. Chem. Soc.*, **126**, 15372–15373.
43. Heredia, K.L., Bontempo, D., Ly, T., Byers, J.T., Halstenberg, S., and Maynard, H.D. (2005) *J. Am. Chem. Soc.*, **127**, 16955.
44. Iwasaki, T. and Ishihara, K. (2005) *Anal. Bioanal. Chem.*, **381**, 534–546.
45. Samanta, D., McRae, S., Cooper, B., Hu, Y., Emrick, T., Pratt, J., and Charles, S.A. (2008) *Biomacromolecules*, **19**, 2891–2897.
46. Lewis, A., Tang, Y.Q., Brocchini, S., Choi, J., and Godwin, A. (2008) *Bioconjug. Chem.*, **19**, 2144–2155.
47. Velonia, K., Rowan, A.E., and Nolte, R.J.M. (2002) *J. Am. Chem. Soc.*, **124**, 4224–4225.
48. Dirks, A.J., van Berkel, S.S., Hatzakis, N.S., Opsteen, J.A., van Delft, F.L., Cornelissen, J.J.L.M., Rowan, A.E., van Hest, J.C.M., Rutges, F.P.J.T., and Nolte, R.J.M. (2005) *Chem. Commun.*, 4172–4174.
49. Wang, Q., Chan, T.R., Hilgraf, R., Fokin, V.V., Sharpless, K.B., and Finn, M.G. (2003) *J. Am. Chem. Soc.*, **125**, 3192–3194.

7
Towards Sustainable Solution Polymerization: Biodiesel as a Polymerization Solvent

Marc A. Dubé and Somaieh Salehpour

7.1
Introduction

For over half a century, a number of polymer products with valuable properties have been made using *solution polymerization*. The use of a solvent as a polymerization medium offers first and foremost a reduction in viscosity of the reaction mixture. This translates into improved heat transfer and prevention of thermal runaway by absorbing the heat of polymerization [1]. Solution polymerization may also present the opportunity for chain transfer to solvent reactions, which can have an impact on the molecular weight of the polymer product and the molecular weight distribution. Many solution polymerization products have been developed over time that present final performance properties that cannot easily be met using other polymerization techniques. Nonetheless, solution polymerizations are seen in a negative light due to the use of volatile organic compounds (VOCs) as the solvents, which cause health and environmental problems. An example is ethylbenzene, an anticipated carcinogen [2], which is a common solvent for the commercial production of polystyrene. From US production alone, over 4000 tons of ethylbenzene are released into the atmosphere yearly [3]. In terms of environmental effects, these solvents can cause tropospheric pollution and deplete the ozone layer [4]. Moreover, they can cause cancer, infertility, and genetic disorders in individuals that experience frequent exposure [4].

One of the ways to avoid some of these side effects is to use technologies that do not employ solvents, such as bulk polymerization, although other difficulties can then arise [5]. Emulsion polymerization can be an attractive alternative to solution polymerization although the same level of end-use product performance is not always achieved [6]. Therefore, the superior properties of polymers prepared by solution polymerization suggest that rather than altering the polymerization technology, one could look for more environmentally friendly solvents to replace more harmful conventional solvents [7].

Green Polymerization Methods: Renewable Starting Materials, Catalysis and Waste Reduction
Edited by Robert T. Mathers and Michael A. R. Meier
Copyright © 2011 WILEY-VCH Verlag GmbH & Co. KGaA, Weinheim
ISBN: 978-3-527-32625-9

7.2
Solution Polymerization and Green Solvents

With the need to create a cleaner environment, along with strict regulation and concerns over the VOCs in various polymer production processes, the implementation of *green solvents in polymer processes* has a great potential to contribute to more sustainable production [8].

The choice of polymerization solvent is influenced by a number of issues. It should be non-toxic and reasonably non-hazardous, and also easily removed from the product. Water is the most inexpensive, environmentally benign and relatively plentiful solvent. However, many organic compounds such as some monomers are insoluble in *water*. Another challenge to using water is separating water-soluble products from it [9]. The use of environmentally friendly solvents such as supercritical fluids (e.g., carbon dioxide and water) and room-temperature ionic liquids (RTILs) has emerged as an important development in *polymer production technology* [10].

Supercritical water and carbon dioxide have been investigated as green solvents for the polymerization of various monomers [11]. In view of the fact that water exists in the supercritical state at temperatures above 647 K and pressures above 217 atm (21 987 kPa), it can perform as a non-polar solvent mainly due to the absence of hydrogen bonding and the stretching out of the molecule under these extreme conditions. As an example, supercritical CO_2 has been used in the manufacture of fluoropolymers as a replacement solvent for 1,1,2-trichloro-1,2,2-trifluoroethane [12]. However, implementation of this technology can cause corrosion problems. Considering the economic reasons, one can surmise that supercritical water may not be a convenient solvent alternative [11].

Ionic liquids are another class of non-classical polymerization solvents that are often considered as green solvents capable of replacing traditional organic solvents, usually because of their low vapor pressure. They have been extensively studied in the materials chemistry and catalysis fields [13]. They usually consist of organic cationic species and inorganic anionic species, which are liquids at room temperature with very low vapor pressures [14]. However, they have yet to be applied at the industrial scale as a "green" solvent [15]. Moreover, in many cases the use of ionic liquids does not lower energy expenditures compared with conventional solvents and would result in higher costs [14].

7.3
Biodiesel as a Polymerization Solvent

Recently, *biodiesel* (fatty acid methyl ester or FAME) produced from canola oil has been used as a green polymerization solvent [1]. In the 1960s, the use of methyl oleate, a major component of biodiesel, as a polymerization solvent was explored [16]. However, the high cost of the pure compound prevented its widespread application. The solvating ability of biodiesel has only recently started to be

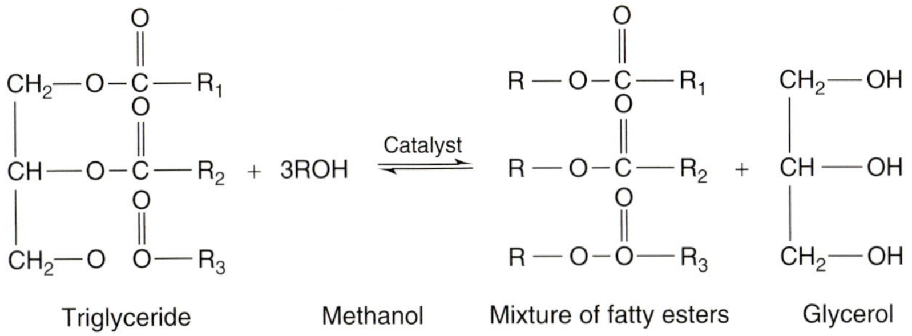

Figure 7.1 General reaction of the transesterification of triglyceride [18].

investigated [7]. Biodiesel is more commonly known as an environmentally friendly alternative to petroleum diesel for use in combustion engines. It has the chemical structure of fatty acid alkyl esters (FAAEs) and is produced by the transesterification of vegetable oils, animal fats, or grease with an alcohol in the presence of a catalyst. Different types of alcohols can be used in the transesterification reaction, such as methanol, ethanol, propanol, and butanol. Usually, the process is carried out with an alkaline catalyst (e.g., NaOH) dissolved in excess methanol under agitation, controlled temperature, and atmospheric pressure [17]. The reaction scheme is shown in Figure 7.1.

Aside from the obvious environmental benefits of using a renewable material such as biodiesel, one can also consider its high boiling-point and its low cost. Biodiesel boils at temperatures above 300 °C (326 °C for canola-based biodiesel) [19]. This means that it will not pose a hazard in the workplace due to evaporation and reactions can be carried out at elevated temperatures without fear of excessive pressure buildup aside from the contributions of the monomers. There has been mounting interest in carrying out *polymerizations at elevated temperatures* [20]. Advantages of using high temperatures include:

- Decrease in concentration of chain transfer agents and initiators; this would reduce costs.
- Increase in the rate of reaction and therefore, an increase in productivity.
- Decrease in the polymer molecular weight without using chain transfer agents.

However, there are disadvantages inherent to running reactions at elevated temperatures, such as increased energy consumption, safety considerations, and possible undesired side reactions that may occur, that is, intramolecular chain transfer and depropagation [21].

Table 7.1 shows the toxicity, the permissible exposure limit (PEL) reported by the Occupational Safety and Health Administration (OSHA) and the price of common polymerization solvents compared with biodiesel and methyl oleate. There are clear indications from both an economic and health and safety perspective, that biodiesel may prove to be a useful industrial solvent for polymerization. Therefore, *application of biodiesel in polymer production* merits thorough investigation. In this

Table 7.1 Comparison of polymerization solvents.

Solvent	Toxicity	OSHA PEL (8 h ppm)	Price (US$ l^{-1})
Toluene[a]	Narcotic, liver and kidney damage in high concentration	200	40
Benzene[a]	Carcinogen	10	52
Xylene[a]	Narcotic at high concentrations	100	45
Methyl oleate (technical grade 70%)[a]	Non-hazardous material	None	46
Biodiesel[b]	Non-hazardous material	None	0.60

[a] Sigma–Aldrich, MSDS. Toxicity information from The Merck Index [2].
[b] Toxicity information from Biodiesel Handbook [23].

chapter, the use of FAME as a green polymerization solvent at the laboratory scale is discussed. It is noted that one should only refer to FAME as biodiesel if it has passed strict ASTM (American Society for Testing and Materials) testing standards [22]. Thus, we will henceforth refer to the use of FAME.

7.4
Experimental Section

Solution polymerizations of four commercially important monomers [i.e., methy methacrylate (MMA), styrene (Sty), butyl acrylate (BA), and vinyl acetate (VAc)] were studied using FAME produced from canola oil as a polymerization solvent at 60 and 120 °C. The same initiator to monomer ratio was utilized for all concentrations of monomers at each temperature. Specifically, we used 0.4 phm (parts per 100 parts by weight of monomer) of 2,2′-azobisisobutyronitrile (AIBN) at 60 °C and 0.2 phm of Trigonox B at 120 °C. Furthermore, to study the effect of biodiesel feedstock, Sty homopolymerizations using 0.4 phm AIBN at 60 °C were performed in various types of FAME. The FAME was produced from three different sources of triglycerides: canola oil, soybean oil, and a 50/50 v/v mixture of yellow grease and canola oil.

7.4.1
Materials

The *monomers* (Sigma–Aldrich, Milwaukee, WI, USA) were received with added inhibitors. To remove the *inhibitor*, VAc was solely distilled under vacuum while BA, MMA, and Sty were washed three times with a 10% (v/v) sodium

Table 7.2 Biodiesel fatty acid alkyl esters composition from different feedstocks [18].

Biodiesel feedstock	FAAE composition[a](wt%)							
	12:0	14:0	16:0	16:1	18:0	18:1	18:2	18:3
Canola	–	–	3–5	–	1–2	55–65	20–26	8–10
Soybean	–	–	11–12	–	3–5	23–25	52–56	6–8
Yellow grease	0.5	1–3	13–25	0–4	5–12	43–52	7–22	0.5–3

[a] Carbon chain length: number of unsaturations.

hydroxide solution, then washed three times with distilled de-ionized water and dried over calcium chloride prior to vacuum distillation. Distillations were completed a maximum of 24 h prior to polymerization, and the monomers were stored at −10 °C. The initiator AIBN (DuPont Chemicals) was recrystallized three times in absolute methanol. Di-t-butyl peroxide (Trigonox B), ≥95% pure, was obtained from the Sigma–Aldrich Chemical Company and used without purification.

FAME was produced in-house by transesterification of triglyceride with methanol (99.85% Reag. Grade, Commercial Alcohols Inc.; Brampton, ON, Canada) using sodium hydroxide catalyst (97 wt% NaOH, EMD Chemicals Inc.; NJ, USA) at 65 °C via a membrane reactor process [24]. A catalyst concentration of 0.5 wt% based on the amount of oil and a methanol/oil molar ratio of 6/1 were used. FAME was produced from three different sources of triglycerides: canola oil (No Name®, Toronto, ON, Canada, purchased at a local food store), soybean oil (Mr. Goudas®, Goudas Food Products; Concord, ON, Canada, purchased at a local food store), and 50% yellow grease (supplied by a local restaurant) – 50% canola oil (No Name®, Toronto, ON, Canada, purchased at a local food store). HPLC analysis of the FAME showed no traces of triglycerides. The FAME was washed with de-ionized water and then distilled under vacuum (25 °C, −30 kPA). FAME compositions are shown in Table 7.2.

Hexanes (Fisher Chemicals, HPLC Grade, UV cutoff 195 nm) and tetrahydrofuran (THF) (Sigma–Aldrich, HPLC Grade, ≥99.9% inhibitor free) were used for polymer characterization and sample work-up without further purification.

7.4.2
Polymerization

Polymerizations were performed in sealed glass ampoules (17-cm length, 0.8-cm outer diameter) in a water bath. The feed was prepared by weighing the monomer, FAME solvent, and initiator into a flask and delivered into a series of 5-ml glass ampoules. The ampoules were degassed by several freeze–pump–thaw cycles under high vacuum, were flame sealed and then submerged in a constant temperature water bath (for runs at 60 °C) or an oil bath (for runs at 120 °C). At the appropriate time interval, ampoules were removed from the bath and quenched in an ice bath to stop the reaction. The ampoules were broken and the contents poured into a pre-weighed beaker. The solution was diluted with 5 ml of acetone. The extraction of polymer was done by precipitation in hexane (for MMA, VAc, and Sty) and in methanol (for BA) at room temperature. The polymer–monomer–FAME solution was added slowly to the non-solvent (i.e., hexane or methanol) while stirring and generally appeared as a fine white fiber or a milky dispersion. After settling (~24 h), decantation and evaporation of the non-solvent, extracted polymer was dried in a vacuum oven at 40 °C until a constant weight was reached.

7.4.3
Characterization

Gravimetry was used to calculate conversion. A Waters Associates gel permeation chromatograph equipped with a Waters Model 410 refractive index detector was used to determine the cumulative number- and weight-average molecular weights of the polymers. Three Waters Ultrastyragel packed columns (10^3, 10^4, and 10^6 Å) were installed in series. Filtered THF circulating at a flow rate of 0.3 ml min^{-1} at 38 °C was used as eluent. The universal calibration method using Mark–Houwink relationships was used with ten standard samples of polystyrene (SHODEX, Showa, Denko, Tokyo, Japan) with peak molecular weights between 1.3×10^3 and 3.15×10^6 g mol^{-1} [1]. Standards and samples were dissolved in THF to a concentration of 0.2% w/v and filtered through 0.45-μm filters before injection to remove any gel or impurities that may have been present. Millennium 32 software (Waters) was used for data acquisition.

^{1}H-NMR (nuclear magnetic resonance) and ^{13}C-NMR spectra were recorded at room temperature for all samples dissolved in CDCl$_3$ (Sigma–Aldrich) with a Bruker 400-MHz spectrometer. To refine the assignments of signals in the ^{1}H-NMR and ^{13}C-NMR spectra, a 2D hetero-nuclear multiple quantum coherence (HMQC) experiment was used to clarify correlations between the chemical shifts of carbon nuclei and their attached protons.

7.5
Effect of FAME Solvent on Polymerization Kinetics

Solvent effects on the *free radical polymerization* of many monomers have been widely documented [25]. The presence of solvent is known to present some effect on kinetic reaction rate constants and therefore on the rate of polymerization.

7.5.1
Chain Transfer to Solvent Constant

The chain transfer to solvent constant, C_{fs}, is an important parameter that describes the effect of solvent on polymerization kinetics, where

$$C_{fs} = \frac{k_{fs}}{k_p} \tag{7.1}$$

where

k_{fs} is the chain transfer to solvent rate parameter
k_p is the propagation rate parameter

Various methods can be employed to determine the value of C_{fs} [26]. Traditionally, the *Mayo method*, which shows the quantitative effect of various transfer reactions on the number-average degree of polymerization, has been used. This method is based on overall chain-growth and chain-stopping rates under the assumption of steady state and uses the long-chain approximation [26]. An alternative is the chain-length distribution (CLD) method, which has been developed by Clay and Gilbert [27]. This latter procedure is based on the calculation of the high molecular weight slope for the number molecular weight distribution. Both procedures yield very similar results, especially in chain transfer dominated systems [26]. In the work presented here, transfer constants were obtained using the Mayo equation [28]:

$$\frac{1}{\overline{X}_n} = \frac{k_t R_p}{k_p^2 [M]^2} + C_{fm} + C_{fs}\frac{[S]}{[M]} + C_{fi}\frac{k_t R_p^2}{k_p^2 f k_d [M]^3} \tag{7.2}$$

where

$\overline{X}_n$ is the number-average degree of polymerization
R_p is the rate of polymerization
k_t, k_p, and k_d are the termination, propagation, and initiator decomposition rate coefficients, respectively
f is the initiator efficiency
$[M]$ and $[S]$ are the concentration of the monomer and solvent, respectively
C_{fm}, C_{fs}, and C_{fi} are the transfer constants to monomer, solvent, and initiator, respectively

The first term on the right-hand side of Equation (7.2) represents the contribution of the rates of polymerization and termination to the polymer chain size. The second term corresponds to the effect of transfer to monomer, the third term describes the role of transfer to solvent reactions, and the last term expresses contributions of the initiator. For the present case, the third term makes the biggest contribution to the *degree of polymerization*. By using low concentrations of initiator or initiators with very small C_{fi} values (e.g., AIBN), the last term in Equation (7.2) becomes negligible and rate retardation becomes minimal [28]. In addition, by keeping $k_t R_p/[M]^2$ constant, the first term on the right-hand side may be kept constant

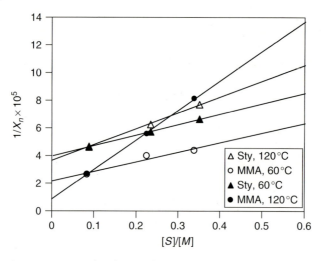

Figure 7.2 Mayo plot showing the reciprocal number-average degree of polymerization versus the solvent/monomer ratio for Sty and MMA polymerized in canola-based FAME at 60 and 120 °C.

by adjusting the initiator concentration over a series of separate polymerizations. Nonetheless, this term will change with changes in [S]/[M] because the termination rate parameter, k_t, is chain-length dependent and will vary due to the effect of [S]/[M] on the average chain length. This is a weakness of the Mayo method; however, the effect is not significant for all practical purposes [26]. Under these conditions, Equation (7.2) is reduced to

$$\frac{1}{\overline{X}_n} = \frac{1}{\overline{X}_{n0}} + C_{fs}\frac{[S]}{[M]} \qquad (7.3)$$

where

$(1/\overline{X}_{n0})$ is the value of $(1/\overline{X}_n)$ in the absence of solvent.
C_{fs} is obtained from the slope of the line by plotting $(1/\overline{X}_n)$ versus [S]/[M]

Various solvent concentrations were used and *chain transfer to solvent rate constants* were determined using the Mayo method. A Mayo plot for the homopolymerizations of MMA and Sty is presented in Figure 7.2 [1, 19]. All estimated values of transfer to solvent constants were higher at 120 °C compared with at 60 °C. It should be noted that at 120 °C, all polymers were soluble in the FAME solvent while precipitation polymerization of MMA was observed at 60 °C. At 120 °C chain transfer constants for MMA and BA were comparable and the chain transfer constant for any compound increased in the order of increasing radical reactivity in the chain transfer reaction. However, MMA presented the lowest transfer constant at 60 °C, which is expected due to the precipitation polymerization mechanism it underwent. Chain transfer to solvent data for various polymerization solvents are shown in Table 7.3.

Table 7.3 Comparison of chain transfer to solvent data for different polymerization solvents.

Solvent	$C_{fs} \times 10^4$ for polymerization of:			
	MMA	Sty	VAc	BA
Benzene[a]	0.83	0.023	6	0.4
Toluene[a]	0.96	0.125	35	1.8
Methyl oleate[a]	1.68	3.52	217	–
Xylene[a]	0.50	0.78	140	–
FAME at 60 °C[a]	68	74	190	110
FAME at 120 °C[b]	173	114	246	153

[a] Data from polymerization at 60 °C [1].
[b] Data from polymerization at 120 °C [19].

Transfer to solvent effectiveness depends on the solvent's amount, structure, strength of the breaking bond, and the stability of the solvent radical formed. Transfer to solvent phenomena relate to propagating radical reactivity [28]. A comparison of transfer to solvent constants for FAME and some common solvents at 60 °C has been shown previously [1]. Accordingly, FAME would be expected to have higher transfer constants compared with aliphatic and aromatic hydrocarbons because of C–H breakage and stabilization of the radical by an adjacent carbonyl group. This is consistent with evidence that chain transfer to FAME was significantly greater than benzene, toluene, and xylene [1]. In addition, transfer to internal double bonds of FAME compounds is possible. FAME from canola oil consists of 55–65 wt% methyl oleate with one internal double bond at the 9-position, 20–26 wt% methyl linoleate with two internal double bonds at the 9- and 12-positions, and 8–10 wt% linolenate with three internal allylic unsaturations at the 9-, 12-, and 15-positions. In an early study on chain transfer constants for vinyl monomers polymerized in methyl oleate and methyl stearate, estimated C_{fs} values for methyl oleate, with an 18-carbon structure and one internal double bond, were considerably higher at 60 °C compared with that for methyl stearate, with an 18-carbon structure and no internal double bonds [16]. Therefore, it is possible that internal allylic double bonds play a considerable role in the transfer to solvent reaction, which leads to high C_{fs} values. However, no significant differences between chain transfer constants for polymerization in FAME derived from different biodiesel feedstocks were observed in this work. Therefore, chain transfer to an internal double bond is not likely to occur. This was also confirmed by ^{13}C-NMR results [18].

7.5.2
Rate Constant

In general, according to the *rate of polymerization*, R_p, [see Equation (7.4)] the dilution effect of the solvent decreases the concentration of monomer, thereby decreasing the rate of polymerization. In addition, the presence of solvent may affect the propagation and termination steps in free radical polymerization [29]. The overall rate of polymerization is expressed as follows [28]:

$$R_p = \frac{k_p}{k_t^{1/2}}[M]\left(fk_d[I]\right)^{1/2} \tag{7.4}$$

The term $k_p/(k_t)^{1/2}$ is a "lumped" parameter (units are l mol^{-1} s^{-1}) and it can be obtained directly from experimental measurements, using simple techniques such as gravimetry [1]. This *lumped parameter* can be used in free radical polymerization modeling as described in Dubé et al. [30] and Gao and Penlidis [31]. The collected kinetic experimental data for the polymerization in FAME solvent were modeled using a comprehensive polymerization model (WATPOLY polymerization simulator) developed by Gao and Penlidis for bulk/solution free radical polymerization [31].

Physical property data of the FAME and the chain transfer to solvent rate parameters were incorporated into the *WATPOLY polymerization simulator database* and all experimental data were modeled [31]. The "lumped" kinetic rate parameter $(k_p/k_t^{1/2})$ for each monomer was modified to fit the experimental conversion data at each concentration, where k_p is the propagation rate parameter and k_t is the termination rate parameter. The model demonstrated reliable predictions of reaction lumped rate constant and cumulative average molecular weight for all monomers. At 120 °C, the value of the modified parameters $(k_p/k_t^{1/2})$ were 1.52, 1.62, and 2.35 for MMA polymerization, 0.16, 0.22, and 0.33 for VAc and 0.37, 0.56, and 0.63 for BA at 50, 40, and 20 wt% concentration of solvent, respectively, while it was kept constant at 2.46 for Sty at all solvent concentrations. At 60 °C, the value of the modified parameters were 0.2, 0.24, and 0.31 for VAc and 1.09, 1.16, and 1.22 for BA at 50, 40, and 20 wt% concentration of solvent, respectively, while it was kept constant at 0.39 for Sty and 1.16 for MMA at all solvent concentrations. This indicates different effects of solvent on propagation and termination rate constants for each individual monomer. The lumped rate constants were found to be related in the increasing order: Sty < BA < MMA < VAc at 120 °C and MMA < Sty < BA < VAc at 60 °C [1, 19].

The presence of FAME solvent was observed to affect the kinetic reaction rate parameters and therefore the rates of polymerization at these temperatures. Pulsed laser polymerization (PLP) data show that changing solvent viscosity and polarity has no significant effect on the propagation rate coefficient of Sty and MMA. However, preferential solvation of the radical site seems to occur when polymer precipitation takes place, thus increasing the monomer concentration at the free radical site and increasing the apparent k_p [32]. There is evidence of either a radical–solvent or radical–monomer complex, which participates in propagation reactions and modifies the propagation rate parameter. The stability and reactivity of these complexes determines the effect on k_p [33]. Regarding the effect on k_t, it is known that k_t is nearly always diffusion controlled [34]. By increasing the solvent concentration, more transfer to solvent reactions are possible, resulting in faster termination reactions due to the generation of shorter and more mobile chains. This leads to an increase in k_t and therefore, to a decrease in the rate of polymerization [5]. Regardless of the effect of solvent on the individual rate parameters, it is the *lumped rate parameter*, $k_p/k_t^{1/2}$, which should be manipulated because the individual parameters are coupled.

A considerable difference in the *polymerization rate* was observed for all monomers by changing the concentration of the solvent. Therefore, the solvent must be a major contributing factor to the change in rate, in addition to its dilution effect. The lowest polymerization rate was at the monomer/FAME ratio of 50/50 wt%, whereas the highest was observed at a ratio of 80/20 wt% for each polymer system. In general, the lowest rate of polymerization was observed for VAc, which is not surprising given its high transfer to solvent coefficient compared with other monomers. At 60 °C, no thermal polymerization was noted and all polymers were soluble except for pMMA [poly(methyl methacrylate)], which precipitated in the FAME. The same behavior was observed for pMMA in methyl oleate [16]. At 120 °C, the highest reaction rate was observed for Sty homopolymerization. Pure thermal initiation or self-initiation is known to occur for Sty at elevated temperatures due to the thermal production of radicals from the monomer [28]. In addition to the presence of thermal initiation, thermal homolysis of an initiator would result in higher rates of polymerization and lower molecular weights. This is consistent with our experimental results for Sty. Sty had a lower rate of polymerization compared with MMA and BA at 60 °C, while it had the highest reaction rate at 120 °C.

The results for polymerization at 120 °C demonstrate that FAME from canola oil fulfills the requirements of a good high boiling solvent for solution polymerization at elevated temperatures. Polymerization at elevated temperature leads to higher reaction rates, and consequently shorter reaction times and lower viscosity, which can dissipate auto-acceleration effects [20]. Moreover, *high temperature polymerization* enables the production of low molecular weight polymers without the use of chain transfer agents.

In systems where the formation of complexes may lead to stabilization, radical–solvent complexes are expected to be favored. For instance, the solvent effect on the *homopolymerization* of VAc results in a reduced k_p. As VAc is an unstable monomer with low reactivity, it may form stable radical–solvent complexes during the propagation step. This can cause a decrease in the rate of propagation. Therefore, in this instance, the effect of solvent was mainly ascribed to the variation in k_p with solvent concentration rather than that in k_t [35]. *Retardation* was reported for VAc polymerization in methyl stearate and methyl oleate as the solvent at 60 °C [16]. Methyl oleate (a major component of FAME) with one internal double bond was more rate-reducing at that temperature compared with methyl stearate, which has no internal unsaturation. Moreover, an even higher C_{fs} value and more retardation were observed for methyl oleate compared with methyl stearate at 90 °C [16]. Hence, it can be hypothesized that the internal allylic bond of FAME makes a considerable contribution to radical–solvent complex formation. Therefore, reduction in the propagation rate constant (k_p) may occur in the polymerization of monomers with low reactivity. This is consistent with our results for VAc polymerization, where a significant effect of solvent on the rate of polymerization was observed. Nonetheless, further studies on the existence of radical–solvent complexes and their influence on free radical polymerization at elevated temperatures are necessary, as the complex might dissociate as temperature increases.

As a representation of typical results, *model predictions* of the reaction rates and cumulative weight-average molecular weights along with the experimental data are presented in Figures 7.3–7.6 for Sty and MMA polymerization at 60 and 120 °C.

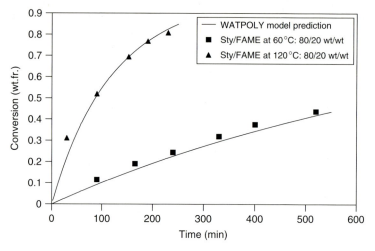

Figure 7.3 Sty homopolymerization in FAME: conversion versus time at different temperatures.

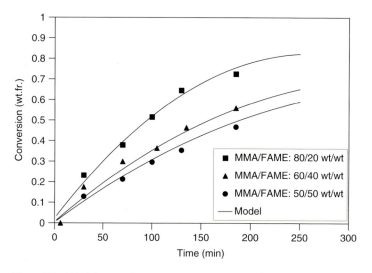

Figure 7.4 MMA homopolymerization in FAME at 120 °C: conversion versus time at different solvent concentrations.

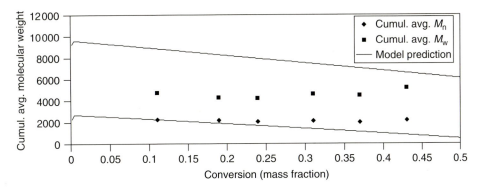

Figure 7.5 Cumulative average molecular weight versus conversion for Sty solution homopolymerization in FAME at 60 °C (80/20 wt%).

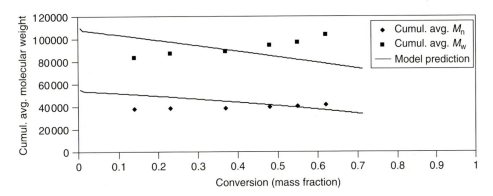

Figure 7.6 Cumulative average molecular weight versus conversion for MMA solution homopolymerization in FAME at 60 °C (80/20 wt%).

The model predictions correctly describe the trends in the data. Molecular weight data for MMA polymerization at 120 °C are shown in Table 7.4. As has been shown in Figure 7.7, by decreasing the solvent concentration, molecular weight increases and also molecular weight distribution becomes broader.

7.6
Effect of Biodiesel Feedstock

Biodiesel can be produced from a variety of different *feedstocks*. The most common feedstock worldwide is rapeseed oil (or the genetically modified form of rapeseed: canola), while that for the USA is soybean oil [23]. In recent years, the focus has been shifting towards the use of multiple feedstocks comprised of animal fats and waste frying oils (yellow grease), so as not to compete with food supplies, in addition

Table 7.4 Molecular weight data for polymerization at 120 °C.

$\overline{M}_n$	[S]/[M]	$C_{fs} \times 10^4$
Methyl methacrylate		
14 200	0.3383	
20 900	0.2255	173
36 900	0.0846	
Styrene		
15 400	0.3518	
19 700	0.2346	114
28 200	0.0879	
Vinyl acetate		
10 500	0.3010	
16 400	0.2006	246
30 100	0.0752	
Butyl acrylate		
10 400	0.4329	
12 500	0.2886	153
17 300	0.1082	

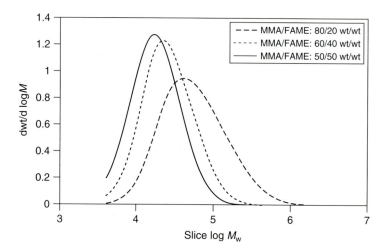

Figure 7.7 Molecular weight distributions for PMMA polymerizations at 120 °C.

to economic reasons [18]. Biodiesel is comprised primarily of C_{16} and C_{18} methyl esters and varies in composition depending on its source. Biodiesel derived from soybean oil has a higher percentage of FAME 18 : 2 (carbon chain length : number of unsaturations) and thus, has more double bonds than biodiesel from other feedstocks. In contrast, biodiesel derived from animal fats or yellow grease tends to

7.6 Effect of Biodiesel Feedstock

have less double bonds and shorter fatty acid chain lengths than that from vegetable oil. To measure the degree of unsaturation or number of double bonds per mass of sample, iodine is introduced to the oil and it will react with carbon-carbon double bonds to form single bonds. The amount of iodine in grams that is taken up by 100 g of the biodiesel is known as the *iodine value*, which is 94–120 for canola, 120–143 for soybean, and 80–100 for yellow grease biodiesel [36]. A higher iodine value indicates a higher degree of unsaturation. Therefore, to investigate the effect of various feedstocks used in FAME production, Sty polymerization at 60 °C in FAME from different feedstocks was studied. FAME produced from soybean oil and 50% yellow grease-50% canola oil were used as polymerization solvents and results were compared with runs at 60 °C in canola-based FAME [18].

7.6.1
Polymerization Kinetics

Chain transfer to solvent constants were measured for Sty homopolymerization in each FAME type and no statistically significant differences were observed between the values obtained from each feedstock. In contrast, the *rates of polymerization* were influenced by the degree of saturation of each FAME solvent. FAME from a soybean oil feedstock was found to have the greatest effect on the conversion of styrene, while FAME from canola oil and that from a mixture of 50% yellow grease and 50% canola oil yielded comparable results. The observed difference between the feedstocks is most pronounced at the highest concentration of solvent (i.e., at a styrene/FAME ratio of 50/50 wt%). As the concentration of monomer was the same for all FAME feedstocks, the observed difference between polymerization rates cannot be explained by the dilution effect alone. Thus, the solvent itself must have been a major contributing factor, and the FAME feedstock must have affected the rate parameters. The significant differences in rates of polymerization, molecular weights, and chain transfer constants between the experiments using different FAME solvents may be attributed to variations in their composition and consequently to their degree of unsaturation or iodine value. As mentioned above, FAME derived from soybean oil has a higher percentage of double bonds in the fatty acid chains of the esters. The presence of these double bonds is hypothesized to play a role in decreasing the rate of polymerization due to the formation of monomer–solvent or radical–solvent complexes. These complexes are known to inhibit the polymerization propagation reaction [25] and this is consistent with our experimental results. In Figure 7.8, the same conversion versus time data as above are shown for the three FAME types at 40 wt% concentrations of solvent to highlight their differences.

In most of the experiments, the *average molecular weight of the polymer* remained roughly constant with increasing conversion. As expected, owing to chain transfer potential, M_w increased with decreasing solvent concentration. The less solvent that was present, the less there was potential for chain transfer to solvent. Growth of the polymer chains was consequently less inhibited, yielding polymers of greater average molecular weight. Model predictions of the cumulative weight-average

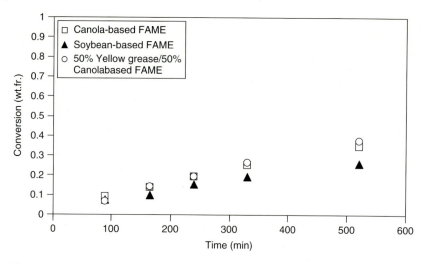

Figure 7.8 Conversion versus time for 60/40 wt% monomer/FAME.

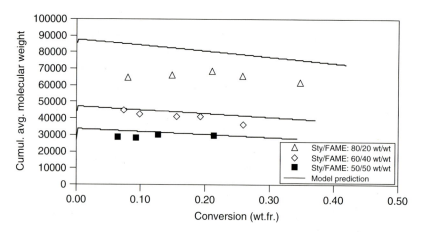

Figure 7.9 Weight-average molecular weight versus conversion for 50% yellow grease-50% canola FAME.

molecular weights along with the experimental data are presented in Figure 7.9 for polymerization in 50% yellow grease-50% canola FAME.

7.6.2
Polymer Composition

One important question that can arise relates to how the *FAME solvent* is involved in the chain transfer reaction mechanism. Thus, the *microstructure* of the final polymer products would be of interest and this could be studied using spectroscopic methods. A ^{1}H-NMR investigation for determining the effect of various FAME on the final polymer product was performed for all polystyrene samples, including samples from polymerization in canola-based FAME [18]. The acquisition of spectra by this method is fast but diagnostic signals for individual compounds may overlap. Therefore, supplementary analyses were made using ^{13}C-NMR for samples generated at 50 wt% solvent concentration. To achieve a better comparison and distinguish any contribution of solvent to the polymer microstructure, ^{1}H-NMR and ^{13}C-NMR analyses were performed on all FAME samples and a sample of polystyrene, which was polymerized in a solvent-free medium under similar experimental conditions. In addition, selected samples were examined by HMQC, which is a two-dimensional NMR analysis [37]. This 2D experiment was used to clarify correlations between the chemical shifts of carbon nuclei and their attached protons.

There were other small but distinguishable peaks in all *polymer spectra*, which are not overlapped with the polymer chain signals and can be attributed to residual FAME in the samples [18]. Resonances of hydrogen nuclei attached to internal double bonds of FAME appear at about 5.5 ppm in the ^{1}H-NMR spectra. Considering the ^{13}C-NMR spectra [14], a signal at about 14 ppm was assigned to the methyl groups at the end of each chain in saturated or unsaturated FAME chains, which is well separated from other signals. Another well-defined resonance was assigned to the carbonyl group at about 173 ppm in the ^{13}C-NMR spectra. Moreover, there is a resonance for the carbon, which is attached to the acyl group, at about 51 ppm. This can be confirmed by a signal in the HMQC spectra for all samples and arises at the intercept of the signal at about 51 ppm on the ^{13}C axis and 3.8 on the ^{1}H axis, confirming the presence of the OCH_3 chain endgroups. Hence, assignment of this signal to any tertiary carbon or carbon of a methylene group is thwarted. Tertiary carbons without an aromatic attachment can appear in our system as a result of transfer to internal double bonds of FAME and initiation of new polymer chains by solvent radicals. The resonances for tertiary carbons are expected to be at 30–50 ppm. On the other hand, methylene groups attached to acyl groups can appear when transfer to solvent takes place at a terminal methyl group and initiation of new polymer chains starts from that point. The signals for this carbon are expected to be at 50–90 ppm. The lack of these signals means that, if even present, these carbons have a concentration lower than the signal-to-noise ratio or overlap with other resonances. However, the ratio of the intensity of the mentioned peaks assigned to FAME in each polymer sample spectrum is the same as the ratio of the intensity of these peaks in the corresponding FAME spectrum, within the range of experimental error. Therefore, it can be concluded that these small peaks are only related to residual FAME solvent in the polymer samples and are not due to copolymerization or any contribution of solvent as an initiator in the polymer chain. From our study of the spectra of polystyrene polymerized in FAME from different feedstocks, no evident contribution of FAME in the long chain polymer samples was noted. It is entirely possible; however, that oligomers or very short chain polymers could have existed, but might have been lost during the extraction procedure.

7.7
Conclusion

Biodiesel fulfills the requirements of a good polymerization solvent; it is environmentally benign and has low volatility, low viscosity, and good solubility. Hence, in an effort to achieve cleaner technologies by reducing VOC of various polymer processes, FAME (or biodiesel) can be used as an alternative to traditional polymerization solvents. The use of FAME or biodiesel as a "green" polymerization solvent with good solubility, low viscosity, and high boiling-point fulfills the requirements of a solution polymerization solvent in various cases. Therefore, the *"environmental friendliness"* of FAME, coupled with its effectiveness as a polymerization medium, makes it an attractive alternative to traditional polymerization solvents.

As with the use of any high boiling solvent, the extraction of the finished polymer from the reaction mixture is an important step. The work discussed here employed *solvent recovery techniques* more suited to the laboratory scale. For a larger scale industrial process, one could revert to a variety of solvent recovery methods. These application-dependent techniques include spray drying, solvent–solvent extraction, and evaporation. Obviously, in the broader context of sustainable processes, one would have to consider the energy costs and environmental safety impacts of each of these methods.

References

1. Salehpour, S. and Dubé, M.A. (2008) *Green Chem.*, **10**, 329–334.
2. Budavari, S.M. and Heckelman, P.E. (1989) *The Merck Index*, Merck and Company, Rahway.
3. Benton, M.G. and Brazel, C.S. (2002) *ACS Symp. Ser.*, **818**, 125–133.
4. Sherman, J., Chin, B., Huibers, P.D.T., Garcia-Valls, R., and Hatton, T.A. (1998) *Environ. Health Perspect.*, **106**, 253–271.
5. Jovanović, R. and Dubé, M.A. (2004) *J. Appl. Polym. Sci.*, **94**, 871–876.
6. Jovanović, R. and Dubé, M.A. (2005) *J. Macromol. Sci., Part C: Polym. Rev.*, **44**, 1–51.
7. Hu, J., Du, Z., Tang, Z., and Min, E. (2004) *Ind. Eng. Chem. Res.*, **43**, 7928–7931.
8. Leitner, W. (2009) *Green Chem.*, **11**, 603.
9. Li, C.J. and Trost, B.M. (2008) *Proc. Natl. Aacd. Sci.*, **105**, 13197–13202.
10. Duan, J., Shim, Y., and Kim, H.J. (2006) *J. Chem. Phys.*, **124**, 204504-1–204504-13.
11. DeSimone, J.M. (2002) *Science*, **297**, 799–803.
12. Romack, T.J., DeSimone, J.M., and Treat, T.A. (1995) *Macromolecules*, **28**, 8429–8431.
13. Mallakpour, S. and Kolahdoozan, M. (2008) *Polym. J.*, **40**, 513–519.
14. Nelson, W.M. (2002) *ACS Symp. Ser.*, **818**, 30–41.
15. Sheldon, R.A. (2005) *Green Chem.*, **7**, 267–278.
16. Jordan, E.F. Jr. and Artymyshyn, B. (1969) *J. Appl. Polym. Sci.*, **7**, 2605–2611.
17. Cao, P., Tremblay, A.Y., and Dubé, M.A. (2009) *Ind. Eng. Chem. Res.*, **48**, 2533–2541.
18. Salehpour, S., Dubé, M.A., and Murphy, M. (2009) *Can. J. Chem. Eng.*, **87**, 129–135.
19. Salehpour, S. and Dubé, M.A. (2008) *Polym. Int.*, **57**, 854–862.
20. McManus, N.T., Hsieh, G., and Penlidis, A. (2004) *Polymer*, **45**, 5837–5845.

21. Quan, C., Soroush, M., Grady, M.C., Hansen, J.E., and Simonsick, W.J. Jr. (2005) *Macromolecules*, **38**, 7619–7628.
22. Cao, P., Dubé, M.A., and Tremblay, A.Y. (2008) *Biomass Bioenergy*, **32**, 1028–1036.
23. Mittelbach, M. and Remschmidt, C. (2004) *Biodiesel: The Comprehensive Handbook*, Martin, Graz, p. 330, ISBN: 3-200-00249-2
24. Dubé, M.A., Tremblay, A.Y., and Liu, J. (2007) *Bioresour. Technol.*, **98**, 639–647.
25. Coote, M.L., Davis, T.P., Klumperman, B., and Monteiro, M.J. (1998) *J. Macromol. Sci., Rev. Macromol. Chem. Phys.*, **38**, 567–593.
26. Heuts, J.P.A., Davis, T.P., and Russell, G.T. (1999) *Macromolecules*, **32**, 6019–6030.
27. Clay, P.A. and Gilbert, R.G. (1995) *Macromolecules*, **28**, 552–569.
28. Odian, G. (2004) *Principles of Polymerization*, 4th edn, John Wiley & Sons, Inc., Hoboken
29. McKenna, T.F., Villanueva, A., and Santos, A.M. (1999) *J. Polym. Sci., Part A: Polym. Chem.*, **37**, 571–588.
30. Dubé, M.A., Soares, J.B.P., Penlidis, A., and Hamielec, A.E. (1997) *Ind. Eng. Chem. Res.*, **36**, 966–1015.
31. Gao, J. and Penlidis, A. (1996) *J. Macromol. Sci., Rev. Macromol. Chem. Phys.*, **36**, 199–404.
32. Olaj, O.F., Zoder, M., and Vana, P. (2001) *Macromolecules*, **34**, 441–446.
33. Zammit, M.D., Davis, T.P., Willett, G.D., and O'Driscoll, K.F. (1997) *J. Polym. Sci., Part A: Polym. Chem.*, **35**, 2311–2321.
34. Buback, M., Egorov, M., Gilbert, R.G., Kaminsky, V., Olaj, O.F., Russell, G.T., Vana, P., and Zifferer, G. (2002) *Macromol. Chem. Phys.*, **203**, 2570–2582.
35. Kamachi, M., Liaw, D.J., and Nozakura, S. (1979) *Polym. J.*, **11**, 921–928.
36. Knothe, G. (2002) *J. Am. Oil Chem. Soc.*, **79**, 847–854.
37. Golotvin, S.S., Vodopianov, E., Pol, R., Lefebvre, B.A., Williams, A.J., Rutkowske, R.D., and Spitzer, T.D. (2007) *Magn. Reson. Chem.*, **45**, 803–813.

Part IV
Catalytic Processes

8
Ring-Opening Polymerization of Renewable Six-Membered Cyclic Carbonates. Monomer Synthesis and Catalysis

Donald J. Darensbourg, Adriana I. Moncada, and Stephanie J. Wilson

8.1
Introduction

Ring-opening polymerization (ROP) reactions of *six-membered cyclic carbonates* provide facile routes to biodegradable polycarbonates, materials which currently have numerous applications in biomedicine [1]. The most well studied or prototypical cyclic carbonate monomer for this process is trimethylene carbonate (TMC) or 1,3-dioxan-2-one [Equation (8.1)]. This monomer affords copolymers with L- and D-lactides, which are used in specialty orthopedics, especially for internal fixation devices for the repair of small bones. These implanted devices degrade over time by hydrolysis, subsequent to being metabolized via natural processes in the body and released as water and carbon dioxide. In general, when employing anionic initiators, high molecular weight polymers are afforded that contain no ether units.

$$(8.1)$$

The synthesis of six-membered cyclic carbonates is achieved by *transesterification* of 1,3-propanediol and derivatives thereof with various reagents including phosgene and its derivatives (di- and triphosgene), dialkylcarbonates, and ethyl chloroformate. Because phosgene and diphosgene are highly poisonous gases, these reagents are to be avoided for health and safety concerns. Indeed, a focus of this chapter is to synthesize these important cyclic carbonates via routes that are environmentally benign and sustainable. In order to achieve this goal, we will describe synthetic methods for preparing 1,3-propanediol from renewable resources and also dialkylcarbonates from carbon dioxide [Equation (8.2)].

$$(8.2)$$

Green Polymerization Methods: Renewable Starting Materials, Catalysis and Waste Reduction
Edited by Robert T. Mathers and Michael A. R. Meier
Copyright © 2011 WILEY-VCH Verlag GmbH & Co. KGaA, Weinheim
ISBN: 978-3-527-32625-9

Subsequent to a brief consideration of the thermodynamic properties associated with ROP reactions, a detailed discussion of the various catalytic processes for the ROP of six-membered cyclic carbonates will ensue. Because these polymers have numerous applications in biomedicine, for example, sutures, drug delivery systems, and tissue engineering, we will focus our attention on an array of *biocompatible catalytic processes*.

8.2
Preparation of 1,3-Propanediol from Renewable Resources

Prior to examining routes to 1,3-propanediol from renewable resources, it is worthwhile mentioning current synthetic methods originating from hydrocarbon precursors. The two major routes to producing 1,3-propanediol based on petrochemicals are the Shell and Degussa–DuPont processes. These processes are described in Scheme 8.1. As indicated in Scheme 8.1, the Shell process involves the hydroformylation of ethylene oxide to 3-hydroxypropanal with subsequent hydrogenation to product, whereas the Degussa–DuPont procedure initially converts propylene into acrolein followed by hydration to 3-hydroxypropanal and hydrogenation [2]. The two processes converge at 3-hydroxypropanal. High pressure is required for both the hydroformylation and hydrogenation steps (15 MPa). The production of 1,3-propanediol from renewable resources, that is, from glycerol (Henkel) or glucose (Dupont–Genencor) represents very attractive alternative routes to this valuable *polymer precursor*. It should also be noted that ethylene can be obtained from renewable resources (sugar cane), thereby, the Shell process could potentially be made more sustainable. Dow currently has a joint venture with Crystalsev in Brazil to produce polyethylene (PE) from ethanol-based ethylene. Also, Braskem and Solvay in Brazil have plans to employ ethanol-based ethylene for the production of bio-derived PE and bio-derived PVC [poly(vinyl chloride)].

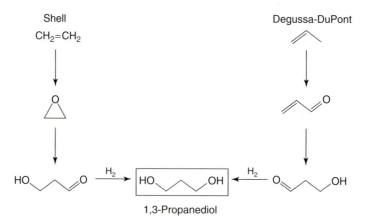

Scheme 8.1 Preparation of 1,3-propanediol from petrochemicals.

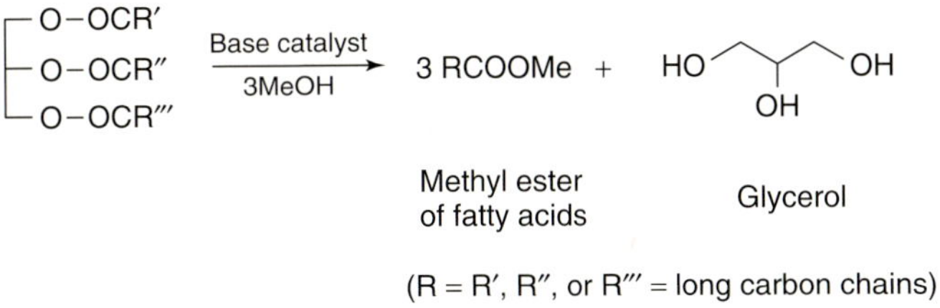

8.2 Preparation of 1,3-Propanediol from Renewable Resources

$$\begin{bmatrix} O-OCR' \\ O-OCR'' \\ O-OCR''' \end{bmatrix} \xrightarrow[3MeOH]{\text{Base catalyst}} 3\ RCOOMe\ +\ \text{HO-CH(OH)-CH}_2\text{OH}$$

Methyl ester of fatty acids   Glycerol

(R = R', R", or R''' = long carbon chains)

Scheme 8.2 Biodiesel production by transesterification of a triglyceride.

The focus for finding a renewable source of 1,3-propanediol is centered on *glycerol*, for there is an excess of crude glycerol on the market. That is, glycerol is a by-product of the rapidly growing biofuels industry, specifically biodiesel production via transesterification of vegetable oils (Scheme 8.2). Although glycerol is a good target for deoxygenation, production of 1,3-propanediol by the intermediacy of 3-hydroxypropanal presents a real challenge. This is in part due to the proficiency of 3-hydroxypropanal to undergo further dehydration to acrolein. As previously mentioned, the hydrogenation of 3-hydroxypropanal also requires very high hydrogen pressure [Equation (8.3)].

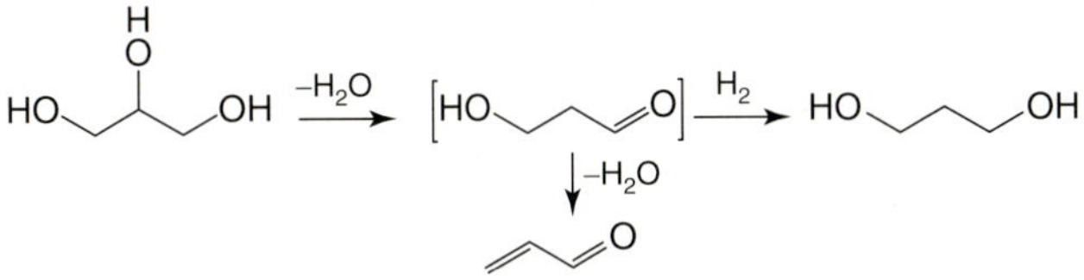

$$\text{HO-CH(OH)-CH}_2\text{OH} \xrightarrow{-H_2O} [\text{HO-CH}_2\text{-CH}_2\text{-CHO}] \xrightarrow{H_2} \text{HO-CH}_2\text{-CH}_2\text{-CH}_2\text{-OH}$$
$$\downarrow -H_2O$$
$$\text{CH}_2=\text{CH-CHO}$$

(8.3)

In general, heterogeneous catalysts are not selective for the transformation of glycerol to 1,3-propanediol [3]. However, there has been limited success employing *homogeneous catalysts* for this process. For example, the cationic ruthenium complex [{Cp*Ru(CO)$_2$}$_2$(μ-H)][O$_3$SCF$_3$] has been shown to be highly selective for the deoxygenation of diols and has exhibited deoxygenation activity for glycerol [4]. When employing this catalyst, equal amounts of 1,3-propanediol and *n*-propanol were produced along with a small amount of 1,2-propanediol at 5.2 MPa of hydrogen and 383 K. On the other hand, Rh(CO)$_2$(acac) has been reported in a patent to be a much better homogeneous catalyst for the production of 1,3-propanediol from glycerol under very mild reaction conditions (ambient temperature and atmospheric pressure) [5]. For a lucid discussion of the numerous obstacles that must be overcome in the chemical transformation of glycerol into 1,3-propanediol, see the overviews by Schlaf and by Behr and coworkers [6].

Much of the current activity in the area of the production of the bulk chemical 1,3-propanediol from renewable resources is centered on the use of *biocatalysts*. It has long been known that glycerol can be converted via anaerobic fermentation into 1,3-propanediol [7]. Forsberg has reported the formation of 1,3-propanediol as the major fermentation product from glycerol using a number of bacteria, including various strains of *Clostridium acetobutylicum* [8]. In principle, the important metabolic reactions in the anaerobic fermentation of glycerol can be separated into

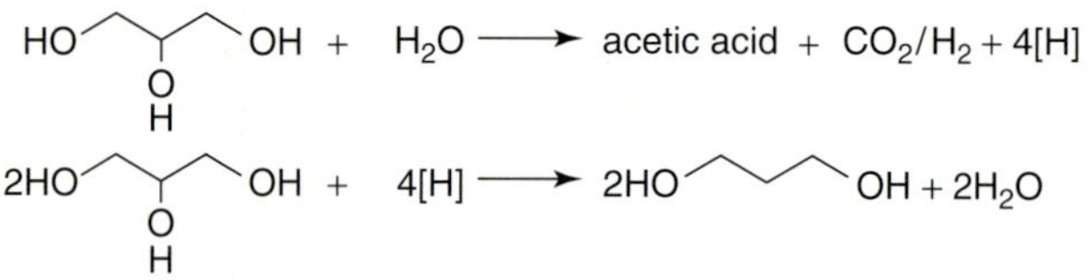

$$\text{HOCH(OH)CH}_2\text{OH} + \text{H}_2\text{O} \longrightarrow \text{acetic acid} + \text{CO}_2/\text{H}_2 + 4[\text{H}]$$

$$2\,\text{HOCH(OH)CH}_2\text{OH} + 4[\text{H}] \longrightarrow 2\,\text{HOCH}_2\text{CH}_2\text{CH}_2\text{OH} + 2\text{H}_2\text{O}$$

Scheme 8.3 Simplified metabolic pathways for glycerol fermentation process.

two steps: the oxidative metabolism, which provides the reducing power ($NADH_2$), and the reductive pathway, which dehydrates glycerol to 3-hydroxypropanal with subsequent reduction by $NADH_2$ to 1,3-propanediol and NAD (nicotinamide adenine dinucleotide) (Scheme 8.3). This microbial production of 1,3-propanediol has been developed on an industrial scale by Henkel, with an expected 67% theoretical yield of 1,3-propanediol [9]. For a comprehensive coverage of the numerous engineering efforts that have been made to optimize this process, see the excellent review by Zeng and Biebl [10].

Owing to the highly desirable properties of poly(trimethylene terephthalate) (PTT), a copolymer produced from 1,3-propanediol and terephthalic acid, Genencor and DuPont have collaborated to develop a biotechnological pathway to 1,3-propanediol from *fermentable carbohydrates*, particularly glucose [Equation (8.4)] [11]. It should be noted that in the United States glucose or starch hydrolyzates are significantly less expensive than glycerol. This in turn makes it possible to produce PTT, commercially known as Sorona®, which is used to make apparel, upholstery, and carpet in a partially sustainable manner. The yielded PTT contains 36% 1,3-propanediol by mass. A parenthetically noteworthy point is that the Coca-Cola Co. has announced its intention to introduce into its plastic bottle line poly(ethylene terephthalate) (PET) made from ethylene glycol derived from sugar and molasses. This procedure would provide a product with about 30% biobased content from the biggest player in the beverage industry [12].

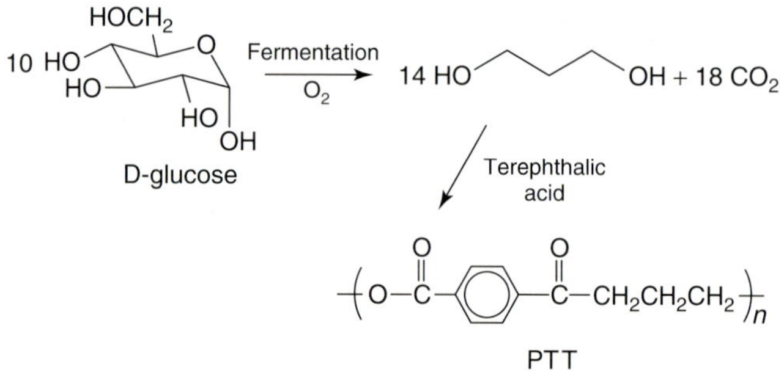

$$10\ \text{D-glucose} \xrightarrow[O_2]{\text{Fermentation}} 14\ \text{HOCH}_2\text{CH}_2\text{CH}_2\text{OH} + 18\ \text{CO}_2$$

$$\xrightarrow{\text{Terephthalic acid}} \text{[O-CO-C}_6\text{H}_4\text{-CO-CH}_2\text{CH}_2\text{CH}_2\text{]}_n \quad \text{PTT}$$

(8.4)

The approach taken in this joint venture by Genencor and DuPont was to metabolically engineer recombinant *E. coli* to carry out the two required processes. Because carbohydrates are not fermented by naturally occurring organisms, the invention involves utilizing genes from natural strains to develop a *recombinant strain* from two organisms [13], that is, one gene converts glucose into glycerol while a second gene ferments glycerol to 1,3-propanediol. The theoretical yield of 1,3-propanediol provided from glucose is 140%, due to the energy and redox requirements to drive the conversion of glucose into 1,3-propanediol provided by transforming some of the glucose into CO_2. In practice, yields of around 120% are obtained. DuPont has also partnered with Tate and Lyle to commercialize the production of 1,3-propanediol at a manufacturing facility in Loudon, TN, USA. A life-cycle assessment of this process reveals that this particular production pathway of 1,3-propanediol is 40% more energy efficient and reduces greenhouse gas emissions by greater than 40% compared with the petroleum-based process [14].

Two national awards have been granted to these groups for the development and commercial production of 1,3-propanediol by a *biological pathway*. These include the Presidential Green Chemistry Award to DuPont in 2003 and the ACS Heroes in Chemistry Award to DuPont, Tate and Lyle, and Genencor International in 2007.

8.3
Preparation of Dimethylcarbonate from Renewable Resources

As alluded to earlier, although TMC can be readily produced from 1,3-propanediol and phosgene or its derivatives, because of environmental and sustainability considerations it would be desirable to utilize other procedures to synthesize this monomer. One such alternative method involves the reaction of dimethylcarbonate (DMC) with 1,3-propanediol to afford TMC and methanol. DMC is an environmentally benign chemical with a broad range of applications including uses in the production of polyurethanes, lubricants, and polycarbonates, and as a solvent for various processes. Also, DMC has good solubility in gasoline, hence it is an ideal gasoline additive to replace methyl *tert*-butyl ether (MTBE). DMC is currently industrially produced via various technologies. These include: phosgene reaction with methanol, reaction of CO and methanol in the presence of an oxidant, reaction of CO_2 and epoxides followed by transesterification, and urea alcoholysis [Equations (8.5–8.8)]

$$\text{Cl-CO-Cl} + 2\text{MeOH} \longrightarrow \underset{\text{DMC}}{\text{MeO-CO-OMe}} + 2\text{HCl} \qquad (8.5)$$

$$2\text{MeOH} + 2\text{CO} \xrightarrow[\text{CuCl}]{\text{O}_2} \text{DMC} + \text{H}_2\text{O} + \text{CO}_2 \qquad (8.6)$$

$$\text{CO}_2 + \text{epoxide} \xrightarrow{\text{Cat}} \text{cyclic carbonate} \xrightarrow{2\text{MeOH}} \text{DMC} + \text{HO-CH}_2\text{CH}_2\text{-OH} \qquad (8.7)$$

$$\text{H}_2\text{N-CO-NH}_2 + 2\text{MeOH} \longrightarrow \text{DMC} + 2\text{NH}_3 \qquad (8.8)$$

The last three processes are all much greener than the production pathway involving phosgene. The reaction described in Equation (8.6) is practiced by Polimeri Europa SpA in Italy. Recently, there has been a major push in China for developing the two processes depicted in Equations (8.7) and (8.8) for the production of DMC. It is important to note for the reaction shown in Equation (8.7) that either ethylene oxide or propylene oxide can be employed depending on the demand for 1,2-propanediol or ethylene glycol. Although this process is usually carried out in two steps, it has been demonstrated to proceed effectively in a one-pot synthesis under supercritical CO_2 conditions [15]. Additional considerations are that ethylene oxide is generally produced by the oxidation of ethylene, a renewable building block, with oxygen, and propylene oxide can be synthesized from 1,2-propanediol derived from glycerol [16]. In Equation (8.8), the released ammonia can be recovered and recycled to produce urea via reaction with CO_2 [Equation (8.9)].

$$2NH_3 + CO_2 \longrightarrow (H_2N)_2CO + H_2O \tag{8.9}$$

The direct synthesis of DMC by the *carboxylation* of methanol has received much attention as an approach for the utilization of carbon dioxide [Equation (8.10)] [17]. However, this process has a ΔG value near zero, thereby resulting in a low yield of DMC at equilibrium (~2%). In order to drive this process toward the products and increase the yield of DMC, *effective water trapping reagents* are needed. This is a very challenging requirement, which might be aided by technological advances in reactor design such as the use of membrane separators for water. Current efficient heterogeneous catalysts for this reaction consist of supported copper and nickel systems [18]. Alternative homogeneous catalysts include transition metal alkoxides of Sn and Nb [19, 20].

$$2MeOH + CO_2 \overset{Cat}{\rightleftharpoons} (MeO)_2CO + H_2O \tag{8.10}$$

The synthetic methodology represented by the reaction in Equation (8.7), which is currently the greener and more sustainable route to those having commercial viability, has been optimized by developments from the Asahi Kasei Corporation. Although the focus of this group's effort is to prepare a green and sustainable source of diphenylcarbonate via the esterification of DMC, their undertaking required major breakthroughs in DMC synthesis. An important issue that these researchers had to overcome was the fact that the ethylene glycol produced by the reaction in Equation (8.7) can easily react with unreacted ethylene carbonate to afford unwanted by-products. For a detailed description of this phosgene-free production of high-performance polycarbonate from bisphenol-A and diphenylcarbonate, see the excellent account by Fukuoka *et al.* [21]. With regard to catalysts for the coupling of CO_2 and epoxides to selectively afford cyclic carbonates, there are numerous choices, including a wide variety of main group and transition metal complexes and also simple alkali and alkaline earth metal halides [22–24]. Noteworthy is an "exceptionally active" bimetallic aluminum salen derivative that effectively operates using flue gases and requires only atmospheric pressure and ambient temperature [25].

8.4
Synthesis of Trimethylene Carbonate

A low yield synthesis of TMC was first reported by Carothers and Van Natta in 1930 starting from 1,3-propanediol and diethylcarbonate (DEC) and catalyzed by sodium ethanolate [26]. This process was later improved by Albertsson and Sjoling, who utilized stannous 2-ethylhexanoate as the transesterification catalyst [27]. Since that time, a high yield, industrially applicable preparation of TMC from 1,3-propanediol and DEC has been developed [28]. Endo and coworkers described the preparation of research amounts of TMC in good yield from 1,3-propanediol and ethylchloroformate in the presence of stoichiometric amounts of triethylamine [29]. In the interest of a green and sustainable source of TMC, synthesis from 1,3-propanediol and DMC or DEC would be highly desirable [Equation (8.11)].

$$HO{-}CH_2CH_2CH_2{-}OH \;+\; RO{-}C(=O){-}OR \;\xrightarrow{Cat}\; \text{(six-membered cyclic carbonate)} \;+\; 2\,ROH \qquad (R = Me \text{ or } Et) \qquad (8.11)$$

8.5
Six-Membered Cyclic Carbonates: Thermodynamic Properties of Ring-Opening Polymerization

Poly(TMC) and its copolymers with other monomer units, for example, D,L-lactides and ε-caprolactone have gained significant attention due to their excellent biocompatibility and good mechanical properties. The size of the ring, its strain energy, and the number and type of its substituents influence the enthalpy and entropy of ROP reactions of cyclic monomers. The free energy change of polymerization (ΔG_p), as determined by both the enthalpy and entropy changes of polymerization (ΔH_p and ΔS_p) and the temperature (T), should be negative [Equation (8.12)]. Cyclic monomers with $\Delta H_p < 0$ and $\Delta S_p > 0$ can be polymerized at all temperatures, while cyclic monomers with $\Delta H_p > 0$ and $\Delta S_p < 0$, cannot be converted into polymeric materials. For monomers with $\Delta H_p > 0$ and $\Delta S_p > 0$ there is a limiting temperature called the *floor temperature* [Equation (8.13)], below which polymerization does not occur. Additionally, for $\Delta H_p < 0$ and $\Delta S_p < 0$, there is another limiting temperature named the *ceiling temperature* [Equation (8.14)], above which polymerization is thermodynamically forbidden [30].

$$\Delta G_p = \Delta H_p - T\Delta S_p \qquad (8.12)$$

$$t_f = \frac{\Delta H_p}{\Delta S_p} \qquad (8.13)$$

$$t_c = \frac{\Delta H_p}{\Delta S_p} \qquad (8.14)$$

Six- and higher membered cyclic carbonates such as TMC can produce polycarbonates with complete retention of their CO_2 contents under certain catalytic conditions [Equation (8.15)]. That is, the $\Delta H_p < 0$ and $\Delta S_p > 0$, and thus the polymerization process is thermodynamically allowed at all temperatures. Additionally, the increased ring-strain of six- and higher membered cyclic carbonates also promotes the ROP reaction [31, 32]. On the other hand, the ROP of five-membered cyclic carbonates is thermodynamically unfavored. Nevertheless, the ROP of five-membered cyclic carbonates has been performed at temperatures higher than their corresponding ceiling temperatures, resulting in the production of polycarbonates with significant amounts of ether linkages that are the consequence of CO_2 loss [Equation (8.16)]. Researchers exploring this area have postulated that when the reactions are performed at high temperatures ($T > 373$ K) the loss of carbon dioxide makes the $\Delta S_p > 0$, and hence, the polymerization process becomes thermodynamically allowed [31, 33–39]. The advantage, however, of producing polycarbonates with none or at least reduced ether linkages is that the physical properties of the resulting polymers are greatly improved.

$$\text{six-membered cyclic carbonate} \xrightarrow{\text{ROP}} {+OCH_2CH_2CH_2OC(O)+}_n \tag{8.15}$$

$$\text{five-membered cyclic carbonate} \xrightarrow{\text{ROP}} {+OCH_2CH_2OC(O)+}_n {+OCH_2CH_2+}_m + CO_2\uparrow \tag{8.16}$$

8.6
Catalytic Processes Using Green Catalysts Methods

Biodegradable aliphatic polycarbonates obtained from the ROP of six-membered cyclic carbonates are an interesting class of polymers that can be employed to produce *non-toxic thermoplastic elastomers*. These elastomers have a variety of potential medical applications including sutures, drug delivery systems, body, and dental implants, and tissue engineering [1, 40]. Because these polymers have numerous biomedical uses, investigating their mechanism is important. Additionally, a full understanding of the reaction pathway will further improve the methodology of synthesis and the biocompatibility with the human body.

The ROP of six-membered cyclic carbonates has been extensively investigated through various mechanisms and utilizing a variety of catalysts including cationic, anionic, enzymatic, coordination–insertion, and organocatalytic type mechanisms. The scope of this section will be to examine the diverse mechanisms for the ROP of six-membered cyclic carbonates. Special attention will be presented to the *biocompatible catalytic systems* that recently have been reported in the literature. These include promising industrial candidates that are organic-based catalysts and biocompatible metal-based catalysts that contain calcium, zinc, potassium, and magnesium.

Prior to reviewing the mechanistic aspects of ROP reactions of six-membered cyclic carbonates, it is beneficial to define two terms that will be referred to in further discussions, namely turnover frequency (TOF) and polydispersity index (PDI). TOF is the moles of cyclic carbonate consumed/(moles of catalyst h). It is important to note that TOFs are highly dependent on the time period that reactions are monitored. That is, the highest TOF values are obtained during the initial period of the polymerization process. The molecular weights of polymers are generally determined by size-exclusion chromatography (SEC) analysis using a polystyrene standard. The molecular weight distribution is referred to as *polydispersity* and is defined as $\overline{M}_w/\overline{M}_n$, where $\overline{M}_w$ is the weight average molecular weight

$$\left(\frac{\sum N_i M_i^2}{\sum N_i M_i}\right)$$

and $\overline{M}_n$ is the number average molecular weight

$$\left(\frac{\sum N_i M_i}{\sum N_i}\right)$$

N_i is the number of chains containing mass M_i and M_i equals the mass of chain. The value of M_w is always $> M_n$.

8.6.1
Cationic Ring-Opening Polymerization

As its name suggests, the term of cationic ROP refers to the fact that the ionic charge of the active propagating species is positive. Detailed mechanistic studies of this process were first reported in the early work of Kricheldorf and coworkers [41, 42]. Initially, the ROP of 5,5-dimethyl-1,3-dioxan-2-one (DTC) catalyzed by methyl triflate, triflic acid, and boron trifluoride was investigated. Polycarbonates with molecular weights $>6000 \text{ g mol}^{-1}$ could not be obtained. As expected for a cationic ROP mechanism, the reactions proceeded faster in polar solvents such as nitrobenzene compared with less polar solvents such as 1,2-dichloroethane [41]. The polymerization reaction mechanism of DTC catalyzed by methyl triflate was elucidated by means of ^{1}H NMR (nuclear magnetic resonance) and IR (infrared) spectroscopy studies (Scheme 8.4). Alkylation of the exocyclic oxygen atom of the cyclic carbonate generates a trioxocarbenium ion [Scheme 8.4, Reaction (a)]. An equilibrium process is established with the covalent triflate after ring-opening by the counter anion [Scheme 8.4, Reaction (b)]. The trioxocarbenium ion attacks another cyclic carbonate molecule cleaving the alkyl–oxygen bond and the exocyclic oxygen of the nucleophile is alkylated [Scheme 8.4, Reaction (c)]. Another possible propagating route is the reaction between the covalent triflate and monomer [Scheme 8.4, Reaction (d)] [41, 42]. The formation of ether linkages in the afforded

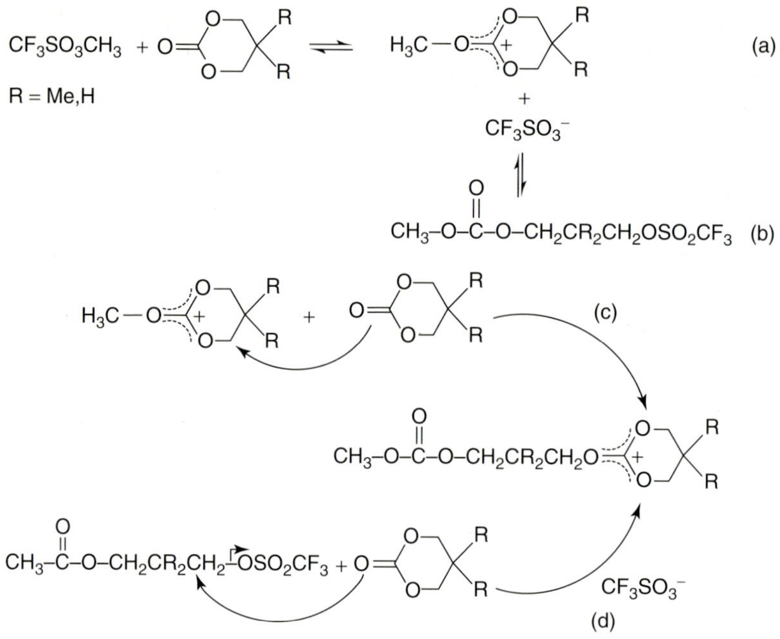

Scheme 8.4 Cationic ring-opening polymerization mechanism [41, 42].

polycarbonate is known as a *side reaction product* of decarboxylation, and this was proposed to occur via intramolecular migration of an alkyl group (Scheme 8.5) [42].

Albertsson and coworkers reported the *melt polymerization* of TMC in the presence of $BF_3 \cdot OEt_2$ as catalyst at temperatures within 353–373 K, yielding poly(TMC) with a molecular weight in excess of 100 000 g mol^{-1}. The polymer produced at higher reaction temperatures contained 2.6% of ether linkages formed by decarboxylation. Polymerization reactions in solutions produced mostly oligomers. Other catalysts such as $AlCl_3$ and $CH_3COO^-K^+$ gave lower molecular weight polymers [27]. Further investigations by Kricheldorf utilizing boron halogenides ($BF_3 \cdot OEt_2$, BCl_3, and BBr_3) demonstrated that $BF_3 \cdot OEt_2$ was an effective initiator, and polymerizations performed in bulk-produced polycarbonates in 95% yield and with M_n in the range of 10 000–60 000 g mol^{-1} at 333 K. Two important factors determined the presence of a cationic-type mechanism. The polymerizations reactions were found to be faster in polar solvents such as nitrobenzene than in the less polar chloroform. Additionally, the polycarbonates contained ether groups, and their mole fraction increased with temperature [43].

Importantly, Endo and coworkers reported the use of alkyl halides as *cationic-based catalysts* for the ROP of six-membered cyclic carbonates, yielding polycarbonates without ether units but of rather low molecular weights [44]. To suppress the decarboxylation side process, Endo proposed to utilize propagating species of lower reactivity according to the *selectivity–reactivity rule* in organic chemistry. Alkyl halides of lower reactivity than that of the triflate were selected, such as methyl iodide and benzyl bromide, which afforded poly(TMC) with $M_n = 1000–3700$ g mol^{-1} and

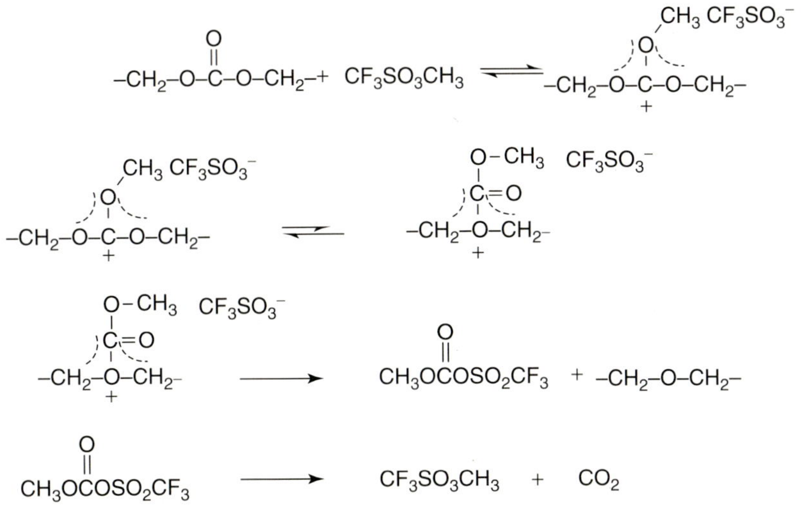

Scheme 8.5 Formation of ether linkages during the cationic ring-opening polymerization process [42].

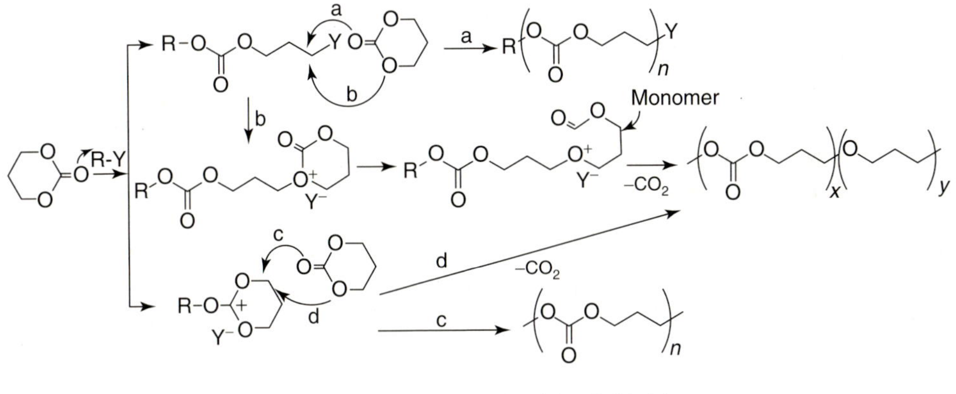

Scheme 8.6 Ring-opening polymerization of TMC catalyzed by alkyl halides [44].

with no ether linkages at 393 K. According to the reactivity–selectivity rule explained by results of molecular orbital calculations, the reaction mechanism with the alkyl halides was proposed to proceed according to pathway **a** (Scheme 8.6). In this instance, the covalent macrohalide is more favored than the carbenium ion species due to the higher nucleophilicity of the halide anion. Nevertheless, in the presence of polar solvents, partial decarboxylation can take place due to shifting of the equilibrium process to the carbenium ion species, which can undergo partial nucleophilic attack along pathway **d**.

Another type of cationic-based polymerization system that has been reported for the ROP of six-membered cyclic carbonates involves the use of alcohols catalyzed by acids. In this respect, Endo and coworkers have reported the ROP

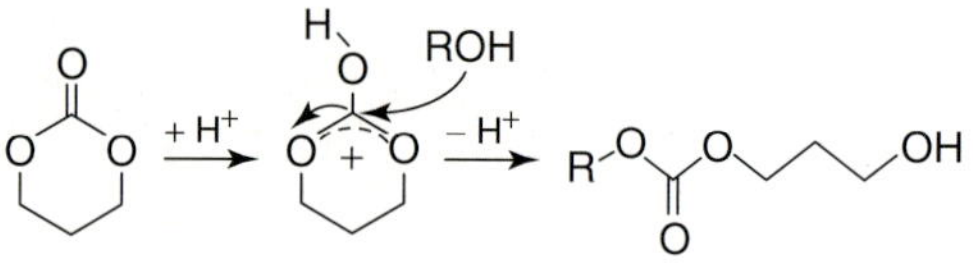

Scheme 8.7 Mechanism for ROP of TMC catalyzed by alcohols in the presence of acid [45].

of TMC by benzyl alcohol or n-butanol in the presence of trifluoroacetic acid to afford the corresponding polycarbonates with M_n = 2500–5300 g mol^{-1} and polydispersities in the range of 1.16–1.24. The molecular weights were found to increase with the percentage conversion to polymer, which demonstrated the *pseudo* living character, and the observed polymerization rate was determined to be 0.8×10^{-6} s^{-1} at 0 °C. The mechanism was shown to proceed by a nucleophilic attack of the hydroxyl endgroup on the carbonyl of the monomer previously activated by trifluoroacetic acid (Scheme 8.7) [45].

Notably, the ROP of DTC and TMC catalyzed by natural amino acids has been reported by Liu *et al.* [46]. The maximum values of M_n of the poly(DTC) and poly(TMC) were 18 900 and 17 800 g mol^{-1} and the corresponding PDIs were 1.67 and 1.65, respectively. Molecular weights were found to increase with the [monomer]/[amino acid] molar ratio. ^{1}H-NMR spectroscopy and titration analysis studies demonstrated that the amino acid was incorporated into the polymer main chain [Equation (8.17)].

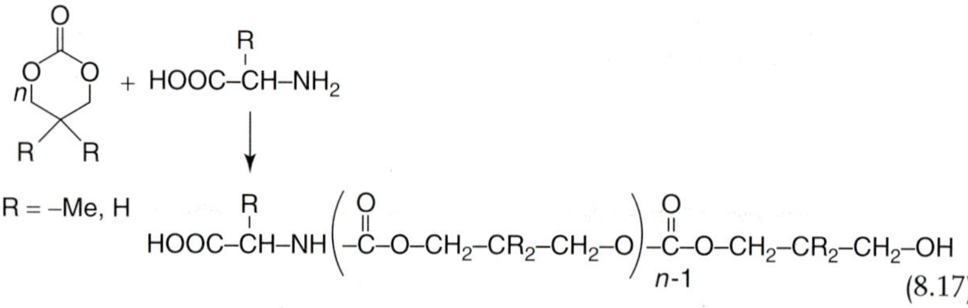

R = –Me, H

(8.17)

8.6.2
Anionic Ring-Opening Polymerization

Contrary to the cationic ROP of six-membered cyclic carbonates, in the anionic ROP mechanism the charge of the active propagating species is negative. In general, this route affords high molecular weight polycarbonates without ether units. The anionic ROP process has been investigated for several different types of catalysts. In 1930, Carothers reported the use of K_2CO_3 as a catalyst for the anionic ROP of TMC [26]. Alkoxide-, alkyllithium-, and alcoholate-based initiators have also been investigated for this process [29, 47–51]. In particular, alkoxide- and alkyllithium-based catalysts have been commonly employed as *anionic initiators*, but their industrial use may be complicated by their instability and high reactivity [32].

Interestingly, it has been demonstrated that the anionic ROP of six-membered cyclic carbonates exhibits an equilibrium process between the monomer and the polymer [52]. In this respect, monomers that undergo equilibrium polymerization are very useful, because this process can be employed in the recycling of polymeric materials [53–55]. As an example of this type of anionic equilibrium polymerization process, Endo and coworkers examined the anionic ROP of several six-membered cyclic carbonates containing two substituents at the 5-position of 1,3-dioxan-2-one [Equation (8.18)], utilizing t-BuOK as catalyst [29]. The conversions into polymer reached a constant value below 100%, which suggested that there was an equilibrium process between the monomer and the polymer. The conversions decreased in the following order $1 > 2 > 3 \geq 4 > 5$. The molecular weights of the polymers ranged from 12 000 to 32 000 g mol^{-1}, and PDIs were around 1.4. The thermodynamic parameters, namely, standard enthalpy and standard entropy of the polymerization, ΔH_{ss} and ΔS_{ss}, were estimated from Dainton's equation [56]. The obtained ΔH_{ss} and ΔG_{ss} decreased in the order of $1 > 2 > 3 \geq 4 > 5$ and reflected their polymerizability.

$$\text{1: } R_1=R_2=H$$
$$\text{2: } R_1=R_2=Me$$
$$\text{3: } R_1=R_2=Et$$
$$\text{4: } R_1=Me, R_2=Ph$$
$$\text{5: } R_1=Et, R_2=Ph$$

(8.18)

Thus, the monomer with the higher percentage conversion into polymer also displayed the higher $|\Delta G_{ss}|$. Molecular orbital calculations on model compounds were performed to investigate their difference in polymerizability. It was demonstrated that the steric repulsion of the substituents on the polymer chain was responsible for the thermodynamic stability of the macromolecules. Two reaction pathways were proposed for the depolymerization mechanism (Scheme 8.8). In pathway 1, the anionic initiator extracts the proton of the –CH$_2$OH endgroup of the polymer, and the formed alkoxide attacks the carbonate of the polymer main chain. On the other hand, in pathway 2, the initiator randomly attacks a carbonate group of the polymer main chain.

Additionally, amine-based initiators have also been reported for the anionic ROP of six-membered cyclic carbonates, such as 1,8-diazabicyclo[5.4.0]undec-7-ene (DBU), 1,4-diazabicyclo[2.2.2]octane (Dabco), 4-(dimethylaminopyridine) (DMAP), N,N-dimethylaniline, triethylamine, aniline, pyridine, and quinuclidine [57]. A six-membered cyclic carbonate containing a norbornene structure (5,5-(bicyclo [2.2.1]hept-2-en-5,5-ylidene)-1,3-dioxan-2-one) (NBC) initiated by DBU at 393 K for 1 h yielded a polycarbonate with $M_n = 6400$ g mol^{-1} and PDI $= 1.48$. No polycarbonate was obtained when triethylamine, aniline, N,N-dimethylaniline, and pyridine were employed. Other amines such as DMAP, Dabco, and quinuclidine could also initiate the polymerization but with lower reactivity than DBU. The lower reactivity of triethylamine was suggested to be due to the greater steric hindrance at the nitrogen atom, whereas aromatic amines displayed low activity because their aromatic resonance effects decreased their nucleophilic character.

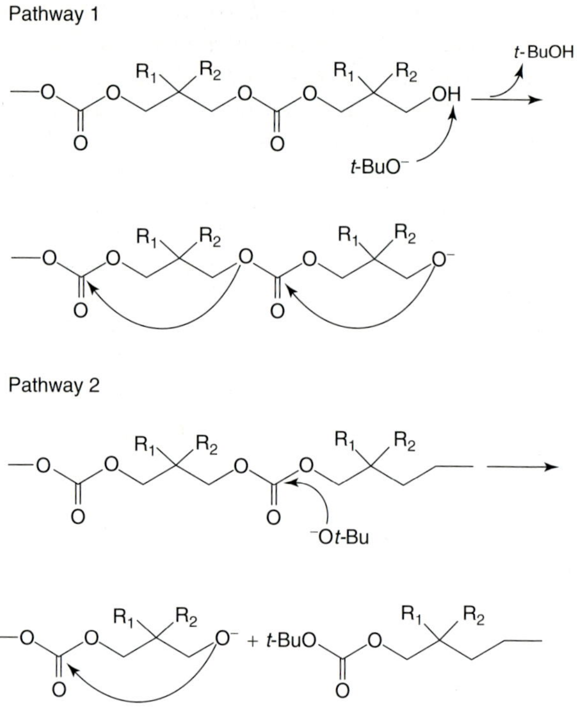

Pathway 1

Pathway 2

Scheme 8.8 Depolymerization mechanisms for polycarbonates [29].

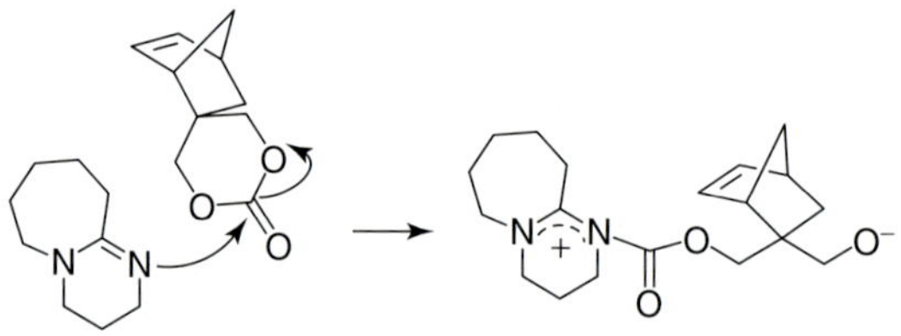

Scheme 8.9 Initiation step of the anionic polymerization reaction [57].

A zwitterionic polymerization mechanism was confirmed by the field desorption mass spectrometry (FD-MS) analysis of the products, where a DBU endgroup could be identified. Hence, the initiation step of the polymerization reaction involves the reaction between DBU and NBC to yield an alkoxide anion (Scheme 8.9).

8.6.3
Enzymatic Ring-Opening Polymerization

The employment of enzymes as catalysts for the synthesis of biodegradable polymers has gained increased attention in the last few years. As opposed to metal-based catalysis that often requires the use of pure monomers, anhydrous conditions, and the need for catalyst removal from the final polymer, *enzymatic catalysis* requires milder conditions, and the enzyme catalysts are safe and can often be recycled [58–66]. Kobayashi *et al.* first demonstrated that lipase derived from *Candida Antarctica* catalyzes the ROP of TMC in bulk yielding the corresponding aliphatic polycarbonate with $M_n = 2500$ g mol^{-1} and PDI $= 3.4$ after 72 h at 348 K, with no decarboxylation side reactions taking place [58].

In 1997, Gross and coworkers reported a comprehensive study on the lipase-catalyzed ROP of TMC in bulk [59]. Seven commercially available lipases derived from different sources were screened and the effects of various reaction conditions on the polymerization reaction were investigated. In polymerizations of TMC at 343 K, the lipase from *Candida Antarctica* gave the highest reaction rate. After 120 h, a monomer conversion of 97% and poly(TMC) with $M_n = 15\,000$ g mol^{-1} and PDI $= 2.2$ with no ether linkages was obtained. A decrease in the molecular weight was observed when the reaction temperature was increased from 328 to 358 K. These workers suggested that side reactions, including hydrolysis, increased chain initiation, and depolymerization reactions, might be responsible for this behavior. On the other hand, as the water content was increased, lower molecular weights and higher polymerization rates were observed. This is most likely due to an increase in the number of propagating chain ends and an enhancement in the enzyme activity. Analysis of low molecular weight polymers led to the proposal of a mechanism for the lipase-catalyzed ROP of TMC (Scheme 8.10). Products obtained from the polymerization reactions carried out in toluene and dioxane were isolated by column chromatography, and these were identified as 1,3-propanediol, α,ω-dihydroxy trimer of trimethylene carbonate (TTMC), and α,ω-dihydroxy dimer of trimethylene carbonate (DTMC), respectively. The initiation step of the polymerization reaction involves the reaction of TMC with the lipase to generate a lipase–TMC, known as an enzyme-activated monomer (EAM). Reaction of EAM and water followed by rapid decarboxylation generates 1,3-propanediol. The propagation step involves the formation of DTMC by reaction of 1,3-propanediol with the EAM, TTMC synthesis by reaction of DTMC with the EAM, and subsequent propagation reactions generate the final polymer.

Importantly, researchers investigating this area have found that the employment of *immobilized lipase enzymes* could greatly improve their catalytic efficiency compared with naked lipase enzymes, due to their thermal activation and thermal stability, and these could be recycled for their re-utilization [60]. Silica microparticules are most commonly used to immobilize the enzymes [60, 61]. Additionally, the enzyme ROP approach to polycarbonates has been successfully extended to other six-membered cyclic carbonates that can be modified by post-polymerization reactions, such as 5-methyl-5-benzyloxycarbonyl-1,3-dioxan-2-one, and 5-benzyloxytrimethylene carbonate. Their copolymerizations with TMC and DTC, respectively, have also been explored [61, 62].

The investigation of a *renewable polycarbonate production and recycling system* has been explored using enzyme catalysis. Toward this objective, Matsumura *et al.* have shown that the polymerization of TMC and depolymerization of the corresponding poly(TMC) can be achieved utilizing lipase derived from *Candida Antarctica*, that is,

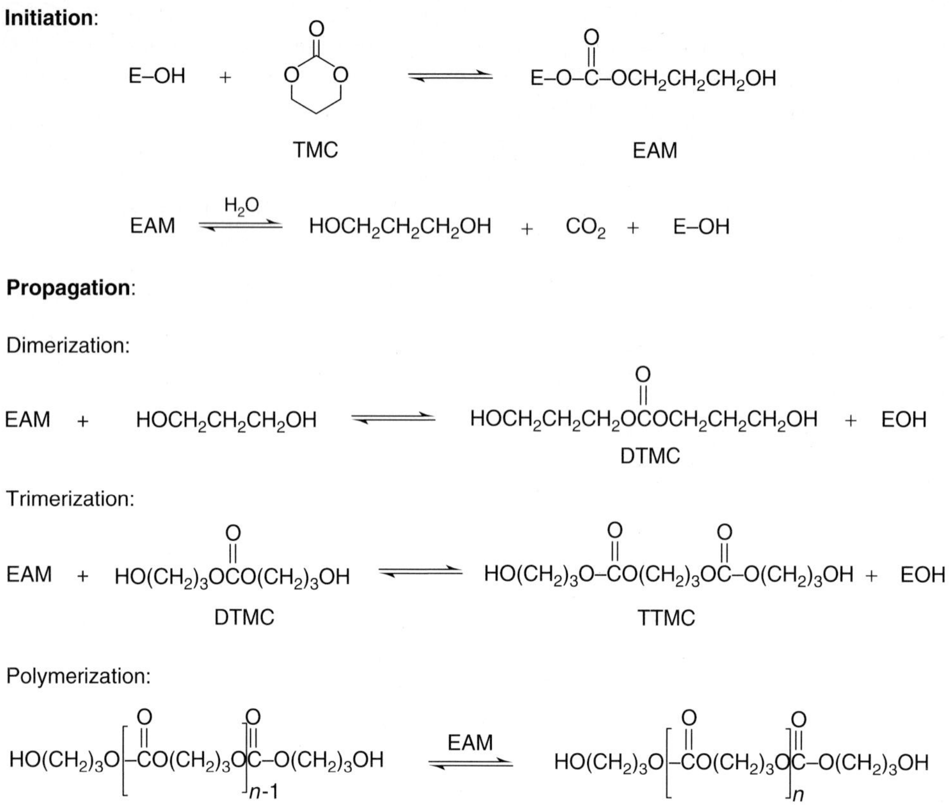

8 Ring-Opening Polymerization of Renewable Six-Membered Cyclic Carbonates

Initiation:

E–OH + TMC ⇌ E–O–C(=O)–OCH$_2$CH$_2$CH$_2$OH (EAM)

EAM + H$_2$O ⇌ HOCH$_2$CH$_2$CH$_2$OH + CO$_2$ + E–OH

Propagation:

Dimerization:

EAM + HOCH$_2$CH$_2$CH$_2$OH ⇌ HOCH$_2$CH$_2$CH$_2$OCOCH$_2$CH$_2$CH$_2$OH (DTMC) + EOH

Trimerization:

EAM + HO(CH$_2$)$_3$OCO(CH$_2$)$_3$OH (DTMC) ⇌ HO(CH$_2$)$_3$O–CO(CH$_2$)$_3$OC–O(CH$_2$)$_3$OH (TTMC) + EOH

Polymerization:

HO(CH$_2$)$_3$O–[CO(CH$_2$)$_3$OC–O(CH$_2$)$_3$O]$_{n-1}$H + EAM ⇌ HO(CH$_2$)$_3$O–[CO(CH$_2$)$_3$OC–O(CH$_2$)$_3$O]$_n$H

Scheme 8.10 Mechanism for the lipase-catalyzed ROP of TMC proposed by Gross [59].

from a poly(TMC) sample with $M_n = 3000-4800$ g mol^{-1} treated at 343 K, the cyclic monomer can be obtained in 80% yield. Hence, the formed TMC readily polymerized using fresh and recovered lipase [63]. Similarly, Matsumura and colleagues reported a successive two-step polymerization process where the polymerization of diethyl carbonate and diols such as 1,3-propanediol, and 1,4-butanediol was catalyzed by lipase at a temperature between 343 and 348 K. These afforded the corresponding aliphatic polycarbonates with weight average molecular weights of 40 000 g mol^{-1} [64].

Interestingly, Matsumura demonstrated that in order to create a completely renewable and recyclable polymer production system, a direct method for synthesizing TMC using dimethyl carbonate and 1,3-propanediol was needed. As we have shown in a previous section of this chapter, dimethyl carbonate and 1,3-propanediol can be obtained from renewable resources. Therefore, the employment of these reagents to synthesize both TMC and poly(TMC) utilizing an enzyme is an attractive alternative. In this case, the obtained TMC can be subsequently polymerized by enzymatic catalysis, and the poly(TMC) formed can be depolymerized to TMC in a recyclable system.

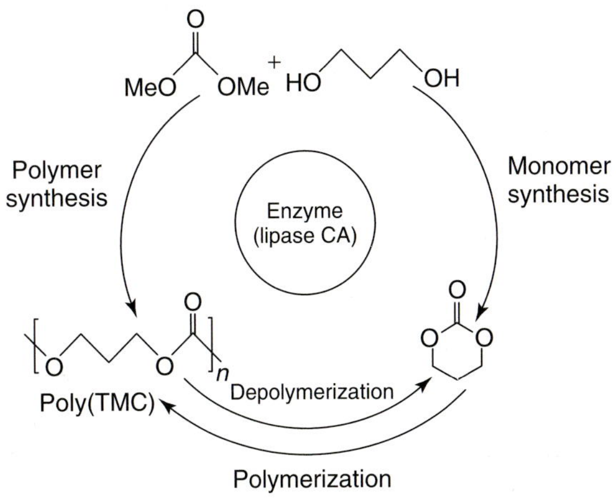

Scheme 8.11 Cycle for the production and recycling of polycarbonate investigated by Matsumura [65].

Indeed, this strategy has been effectively applied in the presence of methyl substituted- and unsubstituted-diols, and dimethyl carbonate catalyzed by immobilized lipase from *Candida Antarctica*, to afford the corresponding six-membered cyclic carbonates in 50–63% yield at 343 K (Scheme 8.11). In addition, the polymerizations of the obtained TMCs were performed at 373 K, catalyzed by lipase, yielding the corresponding polycarbonates with $M_w = 10400$ and $25\,100\,\mathrm{g\,mol^{-1}}$, and PDI = 3.7 and 5.2 for the polymers obtained from Me-TMC and TMC, respectively [65].

8.6.4
Coordination–Insertion Ring-Opening Polymerization

An alternative mechanism for the ROP of six-membered cyclic carbonates is the *coordination–insertion pathway*. This route differs from the cationic and anionic mechanisms involving free ions or ion pairs, in that the charged propagating species and its counterion share a covalent bond. In addition, this method requires the use of a metal-based catalyst. Generally, no ether linkages are observed in the afforded polycarbonates obtained by this pathway, and a better control of the molecular weight of the polymers can be achieved. Depending on the nature of the catalyst employed, two mechanisms have been proposed [32]. That is, when metal halides, oxides, and carboxylates act as Lewis acid-based catalysts in an ROP process initiated by water or alcohol, the polymerization is assumed to proceed through an insertion pathway [Equation (8.19)]. On the contrary, when metal alkoxides containing free p-, d-, or f-orbitals of favorable energy are utilized as catalysts, a two-step coordination–insertion mechanism takes place [Equation (8.20)]. The first step has been proposed to be complexation of the monomer mainly via the carbonyl oxygen of the cyclic carbonate. This complexation enhances the electrophilicity of the monomer and is followed by cleavage of the acyl–oxygen bond of the cyclic

carbonate monomer [32, 67]

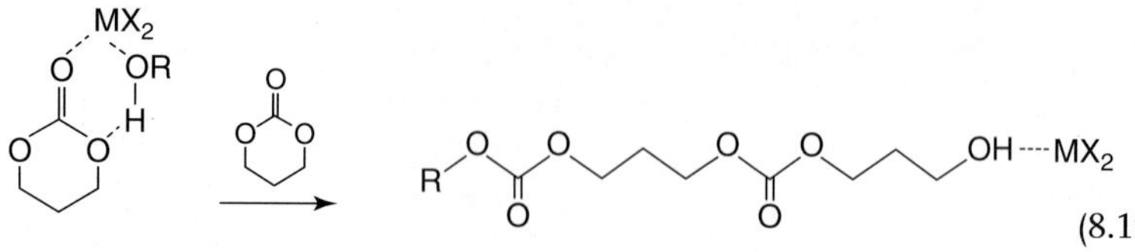

$$(8.19)$$

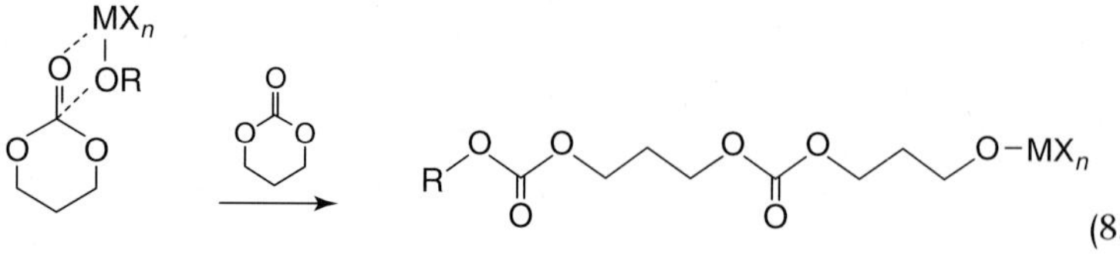

$$(8.20)$$

For the metal catalyzed ROP of six-membered cyclic carbonates, a variety of catalytic systems have been proposed and reported in the literature. To keep within the context of this chapter, we will focus on *catalysts derived from biocompatible metals* that have been investigated, for these represent attractive alternatives to be used in producing polymers for biomedical applications. However, non-biocompatible metal complexes will also be discussed when necessary because of their high reactivity and ease of handling. This section will be divided into metal catalyst groups based on their position on the Periodic Table.

8.6.4.1 Groups 13- and 14 Based Catalysts

Tin(II) bis(2-ethyl-hexanoate), or tin(II) octoate (Figure 8.1), has been shown to be a highly active catalyst for the ROP of cyclic esters [68, 69]. In addition, $Sn(Oct)_2$ is a very efficient catalyst for the ROP of six-membered cyclic carbonates, as was found by Kricheldorf and et al. [67]. Although $Sn(Oct)_2$ *is FDA approved, it is still a rather toxic compound*. Tin(II) octoate has been successfully applied as a catalyst for the ROP of TMC in bulk at 363, 393, and 423 K, producing values for M_w up to approximately 22 000 g mol^{-1}. Despite the high reaction temperatures employed, side reactions such as decarboxylation were not observed. Based on ^{1}H-NMR and IR mechanistic studies, the ring-opening mechanism was suggested to involve the cleavage of the acyl–O bond of the cyclic carbonate. Kricheldorf has also shown that $Bu_2Sn(Oct)_2$ can initiate the ROP of TMC with lower reactivity than $Sn(Oct)_2$. Addition of benzyl alcohol accelerated the polymerization reaction and allowed control of the molecular weight of the polymer by the [monomer]/[alcohol] ratio. Furthermore, benzyl carbonate endgroups were detected in the poly(TMC) samples by ^{1}H-NMR spectroscopy. At temperatures greater than 393 K, transesterification reactions occurred, and octanoate endgroups were also observed. At this high temperature, however,

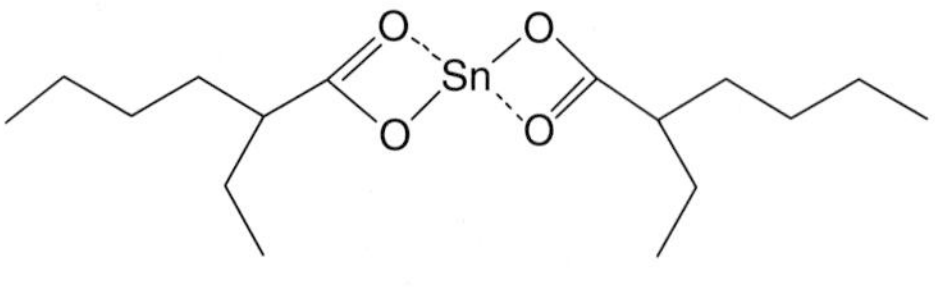

Figure 8.1 Structure of tin(II) bis(2-ethyl-hexanoate), or tin(II)octoate.

an accurate control of the M_n by the [monomer]/[initiator] ratio was not possible due to the backbiting degradation reactions occurring at high conversions [70].

Recently, Kricheldorf and coworkers reported the use of Ph_2BiOEt and Ph_2BiBr as initiators for the ROP of TMC in bulk. The dependence of the molecular weight on the [monomer]/[initiator] ratio suggested that the polymerizations obeyed a coordination–insertion pathway. In the presence of Ph_2BiBr, higher molecular weights were achieved (M_n values up to 340 000 g mol^{-1}), but the molecular weights were higher than the expected values based on the [monomer]/[initiator] ratio. Nevertheless, the addition of tetraethylene glycol as coinitiator resulted in a satisfactory control of the molecular weights. In the instance of Ph_2BiOEt, the molecular weights were lower than the expected values based on the [monomer]/[initiator] ratios, and only a rough control of the molecular weights was possible. Interestingly, both Bi-based complexes were found to be more active than $Sn(Oct)_2$ activated with ethanol, specifically at lower temperatures (e.g., 333 K) [71].

*Aluminum-based complex*es have also been employed as catalysts for the ROP of six-membered cyclic carbonates [72–74]. Early reports on the ROP of 2,2-dimethyltrimethylene carbonate catalyzed by tetraphenylporphyrin–aluminum compounds demonstrated that a halide or a methyl group attached to the aluminum center were inactive for polymerization (Figure 8.2). Only alkoxide nucleophile groups bound to the aluminum center proved effective. The polymerization reaction in methylene chloride was slow at ambient temperature (100 h). As expected, in non-coordinating solvents such as toluene at 323 K, the reaction time was significantly reduced [72]. On the other hand, Cao and coworkers reported on the use of aluminum complexes bearing salen ligands as catalysts for the ROP of six-membered cyclic carbonates and cyclic esters (Figure 8.3) [73]. The two complexes, namely, a monomeric and a dimeric aluminum salen based complex, were synthesized *in situ* from trimethylaluminum, methanol, and the corresponding salen ligand, (1R, 2R)-N,N′-bis(salicylidene)-1,2-diaminocyclohexane, producing the monomeric and dimeric aluminum salen complexes in 85 and 10% yields, respectively. Based on NMR, MS, FTIR (Fourier transform infrared), and single crystal X-ray diffraction data, the molecular structures and coordination states of these two complexes were defined. ^{27}Al-NMR and X-ray crystallography revealed that the dimeric compound appeared in the six-coordinated state. Furthermore, a dynamic equilibrium transition between the five- and six-coordinated states

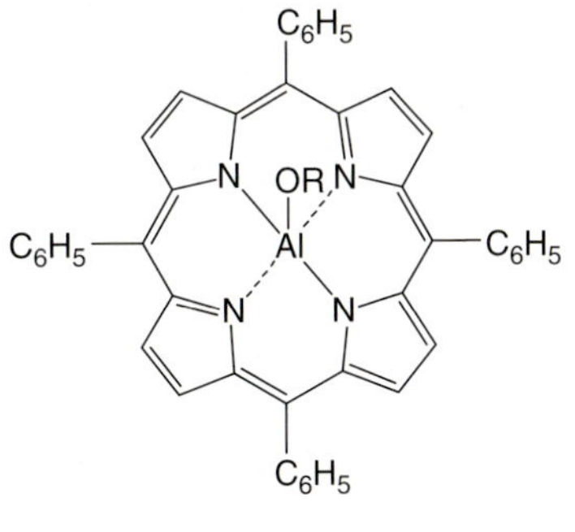

Figure 8.2 Example of tetraphenylporphyrin–aluminum complex (TTP) AlOR, (R=$OCH_2CH_2OC_6H_5$), employed as catalyst for the ROP of DTC [72].

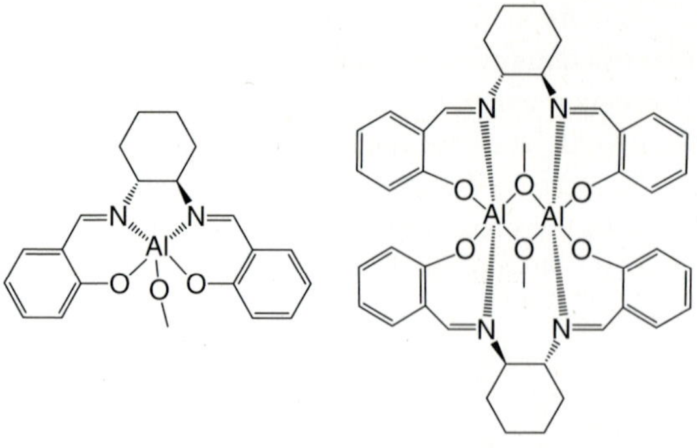

Figure 8.3 Monomeric and dimeric aluminum salen complexes [73].

was detected for the monomeric compound by ^{27}Al-NMR studies in a CDCl$_3$ solution. The two complexes were found to be efficient in catalyzing the ROP of six-membered cyclic carbonates and cyclic esters through a coordination–insertion mechanism, and no decarboxylation side reactions could be detected. For example, the ROP of TMC by the monomeric aluminum complex in anisole at 373 K for 3 h led to poly(TMC) with a PDI of 1.41, >95% conversion into polymer, and $M_n = 13400$ g mol^{-1}, which closely tracked the theoretical value ($M_n = 9800$ g mol^{-1}). Under similar conditions, the ROP of TMC catalyzed by the dimeric aluminum complex led to poly(TMC) with a PDI of 1.26, >95% conversion, and a $M_n = 9200$ g mol^{-1}, which was in excellent agreement with the theoretical value ($M_n = 9100$ g mol^{-1}). These researchers also suggested that the lower PDI observed by the dimeric complex could be due to its steric hindrance, which could diminish competitive chain transfer reactions that would result in a broadening of the polymer's molecular weight distribution.

Initial reports by Kricheldorf and Cao on the ROP of TMC catalyzed by tin(II)-based octoate and aluminum salen complexes, respectively, inspired us to examine the efficiency of Al(III) and Sn(IV) salen complexes as catalysts for the ROP of TMC (Figure 8.4) [74]. Initially, optimization of the catalytic system for the ROP of TMC was carried out. The reactions were performed in bulk at 368 K, maintaining a [monomer]/[initiator] ratio of 350 : 1 for 3 h. Optimization of the (salen)Al(III)Cl system was achieved by utilizing a salen ligand with chloride groups in the 3,5-positions of the phenolate rings and a phenylene backbone for the diimine (TOF = 81 h^{-1}). Further optimization of the aluminum salen catalytic system was attained by varying the initiator of the complexes containing an ethylene backbone for the diimine

M = Al or SnY (Y = Cl or nBu)

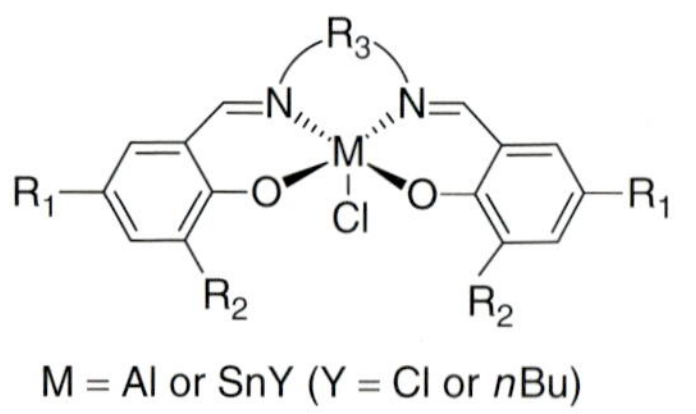

Figure 8.4 Generic diagram of a metal(salen)Cl complex. R_1 and R_2 refer to the 3,5-positions of the phenolate rings, respectively.

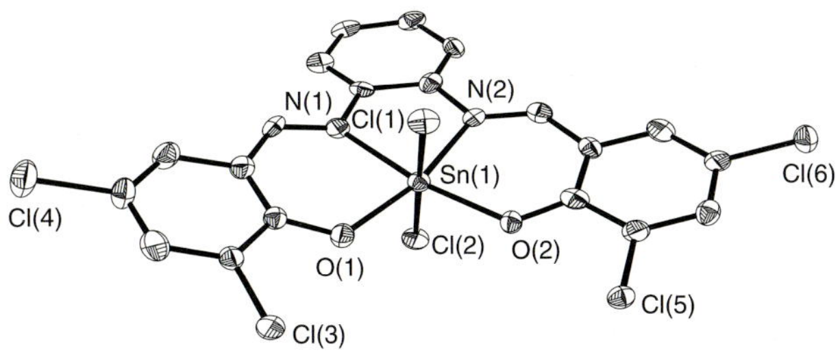

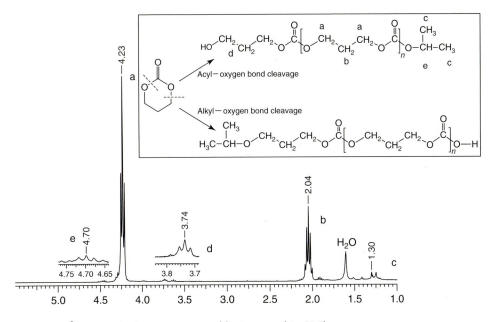

Figure 8.5 Thermal ellipsoid representation of (salen)SnCl$_2$, where the salen ligand contains $-t$-Bu substituents in the 3,5-positions of the phenolates, with a phenylene diamino-backbone.

and *tert*-butyl groups in the 3,5-positions of the phenolate rings. The aluminum salen complex containing an ethoxy group was found to be more active than that with a chloride group, (TOF = 105 h^{-1}, M_w = 24 000 g mol^{-1}, PDI = 1.61). In addition, the activity of the (salen)Sn(IV)(X)(Y) complex was optimized by employing chloride groups in the 3,5-positions of the phenolate rings, a phenylene backbone for the diimine, and chloride groups bound to the tin center (TOF = 22 h^{-1}) (Figure 8.5). The reaction mechanism was further investigated by ^{1}H-NMR spectroscopy. A low molecular weight polymer terminated by 2-propanol exhibited peaks in the ^{1}H-NMR spectrum consistent with monomer insertion into the growing polymer chain by acyl–oxygen bond cleavage (Figure 8.6). Subsequent kinetic investigations were performed in solution employing one of the catalysts explored,

Figure 8.6 ^{1}H-NMR of poly(TMC) terminated by 2-propanol in CDCl$_3$.

namely, N, N′-bis(salicylidene)-1,2-phenylenediimine-aluminum(III) chloride. We found that the reaction mechanism was first order in [catalyst] and [monomer]. The activation parameters, $\Delta H^{\ddagger}$ and $\Delta S^{\ddagger}$, were found to be 51.4 ± 3.2 kJ mol^{-1} and -105 ± 8.4 J mol^{-1} K^{-1}, respectively, which were consistent with a reaction mechanism involving an insertion of the monomer into a metal–nucleophile bond.

8.6.4.2 Groups 4–12 Based Catalysts

As mentioned earlier on in this chapter, a greener method for the synthesis of six-membered cyclic carbonates such as TMC is needed, not only because these are valuable monomers for the production of biodegradable polycarbonates [Equation (8.21)], but also because the current industrial routes to synthesize these cyclic carbonates involve the use of toxic reagents such as chloroformates. Conforming with this objective, the copolymerization of oxetane and carbon dioxide represents an attractive route [Equation (8.22)]. In this instance, the six-membered cyclic carbonate by-product, TMC, can be ring-opened and transformed into the same polymer by way of the reaction shown in Equation (8.21). This reaction has not been widely explored, possibly due to the lower reactivity of oxetane and also because of its limited availability [75].

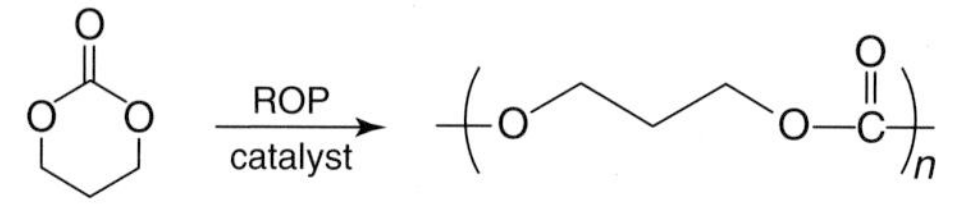

$$\text{TMC} \xrightarrow{\text{ROP catalyst}} \text{Poly(TMC)} \tag{8.21}$$

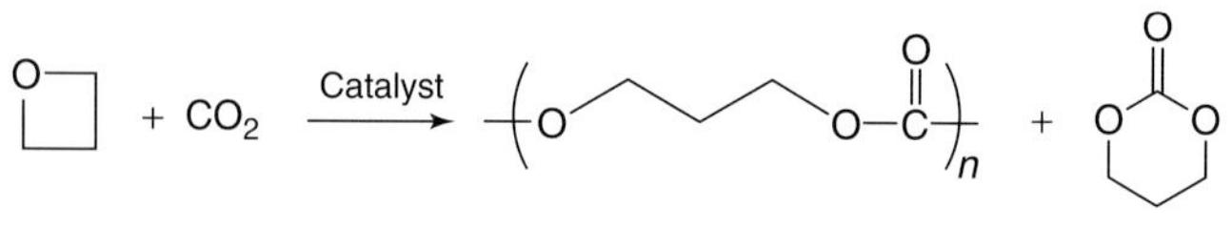

$$\text{oxetane} + CO_2 \xrightarrow{\text{Catalyst}} \text{poly(carbonate)} + \text{TMC} \tag{8.22}$$

Recently, we have investigated the mechanistic aspects of the copolymerization of oxetane and CO_2 catalyzed by (salen)Cr(III)Cl complexes in the presence of n-Bu$_4$NX (X = Cl, N$_3$) as cocatalysts [76, 77]. Optimization of the catalytic systems and catalytic conditions has led us to produce copolymers with 100% selectivity and ~97% carbonate linkages. The mechanistic aspects of the catalytic coupling of oxetane and carbon dioxide have been investigated by monitoring the process by *in situ* infrared spectroscopy. The catalytic system employed in these studies was a (salen)CrCl along with n-Bu$_4$NN$_3$ as cocatalyst (Figure 8.7). The reactions were carried out at 383 K and 3.5 MPa CO_2 pressure. Figure 8.8 illustrates a typical

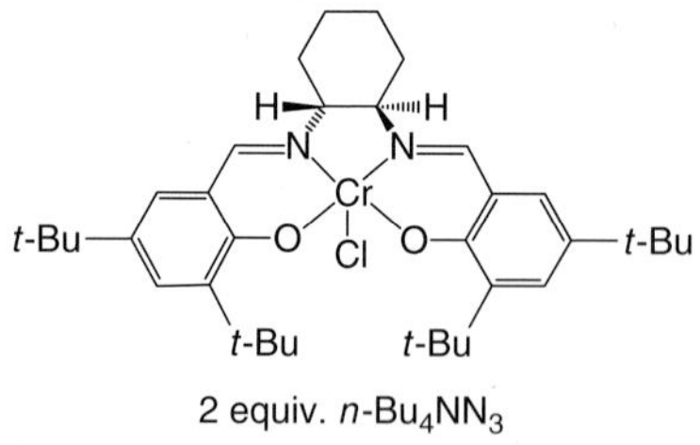

Figure 8.7 Catalytic system employed: (salen)CrCl along with n-Bu$_4$NN$_3$ as cocatalyst.

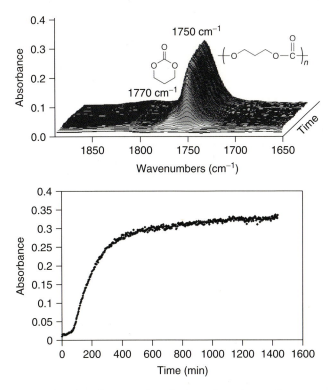

Figure 8.8 Three-dimensional stack plot and reaction profile of the IR spectra collected every 3 min during the copolymerization reaction of oxetane and carbon dioxide. Reaction carried out at 383 K in toluene at 3.5 MPa CO_2 pressure.

reaction profile of the growth of copolymer's $\nu_{C=O}$ band at 1750 cm^{-1} as a function of time. Of importance, an infrared band due to TMC ($\nu_{C=O}$, 1770 cm^{-1}) was detected during the early stages of the reaction, however, it completely vanishes after a few hours. This was also observed by intermittent sampling for ^{1}H-NMR analysis. From temperature-dependent studies, $\Delta H^{\ddagger}$, $\Delta S^{\ddagger}$, and $\Delta G^{\ddagger}$ values of 45.6 ± 3.01 kJ mol^{-1}, −161.9 ± 8.21 J mol^{-1} K^{-1}, and 107.6 kJ mol^{-1}, respectively, were obtained.

A *kinetic study* of the ROP of TMC similarly afforded $\Delta G^{\ddagger}$ of 101.9 kJ mol^{-1} at 383 K, a value very close in energy to that found for the production of poly(TMC) from oxetane and CO_2 (Figure 8.9). Based on our experimental findings, the formation of polycarbonate from the oxetane and CO_2 coupling reaction occurs via two different or concurrent pathways, that is, the intermediacy of TMC formation and subsequent polymerization and/or the direct enchainment of oxetane and carbon dioxide. The presence of small amounts of ether linkages in the copolymer also supports this conclusion. Scheme 8.12 summarizes our proposed mechanism for the coupling of oxetane and carbon dioxide. Following the initial ring-opening step and CO_2 insertion into the resultant chromium–oxygen bond, two pathways

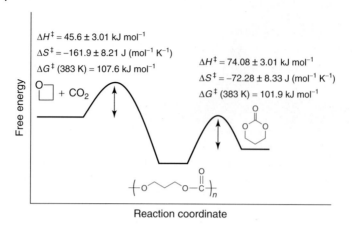

Figure 8.9 Reaction coordinate diagram with activation parameters for the copolymerization of oxetane and CO_2 and for the ROP of TMC.

are open for the intermediate. Pathway (1) involves consecutive additions of oxetane and CO_2 to yield the alternating copolymer, while pathway (2) leads to TMC formation by a backbiting process with ring closure. Once TMC is formed, it can enter the polymer chain by a coordination–insertion mechanism. The portion shown in red is highly dependent on the nature of the anionic leaving group. Indeed, we have noted for bromide and iodide ions, and at a lower reaction

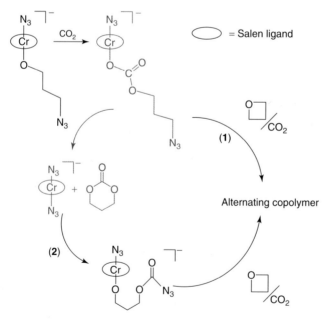

Scheme 8.12 Proposed mechanism for the coupling of oxetane and CO_2 [77].

temperature, that this pathway is competitive with oxetane enchainment and may provide a means for tuning the selectivity of the two pathways for TMC formation and/or for polycarbonate produced from the ROP of preformed TMC. An advantage of proceeding exclusively via ROP of preformed TMC is the absence of ether linkages in the afforded polycarbonate, which results in better physical and thermal properties of the polymer.

The employment of *low toxicity metal complexes* of iron, zinc, and zirconium as catalysts for the production of biodegradable aliphatic polycarbonates has recently been explored. In particular, zinc and iron have roles in human metabolism [78], and zirconium is considered to be inert in human metabolic processes [79]. In this respect, a few reports have been published in the literature, including the use of (salen)Zn(II) complexes and acetylacetonates of iron, zinc, and zirconium as catalysts for the ROP of six-membered cyclic carbonates [80, 81].

Additionally, Guillaume and coworkers reported the application of an "immortal" *solvent-free polymerization of TMC* process by utilizing a simple binary catalyst system [82]. This was a zinc complex supported by a β-diiminate ligand in conjunction with benzyl alcohol as a transfer agent (Scheme 8.13). Prior to showing a summary of Guillaume's work, it is worth noting the definition of the concept of "*immortal polymerization*." This approach allows the growth of several polymer chains per metal center while maintaining the control of the polymerization reaction through the presence of protic sources that act as chain transfer agents. A good control of the molecular weight was obtained with an alcohol/zinc ratio varying from 0 to 50 and a [monomer]/[zinc] loading ranging from 500 up to 50 000. For example, for a bulk polymerization at 383 K with $[TMC]_0 : [BnOH]_0 = 25\,000 : 10$ versus $[Zn]_0$, after 40 min an 83% conversion into polymer could be reached with a TOF of $31125\,h^{-1}$, $M_n = 185\,200\,g\,mol^{-1}$ [$M_{n(theoretical)} = 211\,750\,g\,mol^{-1}$], and a PDI of

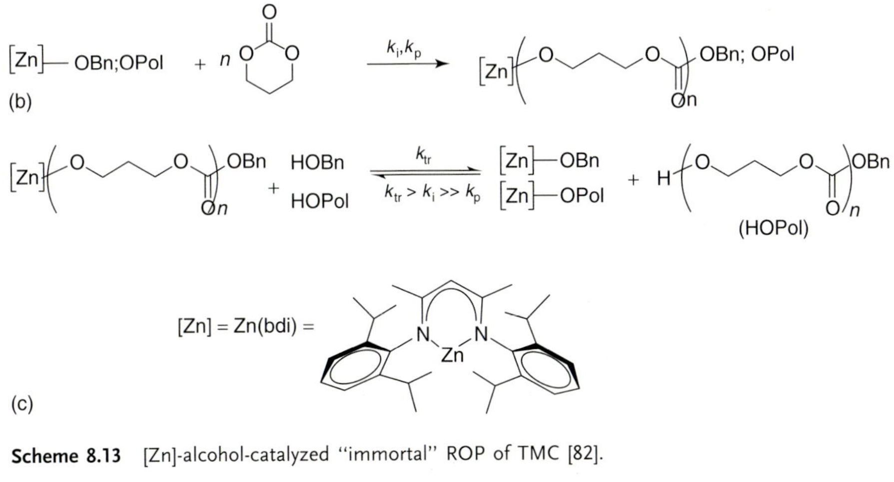

Scheme 8.13 [Zn]-alcohol-catalyzed "immortal" ROP of TMC [82].

1.68. Endgroup analysis of low molecular weight polymers performed by ^{1}H-NMR spectroscopy demonstrated the presence of a benzyloxy and a hydroxyl endgroup resulting from the initiation by the zinc-benzyloxide species and from hydrolysis of the growing species. These results supported a coordination–insertion pathway for polymer formation. Importantly, these analyses showed the absence of ether units.

Based on experimental results, these workers proposed the following mechanism for the immortal ROP of TMC. Treatment of [Zn(bdi){N(SiMe$_3$)$_2$}] with a molecule of BnOH yields a zinc alkoxide initiating species [Zn(bdi)OBn] (Scheme 8.13a). Subsequent addition of TMC leads to the formation of a growing zinc alkoxide (Scheme 8.13b). Excess alcohol molecules take part in a rapid and reversible exchange process with the growing alkoxide, thereby acting as chain transfer agents. This results in the formation of a new growing zinc alkoxide complex along with hydroxy-terminated polymer chains (HOPol) (Scheme 8.13c). All BnOH and HOPol molecules hence act as chain transfer agents resulting in the formation of new growing species from which new polymer chains are formed. It is important to note that the key aspect for this process to be successful is the *rate of the exchange* between the zinc alkoxide and the chain transfer agents. When the rate of the transfer reaction (k_{tr}) is greater than the rate of polymerization (k_p), the molecular weight distribution of the polymer is expected to be narrow, and this was actually observed in this work (PDI < 1.9) for the polymerization reactions performed in bulk.

8.6.4.3 Lanthanide-Based Catalysts

Because lanthanide-based complexes have well-defined structures and can be used as *single-component catalysts*, these have been examined for the ROP of six-membered cyclic carbonates via a coordination–insertion mechanism [83–88]. For example, lanthanide aryloxide complexes of the type (ArO)$_2$Ln(THF)$_3$ (Ln = Sm, Yb) have been shown to be active for the ROP of TMC, where the Sm-based complex was found to be more active than the Yb-based complex [83]. Similarly, single rare earth aryloxide catalysts of the type Ln(OAr)$_3$ (Figure 8.10) have been reported as active catalytic systems for the ROP of DTC. The catalytic activity of these complexes was found to decrease in the following order; La > Nd > Dy ~Y.

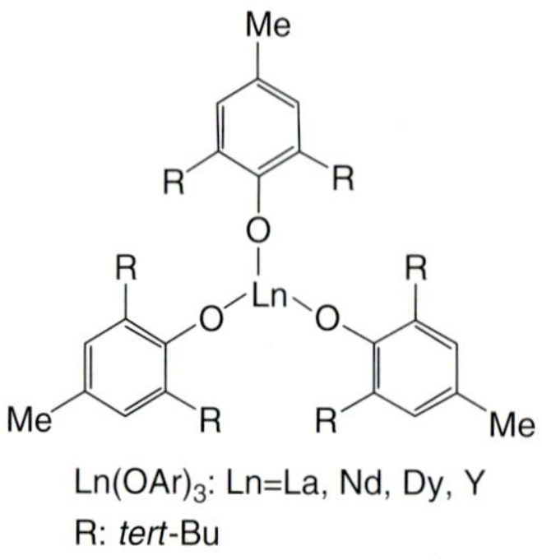

Figure 8.10 General structure of Ln(OAr)$_3$ complexes.

Ln(OAr)$_3$: Ln=La, Nd, Dy, Y
R: *tert*-Bu

Furthermore, poly(DTC) with $M_w = 171\,000$ g mol^{-1} and PDI $= 2.79$ was obtained from the catalyzed ROP of DTC by La(OAr)$_3$. The reaction conditions used were a [DTC]/[initiator] $= 1000$ in toluene at 288 K for 1 h with a 97.9% conversion into polymer obtained [84].

In addition, *homoleptic lanthanide guanidinate complexes* with general structure [Ln{Ph$_2$NC(NCy)$_2$}$_3$]·2C$_7$H$_8$ (Ln = Sm, Yb, Nd) showed high reactivity for the ROP of TMC giving polymers with PDIs ranging from 1.41 to 1.73 [85]. Homoleptic lanthanide amidinate complexes of the general type [CyNC(R)NCy]$_3$Ln have been investigated for the ROP of TMC. In this instance, the substituents on the amidinate ligands and metal centers showed a great effect on the catalytic activity of the complexes. Indeed, the catalytic activity decreased in the following order; Me > Ph, and La > Nd > Sm > Yb. This suggested that the activities for the polymerization of TMC are highly dependent on the metal ionic radius, with the best activities being observed for larger size metals. Among them, [CyNC(Me)NCy]$_3$La displayed the highest catalytic activity producing poly(TMC) with $M_n = 8500$ g mol^{-1} and PDI $= 1.85$. The reaction conditions employed were [TMC]/[initiator] $= 500$ in toluene at 313 K for 30 min with a 100% conversion into polymer achieved. Under the same catalytic conditions, the complex [CyNC(Me)NCy]$_3$Sm yielded poly(TMC) with $M_n = 7000$ g mol^{-1} and PDI $= 2.05$ [86].

8.6.4.4 Groups 1 and 2 Based Catalysts

The development of *non-toxic catalysts* for the production of biodegradable polycarbonates derived from six-membered cyclic carbonates is an area of increasing attention. This is because of the difficulty of removing trace amounts of catalysts residues from the resulting polycarbonates. In addition, this extensive and time consuming procedure increases the cost of medical grades of the final polymer. Therefore, researchers exploring this area are now focusing on the design of safer catalytic systems for this transformation. As we have seen previously in this chapter, organo-based, enzyme-based, and biocompatible metal-based catalysts of zinc, iron, and zirconium are of a great deal of interest. In this section we will present an important pair of non-toxic metal catalysts of the s-block of the Periodic Table, namely, metal complexes of magnesium and calcium.

In 2006, we began efforts to examine the efficiency of *metal salen complexes* of Ca and Mg in the presence of anion initiators for the ROP of TMC (Figure 8.11) [80]. Initially, optimization of various biocompatible complexes for the ROP of TMC was carried out. The common salen ligand employed in this investigation was H$_2$ salen = N, N'-bis(di-*tert*-butylsalicylidene)1,2-ethylene-diimine. The reactions

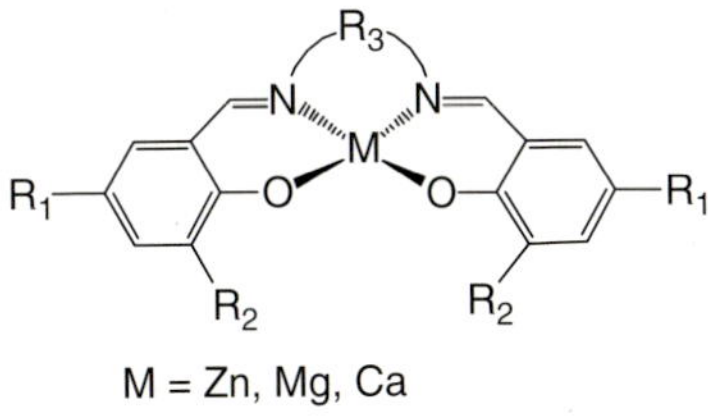

Figure 8.11 General structure of biometal salen complexes utilized as catalysts for the polymerization of trimethylene carbonate.

were performed in bulk at a [monomer]/[catalyst]/[initiator] ratio of 350 : 1 : 1 at 359 K for 15 min under an argon atmosphere. All polymers showed a complete absence of ether linkages. The calcium(II) salen complex was found to be the most active salen derivative, displaying a catalytic activity in the presence of n-Bu$_4$NCl of about twice that of its magnesium analog (TOF = 1123 h^{-1} for the calcium salen complex, and TOF = 541 h^{-1} for the magnesium salen complex). Aluminum and zinc salen complexes were found to display a lower catalytic activity for this process. Subsequent investigations were performed to optimize the catalytic activity of the calcium(II) salen system. This was achieved by employing a salen ligand with *tert*-butyl substituents in the 3,5-positions of the phenolate rings and an ethylene backbone for the diimine grouping, along with PPNN$_3$ acting as an initiator [PPN$^+$ = (Ph$_3$P)$_2$N$^+$] (TOF = 1286 h^{-1}). The molecular weight and PD of poly(TMC) obtained in these studies were measured by gel permeation chromatography. The bulk polymerizations were carried out at 361 K using [N, N'-bis(3,5-di-*tert*-butylsalicylidene) phenylenediimine]Ca(II) the presence of 1 equiv. of PPNN$_3$ as the initiator. In general the molecular weights closely paralleled the theoretical values ($M_n \sim 5600$ g mol^{-1} and PDIs ~ 1.6). Furthermore, a linear increase in M_n with percentage conversion and fairly low polydispersities were obtained. This clearly demonstrated that the level of polymerization control is high with this catalytic system.

Kinetic investigations were performed in TCE (1,1,2,2-tetrachloroethane) solution employing one of the catalytic systems explored, [N, N'-bis(3,5-di-*tert*-butylsalicylidene)-1,2-ethylenediimine]Ca(II) in the presence of 1 equiv. n-Bu$_4$NCl as cocatalyst. We found that the reaction mechanism was first order in [monomer], [catalyst], and [cocatalyst]. From temperature-dependent studies, the activation parameters, $\Delta H^{\ddagger}$ and $\Delta S^{\ddagger}$, were found to be 20.1 ± 1.0 kJ mol^{-1} and -128 ± 3 J mol^{-1} K^{-1} respectively. The $\Delta G^{\ddagger}$ value of 58.2 kJ mol^{-1} observed for the calcium catalyzed ROP of TMC was lower in energy by 24.5 kJ mol^{-1} than that which we previously described in this chapter for the reaction catalyzed by an aluminum derivative (82.7 kJ mol^{-1}). In both instances, these activation parameters were in agreement with a reaction pathway involving the addition of a nucleophile to a metal bound carbonate [Equation (8.23)]. As previously described by us for the ROP of TMC in the presence of (salen)Al(III)Cl catalysts, the mechanism involved an insertion of the monomer into the growing polymer chain by breaking the acyl–oxygen bond instead of the alkyl-oxygen bond. This conclusion was obtained after analyzing a low molecular weight polymer, by ^{1}H-NMR spectroscopy, previously terminated by 2-propanol as described in the literature [84]. Lastly, a study on the initiation step of the reaction, involving ring-opening of TMC by the anion of the cocatalyst, n-Bu$_4$NX (X = N$_3$), was performed in TCE solution by infrared pectroscopy. The equilibrium constant (K_{eq}) shown for the reaction in Equation (8.24) was determined to be 79 ± 21 M^{-1} by monitoring the ν_{N_3} vibrational modes in the metal-bound and free species. The (salen)Ca in the presence of one equivalent of N$_3^-$ exists as an equilibrium mixture of (salen)Ca and (salen)CaN$_3^-$, and upon addition of excess TMC, all the azide ligand is free in solution. Because our kinetic studies were performed using a [monomer]/[initiator] ratio of 350 : 1, the initiation process most probably involves free or weakly bound azide (or other anion, X⁻) at a metal coordinated TMC monomer [Equation (8.23)].

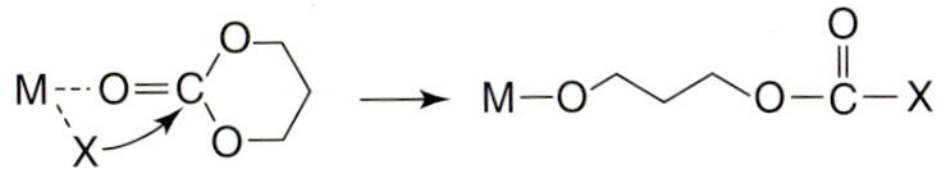

$$M\cdots O=C \cdots \rightarrow M-O\smile\smile O-\overset{O}{\underset{\|}{C}}-X \quad (8.23)$$

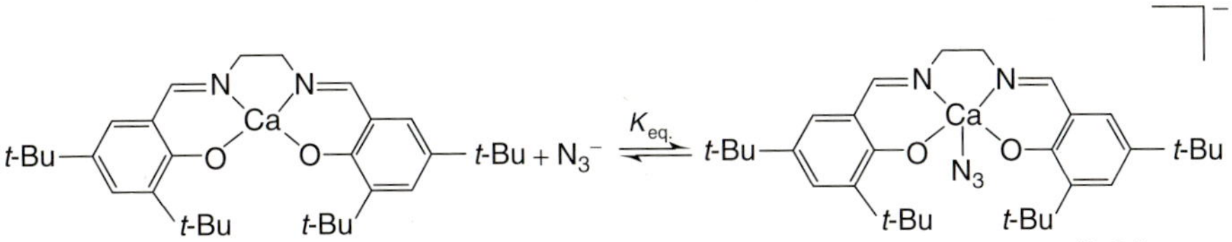

$$(8.24)$$

8.6.5
Organocatalytic Ring-Opening Polymerization

The utilization of organocatalysts for the ROP of six-membered cyclic carbonates is currently receiving more attention due to the versatility of the process and ease of purification compared with metal-based catalyst systems [89]. As an example, the Waymouth research group has extensively investigated the use of various organocatalyts for the ROP of TMC [89a]. Scheme 8.14 shows a series of organocatalysts that were surveyed for the ROP of TMC initiated by benzyl alcohol. These are, two commercially available guanidines, 1,5,7-triazabicyclo-[4.40]dec-5-ene (**6**, TBD) and 7-methyl-1,5,7-triazabicyclo-[4.4.0]dec-5-ene (**7**, MTBD), one amidine base,

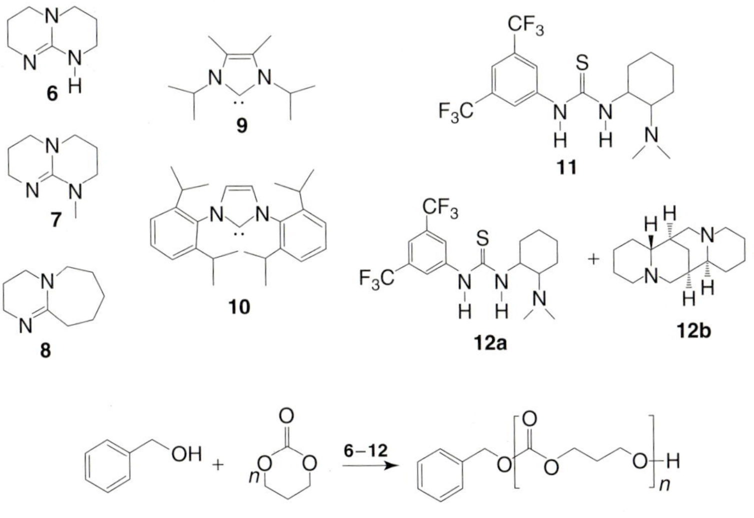

Scheme 8.14 Organocatalysts tested for ROP of TMC initiated by benzyl alcohol.

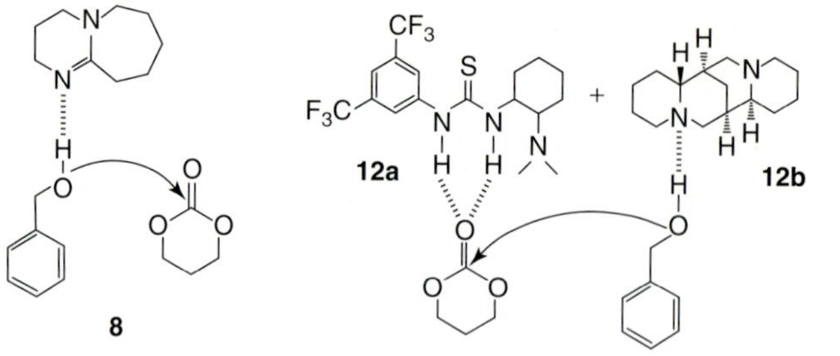

Scheme 8.15 Catalysts **8** and **12** and their mode of action in the polymerization of TMC.

1,8-diazabicyclo[5.4.0]undec-7-ene (**8**, DBU), and two NHCs (*N*-heterocyclic carbene), 1,3-diisopropyl-4,5-dimethyl-imidazol-2-ylidene (**9**) and 1,3-bis(2,6-diisopropylphenyl)-imidazol-2-ylidene (**10**), were investigated as catalysts. Lastly, the bifunctional thiourea–tertiary amine catalyst (**11**) and catalyst mixture (**12**) were also tested as these were found to be successful for catalyzing lactide polymerization. For the solution polymerizations, all catalysts were found to be active, the only difference between them being the polymerization rate. The degree of polymerization and the number average molecular weight closely matched the theoretical values with very low polydispersities for catalysts **8** and **12**. Specifically for catalyst **8**, >99% conversion was reached after 480 min, and polymer with a PDI of 1.04 was obtained. For catalyst **12**, >99% conversion was obtained after 720 min, and polymer with a PDI of 1.07 was achieved. Moreover, analysis of low molecular weight polymers showed no ether linkages and the presence of the anticipated endgroups. Benzyl carbonate was detected at the α-chain position and gave the same integral value as the respective methylene group adjacent to the hydroxyl at the ω-chain position, indicating controlled initiation step and endgroup fidelity. An examination of the control of the polymerization was carried out for catalysts **8** and **12**, and the molecular weight was monitored as a function of monomer conversion. A linear relationship between molecular weight and monomer conversion was observed by ^{1}H-NMR, and the PDI values remained low. Polymerization reactions in bulk using catalyst **8** were performed at 338 K. TMC was polymerized in minutes yielding polymers with controlled molecular weights and endgroup fidelity, and the PDIs were only slightly higher compared with the solution experiments ($\sim$1.09–1.15).

The reaction mechanism was proposed to involve hydrogen-bond activation of monomer and initiator/propagating species. Scheme 8.15 shows the modes of action of catalysts **8** and **12** in the polymerization of TMC.

8.7
Thermoplastic Elastomers and their Biodegradation Processes

Because poly(TMC) produced from the ROP of TMC is a very soft and rubbery amorphous polymer with inappropriate mechanical properties, its use in biomedical applications is limited. However, it is an important component of *thermoplastic*

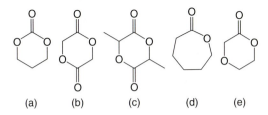

Figure 8.12 Monomers suitable for ring-opening polymerization: (a) TMC, (b) glycolide, (c) lactide, (d) ε-caprolactone, and (e) p-dioxanone.

elastomers obtained by its copolymerization with cyclic esters such as glycolide, lactide, ε-caprolactone, and p-dioxanone (Figure 8.12). These cyclic esters can be polymerized via ROP similarly to ROP of TMC. The copolymerization of TMC with cyclic esters has been thoroughly investigated using metal complexes of Sn [90], Y [91], Ln [83–88, 92], and Ca [93].

As copolymers of TMC and cyclic esters are used in biomedical applications such as sutures, body/dental implants, and tissue engineering, these copolymers should have appropriate mechanical properties and degrade at a specific rate without toxic response. These copolymers normally degrade in two steps (Scheme 8.16) [94]. The first step involves hydrolysis on the hydrolytically unstable backbones of the copolymer. Hydrolysis preferentially occurs in the amorphous phase of the polymer and converts long polymer chains into shorter units. In this step, the physical

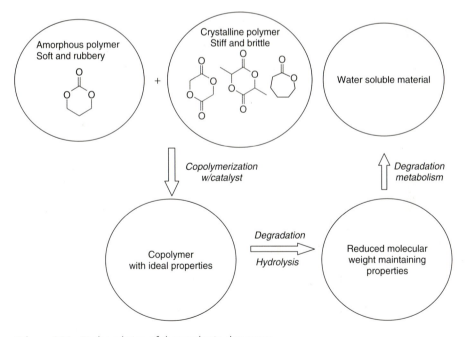

Scheme 8.16 Biodegradation of thermoplastic elastomers.

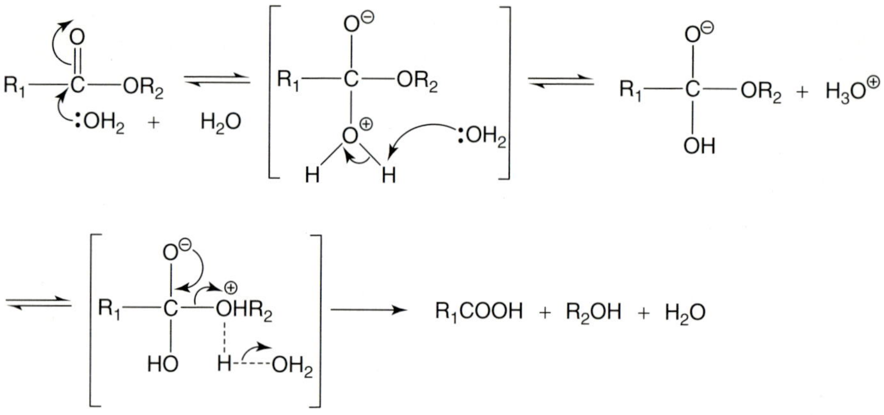

Scheme 8.17 The mechanism for polyester degradation [90a].

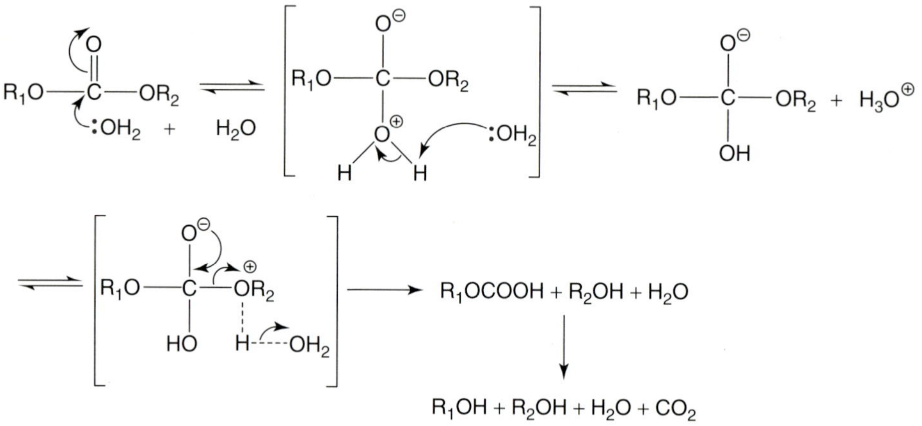

Scheme 8.18 The mechanism for polycarbonate degradation [90a].

properties are maintained by the crystalline regions of the polymer. The hydrolytic degradation mechanism has been discussed by Zhu and coworkers and is shown in Schemes 8.17 and 8.18 for polyester and polycarbonate degradation, respectively [90a]. The second step of biodegradation involves metabolism by enzyme. In this stage, the polymer mass rapidly decreases, and becomes water-soluble material.

Two types of biodegradation are known: bulk erosion and surface erosion [94]. *Bulk erosion* occurs when the rate of water penetration exceeds that at which the polymer is converted into water-soluble materials. The second type of degradation is called *surface erosion*, and occurs when the rate at which the polymer penetrates the medical device is slower than the rate of conversion of the polymer into water-soluble materials.

The need for biomaterials with appropriate mechanical and degradability properties is an area under much current investigation. Mechanical properties are closely related to the hydrophilicity of the polymers, their crystallinity, melt/glass transition temperatures, controlled molecular weights, molecular weight distributions, endgroups, and sequence distribution (random or block) [95]. Therefore, these can be tuned by selecting the appropriate monomer, catalyst (with low toxic response), and reaction conditions. Polymer degradability properties can be tuned by selecting monomers that would give hydrolytically unstable linkages in the final polymers [94].

8.8
Concluding Remarks

In this chapter we have reviewed some of the essential aspects of the synthesis of renewable six-membered cyclic carbonates and the diverse mechanistic aspects of their ROP processes to afford useful biodegradable polycarbonates. The field of the ROP of this important class of cyclic carbonates will continue to flourish, due to the usefulness of the transformation, namely, it is thermodynamically allowed at all temperatures. Additionally, the produced aliphatic polycarbonates are important components of thermoplastic elastomers, which have a wide variety of potential medical applications. Fundamental studies of the transformations presented in this chapter hopefully will provide the knowledge and stimulus for the development of a large scale production of valuable biomaterials via these processes, concomitantly providing safe and stable catalytic systems.

Acknowledgments

The authors' original research on the ROP of cyclic carbonates and the coupling of carbon dioxide and oxetanes has been funded over the years by the U.S. National Science Foundation and the R. A. Welch Foundation of Texas (A.0923).

References

1. (a) Penco, M., Donetti, R., Mendichi, R., and Ferruti, P. (1998) *Macromol. Chem. Phys.*, **199**, 1737; (b) Smith, A. and Hunneyball, I.M. (1986) *Int. J. Pharm.*, **30**, 215; (c) Pêgo, A.P., Siebum, B., Van Luyn, M.J.A., Gallego y Van Seijen, X.J., Poot, A.A., Grijpma, D.W., and Feijen, J. (2003) *Tissue Eng.*, **9**, 981; (d) Marler, J.J., Upton, J., Langer, R., and Vacanti, J.P. (1998) *Adv. Drug Deliv. Rev.*, **33**, 165; (e) Pêgo, A.P., Van Luyn, M.J.A., Brouwer, L.A., van Wachem, P.B., Poot, A.A., Grijpma, D.W., and Feijen, J. (2003) *J. Biomed. Mater. Res. A*, **67A**, 1044.
2. Brossmer, C. and Arntz, D. (2000) US Patent 6, 140, 543.
3. Perosa, A. and Tundo, P. (2005) *Ind. Eng. Chem. Res.*, **44**, 8535.
4. (a) Schlaf, M., Ghosh, P., Fagan, P.J., Hauptman, E., and Bullock, R.M. (2001) *Angew. Chem. Int. Ed.*, **40**, 3887; (b) Schlaf, M., Ghosh, P., Fagan, P.J., Hauptman, E., and Bullock, R.M. (2009) *Adv. Synth. Catal.*, **351**, 789; (c) Ghosh, P., Fagan, P.J., Marshall, W.J., Hauptman, E., and Bullock, R.M. (2009) *Inorg. Chem.*, **48**, 6490.
5. Che, T.M. (1987) US Patent 4, 642, 394.

6. (a) Schlaf, M. (2006) *Dalton Trans.*, 4645 (b) Behr, A., Eilting, J., Irawadi, K., Leschinski, J., and Lindner, F. (2008) *Green Chem.*, **10**, 13.
7. Freund, A. (1881) *Monatsh. für Chem.*, **2**, 636.
8. Forsberg, C.W. (1987) *Appl. Environ. Microbiol.*, **53**, 639.
9. Kretschmann, J., Carduck, F.J., Deckwer, W.D., Tag, C., and Biebl, H. (1993) US Patent 5, 254, 467.
10. Zeng, A.P. and Biebl, H. (2002) *Adv. Biochem. Eng. Biotechnol.*, **74**, 239.
11. Nakamura, C.E. and Whited, G.M. (2003) *Curr. Opin. Biotechnol.*, **14**, 454.
12. Tullo, A. (2009) *Chem. Eng. News*, **87** (21), 9.
13. (a) Zhao, Y.N., Chen, G., and Yao, S.J. (2006) *Biochem. Eng. J.*, **32**, 93; (b) Zheng, P., Sun, J., van den Heuvel, J., and Zeng, A.P. (2006) *J. Biotechnol.*, **125**, 462.
14. http://www.duponttateandlyle.com Dupont Tate and Lyle Bioproducts (accessed 27 Oct. 2010).
15. Cui, H., Wang, T., Wang, F., Gu, C., Wang, P., and Dai, Y. (2003) *Ind. Eng. Chem. Res.*, **42**, 3865.
16. Yu, Z., Xu, L., Wei, Y., Wang, Y., He, Y., Xia, Q., Zhang, X., and Liu, Z. (2009) *Chem. Commun.*, 3934.
17. (a) Aresta, M. and Dibenedetto, A. (2007) *Dalton Trans.*, 2975; (b) Yu, K.M.K., Curcic, I., Gabriel, J., and Tsang, S.C.E. (2008) *ChemSusChem*, **1**, 893.
18. Wu, X.L., Meng, Y.Z., Xiao, M., and Lu, Y.X. (2006) *J. Mol. Catal. A, Chem.*, **249**, 93.
19. Ballivet-Tkatchenko, D., Chambrey, S., Keiski, R., Ligabue, R., Plasseraud, L., Richard, P., and Turunen, H. (2006) *Catal. Today*, **115**, 80.
20. (a) Aresta, M., Dibenedetto, A., Pastore, C., Pápai, I., and Schubert, G. (2006) *Top. Catal.*, **40**, 71; (b) Aresta, M., Dibenedetto, A., Nocito, F., and Pastore, C. (2006) *J. Mol. Catal. A Chem.*, **257**, 149.
21. Fukuoka, S., Tojo, M., Hachiya, H., Aminaka, M., and Hasegawa, K. (2007) *Polym. J.*, **39**, 91.
22. Darensbourg, D.J. and Holtcamp, M.W. (1996) *Coord. Chem. Rev.*, **153**, 155.
23. Coates, G.W. and Moore, D.R. (2004) *Angew. Chem. Int. Ed.*, **43**, 6618.
24. Darensbourg, D.J. (2007) *Chem. Rev.*, **107**, 2388.
25. (a) Meléndez, J., North, M., and Pasquale, R. (2007) *Eur. J. Inorg. Chem.*, 3323; (b) North, M. and Pasquale, R. (2009) *Angew. Chem. Int. Ed.*, **48**, 2946.
26. Carothers, W.H. and Van Natta, F.J. (1930) *J. Am. Chem. Soc.*, **52**, 314.
27. Albertsson, A.C. and Sjoling, M. (1992) *J. Macromol. Sci. A Pure Appl. Chem.*, **29**, 43.
28. Muller, K.R., Buchholz, B., and Hess, J. (1993) US Patent 5, 212, 321.
29. Matsuo, J., Aoki, K., Sanda, F., and Endo, T. (1998) *Macromolecules*, **31**, 4432.
30. Slomkowski, S. and Duda, A. (1993), In: *In Ring-Opening Polymerization* (ed. D.J. Brunelle), Hanser Publisher, New York, p. 87.
31. Clements, J.H. (2003) *Ind. Eng. Chem. Res.*, **42**, 663.
32. Rokicki, G. (2000) *Prog. Polym. Sci.*, **25**, 259.
33. Harris, R.F. (1989) *J. Appl. Polym. Sci.*, **37**, 183.
34. Harris, R.F. and McDonald, L.A. (1989) *J. Appl. Polym. Sci.*, **37**, 1491.
35. Kéki, S., Török, J., Deák, G., and Zsuga, M. (2001) *Macromolecules*, **34**, 6850.
36. Lee, J.C. and Litt, M.H. (2000) *Macromolecules*, **33**, 1618.
37. Soga, K., Tazuke, Y., Hosoda, S., and Ikeda, S. (1977) *J. Polym. Sci.*, **15**, 219.
38. Storey, R.F. and Hoffman, D.C. (1992) *Macromolecules*, **25**, 5369.
39. Vogdanis, L., Martens, B., Uchtmann, H., Hensel, F., and Heitz, W. (1990) *Makromol. Chem.*, **191**, 465.
40. Albertsson, A.C. and Eklund, M. (1994) *J. Polym. Sci. A Polym. Chem.*, **32**, 265.
41. Kricheldorf, H.R., Dunsing, R., and Serra i Albet, A. (1987) *Makromol. Chem.*, **188**, 2453.
42. Kricheldorf, H.R. and Jenssen, J. (1989) *J. Macromol. Sci. Chem.*, **A26**, 631.
43. (a) Kricheldorf, H.R. and Weegen-Schulz, B. (1993) *Macromolecules*, **26**, 5991; (b) Kricheldorf, H.R. and Weegen-Schulz, B. (1995) *J. Polym. Sci. A Polym. Chem.*, **33**, 2193.

44. Ariga, T., Takata, T., and Endo, T. (1997) *Macromolecules*, **30**, 737.
45. Matsuo, J., Nakano, S., Sanda, F., and Endo, T. (1998) *J. Polym. Sci. A Polym. Chem.*, **36**, 2463.
46. Liu, J., Zhang, C., and Liu, L. (2007) *J. Appl. Polym. Sci.*, **107**, 3275.
47. Hovestadt, W., Müller, A.J., Keul, H., and Höcker, H. (1990) *Makromol. Chem. Rapid Commun.*, **11**, 271.
48. Kühling, S., Keul, H., and Höcker, H. (1990) *Makromol. Chem.*, **191**, 1611.
49. (a) Kühling, S., Keul, H., Höcker, H., Buysch, H.J., and Schön, N. (1991) *Makromol. Chem.*, **192**, 1193; (b) Wurm, B., Keul, H., Höcker, H., Sylvester, G., Leitz, E., and Ott, K.H. (1992) *Macromol. Chem. Rapid Commun.*, **13**, 9.
50. Takata, T., Sanda, F., Ariga, T., Nemoto, H., and Endo, T. (1997) *Macromol. Rapid Commun.*, **18**, 461.
51. Weilandt, K.D., Keul, H., and Höcker, H. (1996) *Macromol. Chem. Phys.*, **197**, 3851.
52. Keul, H., Bächer, R., and Höcker, H. (1986) *Makromol. Chem.*, **187**, 2579.
53. Endo, T., Suzuki, T., Sanda, F., and Takata, T. (1996) *Macromolecules*, **29**, 3315.
54. Endo, T., Suzuki, T., Sanda, F., and Takata, T. (1996) *Macromolecules*, **29**, 4819.
55. Höcker, H. and Keul, H. (1994) *Adv. Mater.*, **6**, 21.
56. Dainton, F.S. and Irvin, K. (1958) *Q. Rev.*, **12**, 61.
57. Murayama, M., Sanda, F., and Endo, T. (1998) *Macromolecules*, **31**, 919.
58. Kobayashi, S., Kikuchi, H., and Uyama, H. (1997) *Macromol. Rapid Commun.*, **18**, 575.
59. Bisht, K.S., Svirkin, Y.Y., Henderson, L.A., Gross, R.A., Kaplan, D.L., and Swift, G. (1997) *Macromolecules*, **30**, 7735.
60. Feng, J., He, F., and Zhuo, R. (2002) *Macromolecules*, **35**, 7175.
61. He, F., Wang, Y., Feng, J., Zhuo, R., and Wang, X. (2003) *Polymer*, **44**, 3215.
62. Al-Azemi, T.F., Harmon, J.P., and Bisht, K.S. (2000) *Biomacromolecules*, **1**, 493.
63. Matsumura, S., Harai, S., and Toshima, K. (2001) *Macromol. Rapid Commun.*, **22**, 215.
64. Matsumura, S., Harai, S., and Toshima, K. (2000) *Macromol. Chem. Phys.*, **201**, 1632.
65. Tasaki, H., Toshima, K., and Matsumura, S. (2003) *Macromol. Biosci.*, **3**, 436.
66. Matsumura, S., Tsukada, K., and Toshima, K. (1997) *Macromolecules*, **30**, 3122.
67. Kricheldorf, H.R., Jenssen, J., and Kreiser-Saunders, I. (1991) *Makromol. Chem.*, **192**, 2391.
68. Nijenhuis, A.J., Grijpma, D.W., and Pennings, A.J. (1992) *Macromolecules*, **25**, 6419.
69. Kricheldorf, H.R., Kreiser-Saunders, I., and Boettcher, C. (1995) *Polymer*, **36**, 1253.
70. Krichedrof, H.R. and Stricker, A. (2000) *Polymer*, **41**, 7311.
71. Kricheldorf, H.R., Behnken, G., Schwarz, G., Simon, P., and Brinkmann, M. (2009) *J. Macromol. Sci. A Pure Appl. Chem.*, **46**, 353.
72. Hovestadt, W., Keul, H., and Höcker, H. (1992) *Polymer*, **33**, 1941.
73. Yang, J., Yu, Y., Li, Q., Li, Y., and Cao, A. (2005) *J. Polym. Sci. A Polym. Chem.*, **43**, 373.
74. (a) Darensbourg, D.J., Ganguly, P., and Billodeaux, D. (2005) *Macromolecules*, **38**, 5406; (b) Darensbourg, D.J., Ganguly, P., and Billodeaux, D. (2006) *Macromolecules*, **39**, 2722.
75. (a) Baba, A., Kashiwagi, H., and Matsuda, H. (1985) *Tetrahedron Lett.*, **26**, 1323; (b) Baba, A., Kashiwagi, H., and Matsuda, H. (1987) *Organometallics*, **6**, 137; (c) Baba, A., Meishou, H., and Matsuda, H. (1984) *Makromol. Chem. Rapid Commun.*, **5**, 665; (d) Koinuma, H. and Hirai, H. (1977) *Makromol. Chem.*, **178**, 241; (e) Pritchard, J.G. and Long, F.A. (1958) *J. Am. Chem. Soc.*, **80**, 4162.
76. Darensbourg, D.J., Ganguly, P., and Choi, W. (2006) *Inorg. Chem.*, **45**, 3831.
77. (a) Darensbourg, D.J., Moncada, A.I., Choi, W., and Reibenspies, J.H. (2008) *J. Am. Chem. Soc.*, **130**, 6523;

78. Kricheldorf, H.R., Kreiser-Saunders, I., and Damrau, D.O. (2000) *Macromol. Symp.*, **159**, 247.
79. Ghosh, S., Sharma, A., and Talukder, G. (1992) *Biol. Trace Elem. Res.*, **35**, 247.
80. Darensbourg, D.J., Choi, W., Ganguly, P., and Richers, C.P. (2006) *Macromolecules*, **39**, 4374.
81. Dobrzynski, P., Pastusiak, M., and Bero, M. (2005) *J. Polym. Sci. A Polym. Chem.*, **43**, 1913.
82. Helou, M., Miserque, O., Brusson, J.M., Carpentier, J.F., and Guillaume, S.M. (2008) *Chem. Eur. J.*, **14**, 8772.
83. Zhao, B., Lu, C.R., and Shen, Q. (2007) *J. Appl. Polym. Sci.*, **106**, 1383.
84. Ling, J., Shen, Z., and Huang, Q. (2001) *Macromolecules*, **34**, 7613.
85. Zhou, L., Yao, Y., Zhang, Y., Xue, M., Chen, J., and Shen, Q. (2004) *Eur. J. Inorg. Chem.*, 2167.
86. Li, C., Wang, Y., Zhou, L., Sun, H., and Shen, Q. (2006) *J. Appl. Polym. Sci.*, **102**, 22.
87. Agarwal, S. and Puchner, M. (2002) *Eur. Polym. J.*, **38**, 2365.
88. Palard, I., Schappacher, M., Belloncle, B., Soum, A., and Guillaume, S.M. (2007) *Chem. Eur. J.*, **13**, 1511.
89. (a) Nederberg, F., Lohmeijer, B.G.G., Leibfarth, F., Pratt, R.C., Choi, J., Dove, A.P., Waymouth, R.M., and Hedrick, J.L. (2007) *Biomacromolecules*, **8**, 153; (b) Kamber, N.E., Jeong, W., Waymouth, R.M., Pratt, R.C., Lohmeijer, B.G.G., and Hedrick, J.L. (2007) *Chem. Rev.*, **107**, 5813; (c) Mindemark, J., Hilborn, J., and Bowden, T. (2007) *Macromolecules*, **40**, 3515.
90. (a) Cai, J., Zhu, K.J., and Yang, S.L. (1998) *Polymer*, **39**, 4409; (b) Ruckenstein, E. and Yuan, Y. (1998) *J. Appl. Polym. Sci.*, **69**, 1429; (c) Kricheldorf, H.R. and Stricker, A. (1999) *Macromol. Chem. Phys.*, **200**, 1726; (d) Pospiech, D., Komber, H., Jehnichen, D., Häussler, L., Eckstein, K., Scheibner, H., Janke, A., Kricheldorf, H.R., and Petermann, O. (2005) *Biomacromolecules*, **6**, 439; (e) Tyson, T., Finne-Wistrand, A., and Albertsson, A.C. (2009) *Biomacromolecules*, **10**, 149.
91. Simic, V., Pensec, S., and Spassky, N. (2000) *Macromol. Symp.*, **153**, 109.
92. (a) Tsutsumi, C., Nakagawa, K., Shirahama, H., and Yasuda, H. (2003) *Polym. Int.*, **52**, 439; (b) Nakayama, Y., Yasuda, H., Yamamoto, K., Tsutsumi, C., Jerome, R., and Lecomte, P. (2005) *React. Funct. Polym.*, **63**, 95; (c) Tsutsumi, C., Yamamoto, K., Ichimaru, A., Nodono, M., Nakagawa, K., and Yasuda, H. (2003) *J. Polym. Sci. A Polym. Chem.*, **41**, 3572.
93. (a) Darensbourg, D.J., Choi, W., and Richers, C.P. (2007) *Macromolecules*, **40**, 3521; (b) Darensbourg, D.J., Choi, W., Karroonnirun, O., and Bhuvanesh, N. (2008) *Macromolecules*, **41**, 3493.
94. Middleton, J.C. and Tipton, A. (1998) *Med. Plast. Biomater.*, **2**, 30.
95. Pêgo, A.P., Poot, A.A., Grijpma, D.W., and Feijen, J. (2003) *J. Mater. Sci. Mater. Med.*, **14**, 767.

9
Poly(lactide)s as Robust Renewable Materials
Jan M. Becker and Andrew P. Dove

9.1
Introduction

Polymeric materials are commonly used in a wide range of everyday applications. Since the dawn of the petrochemical age, the decreased weight and manufacturing costs, in conjunction with increased flexibility, durability and lifespan, have led to plastics replacing more traditional materials, such as wood, paper, or metal. Furthermore, the ability to manipulate the physical and mechanical properties of *polymeric materials* through structural changes enables them to fulfill a wide range of applications. Despite these many advantages, both the derivation of these products from depleting resources, typically oil, and their extremely long persistence times compared with useful lifetime [poly(ethylene) packaging is typically used for hours or days yet pervades the environment for decades] will eventually lead to reduced accessibility to monomer feedstocks and needless pollution in land-fill and by release of gases that are harmful to the environment after burning.

Rising oil prices and governmental policies are resulting in a drive towards the investigation of alternate polymeric materials [1]. Of the materials currently under investigation, *poly(lactic acid)/poly(lactide)* (PLA) provides a potential solution. Derived from lactic acid (2-hydroxy propionic acid), a natural product commonly found in the body, the polymer is highly biocompatible and bioresorbable [2–6]. It therefore is not surprising that the original applications of PLA were in the biomedical field as, amongst others, degradable sutures, and fracture fixation pins [4–6]. Improvements in the synthesis of PLA over the past decade, primarily from improved fermentation methodologies to convert corn starch into lactic acid, and the resulting lower costs, have led to the study of PLA for a wider range of applications, such as packaging materials for food and beverages, plastic bags, thin film coatings, and rigid thermoforms [7, 8].

Green Polymerization Methods: Renewable Starting Materials, Catalysis and Waste Reduction
Edited by Robert T. Mathers and Michael A. R. Meier
Copyright © 2011 WILEY-VCH Verlag GmbH & Co. KGaA, Weinheim
ISBN: 978-3-527-32625-9

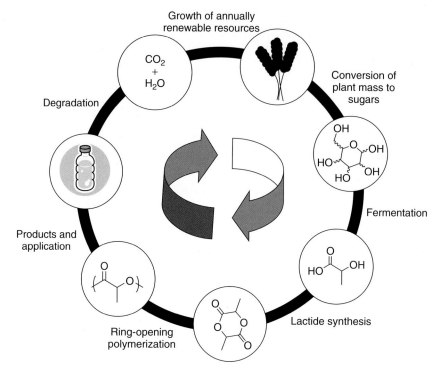

Figure 9.1 The lifecycle of poly(lactide). Adapted from Auras et al. [7].

9.1.1
The Lactide Cycle

As a renewable polymer, the *lifecycle of PLA* (Figure 9.1) can be envisaged as several fermentative/chemical steps to yield polymeric products from renewable resources that will eventually break back down into carbon dioxide and water, which will feed the growth of the next generation of crops, from which more PLA may be derived. Lactic acid can be generated from renewable sources such as corn, sugar beets, and almost any other carbohydrate-containing feedstock by one of two bacterial fermentations methods, classified according to the bacteria used in the process [9–11]. The *heterofermantative process* is the least efficient with commonly <1.8 mol of lactic acid per mol of hexose being produced alongside a large range of other metabolites, such as smaller sugars, acetic acid, and CO_2. In contrast, the *homofermentative process* tends to yield an average of >1.8 mol of lactic acid per mol of hexose with only minor levels of contaminants. In quantitative terms, this equates to >90 g of lactic acid per 100 g sugar, resulting in a commercially viable process [12]. It should be noted that both isomers of the lactic acid [D(+) and L(−)] (Scheme 9.1) can now be readily accessed via this methodology. For use in more demanding applications such as in food or in pharmaceutical grade PLA, the lactic acid synthesized in this manner is further purified via distillation.

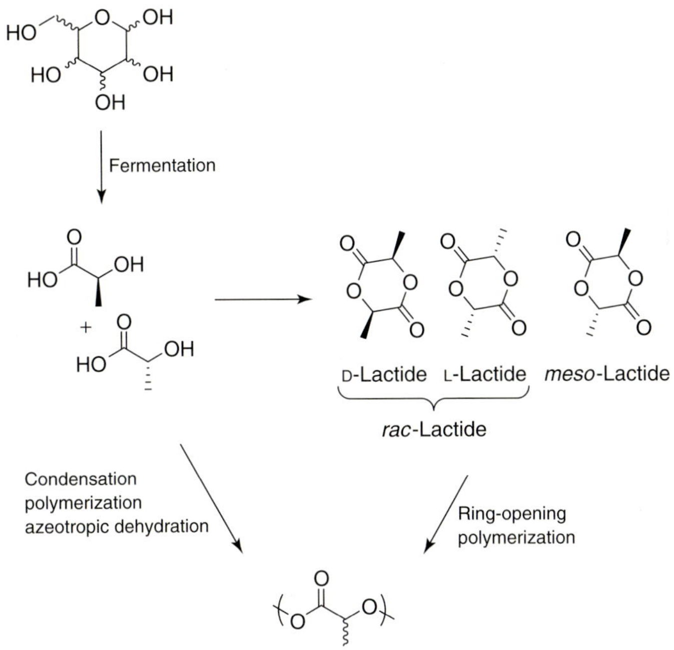

Scheme 9.1 Synthesis of PLA by direct condensation polymerization or azeotropic distillation, or by ring-opening polymerization of lactide.

Synthesis of *high molecular weight PLA* (M_w about 100 000 g mol^{-1}) can be achieved in three ways: direct condensation polymerization; azeotropic dehydration; and ring-opening polymerization (ROP) of lactide (the cyclic diester of lactic acid) (Scheme 9.1). The high energy requirements and the lack of control over the polymer properties in both the direct condensation polymerization and the azeotropic dehydration preclude their use on a large scale [13–18]. PLAs produced using direct condensation polymerization are difficult to obtain as solvent-free high molecular weight polymers. Furthermore, the materials often contain residues of chain transfer agents and other additives, therefore rendering them unusable in food-grade or medical applications. In the case of the azeotropic dehydration polymerization, heating the lactic acid to about 130 °C for several hours before distilling out the bulk of the condensation water under reduced pressure prevents many of these problems. The second step of this process requires addition of catalyst and diphenyl ester before passing the solution through a column packed with activated molecular sieves [7].

Advances in the ROP of lactide have led to this process being the most common *large-scale method for the synthesis of PLAs*. In this process the respective lactic acid is prepolymerized to yield a low molecular weight PLA, which can then be catalytically converted into a mixture of lactide stereoisomers by transesterification [4]. The three stereoisomers of lactide, the cyclic dimer of lactic acid (Scheme 9.1)

L-lactide (S,S), D-lactide (R,R) (commercially available in both enantiopure forms and as a racemic mixture), and *meso*-lactide (R,S), can be synthesized using either homochiral lactic acid or a racemic mixture. The lactides can be purified by distillation or recrystallization before being polymerized via ROP methods using a range of initiators and catalysts (*vide infra*).

PLAs are readily degradable both in the body and in bulk. In physiological media the polymers degrade by an acidic hydrolysis mechanism (of the ester bonds) and the by-products are further removed as CO_2 and water through the Krebs cycle [19]. Industrially, PLAs are compostable, also resulting in *degradation* ultimately into CO_2 and water. PLA degrades via a two-step bulk erosion mechanism, whereby the initial molecular weight of the polymer chains decreases, most probably as a result of random chain scission throughout the polymer matrix [4, 20, 21]. Further degradation is autocatalyzed by the formation of carboxylic acid groups within the polymer matrix [22]. The resultant low molecular weight species can then diffuse to the surface of the bulk and be further degraded biologically into CO_2, water, and humus [7]. This process normally occurs over a period of several months, but can be accelerated by simple acid or base catalysis. Degradation can also take place through the action of a range of enzymes such as proteinase K, pronase, and bromelain, and chemically through both transesterification and hydrolysis in either acidic or basic solutions. The stability and therefore the overall degradation rate of PLA is dependent on many factors, including physical/material properties, such as molecular weight, polymer microstructure, particle size, and shape, and so on, in addition to environmental conditions, such as temperature, pH, humidity, and so on. As may be expected, less crystalline materials (lower molecular weight, non-stereoregular) and more forcing conditions increase the degradation rate and vice versa.

Despite the ultimate release of CO_2 into the environment, the derivation from a renewable feedstock, which consumes CO_2 during the growth period, is argued to make the materials "*carbon neutral*" [23]. Additionally, Tsuji *et al.* have shown that poly(L-lactide) (PLLA) can be hydrolyzed within a period of 30 min at a temperature between 180 and 350 °C to yield L-lactic acid of high enantiopurity in excellent yield, thus showing a potential methodology for recycling PLA without the release of CO_2 during the process [24].

9.2
Ring-Opening Polymerization of Lactide

Ring-opening polymerization of lactide enables the control over many different physical, thermal, and mechanical properties of PLA. ROP enables high levels of control to be exerted over the polymerization, facilitating the synthesis of PLAs with predictable molecular weights (based on the [monomer]/[initiator]), narrow polydispersities and endgroup fidelity, in addition to the ability to synthesize block copolymers. Furthermore, the control of polymer *tacticity* by addition of monomer units based on their stereochemistry has also been realized using this methodology.

While enzymatic ROP of lactide has led to limited success, with poor control over the polymerization being reported, the advancement in metal and organic catalysis of the ring-opening process has led to a range of methods for the synthesis of PLA being reported [19, 25, 26].

9.2.1
Coordination–Insertion Polymerization

The use of *metal complexes in polymerization chemistry* has an extensive history. The most commonly used catalyst in both small-scale synthesis and large-scale manufacture of PLA is tin(II) bis(2-ethylhexanoate), more commonly referred to as *tin(II) octanoate*, $Sn(Oct)_2$. Many other metal salts including aluminum complexes, for example, $Al(O^iPr)_3$, or zinc complexes, such as zinc lactate, have also been used for industrial-scale preparations [19]. ROP mediated by these metal complexes operates via a *coordination–insertion mechanism* in which coordination of the carbonyl oxygen of the lactide monomer to the Lewis acidic metal center is followed by insertion into the existing metal-alkoxide bond, with subsequent ring-opening resulting in a chain extended alcohol (Scheme 9.2).

The high solubility of $Sn(Oct)_2$ in molten monomer/polymer along with its highly robust nature, regularity approval, good activity, and easy to use nature, have led to it become a commonly applied system. Control over the polymerization is highly dependant on the molar ratio of $Sn(Oct)_2$ to initiating alcohol and reaction temperature, such that at temperatures below 120 °C and with molar ratios of initiating alcohol relative to $Sn(Oct)_2$ in the range of 400 : 150, polymers can be produced with excellent molecular weight control and narrow polydispersities [27, 28]. Under less carefully controlled conditions, the *polydispersities of the resultant polymers* are broader as a result of notable inter- and intra-molecular transesterification side reactions. Notably, the mechanism of lactide ROP mediated by $Sn(Oct)_2$ has been the subject of much debate, especially as the control and activity of this system can be increased by the presence of protic additives such as alcohols.

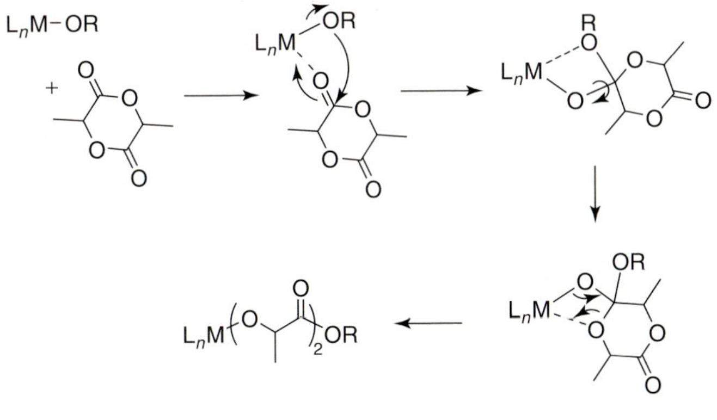

Scheme 9.2 General coordination–insertion mechanism for the ring-opening polymerization of lactide.

Both cationic or activated-monomer mechanisms have been proposed [29, 30], however most of the evidence suggests that ROP mediated by Sn(Oct)$_2$ operates via a coordination–insertion mechanism [31–33]. While in the presence of excess alcoholic initiator it is generally accepted that through reaction of Sn(Oct)$_2$ the corresponding metal alkoxide species Sn(OR)$_2$ will be formed [34, 35], this process is very much dependent on the environmental factors (temperature, solvent, reagent ratios, etc.). Nonetheless, in the presence of these initiating species, the polymer chain-ends will display those of the initiating alcohol.

Both coordination of the alcohol to the metal via an associative [36] or a dissociative [37–40] mechanism with respect to the octanoate ligand has been proposed. As a consequence of a theoretical study for a model system of MeOH and a tin biscarboxylate (SnOct′), in which the 2-ethylhexanoate ligands were replaced by acetoxy groups, a mechanism in which the lactide monomer is activated by coordination to the metal center, with the alcoholic initiator/propagating chain-end being coordinated to both the metal center, and the octanoato ligand is favored (Scheme 9.3) [41]. The mechanism predicts *facile chain transfer* and while the involvement of the octanoato ligand is implicated, octanoic acid is noted to be able to dissociate under polymerization conditions.

Quenching of the polymerization with a protic source results in the realization of a polymer with a hydroxyl chain-end, although other species including acid chlorides or isocyanates have been applied to further manipulate the chain-end chemistry [42–44]. Despite the release of the metal complex, metal residues typically remain trapped inside the polymer matrix, thus leading to environmental concerns both in disposal of waste from the process and after disposal of the polymer, due to appreciable leaching of the metal into the surrounding water table. Additionally, release of metal residues in medical applications is undesirable. While low levels

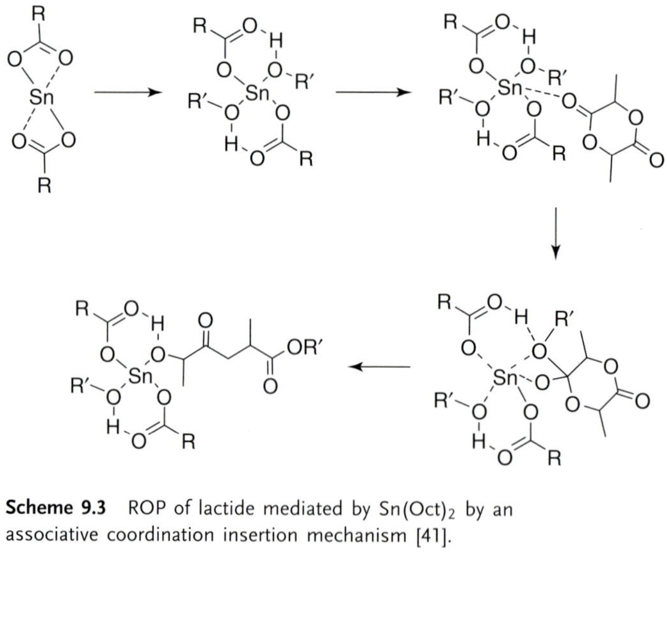

Scheme 9.3 ROP of lactide mediated by Sn(Oct)$_2$ by an associative coordination insertion mechanism [41].

of Sn(Oct)$_2$ present in PLAs are FDA (the U.S. Food and Drug Administration) approved for use in *in vivo* applications [19], concerns remain about the biological and environmental impact of tin(II). To this end, several studies are investigating less toxic metal salts and complexes including zinc, iron, and calcium and also the development of ultra-high productivity single-site metal-based coordination compounds to both improve activities and realize stereoregular polymers from *rac*- and *meso*-lactide [19, 25, 45].

Many single-site *metallo-organic catalysts* have been devised for the stereospecific ROP of lactide [45]. (Salen/salan)Aluminum, (β-diketiminatato)zinc [46–48], tris(pyrazolyl)calcium [49–51], amino-alkoxy(bisphenolate)yttrium [52–56], bis(phenolato)scandium [57–59] complexes (Figure 9.2) under solution conditions

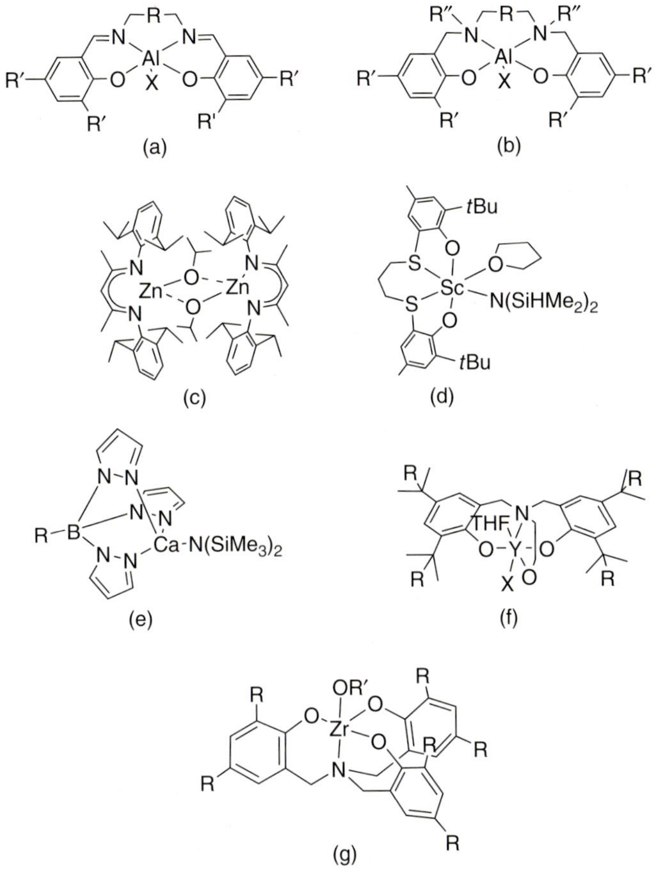

Figure 9.2 Metal-based catalysts for the stereospecific ring-opening polymerization of lactide: (a) (salen)aluminum; (b) (salan)aluminum; (c) (β-diketiminatato)zinc; (d) bis(phenolato)scandium; (e) tris(pyrazolyl)calcium; (f) amino-alkoxy(bisphenolate)yttrium; and (g) amine(trisphenolato)zirconium complexes.

resulted in the highly heterospecific ROP of *rac*-lactide, and several other (salen/salan)aluminum complexes have been reported to be highly efficient for isospecific polymerization to produce stereoblock "isotactic" PLA. While many of these complexes are only effective in solution at temperatures around or below 70 °C, Feijen and coworkers demonstrated that an aluminum complex supported by a single enantiomer of Jacobsens ligand would effectively mediate the isospecific ROP of lactide under industrially relevant melt conditions (*probability of isotactic enchainment*, $P_m = 0.88$) [60]. Further studies from Nomura et al. revealed similar catalysts with flexible aliphatic propyl imine linkers and sterically demanding *ortho-tert*-butyldimethylsilyl phenoxy substituents, resulting in an isoselectivity of 98% in the ROP of *rac*-lactide under melt conditions [61]. Davidson and coworkers have also demonstrated that a range of germanium, zirconium, and hafnium complexes supported by a *C*3-symmetric amine(trisphenolate) ancillary ligand are able to mediate the heterospecific ROP of *rac*-LA (lactide) under both solution and melt conditions at ~130 °C [62, 63]. While in all instances the polymers produced are of predictable molecular weight and narrow polydispersity, the highest levels of stereocontrol are achieved with the Zr analogue (*probability of heterotactic enchainment*, $P_r = 0.98$).

Considerable effort has been dedicated to the development of *highly active catalysts* in order to increase productivities in PLA synthesis and hence reduce the requirement for potentially toxic catalysts. One of the most notable metal-based systems is zinc complexes supported by a trifunctional phenoxyamine ligand [64], converting 650 equiv. *rac*-lactide into atactic PLA, with molecular weights that are slightly lower than predicted based on the monomer : initiator ratio, and polydispersities of about 1.4 within 5 min at room temperature in $CDCl_3$ solution ([catalyst] = 0.7 mM). An alternative approach has seen the investigation of catalysts developed by Carpentier and coworkers. These catalysts have enabled the synthesis of PLAs with high productivities with as little as 5×10^{-4} mol% catalyst [53]. Typically, these metal complexes are based around highly electrophilic metals supported by amino-alkoxy(bisphenolate) ancillary ligands. Furthermore, the yttrium- [53, 54] or lanthanide-based [65] catalysts have additionally demonstrated high levels of stereocontrol.

9.2.2
Organocatalytic Ring-Opening Polymerization

The ultimate reduction in metallic residue can be achieved using organic catalysis. In addition to the elimination of metal-containing impurities, several *organocatalysts* are insensitive to the presence of trace impurities such as water or oxygen, although the presence of water will reduce the control of the polymerization by initiating additional chains [26, 66]. Initial reports of the organocatalyzed ROP of lactide utilized simple nitrogen bases such as 4-dimethylaminopyridine (DMAP) or 4-pyrolidinopyridine (PPY) [67], that were proposed to operate via a "monomer activated" mechanism, although recent theoretical studies have suggested that a general base mechanism may be in operation (Scheme 9.4) [67, 68]. The application

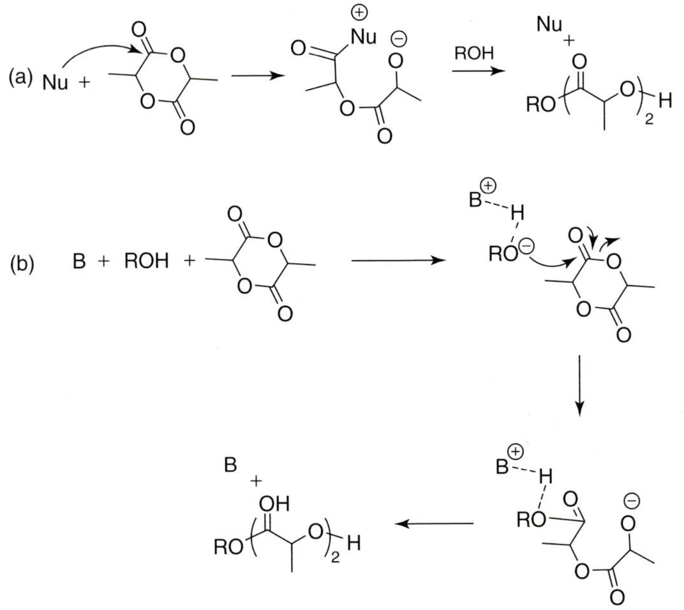

Scheme 9.4 "Monomer activated" (a) and general base (b) mechanisms in organocatalytic ring-opening polymerization of lactide.

of a wide range of nucleophiles has been examined including phosphines [69] and N-heterocyclic carbenes (NHCs) [70, 71], and has enabled the realization of much higher activities, such that NHCs are able to mediate the ROP of LA at ambient temperature in a few minutes [71]. This subsequently enabled the application of sub-stoichiometric amounts of catalyst (relative to initiator) to be applied [71–73]. While NHCs are highly air sensitive, several methods for their *in situ* generation from air-stable precursors have been reported, including silver or acid salts [73, 74], thermally cleavable haloalkane adducts [75], and reversibly cleavable alcohol and amine adducts [72, 76]. Notably the application of hindered NHCs at low temperatures has enabled isoselectivites of up to 90% to be achieved for the ROP of rac-LA at −70 °C in dichloromethane solution [77].

Initial investigations into a *second generation of catalysts* was reported by Dove et al. using a conjoined bifunctional thiourea–tertiary amine compound, which was proposed to mediate ROP by simultaneous activation of both monomer and propagating alcohol by the Lewis acidic and Lewis basic sites, respectively (Scheme 9.5) [78]. Polymerizations demonstrated excellent control and were remarkably resistant to transesterification. Further studies demonstrated that a comparable bimolecular system with an activated thiourea and (−)-sparteine, enabled high DP (degree of polymerization) polymers to be obtained within a few hours [79, 80]. The application of increasingly strong bases resulted in the elimination required for the thiourea cocatalyst with simple commercially available organic compounds, such as 1,8-diazabicyclo[5.4.0]undec-7-ene (DBU) [38],

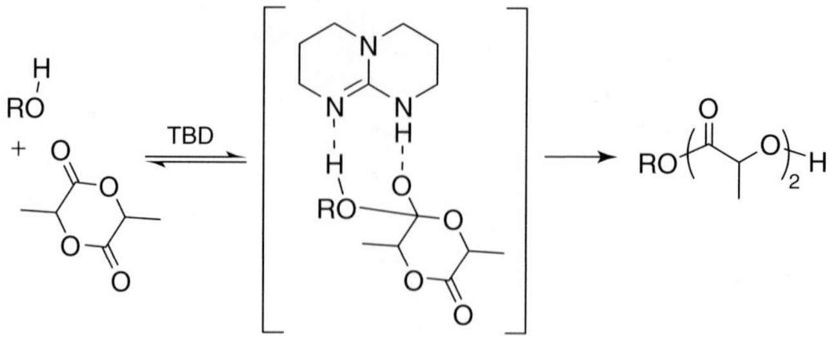

Scheme 9.5 Ring-opening polymerization of lactide catalyzed by 1,5,7-triazabicyclo[4.4.0]dec-5-ene (TBD) [83, 84].

7-methyl-1,5,7-triazabicyclo[4.4.0]dec-5-ene (MTBD) [79], phosphazene bases such as 2-*tert*-butylimino-2-diethylamino-1,3-dimethylperhydro-1,3,2-diazaphosphorine (BEMP), N'-*tert*-butyl-N, N, N', N', N'', N''-hexamethylphosphorimidic triamide (P_1-*t*-Bu) [81], and also the bifunctional 1,5,7-triazabicyclo[4.4.0]dec-5-ene (TBD) [82], being shown to be highly efficient catalysts for the ROP of lactide.

The simple bases (DBU, MTBD, BEMP, and P_1-*t*-Bu) are proposed to operate by a general base catalysis mechanism, whereas the bifunctional guanidine, TBD, activates both monomer and alcohol by stabilization of the transition state of the ring-opening step with hydrogen bonding (Scheme 9.5) [83, 84]. Amongst these catalysts TBD and the analogous P_2-*t*-Bu [1-*tert*-butyl-2,2,4,4,4-pentakis(dimethylamino)-$2\Lambda^5, 4\Lambda^5$-catenadi(phosphazene)] are the most highly active for lactide ROP with TBD consuming 500 equiv. lactide in DCM (dichloromethane) solution, this being complete within 60 s (0.1 mol% TBD to monomer) [82], and P_2-*t*-Bu, in which, with an initial monomer concentration $[M]_0 = 0.32 \, \text{mol l}^{-1}$, 1 mol% catalyst and a target DP $= 100$, LLA was quantitatively polymerized in 10 s [85]. The high activity and sterically hindered nature of P_2-*t*-Bu enabled the stereocontrolled ROP of *rac*-LA with a 95% isoselectivity to be achieved at $-75\,°\text{C}$ [85]. More recently, a further new metal-free approach to the ROP of lactide has been reported using halogen bonding [86].

9.3
Poly(lactide) Properties

PLA has many desirable properties that make it a very interesting material for commodity applications, such as packaging, which is a major consumables market. While the physical properties of PLA are highly dependant on a number of factors (*vide infra*, Section 9.3.1), PLLA possesses many physical and mechanical properties comparable to other major commodity polymers, namely poly(ethylene terephthalate) (PET) and poly(styrene) (PS) (Table 9.1). This makes it possible to market PLA in a wide variety of *packaging and fiber products* [87], as replacements for PET components. The fibrous materials exhibit low odor retention, very good moisture wicking properties [88, 89], and the films, bottles, and thermoformed

Table 9.1 Physical and mechanical properties of poly(L-lactide) (PLLA), poly(styrene) (PS), and poly(ethylene terephthalate) (PET) [7, 88, 90–93].

Property	PLLA	PS	PET
Density (kg m^{-3})	1.26	1.05	1.40
Tensile strength (MPa)	47–1800	45	57
Elastic modulus (GPa)	3.6–7.5	3.2	2.8–4.1
Elongation at break (%)	2–48	3	300
Notched izod (J m^{-1})	19–70	21	59
Heat deflection (°C)	55–66	75	67
Glass transition temperature (°C)	55–60	100	75
Melting temperature (°C)	180	240	260

containers are resistant to oils and fats (hydrophilicity of the polymer), and also display the ability to "block" flavors and aromas [90].

The limited *thermal performance* [measured by parameters such as the glass transition temperature (T_g) or melting temperature (T_m)] and the brittleness/low impact strength of PLA limit its applications where thermal stability, mechanical strength or high elongation are required [7, 87, 94]. In part these properties can be modified by adaptation of the processing conditions to the materials, but several other methods to address these problems are being examined, including reactive extrusion, blending, plasticization, and synthesis of new functional lactide-related cyclic esters or copolymers [90, 95].

9.3.1
PLA Properties and Processing Effects

Several *molecular parameters* have a notable effect on the properties of PLA, including molecular weight, architecture, and microstructure/tacticity. Stereochemically pure (D- or L-) PLAs result in semi-crystalline materials with a T_m of about 180 °C and a T_g of between 55 and 60 °C [8, 19, 25, 96]. Reduction of the stereoregularity, even by the introduction of only 15% *meso*-lactide, results in an amorphous "atactic" polymer with no discernable T_m and a lower T_g [8, 97]. The stereopurity, and hence degree of crystallinity, of the samples also significantly affects the mechanical properties of the materials [92]. Perego *et al.* demonstrated that PLLAs with molecular weights (measured by viscosity) of from 20 to 66 kg mol^{-1} and PDLLAs [poly(D,L-lactic acid)] of between 47.5 and 114 kg mol^{-1} displayed different tensile properties, such that the tensile strengths ranged from 55 to 59 MPa for the PLLA samples, with Young's modulus values of around 3.6–4.2 GPa, and elongations at break of between 1.3 and 7%. In comparison, the PDLLA samples displayed tensile strengths of between 40 and 44 MPa, elongations at break of between 4.8 and 7.5%, but with comparable Young's modulus values. Annealing the samples results in increased crystallinity, and hence for PLLA the tensile strength measurements are between 47 and 66 MPa with increased modulus values of up to 4.1 GPa, although elongation at break reduces slightly to 1.3–4% [92].

Of the other stereoregular PLA microstructures, the most pronounced changes to polymer materials properties occur upon *stereocomplex formation*, either from a 1 : 1 mixture of PLLA and PDLA, or by the synthesis of stereoblock copolymers from *rac*-lactide. The complimentary supramolecular interactions between the two helical homochiral polymers results in increased crystallinity in the polymer and a resultant increase in T_m of about 50 °C (to about 220–230 °C) [96]. Notably, further enhanced melt stability of PLA stereocomplexes can be achieved by modifying the architecture. Biela *et al.* demonstrated that star-shaped PLA stereocomplexes with a greater number of arms were able to survive melting, whereas those with fewer arms displayed homochiral and stereocomplex melting-points upon post-melting thermal analysis [98]. The greater levels of crystallinity of the stereocomplex also lead to enhanced mechanical properties, such that stereocomplex materials display Young's modulus values of about 8.6 GPa, tensile strength = 0.88 GPa, and elongations at break of up to 30%, with rheological studies demonstrating that the stereocomplex crystallites enhanced the heat-resistance of the material [96]. The increased crystallinity also leads to higher hydrolysis resistance and thus retention of its tensile properties for much longer than the homochiral polymers.

The *processing conditions* also have notable effects on the resulting material properties. Studies have shown that by using orientation-inducing processing conditions, such as drawing–injection moulding of amorphous PLA, the resulting samples show a higher tensile strength and an increased impact strength when compared with samples prepared by conventional *injection moulding* [91, 93, 99]. Bigg demonstrated that compared with injection moulded samples, samples that were biaxially oriented displayed increased tensile strength, modulus, and elongation at break, such that a sample of PLA with a weight average molecular weight (M_w) = 120 kg mol^{-1} and a composition of 95% LLA to 5% *rac*-LA displayed a tensile strength of 68.6 MPa, Young's modulus of 1.88 GPa, and elongation at break of 56.7%. Further heat setting of these polymers led to moderate reductions in tensile strength and modulus as the process oriented amorphous section of the PLAs relaxed.

The mechanical properties of the *heat-set polymers* were even higher than those that had been moulded [91]. Gripjma *et al.* also showed that with orientation by hot drawing, increased tensile strengths of the PLLA up to 1800 MPa were possible for the samples, with a high draw ratio of 12–18.5 at temperatures around 204 °C [93]. A similar study with PDLLA drawing at relatively low temperatures (37 and 40 °C) resulted in PLAs that were much less brittle than non-oriented samples. Orienting the samples by drawing with a draw ratio of 2.5 realized notable enhancements in the Young's modulus (4.5 GPa compared with 3.7 GPa for the un-oriented material), tensile strength (73.3 MPa compared with 47 MPa for the un-oriented material), and elongation at break (48.2% compared with 1.5% for the un-oriented material) [99]. Furthermore, the use of additives, for example cross-linkers such as spiro-bis-dimethylencarbonate [100] or multifunctional monomers [e.g., 5,5′-bis(oxepane-2-one)] [101], can also significantly affect the properties of the bulk materials, typically increasing the impact strengths of stereopure PLLA networks. Additionally, plasticizers such as lactide added post-polymerization [102], citrate esters [103], and most notably poly(ethylene glycol)s (PEGs) [102, 104], can be used, with the last showing up to 500% increased elongation at break values over untreated PLLA [104].

9.3.2
Polymer Blends

Several other approaches have been examined to alter the thermal and mechanical properties of PLA [90]. Amongst these, *blending* of PLA with other materials has been applied to achieve toughening by rubber-toughening using binary or ternary blends of compatibilizers [90, 105]. One of the major issues with this approach is maintaining the inherent biological compatibility and degradability of the blending components. Thus, many studies have focused on the use of other poly(ester)s with blending conditions carefully controlled to prevent polymer degradation and/or transesterification [106]. Polymer blending typically involves the formation of an emulsion from a number of immiscible homopolymers, which is then stabilized by the addition of a block copolymer of the individual parts [107]; the resultant blend incorporates properties from both components. While many polymer combinations have been examined with PLA, this section will focus on the maintenance of renewability and degradability in the choice of conditions and blending partners.

9.3.2.1 Poly(Lactide)/Poly(ε-Caprolactone) Blends

The most common *blending partner* in applications where degradability is desired is undoubtedly poly(ε-caprolactone) (PCL). This *biodegradable polyester* displays rubbery characteristics, has a low T_g ($-60°C$) and a very good elongation at break [108]. While binary blended materials provide insignificant increases in the overall mechanical properties, potentially a consequence of the formation of macroscopic phase domains in the material [109–111], the addition of up to ~10 wt% of a preformed discreet diblock PLA-*b*-PCL or triblock PLA-*b*-PCL-*b*-PLA copolymer results in an up to 300% increase in elongation strength and in impact strength by up to a factor of 5 [110, 112, 113]. A similar, albeit diminished, effect has also been obtained by the addition of a transesterification catalyst to PLA–PCL binary blends. Here, transesterification of the poly(ester)s results in reactive blending of the PLA and PCL leading to the *in situ* formation of block copolymers and a concurrent increase in elongation at break up to 130% [108]. Similar results to this have been achieved by employing additional cross-linkers, such as dicumyl peroxide (DCP) [114], or diisocynantes and PLA triols to form semi-interpenetrating polyurethane–PLA networks [115].

9.3.2.2 Other Biodegradable/Renewable Polyesters

A range of other *biodegradable/renewable materials* have been blended with PLA in recent years. Blending PLA with poly(*para*-dioxanone) (PPD) [116] or poly(propylene carbonate) (PPC) [117] results in very limited changes in the mechanical properties of the resultant blends. However biodegradable poly(ester)s based on statistical adipate–terephthalate copolymers have resulted in the development of a range of interesting materials. Poly(tetramethylene adipate-*co*-terephthalate) (PTAT) and poly(butylene adipate-*co*-terephthalate) (PBAT) result in blends that display improved tensile and impact strength, although these improvements are very dependent on the individual blending techniques and blend compositions [118–121]. The same general trend also applies to poly(ester) materials synthesized from succinates [e.g., poly(butylenes succinate)] and poly(hydroxyalkanoates) such as poly[(*R*)-3-hydroxybutyrate] (PHB) [122–127].

9.4
Thermoplastic Elastomers

Thermoplastic elastomers (TPEs) are typically ABA block copolymers that contain an inner soft amorphous segment with a low T_g and hard, often crystalline, outer segments. TPEs display the properties of rubber and offer a potential method for the synthetic enhancement of the mechanical properties of PLA [128]. Many mid-blocks have been investigated to synthesize PLA-based TPEs, including poly(ethylene) oligomers [129], poly(isobutylene) [130], poly(isoprene) [131–133], and poly(cyclooctadiene) [134]. Several entirely renewable/degradable ABA triblock copolymers have also been reported including poly(butylene succinate) (PBS), obtained by condensation polymerization [135, 136], poly(β-butyrolactone) (PBL), PCL, poly(1,5-dioxepan-2-one) (PDXO), poly(trimethylene carbonate) (PTMC), and poly(menthide) as a soft intermediate block [137–143].

Low molecular weight PDLLA-*b*-PTMC-*b*-PDLLA triblock copolymers do not display improved mechanical properties over high molecular weight poly(*rac*-LA-*co*-TMC) [143]; however, replacement of the amorphous PDLLA with homochiral PLLA or PDLA and increase in the mol% PLA led to a significant increase in the mechanical performance with Young's modulus values of up to 470 MPa, yield strength ~15 MPa, and ultimate tensile strength ~20 MPa. PLLA-*b*-poly(1,5-dioxepan-2-one)-*b*-PLLA triblock copolymers displayed high elasticities and very high elongations at break compared with other hydrolyzable poly(ester)s. An elongation at break of up to 910% for a 50 : 300 : 50 triblock composition (ratio of degree of polymerization) was measured with a corresponding stress at break equal to 42 MPa. Reducing the poly(DXO) block length led to the materials becoming more stiff and inflexible [139]. The synthesis of triblock copolymers with a poly(DXO-*co*-TMC) mid-section led to more notable phase separation and had significantly higher strain at break (up to 1089%) than simple triblock copolymers containing either DXO or TMC with PLLA; the Young's modulus was also increased slightly upon the introduction of DXO into the TMC

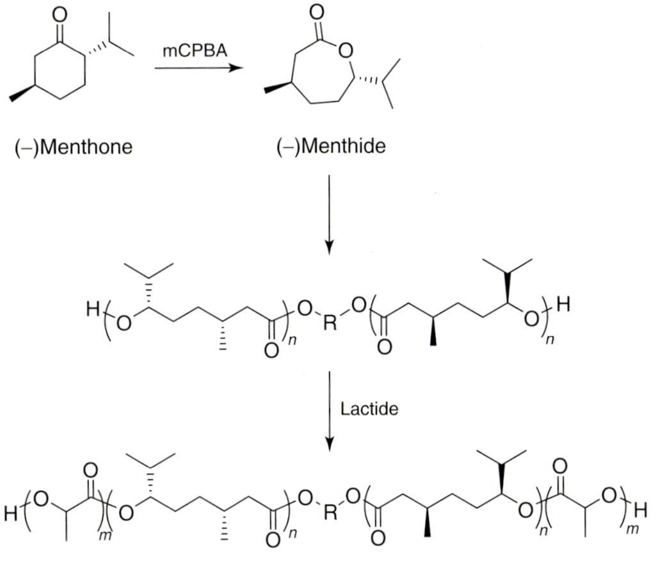

Scheme 9.6 Synthesis of poly(lactide)-b-poly(menthide)-b-poly(lactide) thermoplastic elastomers [140–142, 144].

block [137]. PLLA-b-PBL-b-PLLA triblock copolymers, through the sequential ROP of β-butyrolactone and LA from a 1,4-butanediol initiator, showed increasing Young's modulus values with increasing PLLA content (from 30 to 160 MPa, 44–69% PLLA), and decreasing elongations at break from 200 to 86% [138].

Hillmyer, Tolman, and coworkers have reported the synthesis and properties of a triblock copolymer comprised of PLA outer blocks and an inner block of poly(menthide) (Scheme 9.6) [140–142]. Poly(menthide) is obtained by ROP of menthide that can in turn be realized by the oxidation of (−)menthone [144]. Sequential ROP of menthide and lactide from a bifunctional initiator resulted in triblock copolymers that displayed microphase separation, evidenced by the observation of two glass transition temperatures and small angle X-ray scattering experiments (SAXS). PLA-b-PM-b-PLA (M_n of respective blocks 7.8–29-7.8 kg mol^{-1}) displaying a hexagonally packed cylindrical morphology provided a Young's modulus of ∼1.4 MPa, an ultimate tensile strength of 1.7 MPa, and an elongation of 960%, with an increasing weight fraction of PLA (to 42%) leading to a threefold increase in these parameters [140]. Application of PLLA or PDLA blocks revealed a further threefold increase in tensile strength and a twofold increase in the Young's modulus compared with the amorphous atactic PLA-containing triblocks, although the ultimate tensile strength remained largely invariant. Blending equal amounts of PLLA-b-PM-b-PLLA and PDLA-b-PM-b-PDLA block copolymers to obtain stereocomplexation in the PLA blocks resulted in a further marked enhancement in the mechanical strength of the copolymers, such that modulus values of about 30 MPa, tensile strengths of ∼20 MPa, and ultimate elongations of ∼1000% could be achieved [140].

9.5
Future Developments/Outlook

PLA has already become an established renewable polymer for many applications. The possibility of a full lifecycle analysis and the use of renewable feedstocks in its manufacture are also compatible with the current economic and regulatory demands, and therefore this sector seems likely to grow over the coming decades. The improvements in synthesis of PLA and modification of its properties through, amongst other things, blending and block copolymer synthesis, have already led to notable improvements in the performance of the resultant materials. Future research will no doubt focus on improvements to all areas of PLA research to deliver better fermentation technology, in order to more efficiently synthesize lactic acid from an increasing range of sources, discover more active and selective ROP catalysts, and find new ways of improving the thermal and mechanical properties of bulk renewable materials based on PLA.

References

1. Ragauskas, A.J., Williams, C.K., Davison, B.H., Britovsek, G., Cairney, J., Eckert, C.A., Frederick, W.J., Hallett, J.P., Leak, D.J., Liotta, C.L., Mielenz, J.R., Murphy, R., Templer, R., and Tschaplinski, T. (2006) *Science*, **311**, 484.
2. Mochizuki, M. (2002) *Biopolymers. Polyesters III. Applications and Commercial Products*, 1st edn, Wiley-VCH Verlag GmbH, Weinheim, p. 1.
3. Tsuji, H. (2002) *Biopolymers. Polyesters III. Applications and Commercial Products*, 1st edn, Wiley-VCH Verlag GmbH, Weinheim, p. 129.
4. Albertsson, A.C. and Varma, I.K. (2002) *Adv. Polym. Sci.*, **157**, 1.
5. Uhrich, K.E., Cannizzaro, S.M., Langer, R.S., and Shakesheff, K.M. (1999) *Chem. Rev.*, **99**, 3181.
6. Lipinsky, E.S. and Sinclair, R.G. (1986) *Chem. Eng. Prog.*, **82**, 26.
7. Auras, R., Harte, B., and Selke, S. (2004) *Macromol. Biosci.*, **4**, 835.
8. Gruber, P.R., Drumright, R.E., and Henton, D.E. (2000) *Adv. Mater.*, **12**, 1841.
9. Bogaert, J.-C. and Coszach, P. (2000) *Macromol. Symp.*, **153**, 287.
10. Gewin, V. (2003) *PLoS Biol.*, **1**, 15.
11. Holland, S.J. and Tighe, B.J. (1992) *Biodegradable Polymers*, 1st edn, Academic Press, London, p. 101.
12. Kharas, G.B., Sanchez-Riera, F., and Severson, D.K. (1994) *Polymers of Lactic Acid*, Hanser Publishers, Munich, p. 93.
13. Dunsing, R. and Kricheldorf, H.R. (1985) *Polym. Bull.*, **14**, 491.
14. Hartmann, M.H. (1998), In: *Biopolymers from Renewable Resources*, 1st edn (ed. D.L. Kaplan), Springer-Verlag, Berlin, Heidelberg, p. 364.
15. Hyon, S.H., Jamshidi, K., and Ikada, Y. (1997) *Biomaterials*, **18**, 1503.
16. Kricheldorf, H.R., Kreiser-Saunders, I., Jurgens, C., and Wolter, D. (1996) *Macromol. Symp.*, **103**, 85.
17. Leenslag, J.W. and Pennings, A.J. (1987) *Makromol. Chem.*, **188**, 1809.
18. Zhao, Y., Wang, Z., Wang, J., Mai, H., Yan, B., and Yang, F. (2004) *J. App. Polym. Sci.*, **91**, 2413.
19. Dechy-Cabaret, O., Martin-Vaca, B., and Bourissou, D. (2004) *Chem. Rev.*, **104**, 6147.
20. Hakkarainen, M. (2002) *Adv. Polym. Sci.*, **157**, 113.
21. Li, S.M. (1999) *J. Biomed. Mater. Res.*, **48**, 342.

22. Zhang, X., Wyss, U.P., Pichora, D., and Goosen, M.F.A. (1994) *J. Bioact. Compat. Polym.*, **9**, 80.
23. Dorgan, J.R., Lehermeier, H.J., Palade, L.-I., and Cicero, J. (2001) *Macromol. Symp.*, **175**, 55.
24. Tsuji, H., Daimon, F., and Fujie, K. (2003) *Biomacromolecules*, **4**, 835.
25. Dove, A.P. (2008) *Chem. Commun.*, 6446.
26. Kamber, N.E., Jeong, W., Waymouth, R.M., Pratt, R.C., Lohmeijer, B.G.G., and Hedrick, J.L. (2007) *Chem. Rev.*, **107**, 5813.
27. Trollsas, M., Hedrick, J.L., Mecerreyes, D., Dubois, P., Jerome, R., Ihre, H., and Hult, A. (1998) *Macromolecules*, **31**, 2756.
28. Kricheldorf, H.R., Boettcher, C., and Tonnes, K.U. (1992) *Polymer*, **33**, 2817.
29. Du, Y.J., Lemstra, P.J., Nijenhuis, A.J., van Aert, H.A.M., and Bastiaansen, C. (1995) *Macromolecules*, **28**, 2124.
30. Schwach, G., Coudane, J., Engel, R., and Vert, M. (1997) *J. Polym. Sci., Part A: Polym. Chem.*, **35**, 3431.
31. Kowalski, A., Duda, A., and Penczek, S. (2000) *Macromolecules*, **33**, 7359.
32. Kricheldorf, H.R., Kreiser-Saunders, I., and Stricker, A. (2000) *Macromolecules*, **33**, 702.
33. von Schenk, H., Ryner, M., Albertsson, A.-C., and Svensson, M. (2002) *Macromolecules*, **35**, 1556.
34. Duda, A., Penczek, S., Kowalski, A., and Libiszowski, J. (2000) *Macromol. Symp.*, **153**, 41.
35. Kowalski, A., Libiszowski, J., Duda, A., and Penczek, S. (2000) *Macromolecules*, **33**, 1964.
36. Kricheldorf, H.R., Kreiser-Saunders, I., and Boettcher, C. (1995) *Polymer*, **36**, 1235.
37. Kricheldorf, H.R. (2000) *Macromol. Symp.*, **153**, 55.
38. Majerska, K., Duda, A., and Penczek, S. (2000) *Macromol. Rapid Commun.*, **21**, 1327.
39. Pack, J.W., Kim, S.H., Park, S.Y., Lee, Y.-W., and Kim, Y.H. (2003) *Macromolecules*, **36**, 8923.
40. Zhang, X., MacDonald, D.A., Goosen, M.F.A., and McCauley, K.B. (1994) *J. Polym. Sci., Part A: Polym. Chem.*, **32**, 2965.
41. Ryner, M., Stridsberg, K., Albertsson, A.C., von Schenck, H., and Svensson, M. (2001) *Macromolecules*, **34**, 3877.
42. Dubois, P., Jerome, R., and Teyssie, P. (1991) *Makromol. Chem. Macromol. Symp.*, **42-43**, 103.
43. Dubois, P., Jerome, R., and Teyssie, P. (1991) *Macromolecules*, **24**, 977.
44. Stanford, M.J. and Dove, A.P. (2009) *Macromolecules*, **42**, 141.
45. Stanford, M.J. and Dove, A.P. (2010) *Chem. Soc. Rev.*, **39**, 486.
46. Chamberlain, B.M., Cheng, M., Moore, D.R., Ovitt, T.M., Lobkovsky, E.B., and Coates, G.W. (2001) *J. Am. Chem. Soc.*, **123**, 3229.
47. Chisholm, M.H., Gallucci, J., and Phomphrai, K. (2002) *Inorg. Chem.*, **41**, 2785.
48. Marshall, E.L., Gibson, V.C., and Rzepa, H.S. (2005) *J. Am. Chem. Soc.*, **127**, 6048.
49. Chisholm, M.H., Gallucci, J., and Phomphrai, K. (2003) *Chem. Commun.*, 48.
50. Chisholm, M.H., Gallucci, J.C., and Phomphrai, K. (2004) *Inorg. Chem.*, **43**, 6717.
51. Chisholm, M.H., Gallucci, J.C., and Yaman, G. (2006) *Chem. Commun.*, 1872.
52. Alaaeddine, A., Amgoune, A., Thomas, C.M., Dagorne, S., Bellemin-Laponnaz, S., and Carpentier, J.F. (2006) *Eur. J. Inorg. Chem.*, 3652.
53. Amgoune, A., Thomas, C.M., and Carpentier, J.F. (2007) *Macromol. Rapid Commun.*, **28**, 693.
54. Amgoune, A., Thomas, C.M., Roisnel, T., and Carpentier, J.F. (2006) *Chem. Eur. J.*, **12**, 169.
55. Castro, P.M., Zhao, G., Amgoune, A., Thomas, C.M., and Carpentier, J.F. (2006) *Chem. Commun.*, 4509.
56. Heck, R., Schulz, E., Collin, J., and Carpentier, J.F. (2007) *J. Mol. Catal. A: Chem.*, **268**, 163.
57. Ma, H., Spaniol, T.P., and Okuda, J. (2008) *Inorg. Chem.*, **47**, 3328.

58. Ma, H.Y., Spaniol, T.P., and Okuda, J. (2003) *J. Chem. Soc., Dalton Trans.*, 4770.
59. Ma, H.Y., Spaniol, T.P., and Okuda, J. (2006) *Angew. Chem. Int. Ed. Engl.*, **45**, 7818.
60. Zhong, Z.Y., Dijkstra, P.J., and Feijen, J. (2002) *Angew. Chem. Int. Ed. Engl.*, **41**, 4510.
61. Nomura, N., Ishii, R., Yamamoto, Y., and Kondo, T. (2007) *Chem. Eur. J.*, **13**, 4433.
62. Chmura, A.J., Chuck, C.J., Davidson, M.G., Jones, M.D., Lunn, M.D., Bull, S.D., and Mahon, M.F. (2007) *Angew. Chem. Int. Ed. Engl.*, **46**, 2280.
63. Chmura, A.J., Davidson, M.G., Frankis, C.J., Jones, M.D., and Lunn, M.D. (2008) *Chem. Commun.*, 1293.
64. Williams, C.K., Breyfogle, L.E., Choi, S.K., Nam, W., Young, V.G., Hillmyer, M.A., and Tolman, W.B. (2003) *J. Am. Chem. Soc.*, **125**, 11350.
65. Ajellal, N., Lyubov, D.M., Sinenkov, M.A., Fukin, G.K., Cherkasov, A.V., Thomas, C.M., Carpentier, J.-F., and Trifonov, A.A. (2008) *Chem. Eur. J.*, **14**, 5440.
66. Dove, A.P. (2009), In: *Handbook of Ring-Opening Polymerization*, 1st edn (eds P. Dubois, O. Coulembier, and J.-M. Raquez), Wiley-VCH Verlag GmbH, Weinheim, p. 357.
67. Nederberg, F., Connor, E.F., Moller, M., Glauser, T., and Hedrick, J.L. (2001) *Angew. Chem. Int. Ed. Engl.*, **40**, 2712.
68. Bonduelle, C., Martin-Vaca, B., Cossio, F.P., and Bourissou, D. (2008) *Chem. Eur. J.*, **14**, 5304.
69. Myers, M., Connor, E.F., Glauser, T., Mock, A., Nyce, G., and Hedrick, J.L. (2002) *J. Polym. Sci., Part A: Polym. Chem.*, **40**, 844.
70. Connor, E.F., Nyce, G.W., Myers, M., Mock, A., and Hedrick, J.L. (2002) *J. Am. Chem. Soc.*, **124**, 914.
71. Dove, A.P., Pratt, R.C., Lohmeijer, B.G.G., Culkin, D.A., Hagberg, E.C., Nyce, G.W., Waymouth, R.M., and Hedrick, J.L. (2006) *Polymer*, **47**, 4018.
72. Csihony, S., Culkin, D.A., Sentman, A.C., Dove, A.P., Waymouth, R.M., and Hedrick, J.L. (2005) *J. Am. Chem. Soc.*, **127**, 9079.
73. Nyce, G.W., Glauser, T., Connor, E.F., Mock, A., Waymouth, R.M., and Hedrick, J.L. (2003) *J. Am. Chem. Soc.*, **125**, 3046.
74. Sentman, A.C., Csihony, S., Waymouth, R.M., and Hedrick, J.L. (2005) *J. Org. Chem.*, **70**, 2391.
75. Nyce, G.W., Csihony, S., Waymouth, R.M., and Hedrick, J.L. (2004) *Chem. Eur. J.*, **10**, 4073.
76. Coulembier, O., Dove, A.R., Pratt, R.C., Sentman, A.C., Culkin, D.A., Mespouille, L., Dubois, P., Waymouth, R.M., and Hedrick, J.L. (2005) *Angew. Chem. Int. Ed. Engl.*, **44**, 4964.
77. Dove, A.P., Li, H.B., Pratt, R.C., Lohmeijer, B.G.G., Culkin, D.A., Waymouth, R.M., and Hedrick, J.L. (2006) *Chem. Commun.*, 2881.
78. Dove, A.P., Pratt, R.C., Lohmeijer, B.G.G., Waymouth, R.M., and Hedrick, J.L. (2005) *J. Am. Chem. Soc.*, **127**, 13798.
79. Lohmeijer, B.G.G., Pratt, R.C., Leibfarth, F., Logan, J.W., Long, D.A., Dove, A.P., Nederberg, F., Choi, J., Wade, C., Waymouth, R.M., and Hedrick, J.L. (2006) *Macromolecules*, **39**, 8574.
80. Pratt, R.C., Lohmeijer, B.G.G., Long, D.A., Lundberg, P.N.P., Dove, A.P., Li, H.B., Wade, C.G., Waymouth, R.M., and Hedrick, J.L. (2006) *Macromolecules*, **39**, 7863.
81. Zhang, L., Nederberg, F., Pratt, R.C., Waymouth, R.M., Hedrick, J.L., and Wade, C.G. (2007) *Macromolecules*, **40**, 4154.
82. Pratt, R.C., Lohmeijer, B.G.G., Long, D.A., Waymouth, R.M., and Hedrick, J.L. (2006) *J. Am. Chem. Soc.*, **128**, 4556.
83. Chuma, A., Horn, H.W., Swope, W.C., Pratt, R.C., Zhang, L., Lohmeijer, B.G.G., Wade, C.G., Waymouth, R.M., Hedrick, J.L., and Rice, J.E. (2008) *J. Am. Chem. Soc.*, **130**, 6749.
84. Simon, L. and Goodman, J.M. (2007) *J. Org. Chem.*, **72**, 9656.
85. Zhang, L., Nederberg, F., Messman, J.M., Pratt, R.C., Hedrick, J.L., and

Wade, C.G. (2007) *J. Am. Chem. Soc.*, **129**, 12610.
86. Coulembier, O., Meyer, F., and Dubois, P. (2010) *Polym. Chem.* doi: 10.1039/c0py00013b
87. Tullo, A. (2002) *Chem. Eng. News*, **80**, 13.
88. Gross, R.A. and Kalra, B. (2002) *Science*, **297**, 803.
89. Tullo, A. (2000) *Chem. Eng. News*, **78**, 13.
90. Anderson, K.S., Schreck, K.M., and Hillmyer, M.A. (2008) *Polym. Rev.*, **48**, 85.
91. Bigg, D.M. (2005) *Adv. Polym. Technol.*, **24**, 69.
92. Perego, G., Cella, G.D., and Bastioli, C. (1996) *J. App. Polym. Sci.*, **59**, 37.
93. Grijpma, D.W., Penning, J.P., and Pennings, A.J. (1994) *Colloid Polym. Sci.*, **272**, 1068.
94. Williams, C.K. and Hillmyer, M.A. (2008) *Polym. Rev.*, **48**, 1.
95. Baker, G.L., Vogel, E.B., and Smith, M.R. (2008) *Polym. Rev.*, **48**, 64.
96. Tsuji, H. (2005) *Macromol. Biosci.*, **5**, 569.
97. Garlotta, D. (2001) *J. Polym. Environ.*, **9**, 63.
98. Biela, T., Duda, A., and Penczek, S. (2006) *Macromolecules*, **39**, 3710.
99. Grijpma, D.W., Altpeter, H., Bevis, M.J., and Feijen, J. (2002) *Polym. Int.*, **51**, 845.
100. Grijpma, D.W. and Pennings, A.J. (1994) *Macromol. Chem. Phys.*, **195**, 1649.
101. Grijpma, D.W., Nijenhuis, A.J., and Pennings, A.J. (1996) *Polymer*, **37**, 2783.
102. Jacobsen, S. and Fritz, H.G. (1999) *Polym. Eng. Sci.*, **39**, 1303.
103. Labrecque, L.V., Kumar, R.A., Dave, V., Gross, R.A., and McCarthy, S.P. (1997) *J. App. Polym. Sci.*, **66**, 1507.
104. Nijenhuis, A.J., Colstee, E., Grijpma, D.W., and Pennings, A.J. (1996) *Polymer*, **37**, 5849.
105. Anderson, K.S., Lim, S.H., and Hillmyer, M.A. (2003) *J. App. Polym. Sci.*, **89**, 3757.
106. Schreck, K.M. and Hillmyer, M.A. (2007) *J. Biotechnology*, **132**, 287.
107. Datta, S. and Lohse, D.J. (1996) *Polymeric Compatibilizers: Uses and Benefits in Polymer Blends*, 1st edn, Hanser Publishers, Munich.
108. Wang, L., Ma, W., Gross, R.A., and McCarthy, S.P. (1998) *Polym. Degrad. Stab.*, **59**, 161.
109. Broz, M.E., VanderHart, D.L., and Washburn, N.R. (2003) *Biomaterials*, **24**, 4181.
110. Hiljanen-Vainio, M., Varpomaa, P., Seppala, J., and Tormala, P. (1996) *Macromol. Chem. Phys.*, **197**, 1503.
111. Tsuji, H. and Ikada, Y. (1996) *J. Appl. Polym. Sci.*, **60**, 2367.
112. Maglio, G., Migliozzi, A., Palumbo, R., Immirzi, B., and Volpe, M.G. (1999) *Macromol. Rapid Commun.*, **20**, 236.
113. Tsuji, H., Yamada, T., Suzuki, M., and Itsuno, S. (2003) *Polym. Int.*, **52**, 269.
114. Semba, T., Kitagawa, K., Ishiaku, U.S., and Hamada, H. (2006) *J. App. Polym. Sci.*, **2006**, 1861.
115. Yuan, Y. and Ruckenstein, E. (1998) *Polym. Bull.*, **40**, 485.
116. Pezzin, A.P.T., Alberda van Eckenstein, G.O.R., Zavaglia, C.A.C., ten Brinke, G., and Duek, E.A.R. (2003) *J. App. Polym. Sci.*, **88**, 2744.
117. Ma, X., Yu, J., and Wang, N. (2006) *J. Polym. Sci., Part B: Polym. Phys.*, **44**, 94.
118. Jiang, L., Wolcott, M.P., and Zhang, J. (2006) *Biomacromolecules*, **7**, 199.
119. Lui, T.-Y., Lin, W.-C., Yang, M.-C., and Chen, S.-Y. (2005) *Polymer*, **46**, 12586.
120. Lui, X., Dever, M., Fair, N., and Benson, R.S. (1997) *J. Environ. Polym. Degrad.*, **5**, 225.
121. McCarthy, S.P., Ranganthan, A., and Ma, W. (1999) *Macromol. Symp.*, **144**, 63.
122. Chen, G.-X., Kim, H.-S., Kim, E.-S., and Yoon, J.-S. (2005) *Polymer*, **46**, 11829.
123. Chen, G.-X. and Yoon, J.-S. (2005) *Polym. Degrad. Stab.*, **88**, 206.
124. Chen, G.-X. and Yoon, J.-S. (2005) *J. Polym. Sci., Part B: Polym. Phys.*, **43**, 478.
125. Noda, I., Bond, E.B., Green, P.R., Melik, D.H., Narasimhan, K.,

Schechtman, L.A., and Satkowski, M.M. (2005), In: *Polymer Biocatalysis and Biomaterials*, ACS Symposium Series, vol. 900 (eds H.N. Cheng and R.A. Gross) American Chemical Society, Washington, DC, p. 280.

126. Noda, I., Green, P.R., Schechtman, L.A., and Satkowski, M.M. (2005) *Biomacromolecules*, **6**, 580.

127. Shibata, M., Inoue, Y., and Miyoshi, M. (2006) *Polymer*, **47**, 3557.

128. Spontak, R.J. and Patel, N.P. (2000) *Curr. Opin. Colloid Interface Sci.*, **5**, 334.

129. Abayasinghe, N.K., Glaser, S., Prasanna, K., Perera, U., and Smith, D.W. (2005) *J. Polym. Sci., Part A: Polym. Chem.*, **43**, 5257.

130. Sipos, L., Zsuga, M., and Deak, G. (1995) *Macromol. Rapid Commun.*, **16**, 935.

131. Frick, E.M. and Hillmyer, M.A. (2000) *Macromol. Rapid Commun.*, **21**, 1317.

132. Frick, E.M., Zalusky, A.S., and Hillmyer, M.A. (2003) *Biomacromolecules*, **4**, 216.

133. Schmidt, S.C. and Hillmyer, M.A. (1999) *Macromolecules*, **32**, 4794.

134. Pitet, L.M. and Hillmyer, M.A. (2009) *Macromolecules*, **42**, 3674.

135. Ba, C.Y., Yang, J., Hao, Q.H., Liu, X.Y., and Cao, A. (2003) *Biomacromolecules*, **4**, 1827.

136. Jia, L., Yin, L.Z., Li, Y., Li, Q.B., Yang, J., Yu, J.Y., Shi, Z., Fang, Q., and Cao, A. (2005) *Macromol. Biosci.*, **5**, 526.

137. Andronova, N. and Albertsson, A.C. (2006) *Biomacromolecules*, **7**, 1489.

138. Hiki, S., Miyamoto, M., and Kimura, Y. (2000) *Polymer*, **41**, 7369.

139. Ryner, M. and Albertsson, A.C. (2002) *Biomacromolecules*, **3**, 601.

140. Wanamaker, C.L., Bluemle, M.J., Pitet, L.M., O'Leary, L.E., Tolman, W.B., and Hillmyer, M.A. (2009) *Biomacromolecules*, **10**, 2904.

141. Wanamaker, C.L., O'Leary, L.E., Lynd, N.A., Hillmyer, M.A., and Tolman, W.B. (2007) *Biomacromolecules*, **8**, 3634.

142. Wanamaker, C.L., Tolman, W.B., and Hillmyer, M.A. (2009) *Biomacromolecules*, **10**, 443.

143. Zhang, Z., Grijpma, D.W., and Feijen, J. (2004) *Macromol. Chem. Phys.*, **205**, 867.

144. Zhang, D.H., Hillmyer, M.A., and Tolman, W.B. (2005) *Biomacromolecules*, **6**, 2091.

10
Synthesis of Saccharide-Derived Functional Polymers
Julian Thimm and Joachim Thiem

10.1
Introduction

Current estimations of the annual formation of *biomass*, as shown in the general survey in Scheme 10.1, give numbers of about 170×10^9 tons. The majority (over 95%) consists of carbohydrates in the form of polysaccharides, higher and lower complex oligosaccharides and an enormous amount of glycoconjugates, which combine saccharides functionalized with other natural product components.

Approximately 4.4×10^9 tons, equivalent to 3% of the total mass, are utilized by mankind for animal feed, food, clothing, construction, and other purposes. Over recent decades, numerous attempts have been supported by initiatives in many countries for the study and subsequent promotion of the use of renewable resources. These considerable efforts have provided some impressive results, most of which have not yet left the research laboratory level.

In particular, hydrophilic and also water-soluble polymers are of interest, and a large body of early research addressed carbohydrate-derived precursors and intermediates [1–3]. Of these, one of the attractive goals was *hydrophilic surgical materials*, and thus the formation and use of polylactides has been studied extensively [4, 5]. Previously, US companies (Upjohn, Thermedix) had developed polyoxybutylene-based polyurethanes termed "*pellethen*®" (**1**) or "*tecoflex*®" (**2**) for use as "biomaterials" in artificial organs. These, however, were derived solely from mineral oil sources (Scheme 10.2).

In contrast, the fairly clever and facile condensation of gluconic acid and polyethylene amine resulted in the highly water soluble compound **3** obtained from renewables, the properties of which have not as yet been determined (Scheme 10.3) [6].

At this time, it would appear attractive and worthwhile to follow pathways toward polyhydroxy- or polyoxy-containing scaffolds based on saccharide derived building units. Initially the synthetic approach must be solved and then the basic properties of the novel components established.

Green Polymerization Methods: Renewable Starting Materials, Catalysis and Waste Reduction
Edited by Robert T. Mathers and Michael A. R. Meier
Copyright © 2011 WILEY-VCH Verlag GmbH & Co. KGaA, Weinheim
ISBN: 978-3-527-32625-9

10 Synthesis of Saccharide-Derived Functional Polymers

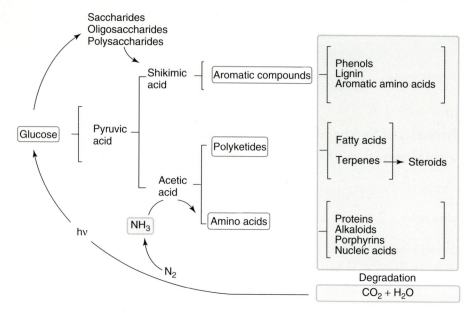

Scheme 10.1 Lifecycle of biomass buildup and degradation.

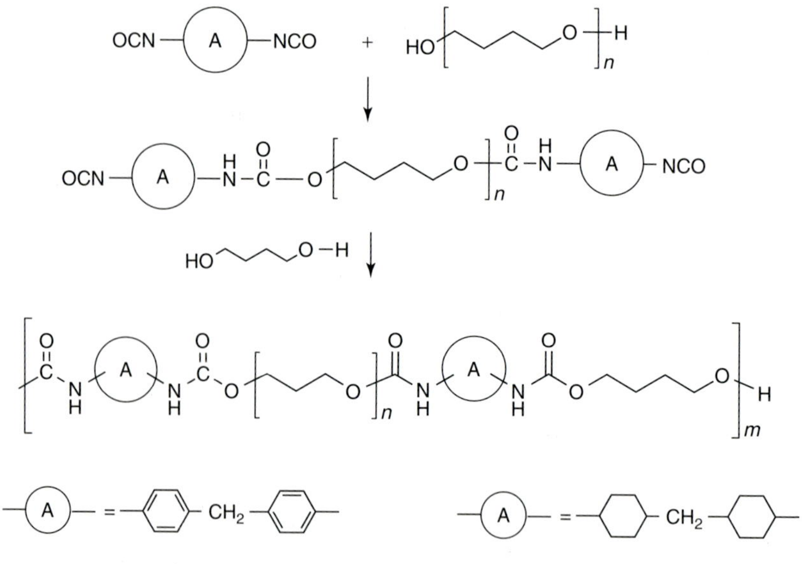

Scheme 10.2 Synthesis of polyoxybutylene-based polyurethanes as biocompatible materials.

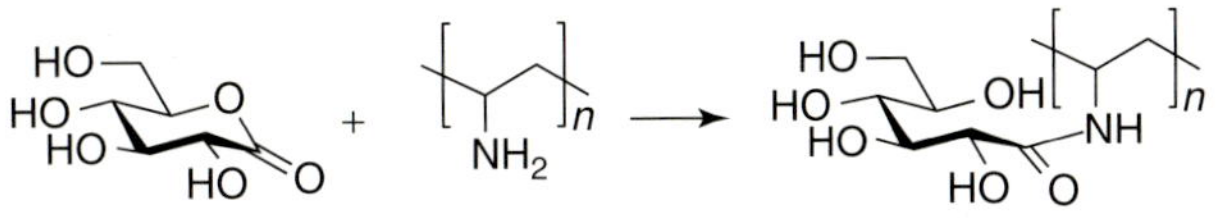

Scheme 10.3 Water soluble polymer from gluconic acid and polyethylene amine.

10.2 Polyethers

The acid-catalyzed ring opening of tetrahydrofuran to give *polyoxybutylene*, observed in 1939 by Meerwein and coworkers [7, 8], is commercially a very important reaction (Scheme 10.4). Polyoxybutylene (**4**) is produced by a number of large chemical

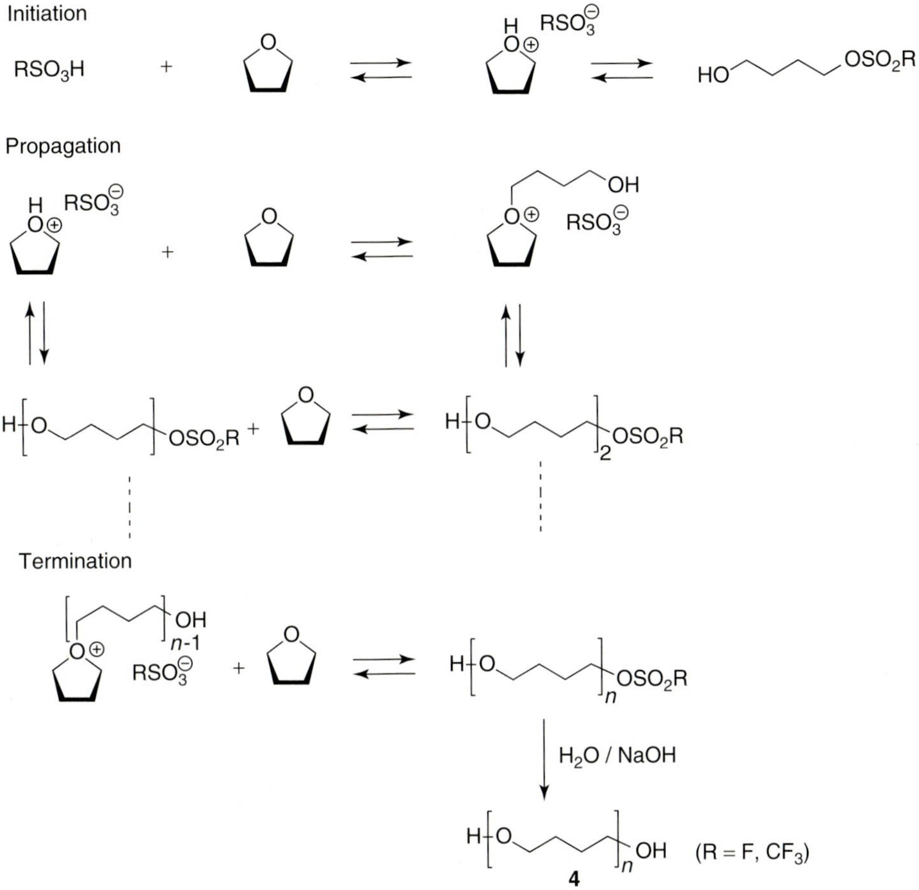

Scheme 10.4 Synthesis of polyoxybutylene by acid-catalyzed ring opening of tetrahydrofuran.

companies in the form of low, medium, and high molecular weight polymeric material and is often employed as a softening segment in polyurethanes.

Early experiments on the *acid-catalyzed ring opening reactions of glycosides* could not be controlled in order to provide oligomeric material. However, starting from starch, enzymatic synthesis gives glucose, which leads to sorbitol (**5**), through catalytic hydrogenation. Acid-catalyzed dehydration then leads to 1,4-monoanhydro sorbitol (**6**), and further elimination of water results in the formation of the thermodynamically very stable 1,4:3,6-dianhydrosorbitol (**7**). According to recent figures, sorbitol (**5**) is produced at a rate of around 100 000 tons per year at a cost of €0.40 per kg. The dianhydrosorbitol (**7**, often known as "isosorbide") is produced at a rate of 1000 tons per year, costing €6.00 per kg.

1,4:3,6-Dianhydro-D-sorbitol (**7**) shows a more reactive secondary *exo*-hydroxy group at the 2-position and another less reactive *endo*-hydroxy group at the 5-position, which allows for regioselective transformations. Structurally related 1,4:3,6-dianhydroalditols of the L-ido type **8** with two *exo*-hydroxy groups and the D-manno type **9** with two *endo*-hydroxy groups can be obtained correspondingly from L-idose or D-mannose, respectively. However, compound **8** may be accessed much more readily from **7** by inversion of the configuration at the C-5 position. In addition, in the tetrose series, structurally further simplified 1,4-anhydro alditols **10** and **11** with D-erythro (meso) and D,L-threo configurations, respectively, could be obtained via corresponding approaches (Scheme 10.5).

The first successful acid-catalyzed ring opening reactions were performed with the 2,3-dialkylethers of 1,4-anhydro-D-erythritol **12–14**. With fluorosulfonic acid (pK_a 12.6) or trifluoromethane sulfonic acid (pK_a 19.1) the 2-*O*-methyl-3-*O*-ethyl derivative **12** gave the corresponding functionalized polyoxybutylene **15** with

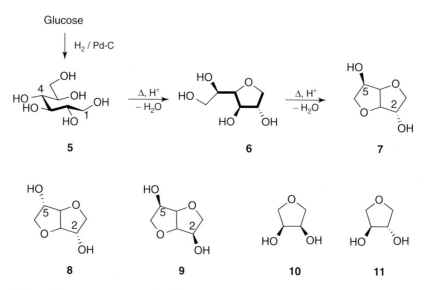

Scheme 10.5 Route to 1,4:3,6-dianhydrosorbitol (**7**) and further dianhydro- and monoanhydroalditols **8–11**.

$\overline{DP_n} = 10\text{--}25$. Under similar conditions the increasing ring tension of the 2,3-di-O-ethyl compound **13** led to the corresponding polyoxybutylene derivative **16** with $\overline{DP_n} = 15\text{--}35$. In both instances the derivatives were esterified as benzoates or acetates and the $\overline{DP_n}$ established by determination of the terminal group. Correspondingly, the 2-O-benzyl-3-O-ethyl derivative **14** gave the functionalized polyoxybutylene **17** with a $\overline{DP_n} \sim 10$. In this case, the acetates could also be obtained, and after hydrogenolysis treatment with cholesteryl-3-oxycarbonyl chloride led to the liquid crystalline material **18**. At this stage $\overline{M_n}$ measurements by ^{1}H-NMR ($\overline{M_n} \sim 2.100$), by GPC (gel permeation chromatography) ($\overline{M_n} \sim 2.100$) and by osmometry ($\overline{M_n} \sim 2.170$) could be compared, and they matched fairly well and documented the correct assignment of the linear functionalized polyoxybutylene derivatives **15–18** (Scheme 10.6) [9–11].

Apparently, reaching the ceiling temperature with these 2,3-alkoxy-functionalized tetrahydrofurans is more difficult, and polymerization led only to modest $\overline{M_n}$ values. Consequently, ring-opening copolymerizations between the diethoxy derivate **13**

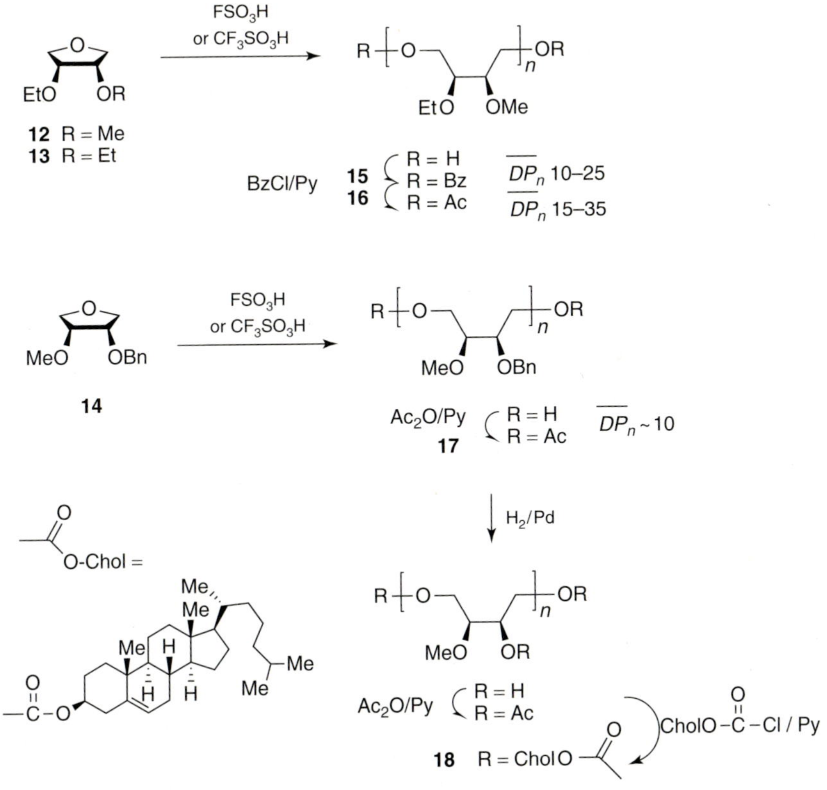

Scheme 10.6 Acid-catalyzed ring opening of 2,3-dialkyl-1,4-anhydro-D-erythritols **12–14** to give functionalized polyoxybutylenes **15–18**.

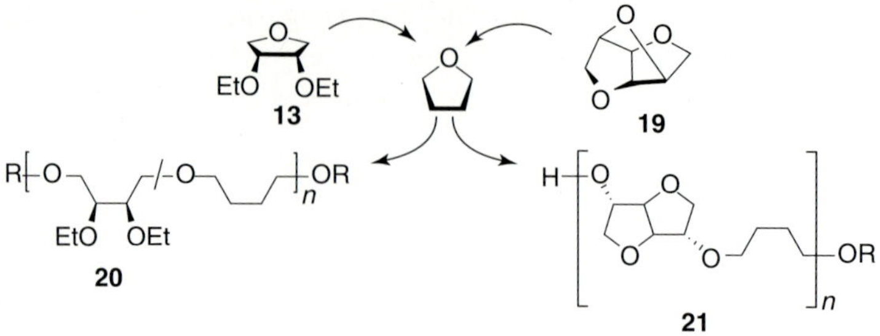

Scheme 10.7 Copolymerization of functionalized anhydroalditols and tetrahydrofuran leading to saccharide-doted polyoxybutylenes.

and THF (tetrahydrofuran) were studied. Polymerization with a co-monomer ratio of **13**/THF = 1 : 2 and 10 mol% CF$_3$SO$_3$H resulted in a 50% yield of the functionalized polyoxybutylene copolymer **20**, showing $\overline{M_n}$ = 14.770 with a ratio of **13**/THF = 1 : 7.7. By copolymerization of THF and the highly strained 1,4:3,6;2,5-trianhydro-D-mannitol (**19**) [12] (THF/**19** = 1 : 1) and 5 mol% CF$_3$SO$_3$H the L-iditol-doted polyoxybutylene **21** was obtained in 63% yield. This showed an $\overline{M_n}$ = 9.390 with a ratio of **19**/THF = 1 : 16.2 and an optical rotation of $[\alpha]_D^{20}$ = 3.1 (c 1.0, THF). The molecular weights of these carbohydrate-containing copolyether polyols are acceptable and demonstrate the principal validation of such approaches toward functionalized polyether structures (Scheme 10.7) [13].

10.3
Polyamides

Following the basic research of Carothers [14], *polyamides* became large scale commercial products around 1950. They displayed the novel properties of cold ductile materials, which allowed the production of fibers and threads of uncommon strength. Thus, polyamides are among the most versatile and important synthetic polymers [15]. Polycondensation of α,ω-diamino compounds with α,ω-dicarboxylic acid derivatives will lead to AA/BB-type Nylon, and the corresponding treatment of ω-amino carboxylic acid derivatives will result in the AB-type Nylon. The latter type can be also accessed by ring-opening polymerization of lactams leading to Perlon (Scheme 10.8).

Facile pathways lead from dianhydroalditols to all stereoisomers of the diamino derivatives via sulfonation of the diols in pyridine, inverting treatment with azide in DMF (N,N'-dimethylformamide), and final hydrogenation. Thus, the D-*gluco* diol **7** will give the diamino L-*gulo* derivative **22**. Correspondingly, the L-*ido* diol **8** will lead to the diamino D-*manno* compound **23**, and from the D-*manno* diol **9** in turn the diamino L-*ido* derivative **24** is obtained. In the tetrose series, the D-*erythro* diol **10** will yield the diamino anhydro-D-*erythritol* **25**, and the D/L-*threo* diol **11** leads to the diamine D/L-*threo* compound **26** (Scheme 10.9) [16, 17].

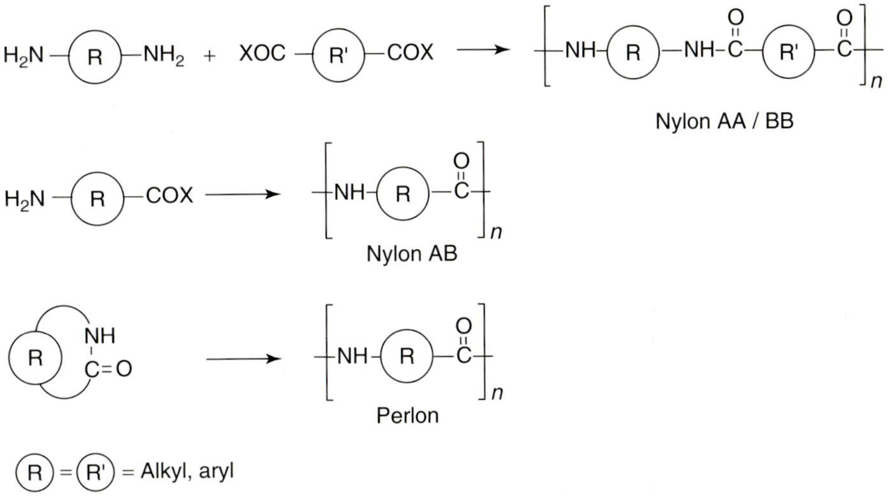

Scheme 10.8 Formation of Nylon and Perlon.

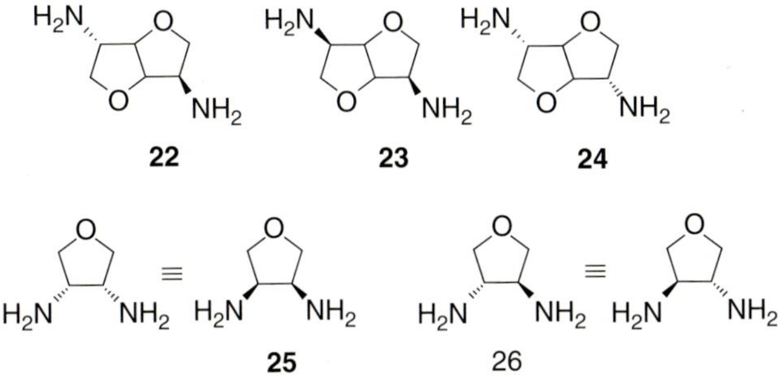

Scheme 10.9 Stereoisomers of diamino di- and monoanhydroalditols.

All the reported diamino mono- and dianhydroalditols **22–26** are thermodynamically very stable compounds. Their bisammonium salts could be employed for both interfacial or solution polymerization with a variety of aromatic and aliphatic diacyl chlorides (Scheme 10.10a–f) to give the corresponding chiral polyamides **27–31**. These polymers are insoluble in most organic solvents except for DMSO (dimethyl sulfoxide) and NMP (N-methylpyrrolidone), with those derived from aromatic dicarboxylic acids (a and b) being slightly more soluble than those from aliphatic dicarboxylic acid precursors (c–f). In all polymers the characteristic NH···O=C bond at 1630 cm^{-1} is observed by IR (infrared) spectroscopy. The molecular weights $\overline{M_n}$ = 5000–20 000 ($\overline{DP_n}$ = 20–40, M_w = 20 000–50 000) are comparable to polyamides based on classical precursors. The structures of polymers **27–31** were confirmed by ^{13}C- and ^{1}H-NMR spectra in DMSO-d$_6$ (Scheme 10.10) [16, 17].

Other work that focused on saccharide-based polyamides [18] has been reviewed, in part, previously [3, 19].

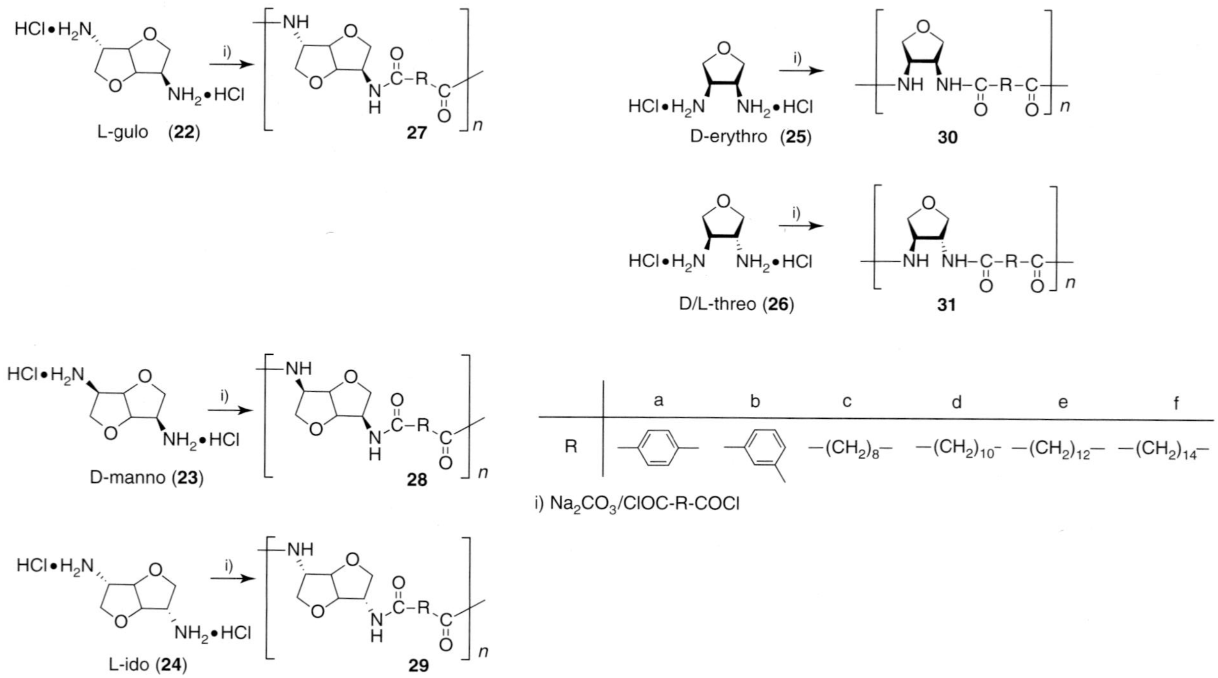

Scheme 10.10 Chiral polyamides 27–31 formed by polymerization of diamino mono- and dianhydroalditols 22–26 with diacyl chlorides.

10.4
Polyurethanes and Polyureas

In addition to their use as precursors to polyamides, diamino dianhydro alditols 22–24 can be employed for the formation of *polyurethanes* by reaction with the bis(chloroformate) of dianhydro-D-sorbitol [20]. Subsequently more efficient pathways to polyurethanes have been developed, which begin with treatment of diamino dianhydro derivatives 22–24 with phosgene to give the corresponding 2,5-diisocyanato compounds 32–34 with L-*gulo*, D-*manno*, and L-*ido* configurations in 64, 41, and 80% yield, respectively. Also, the isothiocyanates could be prepared correspondingly, characterized, along with exemplary transformation into polymers. Reaction of the L-*gulo* diisocyanate 32 with butylene glycol resulted in formation of the polyurethane 35. On treatment with *m*-phenylene diamine, the polyurea compound 36 resulted. Finally, the polyurea 37, completely constructed from dianhydro alditol units, could be obtained from 32 and the L-*gulo* diamine 22 (Scheme 10.11) [21, 22].

In more recent studies following the above mentioned routes, two alternative pathways could be elaborated to form a sugar-derived AB-type *polyurethane*. Starting with dianhydrosorbitol (7) a regioselective tosylation gave the *endo*-tosylate

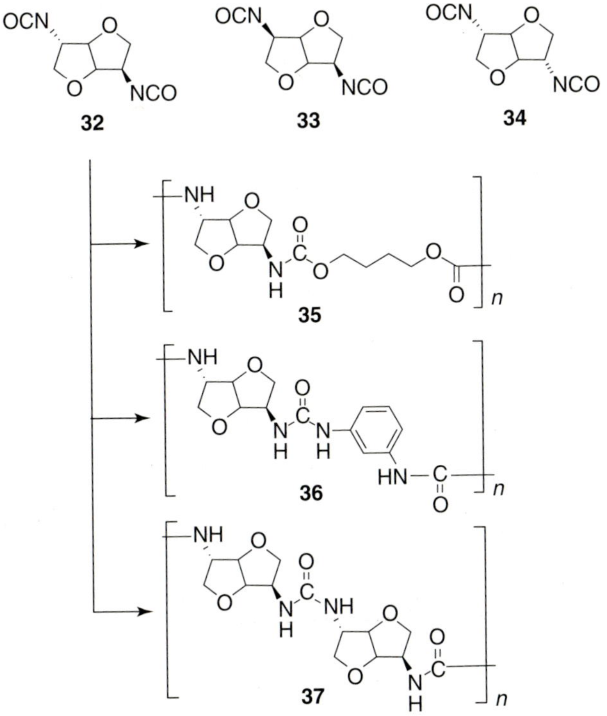

Scheme 10.11 From 2,5-diisocyanato derivative 32, to chiral polyurethane 35, and polyureas 36–37.

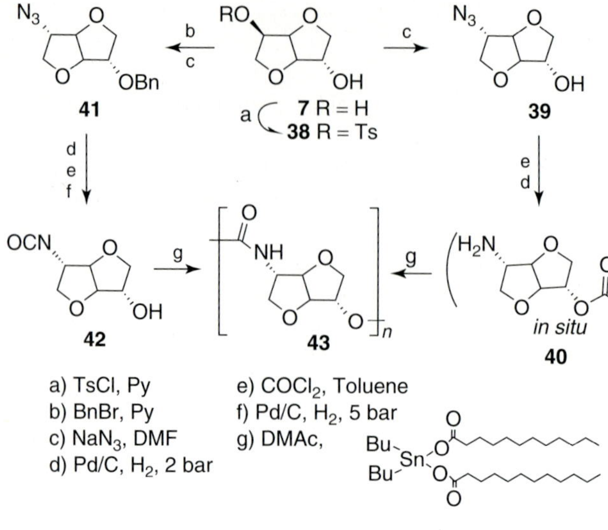

Scheme 10.12 Polycondensation and polyaddition routes to sugar-derived AB-type polyurethane **43**.

a) TsCl, Py
b) BnBr, Py
c) NaN$_3$, DMF
d) Pd/C, H$_2$, 2 bar
e) COCl$_2$, Toluene
f) Pd/C, H$_2$, 5 bar
g) DMAc, Bu$_2$Sn(laurate)$_2$

38 exclusively. By treatment with azide, the 5-azido-1,4:3,6-dianhydro-L-iditol (**39**) was obtained. Phosgenation to the 2-O-chloroformate and subsequent mild hydrogenation led to compound **40**, which *in situ* underwent polycondensation to give the AB-polyurethane **43**. Alternatively, **38** could be benzylated and transferred into the 5-azido-2-O-benzyl-1,4:3,6-dianhydro-L-iditol **41**. By further azide hydrogenolysis and workup, the 5-amino hydrochloride was obtained, and with phosgene the formation of the 5-isocyanato derivative could be achieved. Hydrogenolysis of the benzyl ether gave compound **42**. Under polyaddition conditions, employing dibutyltin dilaurate as a mild Lewis acid catalyst in N,N-dimethylacetamide, the polyurethane **43** was obtained (Scheme 10.12).

Whereas the *polycondensation route* over five steps gave the polyurethane in a 25% overall yield, the *polyaddition route* required seven steps, yet gave a 28% overall yield. $\overline{M_n}$ in the polycondensation was 7900 ($\overline{DP_n} = 46$) and for the polyadduct $\overline{M_n} = 12\,000$ ($\overline{DP_n} = 70$). The much purer and partly crystalline polyadduct showed $T_g = 118.2\,°C$ and $T_{max} = 194.1\,°C$, whereas the polycondensation polymer contained some water and showed a $T_{max} = 145.5\,°C$ [23].

10.5
Glycosilicones

Silicone surfactants are *hydrophilically modified silicones*, which find applications as stabilizing and wetting agents, antistatics, emulgators, and various other uses [24]. Forming carbohydrate-based silicone structures by hydrosilylation is of interest. The hydrosilylation of terminal olefins is catalyzed by hexachloroplatinic acid (Speier's catalyst). Side reactions arising from alcohol impurities led to alkoxy

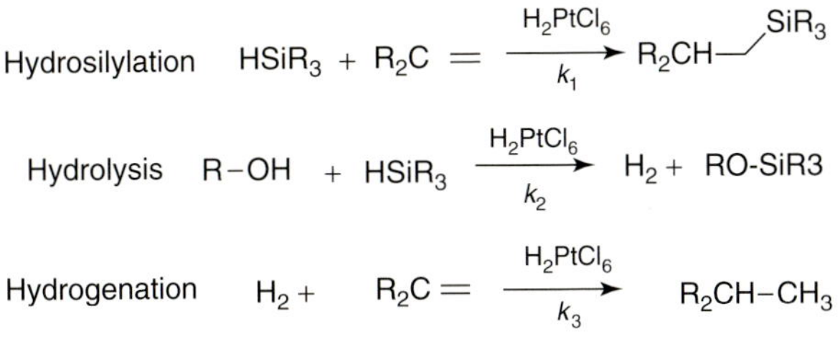

10.5 Glycosilicones

Hydrosilylation $HSiR_3 + R_2C= \xrightarrow[k_1]{H_2PtCl_6} R_2CH-SiR_3$

Hydrolysis $R-OH + HSiR_3 \xrightarrow[k_2]{H_2PtCl_6} H_2 + RO-SiR_3$

Hydrogenation $H_2 + R_2C= \xrightarrow[k_3]{H_2PtCl_6} R_2CH-CH_3$

Scheme 10.13 Hydrosilylation and undesired side reactions.

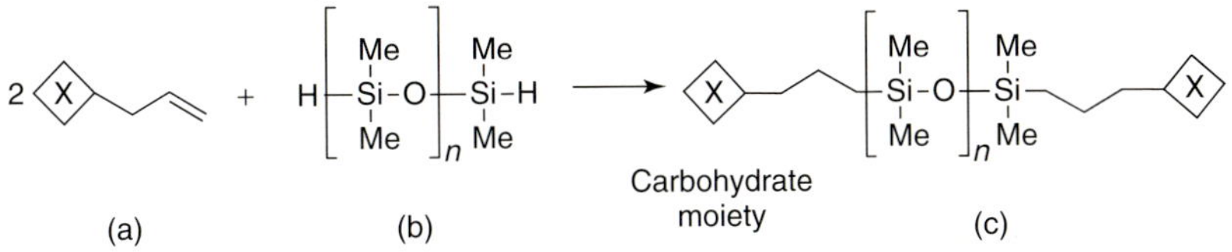

Scheme 10.14 Exemplary hydrosilylation using terminal olefinated hydroxyl protected glyco moieties.

silanes and the liberation of hydrogen, which in turn results in alkane formation. Thus, hydrolysis has to be strictly avoided and a protected saccharide structure must be employed (Scheme 10.13).

The general reaction pattern (Scheme 10.13) requires a hydroxy-protected saccharide structure with a terminal olefine (a), which, through hydrosilylation with the linear polydimethylsiloxane (PDMS-H$_2$, (b), $\overline{M_n} = 592$) results in the formation of the novel glycoconjugates (c) (Scheme 10.14).

Through treatment of the peracetylated alkyl ß-D-glucopyranoside **44** with PDMS-H$_2$ the coupling product **45** results quantitatively, which in turn, after Zemplén cleavage of the ester groups, gives the amphiphilic, methanol soluble saccharide silicone **46**. Correspondingly, the three-allyl ether derivatives of glucose **47** and **50** could be treated to give compounds **48** and **51**, respectively. Again Zemplén transesterification of **48** led to **49**, and reaction of **51** with HBr in methanol resulted in formation of the anomeric mixture of methyl glucopyranoside silicone conjugate **52** (Scheme 10.15) [25].

Reaction of a carbohydrate segment equipped with two olefin functionalities (d) with PDMS (b) means the hydrosilylation can be used to give novel functionalized linear glycopolymers (e) (Scheme 10.16).

Thus, di-O-allylated sugar derivatives were prepared and could be transformed accordingly to give novel glycostructure-segmented olydimethylsiloxanes [26]. For example, allylation of dianhydro sorbitol (**7**) gave derivate **53**, which in turn resulted in formation of polymer **54**. Aldaric acids with *gluco* and *galacto* configuration were treated in analogy with an approach for synthesis of polyamides [27, 28], and in the form of their tetra-O-trimethylsilyl bis-N-allyl amides **55/56** could be transferred into the glycosilico polymers **57/58**. Allyl ether allyl glycosides such as **59** and also di-O-allyl ether derivatives **61** could be used to obtain the polymers **60** and **62**, respectively. These novel carbohydrate modified silicones

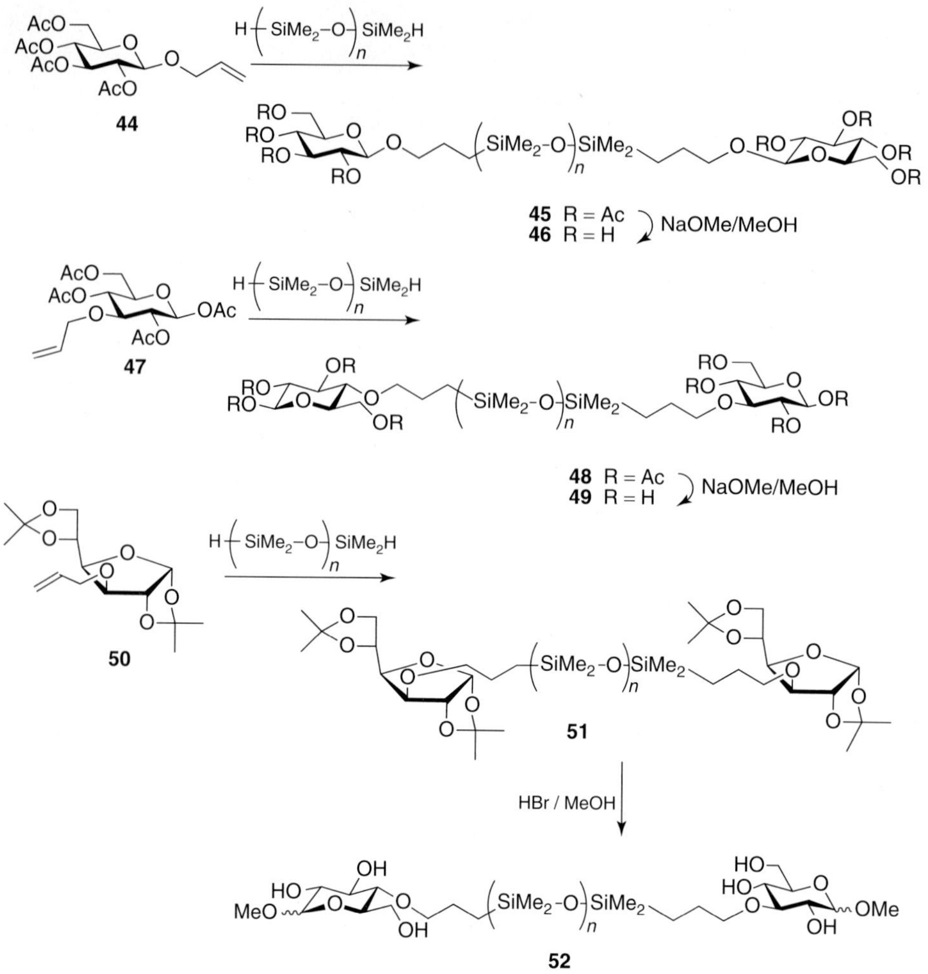

Scheme 10.15 Synthesis of terminal carbohydrate-modified poly(dimethylsiloxane)s.

showed $\overline{M_n}$ = 4000–5500, which should be of interest for further applications (Scheme 10.17).

Some recent investigations studied *glycosilicon-polyamide type structures* [28], another novel class of conjugates. These materials were accessible by employing several different ratios of *gluco : galacto* : PDMS and $\overline{M_n}$ of from 1500 to 5500 with interfacial polycondensation. Those showing a high T_g (120–130 °C) are gum-like, those with a broad T_g around room temperature are wax-like and give transparent foils. Initial studies on their biodegradability showed a considerable decomposition in the presence of proteinase XIV, amounting to about 0.15% after 10 h.

As has been briefly reviewed in this chapter, our interest was to incorporate saccharide-immanent functionalities into various types of polymers. While considering accessible renewable starting materials, the facile formation of adapted

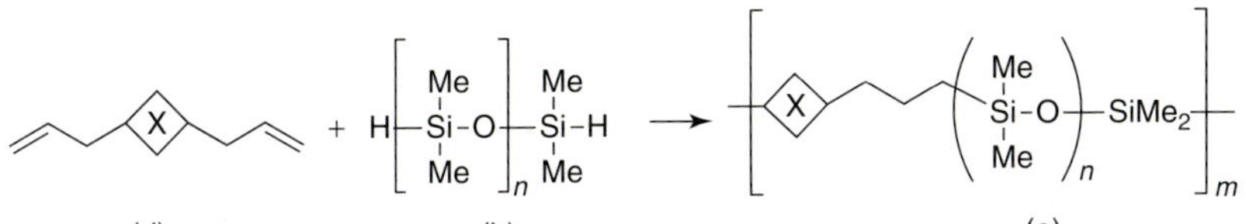

Scheme 10.16 Hydrosilylation of bis-olefin functionalized saccharide derivatives with polydimethylsiloxane.

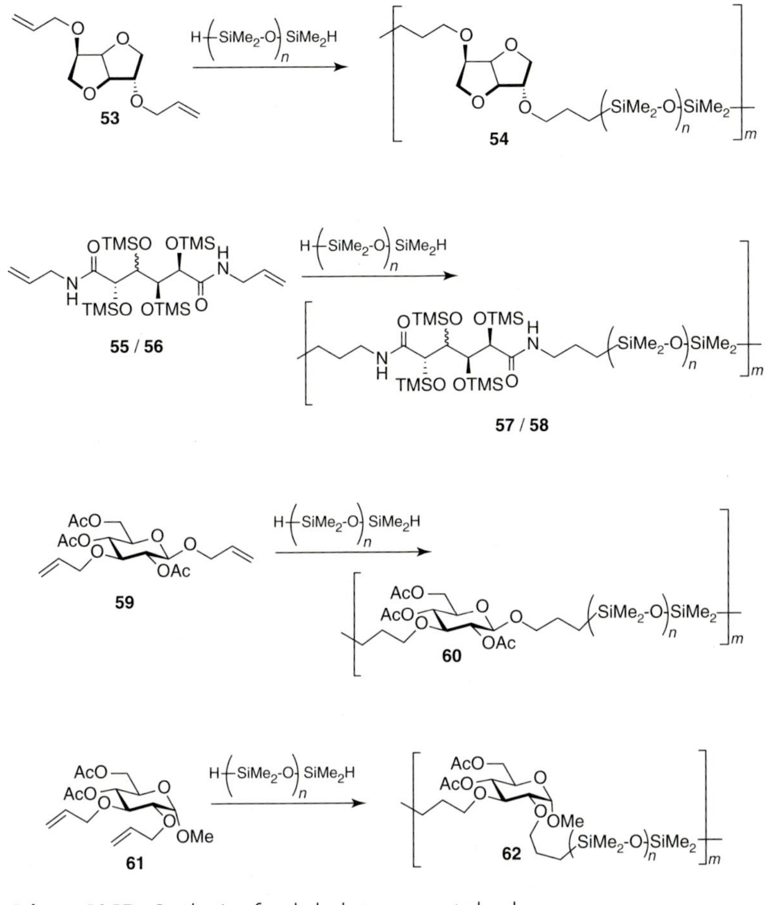

Scheme 10.17 Synthesis of carbohydrate-segmented polydimethylsiloxanes by hydrosilylation.

building units is attractive. Through the use of saccharide precursors, novel materials should become accessible, which may have approximately the same structural stability as polymers presently employed, yet have the advantage of a more rapid degradation via natural cycles. Thus, a number of polyethers, polyamides, polyurethanes, polyureas, and polyglycosilicones could be prepared and characterized. It should be mentioned that at present (oligo)saccharides can be employed in simple turnover processes, such as catalytic hydrogenations, reductive amination, esterifications, dihydrations, and hydrosilylations to give alditols, anhydroalditols, aminoalditols, alkyl polyglycosides, saccharide esters, hydroxy methyl furfural, and saccharide silicones. Some of these components can be used on a larger scale for an enormously wide range of applications, including interesting niche products, while others are still waiting for their first convincing utilizations.

References

1. Werpy, T. and Petersen, G. (2004) Top Value Added Chemicals from Biomass: Vol. I – Results of Screening for Potential Candidates from Sugars and Synthesis Gas. NREL Report No. TP-510-35523; DOE/GO-102004-1992. National Renewable Energy Laboratory, Golden, CO, 76 pp.
2. Werpy, T.A. et al. (2005) Top Value Added Chemicals from Biomass (Vol. 1). NREL Report No. AB-510-39815. 27th Symposium on Biotechnology for Fuels and Chemicals: Program and Abstracts (NREL/BK-510-36826), 1-4 May 2005, Denver, CO. National Renewable Energy Laboratory, Golden, CO, p. 20.
3. Callstrom, M.R. and Bednarski, M.D. (1992) *MRS Bull.*, **17**, 54–59.
4. Kricheldorf, H.R. and Damrau, D.O. (1997) *Macromol. Chem. Phys.*, **198**, 1753–1766.
5. Kricheldorf, H.R. (2001) *Chemosphere*, **43**, 49–54.
6. Pfannemueller, B. and Welte, W. (1985) *Chem. Phys. Lipids*, **37**, 227–240.
7. Meerwein, H., and Battenberg, E. (1939) A.-G. Farbenindustrie, I.G. D.R.P. 741.478.
8. Meerwein, H., Delfs, D., and Morschel, H. (1960) *Angew. Chem.*, **72**, 927–934.
9. Thiem, J. and Haering, T. (1987) *Makromol. Chem.*, **188**, 711–718.
10. Thiem, J., Haering, T., and Strietholt, W.A. (1989) *Starch/Staerke*, **41**, 4–10.
11. Thiem, J., Strietholt, W.A., and Haering, T. (1989) *Makromol. Chem.*, **190**, 1737–1753.
12. Lemieux, R.U. and Innes, A.G.M. (1960) *Can. J. Chem.*, **38**, 136–140.
13. Thiem, J. and Strietholt, W.A. (1995) *Makromol. Chem.*, **196**, 1487–1493.
14. Carothers, W.H. (1929) *J. Am. Chem. Soc.*, **51**, 2548–2559.
15. Kohan, M.I. (2001) *Industrial Polymers Handbook* in (ed. E.C. Wilks) Polyamides, I. Vol. 1, Wiley-VCH Verlag GmbH, Weinheim, pp. 245–289.
16. Thiem, J. and Bachmann, F. (1991) *Makromol. Chem.*, **192**, 2163–2182.
17. Bachmann, F. and Thiem, J. (1992) *J. Polym. Sci. Part A: Polym. Chem.*, **30**, 2059–2062.
18. Thiem, J. and Bachmann, F. (1993) *Makromol. Chem.*, **194**, 1035–1057.
19. Thiem, J. and Bachmann, F. (1994) *Trends Polym. Sci.*, **2**, 425–432.
20. Thiem, J. and Lueders, H. (1986) *Makromol. Chem.*, **187**, 2775–2785.
21. Bachmann, F., Ruppenstein, M., and Thiem, J. (2001) *J. Polym. Sci. A, Polym. Chem.*, **39**, 2332–2341.
22. Bachmann, F., Reimer, J., Ruppenstein, M., and Thiem, J. (2001) *Macromol. Chem. Phys.*, **202**, 3410–3419.
23. Bachmann, F., Reimer, J., Ruppenstein, M., and Thiem, J. (1998) *Macromol. Rapid Commun.*, **19**, 21–26.
24. Gruening, B. and Koerner, G. (1989) *Tenside Surfact. Deterg.*, **26**, 312–317.
25. Henkensmeier, D., Abele, B.C., Candussio, A., and Thiem, J. (2004) *Macromol. Chem. Phys.*, **205**, 1851–1857.
26. Henkensmeier, D., Abele, B.C., Candussio, A., and Thiem, J. (2005) *J. Polym. Sci. A, Polym. Chem.*, **43**, 3814–3822.
27. Kiely, D.E., Chen, L., and Lin, T.H. (1994) *J. Am. Chem. Soc.*, **116**, 571–578.
28. Henkensmeier, D., Abele, B.C., Candussio, A., and Thiem, J. (2004) *Polymer*, **45**, 7053–7059.

11
Degradable and Biodegradable Polymers by Controlled/Living Radical Polymerization: From Synthesis to Application

Nicolay V. Tsarevsky

11.1
Introduction

The kinetic and thermodynamic aspects and also the mechanisms of polymer degradation have been thoroughly covered in the literature [1–4]. This chapter deals specifically with the application of controlled/living radical polymerization (CRP) in the synthesis of well-defined macromolecules with precisely placed degradable or biodegradable links.

The interest in developing methods for the synthesis of *(bio)degradable polymers* has been fuelled both by environmental concerns and by the numerous applications of these materials. The ever-increasing production of polymers has inevitably led to the generation of larger and larger amounts of waste. Although efficient recycling is the most important strategy to deal with the *polymeric waste*, some of the waste does accumulate in the environment and it is important to develop polymers that can slowly degrade when exposed to ambient conditions (sunlight, moisture, bacteria, etc.). Whereas many macromolecules containing heteroatoms in their backbones (polyesters, polyethers, polyamides, and polyurethanes) can generally be degraded by microorganisms [5], high molecular weight (MW) vinyl polymers are not biodegradable, with the notable exception of poly(vinyl alcohol) [6]. It has been well established, however, that microorganisms are capable of degrading polymers with all-carbon backbones, with relatively low MW or degree of polymerization (DP) (usually the DP should be below 30–40, but this depends upon the type of polymer). This is why significant efforts have been made to develop synthetic methods yielding vinyl polymers containing cleavable functionalities randomly distributed along the macromolecular backbone. The breakage of the "labile" groups yields polymers of reduced MWs, which can be degraded by microorganisms.

Many high-tech applications of (bio)degradable polymers, including soil treatment, drug or gene delivery, and tissue engineering, have additionally stimulated the development of methods for their synthesis [7–10]. Biodegradation is important if a polymer is to be used as a *drug carrier*, because high-DP polymers cannot be excreted and tend to accumulate in the body. The renal excretion efficiency of natural and synthetic macromolecules decreases with the increase of the macromolecular

Green Polymerization Methods: Renewable Starting Materials, Catalysis and Waste Reduction
Edited by Robert T. Mathers and Michael A. R. Meier
Copyright © 2011 WILEY-VCH Verlag GmbH & Co. KGaA, Weinheim
ISBN: 978-3-527-32625-9

hydrodynamic volume [11–13], which is related to the MW. For example, it has been determined [14] that for linear copolymers of N-(2-hydroxyethyl) methacrylamide and N-methacryloyltyrosinamide the MW threshold above which glomerular filtration becomes very limited is of the order of 45 000 g mol^{-1}. The *molecular architecture* (linear or branched) also determines the efficiency of renal excretion. For molecules with the same Stokes–Einstein radii, molecules with shapes close to spherical, such as ficoll (cross-linked copolymer of sucrose and epichlorohydrin), are excreted through the kidneys less efficiently than prolate ellipsoid-shaped molecules such as dextran [15]. Additionally, because a variety of densely charged anionic glycosialoproteins are present in the glomerular capillary, negatively charged macromolecules are excreted less efficiently than neutral macromolecules, which in turn are less efficiently excreted than cationic macromolecules of the same hydrodynamic volume [16]. An ideal polymeric drug carrier should have sufficiently high MW prior to delivery in order to avoid too early renal clearance but should degrade to fragments that can be excreted from the body within reasonable time once the drug has been delivered to the targeted location. In this respect, the possibility to control precisely the molecular composition, size, architecture, in addition to the placement of functional groups (degradable links) is crucial for the development of efficient *polymeric drug carriers*. Controlling the efficiency of degradation and the number and location of functional groups in a polymeric material is also important if the material is to be used as a scaffold for tissue regeneration. Here the erosion and eventually the complete degradation of the material should be at a rate that takes into account the rate of tissue regeneration. Also, the ability to attach to the degradable polymer small molecules that improve cell adhesion or promote cell proliferation (for instance, RGD-containing small peptides [17, 18]) is very desirable. All these requirements make CRP techniques very attractive.

A variety of cleavable links are known, including disulfide **1**, ester **2** and **3**, thio- or dithioester **4**, amide or hydrazide **5**, imide (Schiff base) **6**, carbamate (urethane) **7**, and acetal **8**. These functional groups and the modes of their *degradation* (reduction in the case of disulfides and hydrolysis in the other cases) are presented in Figure 11.1. The monomer units in most polycondensation-type polymers, including natural polymers (proteins, nucleic acids, and polysaccharides), are normally connected via these groups. Additionally, there are methods to introduce one or more of them into the backbone of vinyl polymers to make them (bio)degradable [19]. It should be noted in passing that the presence of one or more biodegradable groups in a large polymer molecule does not guarantee that the polymer itself will be biodegradable. It can be argued that for a polymer molecule to be degradable under biological conditions, that is, in the presence of enzymes as catalysts, certain flexibility is required so that the polymer can fit into the active site of the enzyme. Additionally, the solubility or swelling behavior of the polymer, which may be very different in pure solvents or simple solvent mixtures compared with real biological fluids, normally affects dramatically their degradation behavior. In the majority of polymer degradation studies, the processes (hydrolysis, aminolysis, reduction, oxidation) are carried out under conditions that

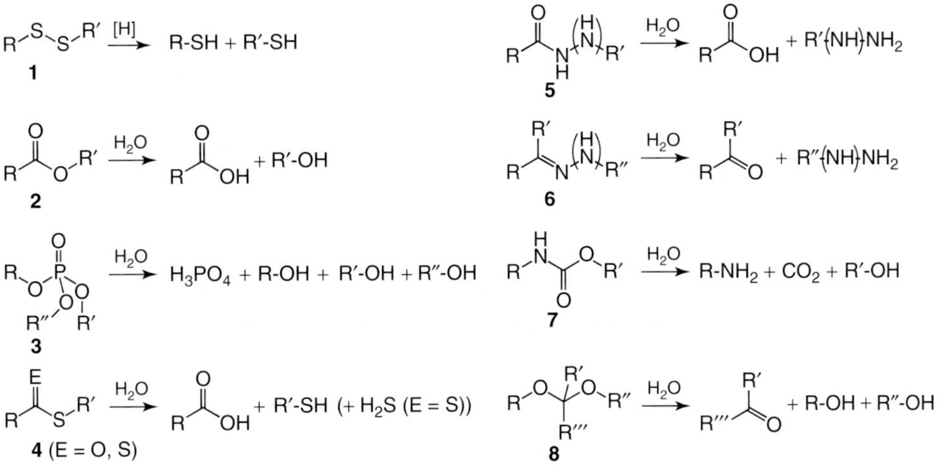

Figure 11.1 Selected (bio)degradable functional groups and products of their cleavage. The parentheses around the NH fragment in **5** and **6** indicate that it may or may not be part of the structures. The hydrolytic cleavage reactions are carried out in the presence of either a base or an acid, and in the former case, the deprotonated forms of the products are formed (e.g., carboxylate, phosphate, or thiolate anions, free amines or hydrazines), while in the latter, the protonated products are formed (e.g., carboxylic or phosphoric acid, thiols, and ammonium or hydrazonium salts).

do not even remotely resemble those in biological systems. Therefore, conclusions about the "biodegradability" (instead of merely "degradability") of a polymer based on experiments conducted *in vitro* should always be taken with a grain of salt.

Radical polymerization is a very attractive synthetic method, both for the small-scale and industrial production of polymers, due to the easy experimental setup (i.e., lack of sensitivity toward moisture, carbon dioxide, or other protic or nucleophilic compounds), the wide range of monomers, including a multitude of compounds with a carbon=carbon double bond and certain heterocyclic compounds, and also the ease of conducting copolymerization reactions. Radical polymerization has been successfully used to synthesize degradable polymers. The incorporation of the cleavable functional group can be achieved by the use of functional monomers or initiators. Further, radical ring-opening polymerization (RROP) [20–23] has a great potential in the preparation of degradable polymers. Selected structures of monomers that undergo RROP, including heterocyclic monomers with exocyclic carbon=carbon double bonds (**9–13**) [24] and cyclic di- or polysulfides (**14**), and that yield polymers with degradable groups along the backbone are presented in Figure 11.2.

One of the drawbacks of conventional radical polymerization is that it cannot be used to synthesize polymers with controlled size, that is, DP, narrow molecular weight distribution (MWD), and well-defined molecular architectures or chain-end functionality. Such control over molecular parameters is crucial in the design of degradable polymers for biomedical applications.

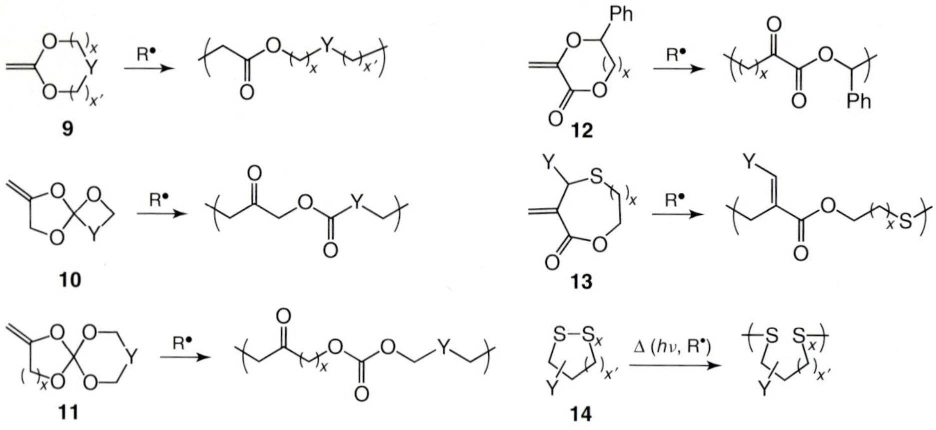

Figure 11.2 Selected heterocyclic monomers undergoing RROP and structures of the polymers prepared from them and containing (bio)degradable functional groups in the backbone. Polymeric esters degrade upon hydrolysis, whereas di- or polysulfides degrade upon reduction. Y represents a functional group and x and x′ are small integers.

11.2
(Bio)degradable Polymers by CRP

To synthesize polymers of desired MW and narrow polydispersity, *living polymerization methods* are generally used [25], in which irreversible chain termination is eliminated. The first living polymerization techniques discovered were ionic processes [26–28]. They, however, suffer from a marked sensitivity to moisture, carbon dioxide, and a number of acidic or basic/nucleophilic compounds, many of which are common impurities in commercial monomers or solvents. Additionally, *ionic polymerizations* can only be applied to a limited range of monomers, and, due to significant differences in the reactivity ratios of the monomers, copolymerization reactions are often challenging. Radical polymerizations, in contrast, are significantly more robust and are applicable to a much wider range of monomers. Owing to the occurrence of inevitable bimolecular radical termination, the development of *living-like radical polymerizations* seemed challenging for a long time. This situation changed in the late 1980s and especially in the 1990s, when several methods of CRP [29–37] were developed that allowed for the preparation of a many previously unknown well-defined polymers. The most widely used CRP methods are: (i) atom transfer radical polymerization (ATRP) [38–41]; (ii) stable free radical polymerization (SFRP), including nitroxide-mediated polymerization (NMP) [42–44], and polymerizations mediated by Co complexes [45, 46] or other organometallic compounds [47]; (iii) degenerative transfer polymerization, such as reversible addition-fragmentation chain transfer (RAFT) polymerization [48–52], iodine-transfer polymerization [53], and polymerizations mediated by

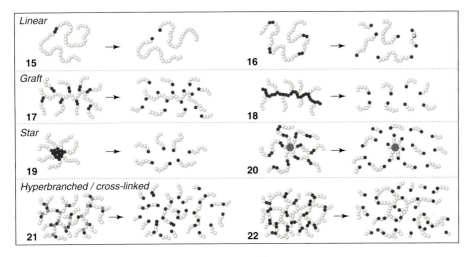

Figure 11.3 Examples of (bio)degradable polymers with controlled architectures and placement of the degradable functionality (represented by two dark solid spheres).

organotellurium, antimony, or bismuth compounds [54, 55], and (iv) reversible chain transfer catalyzed polymerization [56].

All CRP methods allow for the synthesis of polymers of controlled MW, MWD, and architecture, and numerous nanostructured polymeric materials are known [57]. Figure 11.3 shows some examples of degradable polymers and the products of their degradation, including linear (**15** and **16**, with one or multiple degradable functionalities, respectively), graft (**17** and **18**, with degradable side chains or backbone, respectively), star-shaped (**19** and **20**, with degradable core or arms, respectively), (hyper)branched **21**, and cross-linked (network) **22** polymers. All of these macromolecules have been successfully prepared by CRP methods, although their degradation has not always been studied in detail or even demonstrated. The remainder of this work describes the synthesis of degradable polymers by CRP and, where applicable, summarizes the studies on their degradation. The text is organized based on the structure of the polymers.

11.2.1
Linear (Bio)degradable Polymers

11.2.1.1 Polymers with a Degradable Functional Group

A single *degradable functional group* is easily introduced into polymers prepared by ATRP. To achieve this, ATRP is conducted using a difunctional alkyl halide initiator with an internal cleavable link (see Figure 11.1 for examples).

Disulfide is a degradable group that can be cleaved reversibly upon reduction, producing a mixture of the corresponding thiols [58]. Thiols [59, 60], phosphines [61–63], metal hydrides, and various metal–acid combinations are suitable reducing agents. One of the popular reducing agents for disulfides is the dithiol dithiothreitol

(DTT) (Cleland's reagent [59]), which forms a six-membered cyclic disulfide in the redox process. The thiol-containing tripeptide L-γ-glutamyl-L-cysteinylglycine, named glutathione (GSH) [64], one of the most important biological reducing agents, is also efficient at reducing disulfides. Owing to the ease of cleavage of disulfide bonds, they have been widely utilized in conjugates for controlled drug delivery [65]. Polymers containing disulfide or polysulfide groups have attracted significant interest [66, 67], and various procedures for their synthesis have been described [68, 69]. As the reductive degradation of a disulfide leading to two thiol molecules is reversible and the disulfide can be easily regenerated upon oxidation of the thiols, polymers with disulfide links can be viewed as *"responsive"* [70], that is, responding (cleavage or coupling) to changes in the concentration of redox-active compounds.

Well-defined polymers with a disulfide group in the middle of the macromolecule can be synthesized by using one of the initiators **23**, suitable for the polymerization of styrene (Sty) [71] or acrylates [72], or **24–25**, suitable also for the polymerization of methacrylates [73–78] (Figure 11.4). The 2-pyridyldisulfide (2-PyS$_2$-) group-containing 2-bromoisobutyrate initiator **26** ($n = 2$) was used in the ATRP of 2-hydroxyethyl methacrylate (HEMA), yielding the biocompatible [79] polyHEMA with a reactive disulfide bond, which was then reacted with the free thiol groups of a protein, bovine serum albumin (BSA), leading to the formation of a polymer–protein conjugate material [74]. Similarly, ATRP was employed for the preparation of thiol-reactive poly N-isopropylacrylamide (NIPAAm) [80]. A variety of polymers were synthesized by ATRP using initiator **26** ($n = 1$). When a polymer

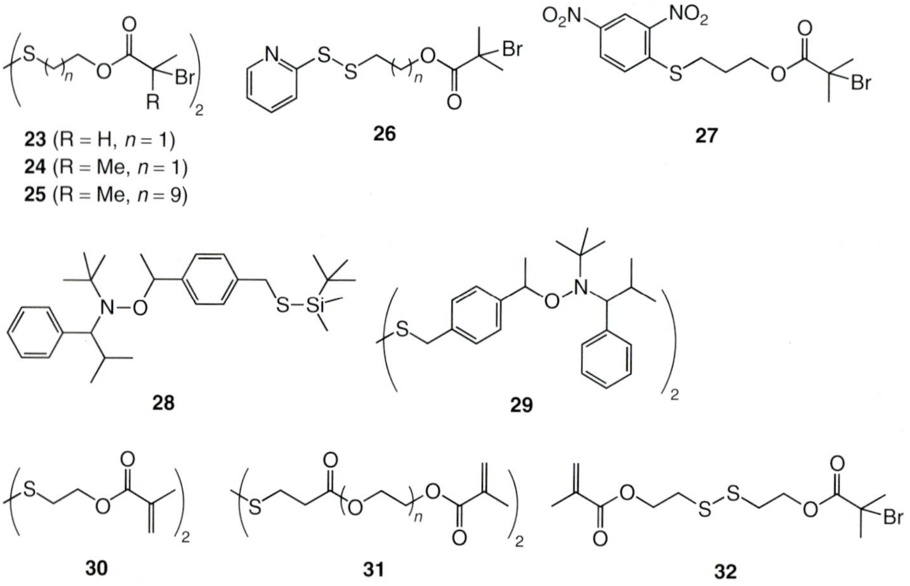

23 (R = H, $n = 1$)
24 (R = Me, $n = 1$)
25 (R = Me, $n = 9$)

Figure 11.4 Initiators and monomers (including cross-linkers) used for preparation of disulfide- or thiol-containing polymers by CRP.

with a 2-PyS$_2$ endgroup, polyA-S$_2$-2-Py, was reacted with DTT, the thiol-terminated polymer polyA-SH was produced. The reaction of this polymer with polyB-S$_2$-2-Py afforded the unsymmetrical disulfide-linked block copolymer polyA-S$_2$-polyB [81].

The synthesis by RAFT was also reported for 2-PyS$_2$-terminated polymers, such as polyNIPAAm [82] and polyoligo(ethylene glycol) acrylate (OEGA) [82, 83], which reacted with BSA [82] or small interfering ribonucleic acid (siRNA) [83]. Micelles formed in aqueous media from a polySty-b-polyOEGA, terminated at the side of the hydrophilic segment with 2-PyS$_2$ groups had numerous thiol-reactive groups on the surface and reacted with BSA [84]. The use of CRP methods, mostly ATRP, for the synthesis of bioconjugates materials has been summarized in a recent review [85].

To prove that a degradable polymer is also biodegradable, it is essential to show that the degradation is successful either *in vivo* or at least under conditions close to the biological environment of living cells. In other words, an aqueous medium with physiological pH, salt concentration, and temperature, should be used in addition to reagents that living cells would use in the degradation. A disulfide-containing block copolymer, polyNIPAAm-b-polyMPC-S$_2$-polyMPC-b-polyNIPAAm (MPC = 2-methacryloyloxyethyl phosphorylcholine), was shown to easily degrade in aqueous medium using GSH [86]. As polyNIPAAm is hydrophobic above 37 °C but becomes hydrophilic below this temperature, the triblock copolymer with disulfide groups could serve as a biodegradable gelator. A well-defined polymer containing multiple pendant 2-PyS$_2$ groups was prepared by the ATRP of the corresponding 2-PyS$_2$-containing monomer, 2-pyridyl 2′-methacryloyloxyethyl disulfide [87]. The polymer could react with various thiols [87], including 3-mercaptopropionic acid [88]. The obtained polymer formed supramolecular aggregates with the cationic surfactant decyltrimethylammonium bromide, which disintegrated upon addition of GSH. This was due to the cleavage of the negatively charged pendant carboxylate groups, each linked to the polymer via a disulfide group. The reductively degradable supramolecular assemblies were loaded with a dye, Nile red, which was released in the presence of GSH [88].

The reductive cleavage of molecules containing a disulfide link to two thiols is reversible, and it was shown [71] that the thiol-terminated polySty obtained from the corresponding disulfide upon reduction with DTT could be quantitatively converted into the starting material via oxidation with FeCl$_3$ (Figure 11.5).

A single thiol group can be introduced at the α-end of a polymer prepared by ATRP by using a protected thiol alkyl halide initiator, such as **23–26** or 2-(2,4-dinitrophenylthio)ethyl 2-bromoisobutyrate **27** [89]. Alternatively, it should be remembered that the polymers prepared by ATRP are structurally polymeric alkyl halides and a variety of functionalization reactions are possible [90]. For instance, the halogen atom can be easily converted into a thiol group in a reaction with a suitable S-nucleophile, such as N,N-dimethylthioformamide [71] or thiourea [91].

RAFT polymerization, which is mediated by trithiocarbonates, xanthates, or dithiocarbamates, is very suitable for the synthesis of thiol-terminated polymers. For this purpose, the endgroup can be reduced [92–94], hydrolyzed [95], or aminolyzed [96–102] to thiol. The produced thiols are easily oxidized in the

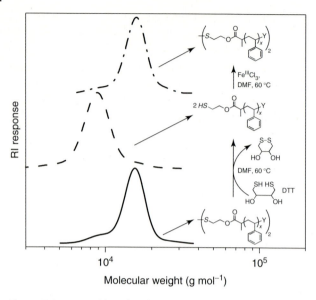

Figure 11.5 Reversible redox cleavage/coupling of polySty with an internal disulfide link prepared by ATRP. (Modified from Tsarevsky and Matyjaszewski [71]. Copyright 2002, American Chemical Society).

presence of air to the corresponding disulfides. If the aminolysis is carried out in the presence of air, the obtained mercapto-terminated polymers often couple to the corresponding disulfides with double the MW. Evidence has been provided that the aminolysis of polymethacrylates synthesized by RAFT affords not only thiol- but also cyclic thiolactone-terminated polymers, the latter being the products of a "backbiting" reaction between the thiol terminal group and the ester group from the penultimate monomer unit [98]. When the aminolysis reaction is carried out in the presence methyl methanethiosulfonate, MeSO$_2$SMe, clean methyldisulfide-terminated polymers are obtained without side products [103]. NMP can also be used for the preparation of polymers with terminal thiol or internal disulfide groups if the alkoxyamines **28** or **29**, respectively, are employed in the polymerization [104].

The great utility of RAFT polymerization in the synthesis of biodegradable polymers, polymeric micelles, or shell–cross-linked micelles, as well as bioapplications, including drug- or nucleic acid delivery, are the subject of a thorough review [105].

CRP functional initiators or transfer agents with other degradable groups (cf. Figure 11.1) can be prepared similarly but all possibilities have not yet been explored fully.

11.2.1.2 Polymers with a Degradable Polymeric Segment

ABA-type block copolymers, in which the middle segment is biodegradable can be synthesized easily by preparing a *macroinitiator*, such as difunctional activated alkyl halide-terminated polyCL, followed by a chain extension via ATRP with one or more vinyl monomers, such as sugar-based methacrylates [106, 107], Sty and acrylonitrile (AN) [108], 2-(N,N-dimethylamino)ethyl methacrylate (DMAEMA) [109], or NIPAAm [110]. Upon degradation, these triblock copolymers should ideally yield the pure A segments and the MW should decrease by at least a factor of two (for the case of long A segments and very short, oligomeric, B segment), and in most instances, the overall MW decrease should be significantly more substantial.

Shell–cross-linked micelles are very useful as nano-sized "capsules" and numerous examples of preparation and uses are known [111]. As an example of a biodegradable block copolymer, polyCL-b-polyAA was prepared by ATRP, and was then assembled in an aqueous solution. The corona of the micelle was cross-linked with a diamine. Thus, nanoparticles with hydrolytically degradable cores (originating from the polyCL segments) were formed [112].

11.2.1.3 Polymers with Multiple Cleavable Groups or Polymeric Segments

Multiple degradable functionalities can be introduced into the backbone of a linear polymer by conducting controlled radical copolymerization of vinyl monomers with a monomer undergoing RROP. To introduce hydrolytically or photodegradable α-ketoester groups, cyclic ester or anhydride monomers with an exocyclic double bond, such as 5-methylene-2-phenyl-1,3-dioxolan-4-one can be employed [113]. The radical homo- or copolymerization of the cyclic monomer, 5,6-benzo-2-methylene-1,3-dioxepane **33** [114], which was successfully carried out under ATRP conditions [115], led to polymers with ester groups (Figure 11.6). The atom transfer copolymerization of **33** with n-butyl acrylate (nBA) proved useful for the synthesis of hydrolytically degradable poly-nBA containing mostly isolated [as confirmed by nuclear magnetic resonance (NMR) and alcoholysis studies], hydrolytically degradable ester links [116]. The same monomer, **33**, was also copolymerized with methacrylate esters of ethylene oxide dimers and oligomers under ATRP conditions [117], or with NIPAAm under both ATRP and RAFT conditions [118], yielding the corresponding biocompatible hydrolytically degradable polymers.

Another approach to introduce degradable groups into polymers prepared by ATRP is to use *difunctional (macro)initiators* containing a cleavable functionality (for instance, one of the groups presented in Figure 11.1) and to couple the prepared well-defined polymer chains. Several efficient methods exist for the

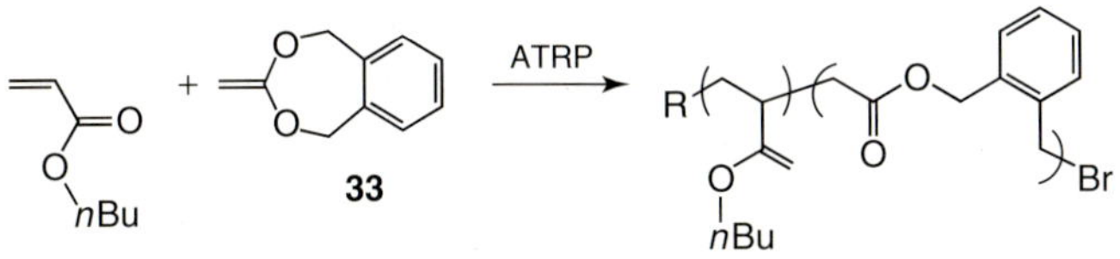

Figure 11.6 Radical ring-opening copolymerization of nBA and 5,6-benzo-2-methylene-1,3-dioxepane (BMDO) under ATRP conditions yielding a copolymer with hydrolytically degradable ester functionality.

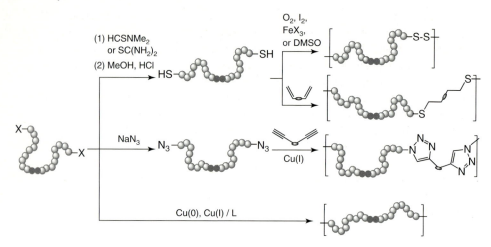

Figure 11.7 Coupling reactions applicable to polymers prepared by ATRP yielding high molecular weight polymers with multiple degradable groups or polymeric segments (represented by two solid dark spheres).

coupling, including the thiol oxidation to disulfide, thiolene coupling [119–121], azide–alkyne click coupling [122], or atom transfer radical coupling [123, 124] (Figure 11.7). In an interesting example, dibromo-terminated polymers prepared by ATRP were reacted with Cu(I) complexes in the presence of Cu(0) and the produced radicals were coupled with various dinitroxides, in which the two nitroxide moieties were connected through an ester or disulfide group. The produced high MW alkoxyamine polymers could be degraded upon heating (with cleavage of the $C_{(polymer)}-O_{(nitroxide)}$ bond), hydrolysis (with cleavage of the ester groups connecting the alkoxyamine moieties), or reduction (disulfide cleavage) [125]. It is envisioned that in the future, the coupling methods depicted in Figure 11.7 will be more widely utilized in the preparation of polymers with multiple degradable groups or polymeric segments.

11.2.2
Degradable Star Polymers

Star-shaped polymeric molecules, consisting of at least three identical or different linear polymer chains connected at a central core, represent the simplest example of *branched polymers* [126, 127]. In many respects, star polymers can be viewed as unimolecular micelles that do not possess a critical micelle concentration. CRP methods allow for the synthesis of star copolymers with independent control over three structural parameters, namely number, size, and composition of the arms. Further, *miktoarm star copolymers* exist, where two (or more) types of arms in the same macromolecule have different composition and/or size. In addition to the "core first" method, which employs the use of multifunctional initiators, "arm first" methods have been developed for the synthesis of star polymers, in which macroinitiator molecules are synthesized first, which are then coupled together, often in the presence of di- or multivinyl cross-linkers [128].

Star-shaped polyCL was synthesized using multifunctional hydroxy-initiators and the obtained polymer was end-functionalized with a bromine-containing ATRP initiator. The polymerization of methyl methacrylate (MMA) [129] or Sty [130] from the star-shaped macroinitiator yielded stars with degradable internal and non-degradable external parts. In a similar approach, hyperbranched poly-HEMA with an average of 18 hydroxy groups was used as a macroinitiator for the ring-opening polymerization (ROP) of ε-caprolactone (CL), the hydroxy endgroups in the obtained star-like molecule with a hyperbranched core were converted into bromoester groups and chain extension reactions were carried out with DMAEMA or *tert*-butyl methacrylate (*t*BMA) under ATRP conditions [131]. The opposite case, namely core–shell stars with a degradable shell was realized by the synthesis of a four-arm polySty star by ATRP, reacting the alkyl bromide chain-ends with diethanolamine leading to a star polymer with eight peripheral hydroxy groups, which was further used as a macroinitiator in the ROP of CL [132]. Miktoarm star copolymers, in which only some of the arms are degradable, for instance polyCL, have also been reported [133]. The products of degradation of these copolymers are still star-shaped copolymers but with lower overall MW.

An example of the *"arm-first" method* for the preparation of degradable star copolymers was the synthesis of polyMMA macroinitiators by ATRP, which were then cross-linked in the presence of **30** (Figure 11.4). Degradation of the star in the presence of a reducing agent led to formation of linear polyMMA. The halogen atoms were preserved at the core of the star and could be successfully used as initiating sites in a chain extension reaction with *n*BA. The produced star copolymers degraded with the formation of polyMMA-b-poly-*n*BA block copolymers (Figure 11.8) [134].

A number of star copolymers, with either degradable arms or cores were prepared by ATRP of vinyl monomers combined with the ROP of monomers with one or two CL groups, via the "arm-first" approach [135, 136]. RAFT of 2-aminoethylmethacrylamide hydrochloride followed by chain extension with NIPAAm and **30** yielded stars with loosely cross-linked cores that could be reductively degraded in the presence of DTT [137].

11.2.3
Degradable Graft Polymers (Polymer Brushes)

There are three general methods to prepare graft copolymers: (i) *grafting through*, where macromonomers are polymerized, containing the side chains of the final graft copolymer; (ii) *grafting from*, in which a multifunctional macroinitiator (the backbone of the final graft copolymer) is prepared first and one or more monomers are polymerized yielding the side chains; and (iii) *grafting to*, where both the graft copolymer backbone and side chains are pre-made and are coupled together using some efficient synthetic methodology.

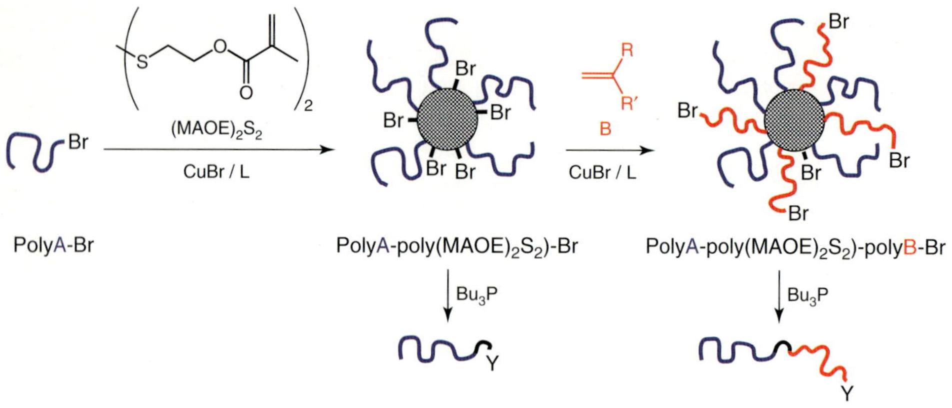

Figure 11.8 Synthesis and reductive degradation of miktoarm star copolymers by the "arm-first" method followed by chain extension with a second monomer. (Adapted from Gao et al. [134]. Copyright 2005, American Chemical Society).

Macromonomers based on the hydolytically degradable polylactide or polyCL have been successfully prepared and polymerized by NMP to yield polymer brushes with degradable side chains [138]. PolyCL-based macromonomers were copolymerized with DMAEMA under ATRP conditions to polymer brushes with hydrolysable side chains [139]. Also, graft copolymers with polyCL side chains were prepared by grafting of polyCL from a hydroxy group-containing polySty backbone, and the side chains were extended (after modification of their hydroxy endgroups) via the ATRP of MMA [140], producing graft copolymers with segmented side chains, the internal (closer to the backbone) segment of which was biodegradable. Polymer brushes with degradable side chains could also be synthesized by a one-pot procedure, in which MMA and HEMA were copolymerized by ATRP initiated by dichloroacetophenone in the presence of CL and a catalyst for the ROP of the latter monomer, aluminum tri(i-propoxide). The hydroxy groups of HEMA served as initiating sites for the polymerization of CL [141]. Copolymers of MMA and HEMA prepared by ATRP were also used as macroinitiators in the enzymatic ROP of CL [142] or in the ROP of dilactide [143].

Graft copolymers with a *(bio)degradable backbone* (typically, polyester-type) have also been prepared by ATRP using "grafting from" techniques. The iBu$_3$Al-initiated ring-opening copolymerization of cyclic phosphates with 2-bromoisobutyrate groups afforded linear polyphosphate esters with pendant ATRP initiating sites (Figure 11.9) that were further used in chain extension reactions with MPC [144]. The hydrolytic degradation of the backbones was studied in buffered solutions and was shown to be rather fast in alkaline media (pH 11). Degradable polymer brushes with polyMPC grafts and containing cholesteryl groups, prepared in a similar manner, were shown to possess practically no cytotoxicity towards v79 cells [145].

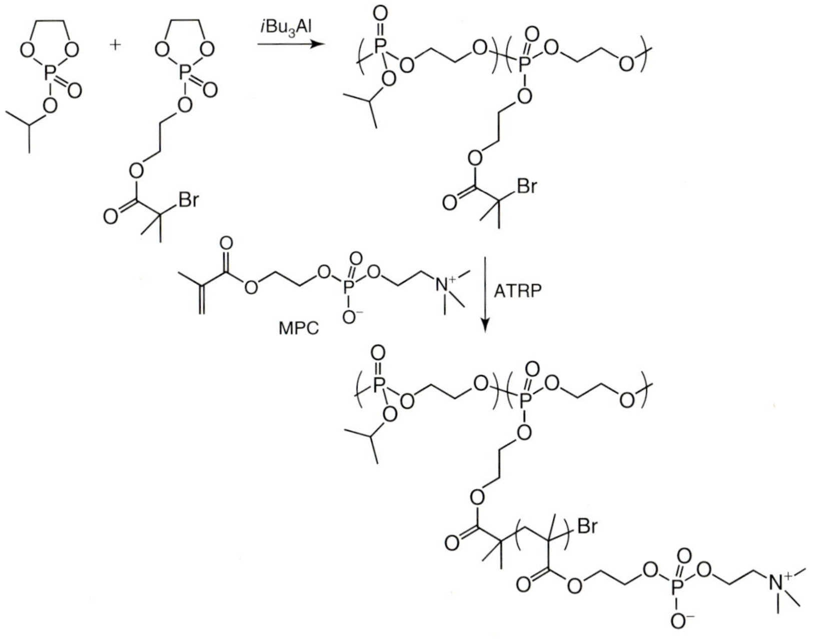

Figure 11.9 Synthesis of degradable graft copolymers (polymer brushes) with polyphosphate backbone [145].

The ROP of γ-2-bromoisobutyryloxyCL [146] or α-chloroCL [147] or α-bromoCL [148] led to a degradable multifunctional ATRP macroinitiator, which was subsequently used in the polymerization of MMA or Sty, yielding brush copolymers with degradable backbones. The polycondensation reactions between 1,4-butanediol and 2-bromoadipic or 2-bromosuccinic acid in the presence of Sc(OTf)$_3$ (Tf = trifluoromethanesulfonyl) yielded bromine-containing polyesters that were used as multifunctional initiators in the ATRP of MMA (Figure 11.10) [149].

Heterograft polymers with a polyCL-type backbone (originating from mixtures of CL and α-chloroCL) and polySty and poly(ethylene oxide) methyl ether methacrylate (polyMePEOMA) side chains have also been reported [150]. Polymer brushes with segmented side chains, the inner segment of which was biodegradable, were prepared by the ATRP of HEMA, followed by ROP of CL, conversion of the hydroxy groups into bromoester and chain extension with *n*BA, as shown in Figure 11.11 [151].

A number of *synthetic–natural polymer bioconjugates* have been synthesized by ATRP. The reason that ATRP is particularly suitable for the synthesis of such hybrid materials is that alkyl halide initiating groups are very easy to attach to a variety of substrates with amino, hydroxy, carboxylate, or mercapto groups, which occur in many biopolymers [57]. Graft copolymers with natural polymer-derived backbones, for instance, polysaccharide backbones, in which the monomer units are linked

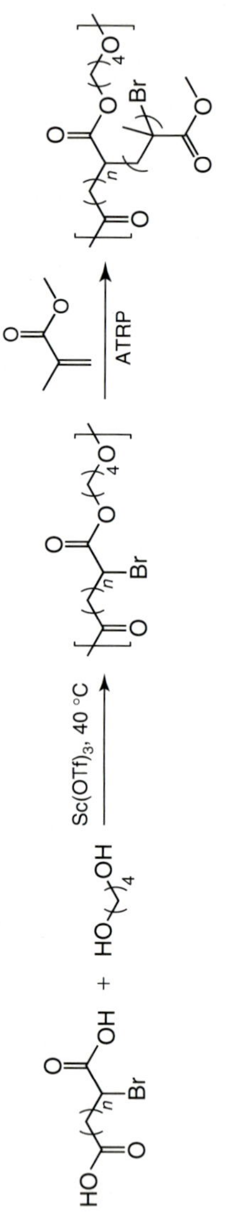

Figure 11.10 Synthesis of polymer brushes with hydrolytically degradable (polyester) backbone [149].

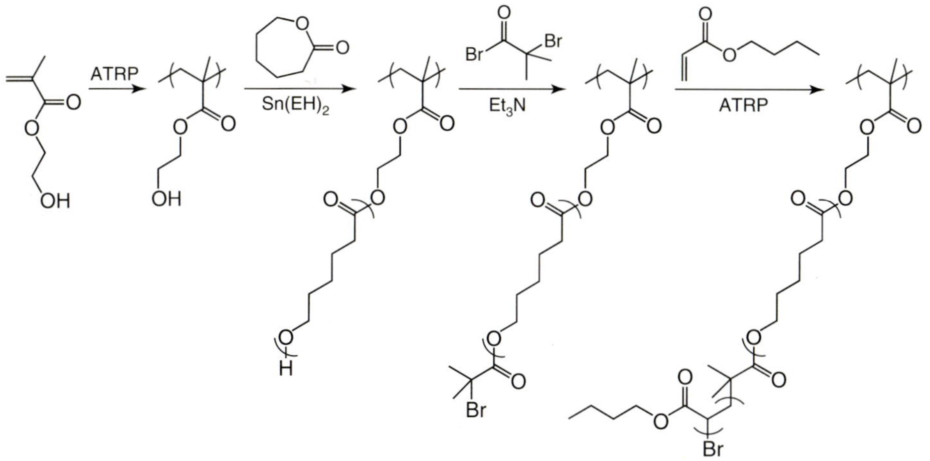

Figure 11.11 Synthesis of graft copolymers with segmented and degradable side chains via combination of ATRP and ROP [Sn(EH)$_2$ = tin(II) 2-ethylhexanoate] [151].

via glycoside groups, are hydrolytically degradable. Examples of polysaccharides modified with synthetic polymer side chains include cellulose [152–155], acetylated cellulose [156], or ethylcellulose [157], chitosan [154, 158–160], pullulan [161], and cross-linked dextran [162] (see Figure 11.12 for an example). Hydroxypropylcellulose was used in the ROP of CL and the hydroxy endgroups of the polyCL chains in the graft copolymer were converted into ATRP initiating sites, followed by the ATRP of *tert*-butyl acrylate (*t*BA) and hydrolysis of the *t*Bu groups. The obtained graft copolymers with cellulose backbone and polyCL-b-polyAA side chains with varying lengths of the polyAA segments were further reacted with diamines, yielding

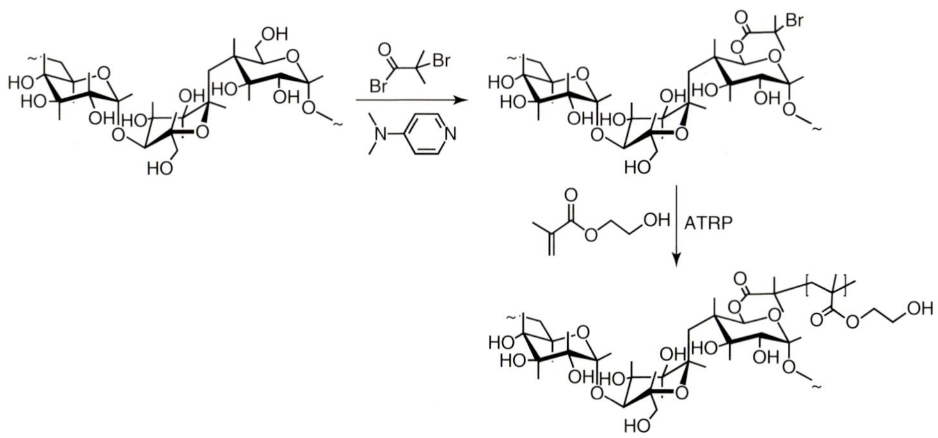

Figure 11.12 Synthesis of a polysaccharide (pullulan)–synthetic polymer bioconjugate by ATRP.

molecular "containers" with cross-linked shells [163]. The pullulan–polyHEMA graft copolymer conjugates obtained at different HEMA conversions were degraded in the presence of trifluoroacetic acid (under conditions at which polyHEMA itself was shown to be stable), and it was shown that the MWs of the yielded free polyHEMA were in agreement with the theoretically predicted ones, and that the polydispersity index (PDI) values were low [161].

11.2.4
Hyperbranched Degradable Polymers

Hyperbranched polymers [164–173] have attracted a great deal of interest as components of materials ranging from low-viscosity multifunctional coatings and adhesives to (drug) delivery formulations [174]. All branched macromolecules are more compact than their linear counterparts with the same MW, and because of their small hydrodynamic radius, combined with the presence of multiple functional groups (the endgroups of each individual branch), they are very promising for biomedical applications. In many respects the structures and applications of hyperbranched polymers are similar to those of dendrimers [175].

There are several approaches to synthesizing hyperbranched polymers from vinyl monomers. (i) Copolymerization of monovinyl with small amounts of di- or multivinyl monomers [176–179] in the presence of an efficient chain transfer agent (in order to avoid gelation). (ii) A similar approach is the *"initiator-fragment incorporation radical copolymerization,"* [180–182] in which di- and multivinyl monomers are (co)polymerized using a very large amount of radical initiator, comparable to that of the monomer, to guarantee that the MWs of the produced polymers will be sufficiently low, thus avoiding cross-linking. (iii) A chain-growth polymerization that results in hyperbranched macromolecules is the *"self-condensing vinyl polymerization"* (SCVP) [183–185], in which compounds are used, known as *inimers*, containing both a polymerizable group [such as styrene or (meth)acrylate moiety] and a group able to initiate polymerization by forming active species (e.g., radicals). All these methods can be used in the synthesis of (bio)degradable polymers by CRP, but the first and the third method have been used extensively.

Hyperbranched polymers with disulfide groups derived from 2-hydroxypropyl methacrylate could also be prepared using bis(2-methacryloyloxyethyl)disulfide [(MAOE)$_2$S$_2$] **30** (Figure 11.4), provided that the disulfide was used at low concentration (on average, less than one branching disulfide unit per polymer chain), in order to avoid formation of a macroscopic gel [186]. In an analogous fashion, highly branched polyHEMA was prepared by the atom transfer copolymerization of HEMA and **30**. The polymer was degraded in the presence of the biological thiol GSH [187]. Highly branched reductively degradable polymers derived from the same disulfide cross-linker, **30**, and MMA have also been synthesized by RAFT [188].

ATRP combined with SCVP is also very useful. For example, the two inimers [2-(2′-bromopropionyloxy)ethyl acrylate], containing the hydrolytically degradable ester functionality, and [2-(2′-bromoisobutyryloxy)ethyl 2″-methacryloyloxyethyl disulfide] **32** with the reductively degradable disulfide bond, have been prepared, polymerized, and copolymerized with Sty and MMA, respectively, under ATRP conditions. The obtained hyperbranched polymers were successfully degraded to lower MW polymers [189].

Hyperbranched polymers with controlled branching and containing polyCL-type segments have been synthesized by polymerizing either sequentially or simultaneously 2-bromoisobutyrate-functionalized CL and HEMA. The former monomer contained the ATRP initiating site, whereas the latter contained an initiating site for the ROP of the functional CL [190].

When inimers containing degradable units are polymerized, the degradation products of the branched polymers can be used to evaluate the degree of branching, provided that a technique that gives the precise MW of the fragments is available and that complete degradation of all cleavable groups is accomplished, as illustrated in Figure 11.13 [189]. In polymers prepared by ATRP, depending on the rate constants of activation of the initiating sites attached to the backbone ($k_{act,b}$) and those belonging to the monomer moiety, that is, the side chains ($k_{act,s}$), the degradation products will be very different. For the case when $k_{act,b} \gg k_{act,s}$, linear polymers of practically the same size as the starting material will be formed upon degradation, while for the case when $k_{act,s} \gg k_{act,b}$, the products of degradation will be small fragments with MWs close to that of the inimer. When truly branched

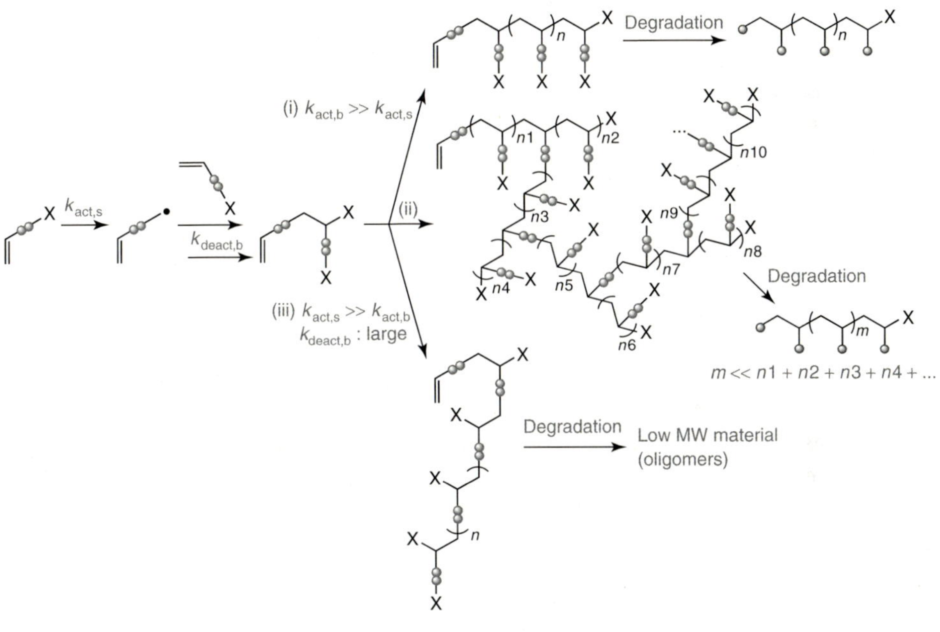

Figure 11.13 Possible structures of polymers derived from an inimer with a degradable functionality and products of their complete degradation. (From Tsarevsky et al. [189]. Copyright 2009, Wiley Interscience).

polymers are formed, the products of degradation will be oligomers with lower DP than the starting material, but with MWs higher than that of the inimer.

11.2.5
Cross-Linked Degradable Polymers

When CRP of a vinyl monomer in the presence of a di- or polyvinyl cross-linker is carried out, the obtained gels are much more homogeneous than those obtained in a conventional radical polymerization system, without formation of microgels, as demonstrated for the case of NMP [191, 192]. Depending on the exact time of introducing the cross-linker (at the beginning of the polymerization or near the end), gels, or star copolymers with various degrees of *core cross-linking* can be obtained [128]. Di-methacryloyloxy-terminated polyCL has been utilized as a hydrolytically degradable cross-linker in the ATRP of DMAEMA [193]. The produced gels were uniform, as expected from a CRP method.

Disulfide cross-linked gels were prepared by ATRP of MMA and **30**. They were successfully degraded in the presence of nBu$_3$P [75]. Each polymer chain in the gels retained its halogen endgroups after the cross-linking and washing, which was demonstrated by the successful use as of the gels as "*supermacroinitiators*" in a subsequent chain extension with a second monomer, Sty (Figure 11.14). The degradation products of the starting gels (polyMMA-based) and of gels with grafted polySty polymer chains were analyzed by 2-dimensional size exclusion chromatography (2D-SEC). It was shown that in the latter case, the degradation products consisted of the block copolymer polyMMA-b-polySty and no homopolymer of MMA could be detected [76].

The high degree of halogen end-functionalization of the cross-linked structures prepared by ATRP could be very useful in biomedical applications, for it would allow the attachment of various functional groups to the degradable gels. For this purpose, use can be made of simple nucleophilic substitution reactions [90]. The ATRP of MMA in the presence of a disulfide-containing cross-linker in a mini-emulsion yielded a stable polymer latex consisting of cross-linked and halogen-functionalized particles that degrade in a reducing environment [76]. It was also demonstrated [194, 195] that degradable nanometer-sized gel particles derived from biocompatible polymers, such as polyMePEGMA, could be prepared by ATRP in an inverse mini-emulsion using the disulfide-based dimethacrylate **31** (Figure 11.4, $n \approx 10$) as the cross-linker. These nanogels released incorporated dyes or drugs in reducing environments [195, 196]. Nanogels, loaded with dyes, with surface polymerizable groups attached through disulfide linkers were incorporated into a matrix comprised of hyaluronic acid modified with polymerizable groups. Photopolymerization yielded gels chemically attached to the hyaluronic acid based macrogel matrix, and the particles could be released upon reduction [197].

RAFT was also utilized for the formation of degradable gels. For instance, micelles formed in an aqueous solution of the block copolymer polyMePEGMA-b-poly (5'-O-methacryloyl uridine) were cross-linked using **30** (Figure 11.4). Efficient degradation in the presence of DTT was demonstrated [198]. The dispersion

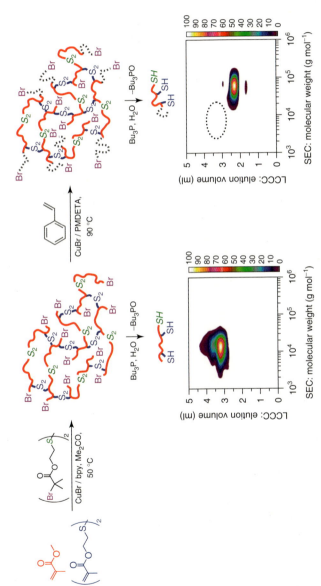

Figure 11.14 Synthesis of polyMMA-based disulfide cross-linked gels by ATRP, and their chemical modification using a chain extension reaction with a second monomer (Sty). The degradation products of the two gels – homopolymer from the "supermacroinitiator" and block copolymer from the modified gel – are also presented. The 2D-SEC plots at the bottom show that the chain extended gel degraded with the formation of polyMMA-b-polySty, and no polyMMA homopolymers were observed. (Adapted from Tsarevsky et al. [76]. Copyright 2006, American Chemical Society).

copolymerization of nBA with an acetal-containing bisacrylate cross-linker under RAFT conditions yielded degradable (acid-sensitive) particles. Similarly to the disulfide gel prepared by ATRP and presented above, the gel particles were successfully used as *"supermacroinitiators"* in a chain extension reaction with OEGA, and degradable cross-linked "hairy" particles with hydrophilic dangling chains were thus prepared [199]. Block copolymer micelles, derived from a poly(acryloyl glucosamine)-b-polyNIPAAm, were formed at 60 °C in an aqueous medium, and the dithioester groups in the polyNIPAAm core served as sites for further reaction with an acetal-containing cross-linker. The reaction yielded core cross-linked particles with acid-sensitive hydrolytically degradable cores [200].

11.3
Conclusions

Controlled radical polymerization techniques developed since the mid-1990s allow unprecedented control over MW, MWD, molecular architecture, and precise placement of functional groups in synthetic polymers. Of particular interest are well-defined functional polymers containing (bio)degradable functional groups or polymeric segments, and numerous examples of synthetic methodologies utilizing CRP and applications have already appeared in the literature. These materials have attracted significant attention due to their potential in the biomedical field (drug delivery, tissue engineering, etc.) and it is envisioned that this area will be developing very rapidly over the next few years. This review, by no means extensive, aims to give examples of (bio)degradable polymers prepared by CRP and serve as a source of ideas and inspiration for future research.

Abbreviations

2-PyS$_2$-	2-Pyridyldisulfide group
2D-SEC	2-Dimensional size exclusion chromatography
AA	Acrylic acid
AN	Acrylonitrile
ATRP	Atom transfer radical polymerization
BMDO	5,6-Benzo-2-methylene-1,3-dioxepane
BSA	Bovine serum albumin
CL	ε-Caprolactone
CRP	Controlled/living radical polymerization
DMAEMA	2-(N,N-dimethylamino)ethyl methacrylate
DP	Degree of polymerization
DTT	Dithiothreitol (Cleland's reagent)
GSH	Glutathione
HEMA	2-Hydroxyethyl methacrylate
L	Ligand

(MAOE)$_2$S$_2$	Bis(2-methacryloyloxy)ethyl disulfide
MePEOMA	Poly(ethylene oxide) methyl ether methacrylate
MMA	Methyl methacrylate
MPC	2-Methacryloyloxyethyl phosphorylcholine
MW	Molecular weight
MWD	Molecular weight distribution
nBA	n-Butyl acrylate
NIPAAm	N-isopropylacrylamide
NMP	Nitroxide-mediated polymerization
NMR	Nuclear magnetic resonance
OEGA	Oligo(ethylene glycol) acrylate
PDI	Polydispersity index (M_w/M_n)
RAFT	Reversible addition-fragmentation chain transfer polymerization
RGD	arginine-glycine-aspartic acid
ROP	Ring-opening polymerization
RROP	Radical ring-opening polymerization
SCVP	Self-condensing vinyl polymerization
SEC	Size exclusion chromatography
SFRP	Stable free radical polymerization
siRNA	Small interfering ribonucleic acid
Sn(EH)$_2$	Tin(II) 2-ethylhexanoate
Sty	Styrene
tBA	tert-butyl acrylate
tBMA	tert-butyl methacrylate
Tf	Trifluoromethanesulfonyl (CF$_3$SO$_2^-$)

References

1. Dainton, F.S. and Ivin, K.J. (1958) *Quart. Rev. Chem. Soc.*, **12**, 61–92.
2. Jellinek, H.H.G. (1966), In: *Encyclopedia of Polymer Science and Technology*, Vol. 4 (eds H.F. Mark, N.G. Gaylord, and N.M. Bikales), John Wiley & Sons, Inc., New York, pp. 740–793.
3. Sawada, H. (1976) *Thermodynamics of Polymerization*, Marcel Dekker, New York.
4. Duda, A. and Kowalski, A. (2009), In: *Handbook of Ring-Opening Polymerization* (eds P. Dubois, O. Coulembier, and J.-M. Raquez), Wiley-VCH Verlag GmbH, Weinheim, pp. 1–51.
5. Kawai, F. (1995) *Adv. Biochem. Eng. Biotechnol.*, **52**, 151–194.
6. Chiellini, E., Corti, A., D'Antone, S., and Solaro, R. (2003) *Prog. Polym. Sci.*, **28**, 963–1014.
7. Schnabel, W. (1981) *Polymer Degradation: Principles and Practical Applications*, Hanser International, Munich.
8. Chandra, R. and Rustgi, R. (1998) *Prog. Polym. Sci.*, **23**, 1273–1335.
9. Hamid, S.H. (ed.) (2000) *Handbook of Polymer Degradation*, Marcel Dekker, New York.
10. Khemani, K. and Scholz, C. (eds) (2006) *Degradable Polymers and Materials. Principles and Practice*, ACS Symposium Series, Vol. 939, ACS, Washington, DC.
11. Blainey, J.D. (1968) *Curr. Probl. Clin. Biochem.*, **2**, 85–100.
12. Chang, R.L.S., Ueki, I.F., Troy, J.L., Deen, W.M., Robertson, C.R., and Brenner, B.M. (1975) *Biophys. J.*, **15**, 887–906.
13. Joergensen, K.E. and Moeller, J.V. (1979) *Am. J. Physiol.*, **236**, F103–F111.
14. Seymour, L.W., Duncan, R., Strohalm, J., and Kopecek, J. (1987) *J. Biomed. Mater. Res.*, **21**, 1341–1358.

15. Bohrer, M.P., Deen, W.M., Robertson, C.R., Troy, J.L., and Brenner, B.M. (1979) *J. Gen. Physiol.*, **74**, 583–593.
16. Bohrer, M.P., Baylis, C., Humes, H.D., Glassock, R.J., Robertson, C.R., and Brenner, B.M. (1978) *J. Clin. Invest.*, **61**, 72–78.
17. Hersel, U., Dahmen, C., and Kessler, H. (2003) *Biomaterials*, **24**, 4385–4415.
18. Perlin, L., MacNeil, S., and Rimmer, S. (2008) *Soft Matter*, **4**, 2331–2349.
19. Okada, M. (2002) *Prog. Polym. Sci.*, **27**, 87–133.
20. Bailey, W.J., Chen, P.Y., Chen, S.-C., Chiao, W.-B., Endo, T., Gapud, B., Lin, Y.-N., Ni, Z., Pan, C.-Y., Shaffer, S.E., Sidney, L., Wu, S.-R., Yamamoto, N., Yamazaki, N., and Yonezawa, K. (1984) *J. Macromol. Sci. Chem.*, **A21**, 1611–1639.
21. Bailey, W.J., Chou, J.L., Feng, P.-Z., Issari, B., Kuruganti, V., and Zhou, L.L. (1988) *J. Macromol. Sci. Chem.*, **25**, 781–798.
22. Sanda, F. and Endo, T. (2001) *J. Polym. Sci., Part A: Polym. Chem.*, **39**, 265–276.
23. Endo, T. and Morino, K. (2009) *ACS Symp. Ser.*, **1023**, 33–48.
24. Klemm, E. and Schulze, T. (1999) *Acta Polym.*, **50**, 1–19.
25. Webster, O.W. (1991) *Science*, **251**, 887–893.
26. Szwarc, M. (1968) *Carbanions, Living Polymers, and Electron Transfer Processes*, John Wiley & Sons, Inc., New York.
27. Matyjaszewski, K. (ed.) (1996) in *Cationic Polymerizations: Mechanisms, Synthesis, and Applications*, Marcel Dekker, New York.
28. Aoshima, S. and Kanaoka, S. (2009) *Chem. Rev.*, **109**, 5245–5287.
29. Matyjaszewski, K. (ed.) (1998) *Controlled Radical Polymerization*, ACS Symposium Series, Vol. 685, ACS, Washington, DC.
30. Matyjaszewski, K. (ed.) (2000) *Controlled/Living Radical Polymerization. Progress in ATRP, NMP, and RAFT*, ACS Symposium Series, Vol. 768, ACS, Washington, DC.
31. Otsu, T. (2000) *J. Polym. Sci.: Part A: Polym. Chem.*, **38**, 2121–2136.
32. Matyjaszewski, K. and Davis, T. P. (eds) (2002) *Handbook of Radical Polymerization*, John Wiley & Sons Inc., Hoboken.
33. Matyjaszewski, K. (ed.) (2003) *Advances in Controlled/Living Radical Polymerization*, ACS Symposium Series, Vol. 854, ACS, Washington, DC.
34. Matyjaszewski, K. (ed.) (2006) *Controlled/Living Radical Polymerization. From Synthesis to Materials*, ACS Symposium Series, Vol. 944, ACS, Washington, DC.
35. Braunecker, W.A., and Matyjaszewski, K. (2007) *Prog. Polym. Sci.*, **32**, 93–146.
36. Matyjaszewski, K. (ed.) (2009) *Controlled/Living Radical Polymerization: Progress in ATRP*, ACS Symposium Series, Vol. 1023, ACS, Washington, DC.
37. Matyjaszewski, K. (ed.) (2009) *Controlled/Living Radical Polymerization: Progress in RAFT and NMP*, ACS Symposium Series, Vol. 1024, ACS, Washington, DC.
38. Matyjaszewski, K. and Xia, J. (2001) *Chem. Rev.*, **101**, 2921–2990.
39. Kamigaito, M., Ando, T., and Sawamoto, M. (2001) *Chem. Rev.*, **101**, 3689–3745.
40. Tsarevsky, N.V. and Matyjaszewski, K. (2007) *Chem. Rev.*, **107**, 2270–2299.
41. Ouchi, M., Terashima, T., and Sawamoto, M. (2009) *Chem. Rev.*, **109**, 4963–5050.
42. Hawker, C.J., Bosman, A.W., and Harth, E. (2001) *Chem. Rev.*, **101**, 3661–3688.
43. Le Mercier, C., Acerbis, S., Bertin, D., Chauvin, F., Gigmes, D., Guerret, O., Lansalot, M., Marque, S., Le Moigne, F., Fischer, H., and Tordo, P. (2002) *Macromol. Symp.*, **182**, 225–247.
44. Sciannamea, V., Jerome, R., and Detrembleur, C. (2008) *Chem. Rev.*, **108**, 1104–1126.
45. Wayland, B.B., Mukerjee, S., Poszmik, G., Woska, D.C., Basickes, L., Gridnev, A.A., Fryd, M., and Ittel, S.D. (1998) *ACS Symp. Ser.*, **685**, 305–315.
46. Debuigne, A., Poli, R., Jerome, C., Jerome, R., and Detrembleur, C. (2009) *Prog. Polym. Sci.*, **34**, 211–239.

47. Poli, R. (2006) *Angew. Chem. Int. Ed.*, **45**, 5058–5070.
48. Moad, G., Chiefari, J., Chong, Y.K., Krstina, J., Mayadunne, R.T.A., Postma, A., Rizzardo, E., and Thang, S.H. (2000) *Polym. Int.*, **49**, 993–1001.
49. Rizzardo, E., Chiefari, J., Mayadunne, R., Moad, G., and Thang, S. (2001) *Macromol. Symp.*, **174**, 209–212.
50. Barner-Kowollik, C., Davis, T.P., Heuts, J.P.A., Stenzel, M.H., Vana, P., and Whittaker, M. (2003) *J. Polym. Sci.: Part A: Polym. Chem.*, **41**, 365–375.
51. Perrier, S. and Takolpuckdee, P. (2005) *J. Polym. Sci.: Part A: Polym. Chem.*, **43**, 5347–5393.
52. Moad, G., Chong, Y.K., Postma, A., Rizzardo, E., and Thang, S.H. (2005) *Polymer*, **46**, 8458–8468.
53. David, G., Boyer, C., Tonnar, J., Ameduri, B., Lacroix-Desmazes, P., and Boutevin, B. (2006) *Chem. Rev.*, **106**, 3936–3962.
54. Yamago, S. (2006) *J. Polym. Sci.: Part A: Polym. Chem.*, **44**, 1–12.
55. Yamago, S. (2009) *Chem. Rev.*, **109**, 5051–5068.
56. Goto, A., Tsujii, Y., and Fukuda, T. (2008) *Polymer*, **49**, 5177–5185.
57. Matyjaszewski, K. and Tsarevsky, N.V. (2009) *Nature Chem.*, **1**, 276–288.
58. Jocelyn, P.C. (1987) *Methods Enzymol.*, **143**, 246–256.
59. Cleland, W.W. (1964) *Biochemistry*, **3**, 480–482.
60. Singh, R., Lamoureux, G.V., Lees, W.J., and Whitesides, G.M. (1995) *Methods Enzymol.*, **251**, 167–173.
61. Humphrey, R.E. and Hawkins, J.M. (1964) *Anal. Chem.*, **36**, 1812–1814.
62. Humphrey, R.E. and Potter, J.L. (1965) *Anal. Chem.*, **37**, 164–165.
63. Burns, J.A., Butler, J.C., Moran, J., and Whitesides, G.M. (1991) *J. Org. Chem.*, **56**, 2648–2650.
64. Schafer, S.Q. and Buettner, G.R. (2001) *Free Radical Biol. Med.*, **30**, 1191–1212.
65. Saito, G., Swanson, J.A., and Lee, K.-D. (2003) *Adv. Drug Delivery Rev.*, **55**, 199–215.
66. Panek, J.R. (1962), In: *Polyethers, Polyalkylene Sulfides and Other Thioethers*, Vol. 3 (ed. N.G. Gaylord), John Wiley & Sons, Inc., New York, London, pp. 115–224.
67. Meng, F., Hennink, W.E., and Zhong, Z. (2009) *Biomaterials*, **30**, 2180–2198.
68. Berenbaum, M.B. (1962), In: *Polyethers, Polyalkylene Sulfides and Other Thioethers*, Vol. 3 (ed. N.G. Gaylord), John Wiley & Sons, Inc., New York, London, pp. 43–114.
69. Kishore, K. and Ganesh, K. (1995) *Adv. Polym. Sci.*, **121**, 81–121.
70. Roy, D., Cambre, J.N., and Sumerlin, B.S. (2010) *Prog. Polym. Sci.*, **35**, 278–301.
71. Tsarevsky, N.V. and Matyjaszewski, K. (2002) *Macromolecules*, **35**, 9009–9014.
72. Van Camp, W., Du Prez, F.E., Alem, H., Demoustier-Champagne, S., Willet, N., Grancharov, G., and Duwez, A.-S. (2010) *Eur. Polym. J.*, **46**, 195–201.
73. Shah, R.R., Merreceyes, D., Husseman, M., Rees, I., Abbott, N.L., Hawker, C.J., and Hedrick, J.L. (2000) *Macromolecules*, **33**, 597–605.
74. Bontempo, D., Heredia, K.L., Fish, B.A., and Maynard, H.D. (2004) *J. Am. Chem. Soc.*, **126**, 15372–15373.
75. Tsarevsky, N.V. and Matyjaszewski, K. (2005) *Macromolecules*, **38**, 3087–3092.
76. Tsarevsky, N.V., Min, K., Jahed, N.M., Gao, H., and Matyjaszewski, K. (2006) *Degradable Polymers and Materials. Principles and Practice*, ACS Symposium. Series, Vol. 939, ACS, Washington, DC, pp. 184–200.
77. Madsen, J., Armes, S.P., Bertal, K., Lomas, H., MacNeil, S., and Lewis, A.L. (2008) *Biomacromolecules*, **9**, 2265–2275.
78. Park, I., Sheiko, S.S., Nese, A., and Matyjaszewski, K. (2009) *Macromolecules*, **42**, 1805–1807.
79. Montheard, J.-P., Chatzopoulos, M., and Chappard, D. (1992) *J. Macromol. Sci. Rev. Macromol. Chem. Phys.*, **C32**, 1–34.
80. Heredia, K.L., Bontempo, D., Ly, T., Byers, J.T., Halstenberg, S., and Maynard, H.D. (2005) *J. Am. Chem. Soc.*, **127**, 16955–16960.

81. Klaikherd, A., Ghosh, S., and Thayumanavan, S. (2007) *Macromolecules*, **40**, 8518–8520.
82. Boyer, C., Liu, J., Wong, L., Tippett, M., Bulmus, V., and Davis, T.P. (2008) *J. Polym. Sci.: Part A: Polym. Chem.*, **46**, 7207–7224.
83. Heredia, K.L., Nguyen, T.H., Chang, C.-W., Bulmus, V., Davis, T.P., and Maynard, H.D. (2008) *Chem. Commun.*, 3245–3247.
84. Liu, J., Liu, H., Bulmus, V., Boyer, C., and Davis, T.P. (2009) *J. Polym. Sci.: Part A: Polym. Chem.*, **47**, 899–912.
85. Heredia, K.L. and Maynard, H.D. (2007) *Org. Biomol. Chem.*, **5**, 45–53.
86. Li, C., Madsen, J., Armes, S.P., and Lewis, A.L. (2006) *Angew. Chem. Int. Ed.*, **45**, 3510–3513.
87. Ghosh, S., Basu, S., and Thayumanavan, S. (2006) *Macromolecules*, **39**, 5595–5597.
88. Ghosh, S., Yesilyurt, V., Savariar, E.N., Irvin, K., and Thayumanavan, S. (2009) *J. Polym. Sci.: Part A: Polym. Chem.*, **47**, 1052–1060.
89. Carrot, G., Hilborn, J., Hedrick, J.L., and Trollsas, M. (1999) *Macromolecules*, **32**, 5171–5173.
90. Coessens, V., Pintauer, T., and Matyjaszewski, K. (2001) *Prog. Polym. Sci.*, **26**, 337–377.
91. Garamszegi, L., Donzel, C., Carrot, G., Nguyen, T.Q., and Hilborn, J. (2003) *React. Funct. Polym.*, **55**, 179–183.
92. Lowe, A.B., Sumerlin, B.S., Donovan, M.S., and McCormick, C.L. (2002) *J. Am. Chem. Soc.*, **124**, 11562–11563.
93. Sumerlin, B.S., Lowe, A.B., Stroud, P.A., Zhang, P., Urban, M.W., and McCormick, C.L. (2003) *Langmuir*, **19**, 5559–5562.
94. Zhu, M.-Q., Wang, L.-Q., Exarhos, G.J., and Li, A.D.Q. (2004) *J. Am. Chem. Soc.*, **126**, 2656–2657.
95. Schilli, C.M., Mueller, A.H.E., Rizzardo, E., Thang, S.H., and Chong, Y.K. (2003) ACS Symposium Series, Vol. 854, ACS, Washington, DC, pp. 603–618.
96. Shan, J., Nuopponen, M., Jiang, H., Kauppinen, E., and Tenhu, H. (2003) *Macromolecules*, **36**, 4526–4533.
97. Motokucho, S., Sudo, A., and Endo, T. (2006) *J. Polym. Sci.: Part A: Polym. Chem.*, **44**, 6324–6331.
98. Xu, J., He, J., Fan, D., Wang, X., and Yang, Y. (2006) *Macromolecules*, **39**, 8616–8624.
99. You, Y.-Z., Zhou, Q.-H., Manickam, D.S., Wan, L., Mao, G.-Z., and Oupicky, D. (2007) *Macromolecules*, **40**, 8617–8624.
100. Qiu, X.-P. and Winnik, F.M. (2007) *Macromolecules*, **40**, 872–878.
101. Vogt, A.P. and Sumerlin, B.S. (2009) *Soft Matter*, **5**, 2347–2351.
102. Setijadi, E., Tao, L., Liu, J., Jia, Z., Boyer, C., and Davis, T.P. (2009) *Biomacromolecules*, **10**, 2699–2707.
103. Roth, P.J., Kessler, D., Zentel, R., and Theato, P. (2008) *Macromolecules*, **41**, 8316–8319.
104. Hill, N.L., Jarvis, J.L., Pettersson, F., and Braslau, R. (2008) *React. Funct. Polym.*, **68**, 361–368.
105. Boyer, C., Bulmus, V., Davis, T.P., Ladmiral, V., Liu, J., and Perrier, S. (2009) *Chem. Rev.*, **109**, 5402–5436.
106. Chen, Y.M. and Wulff, G. (2002) *Macromol. Rapid Commun.*, **23**, 59–63.
107. Narain, R. and Armes, S.P. (2003) *Biomacromolecules*, **4**, 1746–1758.
108. Tsarevsky, N.V., Sarbu, T., Goebelt, B., and Matyjaszewski, K. (2002) *Macromolecules*, **35**, 6142–6148.
109. Motala-Timol, S. and Jhurry, D. (2007) *Eur. Polym. J.*, **43**, 3042–3049.
110. Xu, F.J., Li, J., Yuan, S.J., Zhang, Z.X., Kang, E.T., and Neoh, K.G. (2008) *Biomacromolecules*, **9**, 331–339.
111. O'Reilly, R.K., Hawker, C.J., and Wooley, K.L. (2006) *Chem. Soc. Rev.*, **35**, 1068–1083.
112. Zhang, Q., Remsen, E.E., and Wooley, K.L. (2000) *J. Am. Chem. Soc.*, **122**, 3642–3651.
113. Chung, I.S. and Matyjaszewski, K. (2003) *Macromolecules*, **36**, 2995–2998.
114. Bailey, W.J., Ni, Z., and Wu, S.R. (1982) *Macromolecules*, **15**, 711–714.
115. Yuan, J.-Y., Pan, C.-Y., and Tang, B.Z. (2001) *Macromolecules*, **34**, 211–214.
116. Huang, J., Gil, R., and Matyjaszewski, K. (2005) *Polymer*, **46**, 11698–11706.

117. Lutz, J.-F., Andrieu, J., Uezguen, S., Rudolph, C., and Agarwal, S. (2007) *Macromolecules*, **40**, 8540–8543.
118. Siegwart, D.J., Bencherif, S.A., Srinivasan, A., Hollinger, J.O., and Matyjaszewski, K. (2008) *J. Biomed. Mater. Res. Part A*, **87A**, 345–358.
119. Hoyle, C.E., Lee, T.Y., and Roper, T. (2004) *J. Polym. Sci.: Part A: Polym. Chem.*, **42**, 5301–5338.
120. Dondoni, A. (2008) *Angew. Chem. Int. Ed.*, **47**, 8995–8997.
121. Kade, M.J., Burke, D.J., and Hawker, C.J. (2010) *J. Polym. Sci.: Part A: Polym. Chem.*, **48**, 743–750.
122. Sumerlin, B.S. and Vogt, A.P. (2010) *Macromolecules*, **43**, 1–13.
123. Sarbu, T., Lin, K.-Y., Ell, J., Siegwart, D.J., Spanswick, J., and Matyjaszewski, K. (2004) *Macromolecules*, **37**, 3120–3127.
124. Sarbu, T., Lin, K.-Y., Spanswick, J., Gil, R.R., Siegwart, D.J., and Matyjaszewski, K. (2004) *Macromolecules*, **37**, 9694–9700.
125. Nicolay, R., Marx, L., Hemery, P., and Matyjaszewski, K. (2007) *Macromolecules*, **40**, 9217–9223.
126. Hadjichristidis, N., Iatrou, H., Pitsikalis, M., and Mays, J. (2006) *Prog. Polym. Sci.*, **31**, 1068–1132.
127. Matyjaszewski, K. (2003) *Polym. Int.*, **52**, 1559–1565.
128. Gao, H. and Matyjaszewski, K. (2009) *Prog. Polym. Sci.*, **34**, 317–350.
129. Hedrick, J.L., Trollss, M., Hawker, C.J., Atthoff, B., Claesson, H., Heise, A., Miller, R.D., Mecerreyes, D., Jerome, R., and Dubois, P. (1998) *Macromolecules*, **31**, 8691–8705.
130. Chen, J., Zhang, H., Chen, J., Wang, X., and Wang, X. (2005) *J. Macromol. Sci., Part A: Pure Appl. Chem.*, **42**, 1247–1257.
131. Jia, Z., Zhou, Y., and Yan, D. (2005) *J. Polym. Sci.: Part A: Polym. Chem.*, **43**, 6534–6544.
132. Yuan, W., Yuan, J., Zhou, M., and Pan, C. (2008) *J. Polym. Sci.: Part A: Polym. Chem.*, **46**, 2788–2798.
133. Heise, A., Trollsaas, M., Magbitang, T., Hedrick, J.L., Frank, C.W., and Miller, R.D. (2001) *Macromolecules*, **34**, 2798–2804.
134. Gao, H., Tsarevsky, N.V., and Matyjaszewski, K. (2005) *Macromolecules*, **38**, 5995–6004.
135. Wiltshire, J.T. and Qiao, G.G. (2006) *Macromolecules*, **39**, 9018–9027.
136. Wiltshire, J.T. and Qiao, G.G. (2008) *Macromolecules*, **41**, 623–631.
137. Jiang, X., Liu, S. and Narain, R. (2009) *Langmuir*, **25**, 13344–13350.
138. Hawker, C.J., Mecerreyes, D., Elce, E., Dao, J., Hedrick, J.L., Barakat, I., Dubois, P., Jerome, R., and Volksen, I. (1997) *Macromol. Chem. Phys.*, **198**, 155–166.
139. Mespouille, L., Degee, P., and Dubois, P. (2005) *Eur. Polym. J.*, **41**, 1187–1195.
140. Janata, M., Masar, B., Toman, L., Vlcek, P., Latalova, P., Brus, J., and Holler, P. (2003) *React. Funct. Polym.*, **57**, 137–146.
141. Mecerreyes, D., Moineau, G., Dubois, P., Jerome, R., Hedrick, J.L., Hawker, C.J., Malmstrom, E.E., and Trollsas, M. (1998) *Angew. Chem., Int. Ed.*, **37**, 1274–1276.
142. Villarroya, S., Zhou, J., Thurecht, K.J., and Howdle, S.M. (2006) *Macromolecules*, **39**, 9080–9086.
143. Ydens, I., Degee, P., Dubois, P., Libiszowski, J., Duda, A., and Penczek, S. (2003) *Macromol. Chem. Phys.*, **204**, 171–179.
144. Iwasaki, Y. and Akiyoshi, K. (2004) *Macromolecules*, **37**, 7637–7642.
145. Iwasaki, Y. and Akiyoshi, K. (2006) *Biomacromolecules*, **7**, 1433–1438.
146. Mecerreyes, D., Atthoff, B., Boduch, K.A., Trollsaas, M., and Hedrick, J.L. (1999) *Macromolecules*, **32**, 5175–5182.
147. Lenoir, S., Riva, R., Lou, X., Detrembleur, C., Jerome, R., and Lecomte, P. (2004) *Macromolecules*, **37**, 4055–4061.
148. Wang, G., Shi, Y., Fu, Z., Yang, W., Huang, Q., and Zhang, Y. (2005) *Polymer*, **46**, 10601–10606.
149. Takasu, A., Iio, Y., Mimura, T., and Hirabayashi, T. (2005) *Polym. J.*, **37**, 946–953.
150. Riva, R., Rieger, J., Jerome, R., and Lecomte, P. (2006) *J. Polym. Sci.: Part A: Polym. Chem.*, **44**, 6015–6024.

151. Lee, H., Jakubowski, W., Matyjaszewski, K., Yu, S., and Sheiko, S.S. (2006) *Macromolecules*, **39**, 4983–4989.
152. Carlmark, A. and Malmstroem, E. (2002) *J. Am. Chem. Soc.*, **124**, 900–901.
153. Carlmark, A. and Malmstroem, E.E. (2003) *Biomacromolecules*, **4**, 1740–1745.
154. Lindqvist, J. and Malmstroem, E. (2006) *J. Appl. Polym. Sci.*, **100**, 4155–4162.
155. Chang, F., Yamabuki, K., Onimura, K., and Oishi, T. (2008) *Polym. J.*, **40**, 1170–1179.
156. Vlcek, P., Janata, M., Latalova, P., Dybal, J., Spirkova, M., and Toman, L. (2007) *J. Polym. Sci.: Part A: Polym. Chem.*, **46**, 564–573.
157. Shen, D., Yu, H., and Huang, Y. (2005) *J. Polym. Sci.: Part A: Polym. Chem.*, **43**, 4099–4108.
158. El Tahlawy, K. and Hudson, S.M. (2003) *J. Appl. Polym. Sci.*, **89**, 901–912.
159. Li, N., Bai, R., and Liu, C. (2005) *Langmuir*, **21**, 11780–11787.
160. Liu, P. and Su, Z. (2006) *Mater. Lett.*, **60**, 1137–1139.
161. Bontempo, D., Masci, G., De Leonardis, P., Mannina, L., Capitani, D., and Crescenzi, V. (2006) *Biomacromolecules*, **7**, 2154–2161.
162. Kim, D.J., Heo, J.-Y., Kim, K.S., and Choi, I.S. (2003) *Macromol. Rapid Commun.*, **24**, 517–521.
163. Oestmark, E., Nystroem, D., and Malmstroem, E. (2008) *Macromolecules*, **41**, 4405–4415.
164. Mishra, M.K. and Kobayashi, S. (eds) (1999) *Star and Hyperbranched Polymers*, Vol. 53, Marcel Dekker, New York.
165. Voit, B. (2000) *J. Polym. Sci.: Part A: Polym. Chem.*, **38**, 2505–2525.
166. Sunder, A., Heinemann, J., and Frey, H. (2000) *Chem. Eur. J.*, **6**, 2499–2506.
167. Jikei, M. and Kakimoto, M. (2001) *Prog. Polym. Sci.*, **26**, 1233–1285.
168. Kim, Y.H. and Webster, O. (2002) *J. Macromol. Sci., Polym. Rev.*, **C42**, 55–89.
169. Gao, C. and Yan, D. (2004) *Prog. Polym. Sci.*, **29**, 183–275.
170. Voit, B. (2005) *J. Polym. Sci.: Part A: Polym. Chem.*, **43**, 2679–2699.
171. Taton, D., Feng, X., and Gnanou, Y. (2007) *New J. Chem.*, **31**, 1097–1110.
172. Mori, H., Mueller, A.H.E., and Simon, P.F.W. (2007), In: *Macromolecular Engineering: Precise Synthesis, Materials Properties, Applications*, Vol. 2 (eds K. Matyjaszewski, Y. Gnanou, and L. Leibler), Wiley-VCH Verlag GmbH, Weinheim, pp. 973–1005.
173. Korolev, G.V. and Bubnova, M.L. (2007) *Polym. Sci., Ser. C*, **49**, 332–354.
174. Jiang, G., Chen, W., and Xia, W. (2008) *Designed Monom. Polym.*, **11**, 105–122.
175. Grayson, S.M. and Frechet, J.M.J. (2001) *Chem. Rev.*, **101**, 3819–3867.
176. O'Brien, N., McKee, A., Sherrington, D.C., Slark, A.T., and Titterton, A. (2000) *Polymer*, **41**, 6027–6031.
177. Costello, P.A., Martin, I.K., Slark, A.T., Sherrington, D.C., and Titterton, A. (2002) *Polymer*, **43**, 245–254.
178. Slark, A.T., Sherrington, D.C., Titterton, A., and Martin, I.K. (2003) *J. Mater. Chem.*, **13**, 2711–2720.
179. Camerlynck, S., Cormack, P.A.G., Sherrington, D.C., and Saunders, G. (2005) *J. Macromol. Sci., Part B: Phys.*, **44**, 881–895.
180. Sato, T., Sato, N., Seno, M., and Hirano, T. (2003) *J. Polym. Sci.: Part A: Polym. Chem.*, **41**, 3038–3047.
181. Sato, T., Miyagi, T., Hirano, T., and Seno, M. (2004) *Polym. Int.*, **53**, 1503–1511.
182. Sato, T., Nakamura, T., Seno, M., and Hirano, T. (2006) *Polymer*, **47**, 4630–4637.
183. Frechet, J.M.J., Henmi, M., Gitsov, I., Aoshima, S., Leduc, M.R., and Grubbs, R.B. (1995) *Science*, **269**, 1080–1083.
184. Hawker, C.J., Frechet, J.M.J., Grubbs, R.B., and Dao, J. (1995) *J. Am. Chem. Soc.*, **117**, 10763–10764.
185. Ambade, A.V. and Kumar, A. (2000) *Prog. Polym. Sci.*, **25**, 1141–1170.
186. Li, Y. and Armes, S.P. (2005) *Macromolecules*, **38**, 8155–8162.
187. Wang, L., Li, C., Ryan, A.J., and Armes, S.P. (2006) *Adv. Mater.*, **18**, 1566–1570.

188. Rosselgong, J., Armes, S.P., Barton, W., and Price, D. (2009) *Macromolecules*, **42**, 5919–5924.
189. Tsarevsky, N.V., Huang, J., and Matyjaszewski, K. (2009) *J. Polym. Sci.: Part A: Polym. Chem.*, **47**, 6839–6851.
190. Mecerreyes, D., Trollss, M., and Hedrick, J.L. (1999) *Macromolecules*, **32**, 8753–8759.
191. Ide, N. and Fukuda, T. (1997) *Macromolecules*, **30**, 4268–4271.
192. Ide, N. and Fukuda, T. (1999) *Macromolecules*, **32**, 95–99.
193. Mespouille, L., Coulembier, O., Paneva, D., Degee, P., Rashkov, I., and Dubois, P. (2008) *Chem. Eur. J.*, **14**, 6369–6378.
194. Oh, J.K., Tang, C., Gao, H., Tsarevsky, N.V., and Matyjaszewski, K. (2006) *J. Am. Chem. Soc.*, **128**, 5578–5584.
195. Oh, J.K., Siegwart, D.J., Lee, H.-I., Sherwood, G., Peteanu, L., Hollinger, J.O., Kataoka, K., and Matyjaszewski, K. (2007) *J. Am. Chem. Soc.*, **129**, 5939–5945.
196. Oh, J.K., Siegwart, D.J., and Matyjaszewski, K. (2007) *Biomacromolecules*, **8**, 3326–3331.
197. Bencherif, S.A., Siegwart, D.J., Srinivasan, A., Horkay, F., Hollinger, J.O., Washburn, N.R., and Matyjaszewski, K. (2009) *Biomaterials*, **30**, 5270–5278.
198. Zhang, L., Liu, W., Lin, L., Chen, D., and Stenzel, M.H. (2008) *Biomacromolecules*, **9**, 3321–3331.
199. Chan, Y., Bulmus, V., Zareie, M.H., Byrne, F.L., Barner, L., and Kavallaris, M. (2006) *J. Controlled Release*, **115**, 197–207.
200. Zhang, L., Bernard, J., Davis, T.P., Barner-Kowollik, C., and Stenzel, M.H. (2008) *Macromol. Rapid Commun.*, **29**, 123–129.

Part V
Biomimetic Methods and Biocatalysis

12
High-Performance Polymers from Phenolic Biomonomers
Tatsuo Kaneko

12.1
Introduction

Environmentally friendly polymers originating from renewable starting materials and/or which degrade in the environment are significant materials with respect to the reduction of waste. *Bio-based polyesters* exhibiting smooth degradation have been widely studied as environmentally friendly polymeric materials [1]. However, some aliphatic bio-based polyesters, such as poly(hydroxyalkanoate)s [2], poly(butylene succinate) [3], and so on [4], did not show performances high enough for application in the engineering plastic fields. Improvements in durability and performance have been shown by Ecoflex™ [5] and Biomax™ [6], but the environmental toxicity and availability of terephthalic acid are problematic. Poly(lactic acid)s (PLAs) have been remarkably well developed because of their high mechanical strength [7]. However, it was estimated that these polyesters will only replace a small percentage of the non-degradable plastics currently in use, due to their poor level of thermoresistance. As a result, *high-performance environmentally friendly polymers* originating from and degradable into natural molecules are urgently required to improve human life. Many non-degradable engineering plastics have rigid conjugated rings, such as benzene, benzimide, benzoxazole, benzimidazole, or benzthiazole [8]. Although continuous rigid structures in the polymer backbone do not exist in nature, polysaccharides have many hetero rings. However, the polysaccharides sometimes show too low a degree of processability to be made into useful materials, but they do have a tremendous amount of potential for use as high-performance bio-base polymers, as cellulose shows a very high Young's modulus (more than 10 GPa). In spite of the dramatic increase in plastic properties through the use of a high-strength filler, such as a bacterial cellulose [9] or modified lignin [10], it is the matrix polymers that determine the intrinsic composite performance.

The introduction of an *aromatic component* into a thermoplastic polymer backbone is an efficient method of intrinsically improving material performance [11]. Additionally, the continuous sequence of such aromatic rings could be a mesogenic group. Molding during the thermotropic liquid crystalline (LC) state can induce molecular orientation, giving anisotropy to the mechanical performance, which

Green Polymerization Methods: Renewable Starting Materials, Catalysis and Waste Reduction
Edited by Robert T. Mathers and Michael A. R. Meier
Copyright © 2011 WILEY-VCH Verlag GmbH & Co. KGaA, Weinheim
ISBN: 978-3-527-32625-9

sometimes dramatically increases mechanical strength and the Young's modulus [12]. Researchers have reported the LC properties and *in vitro* degradability of copolymers containing *p*-coumaric acid units [13–21] but they did not confirm the mesogenic role nor the biodegradability of the copolymers or the poly(*p*-coumaric acid) homopolymer. However, more systematic studies of environmentally friendly, degradable materials are needed. In this review, we discuss the preparation of various LC polyarylates derived mainly from phytochemical monomers – in other words, "*phytomonomers*" [22]. We have selected coumarates as the aromatic phytomonomers as their rigidity and photoreactivity may lead to high heat-resistance and photofunctions of the corresponding polymers, respectively.

12.2
Coumarates as Phytomonomers

One of the most well-known coumarates is *p*-coumaric acid (4-hydroxycinnamic acid; 4HCA). In fact, lignin biosynthesis starts with conversion into cinnamate and 4HCA via the enzymatic reaction with phenylalanine ammonia lyase and Cyt P450-dependent monooxygenase, respectively [23]. 4HCA derivatives are used as allelopathic chemicals in plants and widely exist in soil [24]. 4HCA is also present in several photosynthetic bacteria as a protein component [25–27]. *Rhodobacter capsulatus* [25], *Rhodobacter sphaeroides* [26], and so on [27], contain photoactive proteins comprised of a photosensitive 4HCA component. Light irradiation (λ = 375 or 435 nm) results in an (E)–(Z) transformation that enables bacteria to swim away from the light. Furthermore, as the enzymatic conversion of amino acids into

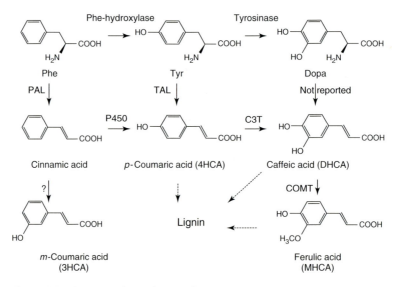

Figure 12.1 Coumarate biosynthetic pathway in an enzymatic reaction based on lignin syntheses.

phytomonomers is well defined and straightforward, scale-up for mass-production is feasible [28]. These phytomonomers were biodegraded by microbial action [29]. Ferulic acid (3-methoxy-4-hydroxycinnamic acid; MHCA), and caffeic acid (3,4-dihydroxycinnamic acid; DHCA), have also been selected as phytomonomers, as these molecules are widely available in various plants with an essential pathway of lignin biosynthesis, as shown in Figure 12.1 [23]. We also used m-coumaric acid (3HCA), the biosynthetic pathway of which was not found but it is contained in acaroids gum [30], for comparison of the polyester properties.

12.3
LC Properties of Homopolymers

12.3.1
Syntheses and Structures

Poly(p-coumaric acid) P4HCA (structure: Figure 12.2a) was obtained by the *thermal polycondensation* of 4HCA phytomonomer as follows [31]. 4HCA was heated at 220 °C for 24 h in the presence of anhydride acetic acid as a condensation reagent and sodium acetate as a catalyst for transesterification. The molten

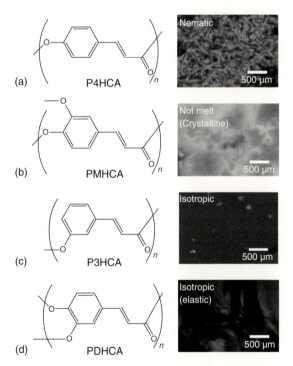

Figure 12.2 Structures of coumarate homopolymers and their polarized microscopic photographs taken above 200 °C.

mixture gradually became viscous during the reaction period. After cooling, the product was dissolved in pentafluorophenol and purified by reprecipitation over methanol, then washed with methanol using the Soxhlet extraction method for 24 h (yield: 90%). The product was a yellowish-white powder that was soluble in pentafluorophenol, giving a viscous solution, but was insoluble in water, ethanol, acetone, tetrahydrofuran (THF), chloroform, and various aprotic amidic solvents. Infrared (IR) and ^{1}H-NMR spectroscopy of P4HCA confirmed the chemical structure of the target polymer. Other homopolymers such as PMHCA [poly(3-methoxy-4-hydroxycinnamic acid)], PDHCA [poly(3,4-dihydroxycinnamic acid)], and P3HCA were prepared and structurally analyzed by the analogous procedures [32].

The average molecular weight of the coumarate homopolymers was measured in DMF (N,N'-dimethylformamide) by gel permeation chromatography (GPC). P3HCA and PDHCA showed a single GPC peak at a molecular weight of 8000 and 70 000, respectively. The polydispersity, M_w/M_n, of P3HCA and PDHCA was 2.0 and 2.4, respectively (M_n: number-average molecular weight, M_w: weight-average molecular weight). We could not measure the molecular weight of P4HCA or PMHCA by GPC because of limitations due to poor solubility. However, using endgroup analysis via ^{1}H-NMR spectroscopy, the M_n value was calculated to be 8000 g mol^{-1}.

Changing the monomer structure from an AB structure to a multifunctional AB$_2$-type monomer, such as DHCA, resulted in high molecular weight polyesters. As described above, it was proved theoretically that the AB$_x$-type multifunctional monomers gave hyperbranching architecture. In fact, PDHCA was soluble in DMF and no cross-linked polymer was detected.

12.3.2
Solubility

The *solubility* of the monomers and homopolymers was investigated (Table 12.1). The monomers dissolved in various organic solvents, but the solubility of all homopolymers was poorer than the corresponding monomers. For example, PMHCA polymerized from the soluble MHCA monomer, having one OH group at the para-position and a methoxy group at the meta-position, could be dissolved in only pentafluorophenol but not any of the other solvents used here. P4HCA polymerized from the 4HCA monomer with an OH group at the para-position could be dissolved in pentafluorophenol and trifluoroacetic acid/dichloromethane (TFA/DCM) (1 : 1, v/v). However, PDHCA polymerized from the DHCA monomer with two OH groups at the para- and meta-positions could be dissolved in NMP (N-methylpyrrolidone), DMF, pentafluorophenol, and TFA/DCM (1 : 1, v/v). In contrast, P3HCA polymerized from the 3HCA monomer with one OH group at the meta-position dissolved in various organic solvents such as THF, chloroform, and various aprotic amidic solvents. This indicated that the position of the OH group strongly affected the solubility of the homopolymers; in the case of monomers with an OH group at the meta-position, the homopolymers showed high solubility.

Table 12.1 Solubility of various coumarates and their corresponding polyesters.[a]

Solvent	4HCA	3HCA	MHCA	DHCA	P4HCA	P3HCA	PMHCA	PDHCA
Water	±	±	±	±	−	−	−	−
Methanol	+	+	+	+	−	−	−	−
Acetone	+	+	+	+	−	−	−	−
Acetonitrile	+	+	+	+	−	−	−	−
N,N′-Dimethylformamide	+	+	+	+	−	+	−	+
N-Methylpyrrolidone	+	+	+	+	−	+	−	+
Dimethylsulfoxide	+	+	+	+	−	+	−	−
Toluene	−	−	−	−	−	−	−	−
Hexane	−	−	−	−	−	−	−	−
Tetrahydrofuran	+	+	+	+	−	+	−	−
Chloroform	−	−	+	−	−	+	−	−
Pentafluorophenol	+	+	+	+	+	+	+	+
DCM/TFA (1 : 1 v/v)	+	+	+	+	+	+	−	+

[a] +, soluble; −, insoluble; ±, soluble at 60 °C.

Homopolymers dissolved in solvents such as pentafluorophenol, suggesting that cinnamoyls in propagating chains were not cross-linked during the polymerization.

12.3.3
Thermotropic Property

In order to investigate thermotropic LC properties of P4HCA, we performed crossed polarizing microscopic observations of samples sandwiched between two glass plates as the temperature was changed. The sample was a birefringent powder at 20 °C. When it was heated at a rate of 10 °C min^{-1}, the sample melted at 215 °C while maintaining its birefringence. In the temperature range 215–280 °C, we observed a schlieren texture with two and four brushes (Figure 12.2a), which was easily transformed into the dark-field view by sliding the cover glass. This finding indicated that P4HCA is macroscopically oriented by the application of shear stress. Therefore, P4HCA exhibits a *nematic phase* where the polymer chains are autonomously oriented but randomly located. The nematic liquid of P4HCA solidified at 200 °C following cooling, and repeatedly appeared on successive changes in temperature. Unexpectedly, the nematic liquid also solidified at 280 °C upon subsequent heating, showing transformation of the microscopic texture from the schlieren to a needle form, which was widely seen in the crystals. The sample solidified at 280 °C and no longer showed any transition with successive heating or cooling. The sample became pale yellow and insoluble in pentafluorophenol or other solvents. This transformation was accompanied by a 9.85% loss in weight, suggesting that the crystallization was due to decomposition.

If the heat solidification phenomenon is utilized, P4HCA may be applied to *thermosetting resins*. In general, the main chain type of a thermotropic LC polymer is composed of more than two monomeric units [33] consisting of poly{4-hydroxybenzoic acid (HBA)-co-5-hydroxynaphthoic acid (HNA)}, where HBA is a phytomonomer but HNA cannot be derived from flora or fauna. Although PHBA can be completely phyto-derived, it is a highly crystallized polymer that shows no melting-point. The HBA co-monomer was incorporated in order to decrease chain rigidity to put the crystalline chain of PHBA into the thermotropic LC state. P4HCA was the first LC synthetic homopolymer derived from a phytomonomer, which was made by molecular weight control.

Melting behaviors of other homopolymers were also investigated by a crossed polarizing microscope. PMHCA with a meta-methoxyl group did not show a melting-point and it indicated the crystalline properties at room temperature (Figure 12.2b). P3HCA showed only a dark field over its melting-point (130 °C), suggesting amorphous properties (Figure 12.2c). Interestingly, when the methoxyl group was changed into a hydroxyl group at the meta-position, the homopolymer (PDHCA) showed a birefringence at even higher temperatures than the softening point (200 °C) (Figure 12.2d). Additionally, when the sample was heated above the birefringence temperature, the sample did not change to a liquid state and showed a rubber-like elasticity. One reason for the specific thermal property exhibited by PDHCA may be caused by the intermolecular entanglement of branched polymer chains. These results indicate that the LC behaviors of the coumaric acid derivative homopolymers are influenced by the number and position of OH groups and the type of substituent.

12.3.4
Ordered Structures

To confirm the crystallinity of coumarate homopolymers, wide angle X-ray diffraction (WAXD) was investigated. Figure 12.3 shows the WAXD patterns of various homopolymers. Samples melted by heating above individual softening temperatures (T_s: P4HCA required 230 °C above the *melting temperature*, T_m, while P3HCA and PDHCA required 130 and 200 °C, respectively, above individual glass transition temperatures, T_g), cooled to room temperature, and WAXD was measured at room temperature. In the case of PMHCA, WAXD was measured without heating because PMHCA does not have a melting-point. These results agreed with those from a crossed polarizing microscope. P4HCA showed sharp peaks with high intensities and thus was a highly-crystallized polymer.

As 4HCA is a photoresponsive molecule showing the photoreactions given in Figure 12.4, the photoreactivity of P4HCA was investigated in terms of its phase behavior. A P4HCA film processed by spin casting (2000 rpm) from its pentafluorophenol solution (1 wt%, 50 μm) was UV-irradiated (wavelength: more than 280 nm) and the structure was investigated by IR spectroscopy and WAXD. UV irradiation in the crystalline state (25 °C) for 1 h did not produce a substantial change in the IR spectra. In contrast, when the P4HCA film was UV irradiated in the LC state (220 °C) for the same period, the IR peaks representing double bonds ($\nu_{C=C}$: 997 and 1283 cm^{-1}) became smaller and carbonyl shoulders appeared at

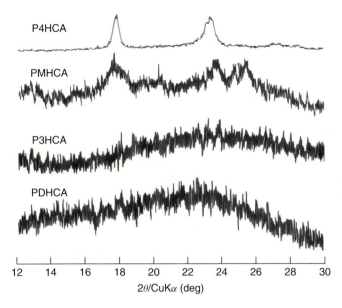

Figure 12.3 Wide angle X-ray diffraction diagrams of coumarate homopolymers.

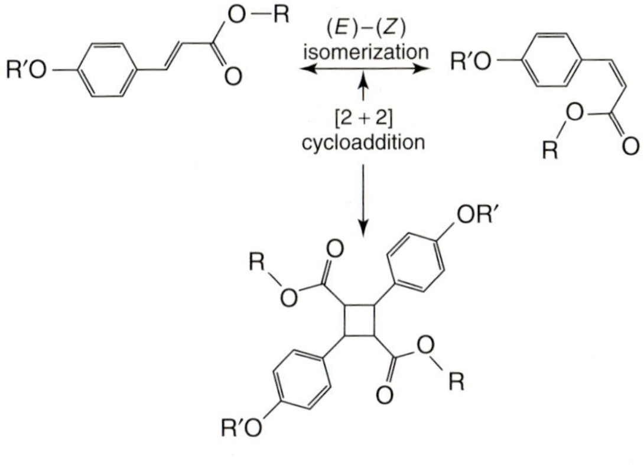

Figure 12.4 Typical photoreactions of cinnamoyl compounds.

about 1700 cm^{-1}, suggesting that the vinylene group reacted [34]. The lack of UV reaction in the crystalline state suggests that the π-electrons of the two vinylene moieties of P4HCA cannot interact, while the vinylenes in the nematic state can react due to the mobility of the polymer rods along the n-direction as shown in Figure 12.5. As the interchain [2 + 2] cycloaddition photoreaction occurred from both the (E)- and (Z)-structures of P4HCA, it was hypothesized that the cyclobutane was effectively formed with UV irradiation in the LC condensed state.

WAXD patterns of other homopolymers were compared with that of 4HCA (Figure 12.3). In PMHCA, several broad WAXD peaks appearing upon a broad halo were observed. These results suggest that its peak intensity was lower than that of

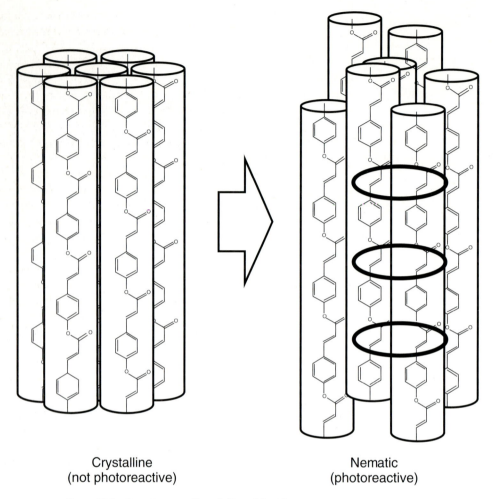

Figure 12.5 Tentative crystalline (left) and liquid crystalline (right) structures where neighboring double bonds are circled.

P4HCA and the crystal size was smaller. The reason seems to be that PMHCA has methoxyl group substituents at the meta-position, indicating that the molecular arrangement of PMHCA may be disturbed by the meta-methoxyl group. P3HCA and PDHCA did not show any peaks. It is thought that the orientation of PDHCA segments is difficult because of its branched structure. From crossed polarizing observations and WAXD analyses, it was understood that the position and type of substituent influences the liquid crystallinity or crystallinity of the homopolymers. It is a well-known fact that, in order to exhibit LC properties, a homopolymer requires molecular rigidity and a high aspect ratio, and so on. The birefringence property of PDHCA seemed to be due to the surface tension being strengthened by molecular entanglement of the branched structures.

12.3.5
Cell Compatibility

We investigated *cell-adhesion properties of the homopolymers* for biomedical applications. Good *cell compatibility* of P4HCA was confirmed by a cell-adhesion test. After L929 fibroblasts were incubated on a P4HCA film for 24 h at 37 °C, they adhered and extended (inset photo of Figure 12.6) as much as cells grown on tissue culture polystyrene (TCPS) did. Therefore, P4HCA showed potential for applications as a novel biopolymer with photoreactivity and liquid crystallinity, which can induce structural transformation and change the molecular orientation. The number of adhered mouse L929 fibroblast cells to various homopolymers is shown in Figure 12.6. It was confirmed that branched PDHCA and crystalline PMHCA have the highest and lowest cell-adhesion properties, respectively. Cellular morphology in these homopolymers was almost the same as that in P4HCA. From the results of cell-adhesion tests, the coumarate homopolymers had cell-adhesion properties

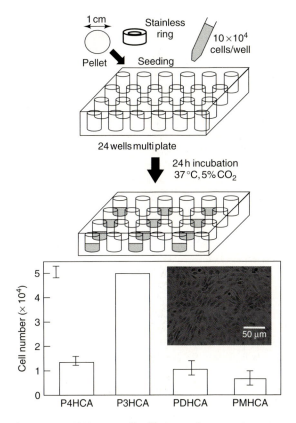

Figure 12.6 L929 mouse fibroblasts seeding on polymer pellets and the number of cells adhered on homopolymer pellets. Inset photograph is of cells extended on the P4HCA pellet.

and the number of adhered cells was influenced by the molecular linearity of the homopolymers.

Thus P4HCA showed some interesting functions but was very brittle because of its low molecular weight. However, P4HCA that was prepared at higher temperature and/or for a longer time was highly crystallized and exhibited no LC phase [14, 35]. Then 4HCA was copolymerized with other hydroxyl carboxylic acid in order to restrict crystallization so as to give higher-molecular-weight polymers.

12.4
LC Copolymers for Biomaterials

12.4.1
Lithocholic Acid as Co-monomer

The 4HCA component in the polymer has a strong preference to exhibit LC phase behavior. As a result, the *copolymerization* of 4HCA with non-mesogenic, flexible, or bent co-monomers will still exhibit some thermotropic properties. Lithocholic acid (LCA), which exists in mammalian bile as a cholesteric derivative hydroxycarboxylic acid, and bile acid were selected as a co-monomer of 4HCA. The copolymerization with LCA having a very stiff steroid plane with a hydroxyl group perpendicular to the plane is expected to prepare the copolymers with a reduced T_m and an increased solubility, keeping the good cell compatibility and the LC properties. P(4HCA-*co*-LCA) [22] was obtained by the thermal polycondensation procedure similar to P4HCA (structure Figure 12.7a) [31]. The purified products were obtained as an ocher powder (yield around 52%). The average molecular weights of P(4HCA-*co*-LCA)s were determined to be 1300–3400 g mol^{-1} in THF

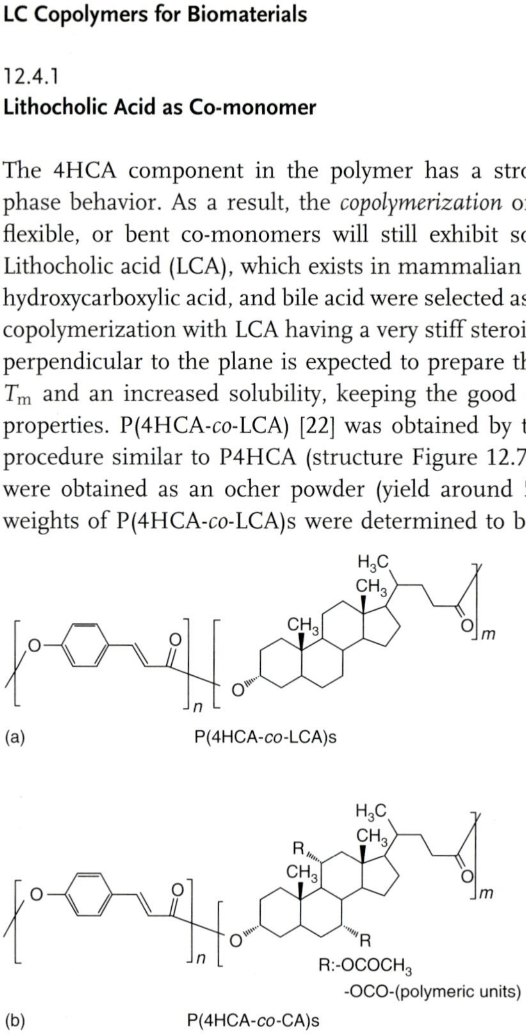

(a) P(4HCA-*co*-LCA)s

(b) P(4HCA-*co*-CA)s

R:-OCOCH$_3$
-OCO-(polymeric units)

Figure 12.7 Chemical structures of the copolymers of 4HCA with bile acid. (a) Lithocholic acid and (b) cholic acid.

by GPC. The low molecular weight may be due to the low chemical reactivity of LCA. The copolymers with an LCA molar ratio of 45 mol% or higher were soluble in pentafluorophenol, TFA/DCM (1 : 5 v/v%), alcohols, acetone, acetonitrile, THF, chloroform, and aprotic amidic solvents.

The thermotropic properties were investigated by crossed polarizing microscopy and differential scanning calorimetry (DSC). The crossed polarizing microscopy showed that P(4HCA-co-LCA) with LCA compositions of 45 mol% and lower in the melting state showed strong birefringence, and sometimes a nematic banded texture, as shown in the representative photograph (Figure 12.8a). The T_g of the copolymers ranged between 160 and 170 °C, independent of the LCA composition. The relationship between the LCA molar composition and T_m is summarized in Figure 12.8b. The T_m of the copolymers increased with increasing LCA molar composition until 7 mol%, and showed its highest value at 250 °C. Above 7 mol%, the T_m decreased. In the case of LCA compositions over 45 mol%, P(4HCA-co-LCA) and PLCA showed no T_m, suggesting that they were amorphous polymers. All of the nematic liquids solidified at 300 °C upon subsequent heating, and no longer

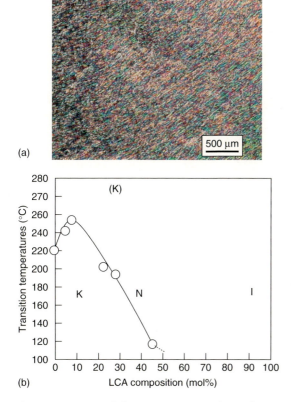

Figure 12.8 (a) Banded texture in nematic phase of P(4HCA-co-LCA). (b) Liquid crystalline phase diagram. K, crystalline; N, nematic; and I, isotropic.

appeared with successive heating and cooling cycles. As a whole, the phase diagram in Figure 12.8b showed that the melting-point of the copolymers could be controlled while retaining the LC properties. In addition, one can see that the high percentage of LCA in the composition resulted in an LC copolymer, although PLCA is not an LC polymer itself.

12.4.2
Cholic Acid as Co-monomer

P(4HCA-co-LCA)s showed a good cell compatibility but too low a molecular weight to process. In order to increase the molecular weight, cholic acid (CA) was selected. CA is a *biomonomer* that is related to bile acid, but contains three hydroxyl groups on the hydrophilic steroid side, methyl groups on the hydrophobic reverse side area, and an aliphatic acidic group at the end. This unique, multifunctional structure of CA can play a role in the branching point of the polymerized chains similarly to with DHCA. Flory showed theoretically that the polymerization of AB_x type multifunctional monomers with one type of functional group (A) and two or more (B) of another type created a hyperbranched architecture without cross-linkage [36]. CA is an AB_3 monomer where A is carboxylic acid and B are the

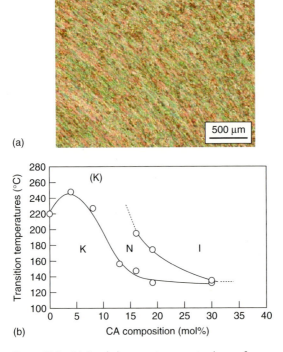

Figure 12.9 (a) Banded texture in nematic phase of P(4HCA-co-CA). (b) Liquid crystalline phase diagram. K, crystalline; N, nematic, and I, isotropic.

hydroxyls, and thus the copolymerization of CA with 4HCA creates a *hyperbranched architecture* rather than cross-linking. The synthesis of P(4HCA-*co*-CA) (structure Figure 12.7b) copolymers was performed in a similar manner to P4HCA [37]. The products were obtained as a yellowish powder. When the CA composition in the feed was increased, the yield decreased from 84 to 7%. This may show a lower reactivity for CA than 4HCA. All the copolymers showed a solubility as high as poly(4HCA-*co*-LCA). The copolymers showed an M_w of 7000–8000 g mol^{-1}, which is unexpectedly low. Based on the assumption of homogeneous arm propagation, CA units play a role of four-way branch points. P(4HCA-*co*-CA)s with a CA composition of less than 30 mol% also showed the nematic banded texture under the crossed polarizing microscopy (Figure 12.9a). Based on the results of the DSC and microscopic observations, the phase diagram can be drawn (Figure 12.9b). The CA composition in P(4HCA-*co*-CA)s showed a narrower range where the nematic phase was exhibited than the LCA composition in the corresponding copolymers. The isotropization temperature, T_i, appeared over a composition range of 16–30 mol%. As CA might make the copolymer chain structure more effectively than LCA, the LC stability of the corresponding polymer was lowered.

On the other hand, the LC melts of P(4HCA-*co*-CA)s were more viscous than P4HCA and P(4HCA-*co*-LCA)s [38]. So we then tried to spin the fibers in the LC state; we picked up the surface of the copolymer melts in the LC state with a pair of tweezers, and pulled them to yield fibers with a thickness of about 100–500 μm.

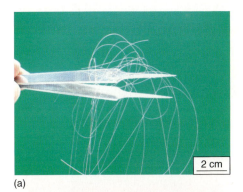

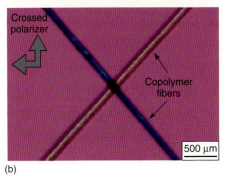

Figure 12.10 (a) Fibers spun from liquid crystalline melt of P(4HCA-*co*-CA). (b) Crossed polarizing microscopic photographs of the fibers under first-order retardation plate (530 nm).

A picture of a representative fiber is shown in Figure 12.10a. Microscopic pictures of the fibers taken under a crossed polarizer are shown in Figure 12.10b. The fibers were bright when they made a ±45° angle to both the polarizer and analyzer, and, furthermore, the fibers clearly showed blue and orange colors in the presence of a sensitive-color plate (λ = 530 nm) depending on the angle sign. This observation indicates that whole fibers were oriented homogeneously, and that the birefringence value was negative, supporting the fact that the polymer chains were oriented along the fiber axis. X-ray diffraction images of the fibers indicated the molecular orientation. The widely studied oriented polyester fiber was derived from the high-speed drawing of high-molecular weight polymer melts. On the other hand, it was quite surprising that the molecularly oriented fibers of the polyester could be formed so easily, even though there was no specific interchain interaction.

The present orientation may be due to the specific structure of the four-way cranked branching point, as shown in Figure 12.11a, suggesting a *hyperbranch architecture*, which effectively increases the extensional viscosity [39]. These arms can behave as a rigid-rod bundle of polymeric chains, thus enabling their LC properties to result in successful spinning of the fibers with a macroscopic

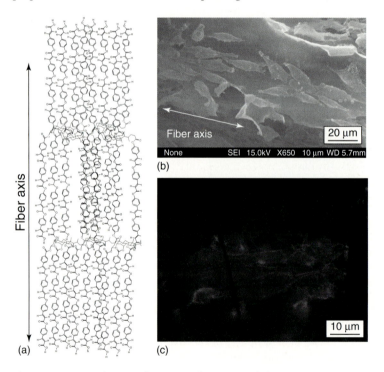

Figure 12.11 (a) Schematic illustration of orientation behavior of P(4HCA-*co*-CA) chains in their fiber. (b) SEM image of the cells adhered on the LC-oriented fiber. (c) CFSM image of cytoskeleton stained by RFP.

molecular orientation. The copolymer showed the good cell compatibility of the copolymers using a cell-adhesion test on the oriented fiber.

The cell morphology was observed by scanning electron microscopy (SEM). One can observe that the extended cells were successfully oriented along the fiber axis (Figure 12.11b), although the substrate fiber was partly broken during the freeze-drying treatment. Confocal fluorescence scanning microscopy (CFSM) demonstrated that their cytoskeletons stained by phalloidin-tetramethylrhodamin B isothiocyanate (RFP) were also oriented (Figure 12.11c). These results can lead to the reconstruction of the oriented tissues, such as blood vessels or nerve fibers, if the P(4HCA-co-CA) fibers were smoothly degraded under the same conditions as when the cells were alive.

12.5
LC Copolymers for Photofunctional Polymers

12.5.1
Syntheses of P(4HCA-co-DHCA)s

We tried to prepare *hyperbranched LC polyarylates* through the copolymerization of 4HCA with DHCA (structure Figure 12.12) [22]. The hyperbranched architecture with oligomeric arms might allow for good degradability from the chain-ends and the high mechanical performance. DHCA units can take the role of a branching point, and the polymeric arms are rigid enough to show an LC phase. ^{1}H-NMR spectroscopy in a mixed solvent of TFA-*d* and DCM-d_2 (1 : 5 v/v) demonstrated that the incorporation of both monomers into the polymer backbone and the copolymer composition, C = [DHCA]/([DHCA] + [4HCA]), can be estimated by the integration ratio of the aromatic proton signals of the individual units. The C value was close to the monomer composition in the feed. The molecular weights of the copolymers were estimated by GPC. The number-average and weight-average molecular weight ranged between $M_n = 1.8$–3.3×10^4 g mol^{-1} and $M_w = 4.4$–9.1×10^4 g mol^{-1}, respectively. The polydispersity ranged between 1.6 and 2.8 although the acetone-soluble fractions ($M_w < 5$ kDa) showed small values of M_w/M_n (<1.3). All of the copolymers were soluble in aprotic polar solvents such as DMF, thus denying the cross-linked network formation.

12.5.2
Phototunable Hydrolyzes

Figure 12.13 shows the photoreactivity of the poly(4HCA-co-DHCA) copolymers. The time course of the *photoreaction conversion* was monitored by UV–visible absorption spectroscopy of the copolymer thin-film cast onto quartz. The absorption with a maximal peak at λ_{max} of 310 nm was reduced as the time of UV irradiation was increased. In contrast, the absorption ($\lambda_{max} = 225$ nm) over a range of less than 250 nm increased with an increase in UV irradiation, resulting in the appearance

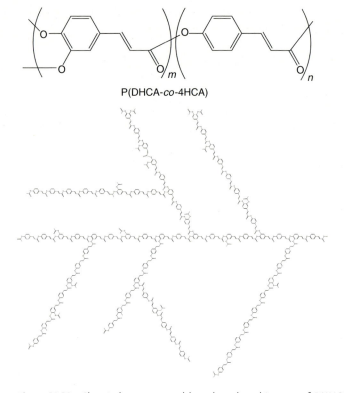

Figure 12.12 Chemical structure and hyperbranch architecture of P(4HCA-co-DHCA).

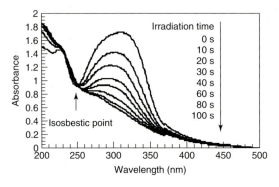

Figure 12.13 Time course of UV-absorption changes of P(4HCA-co-DHCA) membrane by UV irradiation.

of an isosbestic point at $\lambda = 250$ nm. UV irradiation for longer than 20 s made the copolymers insoluble in the solvent. Spectroscopic studies such as ^{1}H-NMR and IR demonstrated that UV irradiation of the copolymers at ambient temperature induced a [2 + 2] cycloaddition of the vinylene units, forming the cyclobutane cross-linkage in the polymeric chains (Figure 12.4).

Figure 12.14 Representative SEM image of P(4HCA-co-DHCA) copolymer particles prepared by mixing of DMF solution with TFA.

The photoreaction effects on *hydrolytic behavior of the copolymers* were investigated by an acceleration test in an alkaline buffer of pH = 10 at 60 °C. The copolymers with a C of 45% or lower showed a slight weight loss of no more than 19% of the initial weight, which demonstrates the high durability. In contrast, the copolymers with a C of 77% and the DHCA homopolymer showed a rapid weight loss, with a half-life of about 15 days. In addition, GPC of the supernatant solution showed only a peak at a maximum of M_n = 2000 and high performance liquid chromatography (HPLC) showed monomer peaks together with other peaks presumably assigned to oligomers. This result indicates that the weight loss was due to hydrolytic degradation; the circular pellets transformed into an amorphous shape accompanying the degradation. An increase in the DHCA composition enhanced the speed of hydrolysis, presumably due to the decreased degree of crystallization and the enhanced degradation from the chain-ends by hyperbranching architecture.

As the copolymer with a C of 77% showed not only high values for the mechanical properties and thermotropic temperatures, but also hydrolytic degradability with a short half-life, it can be regarded as a *hydrolytic engineering polymer*, which could be degraded into the original phytomonomers. Photoconversion from aromatic into aliphatic carboxy ester showed a shortening of the degradation half-life from more than 30 days to only 12 days. In particular, the copolymers with C of 77 and 100% showed a very short half-life of only five days. Furthermore, the period of UV irradiation controlled the speed of hydrolysis. The cyclobutane formation could attenuate the ester carbonyl conjugation with the phenylenevinylene, to make it more polar and easier to degrade randomly, with the cross-linking barely affecting degradation from the chain-ends. As the λ_{max} of the UV absorption peak of the copolymer was 310 nm (UV-B), which is scarcely present in sunlight reaching the ground atmosphere, the hydrolysis was not spontaneously accelerated.

12.5.3
Photoreaction of Nanoparticles

Because the copolymer samples were not transparent, UV light was unable to significantly penetrate the sample. Thus small particles with a size in the nano- to micro-scales were prepared and their photoresponsiveness was investigated. The P(4HCA-co-DHCA) copolymers showed a good solubility in aprotic polar solvents such as DMF and TFA owing to the hyperbranching architecture. Surprisingly when the TFA solution was poured into DMF, the irregular aggregates were formed. Unexpectedly, the mixture became turbid and the turbidity remained for one week. *Monodispersed nanoparticles* with a uniform shape (Figure 12.14) were obtained when the ratio of DMF and TFA was 1 : 1. After mixing DMF and TFA, the electric resistance increased from 10^{-5} Ω cm^{-1} for the pure solvents to 10^{-2} Ω cm^{-1} for the 1 : 1 mixture. As a result, NMR and IR studies indicated that DMF formed a cation while TFA formed an anion. The TFA/DMF ionic pair made the homogeneous solution appear to be an ionic liquid. The ionization of the solvent may be a driving force for the copolymer self-organization to create the nanoparticles because the solubility of copolymers was low in other highly polar solvents such as water and dimethylsulfoxide.

Reversible size changes in P(DHCA-co-4HCA) nanoparticles were observed; the 860 nm diameter decreased to almost a half (420 nm) during UV irradiation over 280 nm and rapidly recovered to 620 nm upon subsequent irradiation at 254 nm. Results from UV-visible and ^{1}H-NMR spectra suggested that *diameter-changing phenomena* corresponded to [2 + 2] cycloaddition formation and cleavage of the cinnamate groups. This significant diameter change was reproduced for at least three cycles, and the reason for this significant size change seemed to be due to the fact that all units of the copolymer contained photochromic groups. Furthermore, these photo-cross-linked bio-based nanoparticles showed various size change behaviors during hydrolytic degradation, depending on the degree of cross-linking. *Photosensitive degradable nanoparticles* may be useful as novel, size-controllable carriers in environmental and biomedical fields.

12.6
LC Copolymers for High Heat-Resistant Polymers

12.6.1
P(4HCA-co-DHCA) Bioplastics

If the poly(4HCA-co-DHCA)s were heated, they melted at specific temperatures to exhibit a thermotropic LC phase, where the schlieren texture was observed by crossed polarizing microscopy (Figure 12.15a). The melting temperatures (T_m) were higher than 220 °C and the glass transition temperatures (T_g) ranged between 115 and 169 °C, which were much higher than the values of the degradable bio-based polymers reported so far [40] and high enough for engineering use

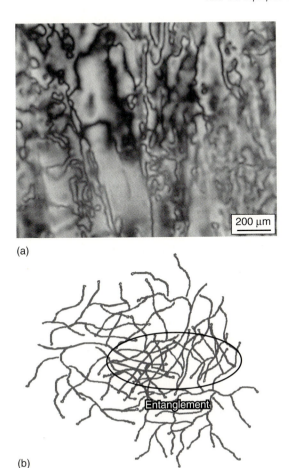

Figure 12.15 (a) Schlieren texture in the nematic phase of P(4HCA-*co*-DHCA). (b) Schematic illustration of the entangled copolymer chains.

{T_g of poly(bisphenol A carbonate), PC, is 145 °C [41]}, as shown in Table 12.1. Further heating induced a LC-crystal transition, which was characteristic of P4HCA [31]. In the LC state, the samples behaved as a very soft elastomer, presumably due to chain entanglement enhanced by the hyperbranching architecture (Figure 12.15b) [39].

The decreased degree of crystallization with an increase in the DHCA composition may also be due to the *hyperbranching architecture*. The copolymers in the nematic state were successfully processed into various shaped molds by pressure (Figure 12.16a) and injection methods (Figure 12.16b). In this nematic state, the samples were molecularly oriented by elongation using a pair of tweezers. The degree of orientation ranged between 0.52 and 0.58 as confirmed by X-ray diffraction, which is relatively low as compared with other LC polymers, presumably due to the hyperbranching architecture.

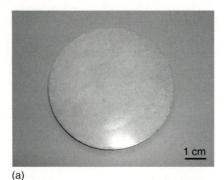

(a)

(b)

Figure 12.16 Compacts of P(4HCA-co-DHCA)s processed from the liquid crystalline melt. (a) Hot-pressed and (b) injected.

A mechanical bending test of the oriented samples showed the mechanical strength, σ, Young's modulus, E, and maximal strain, ε, ranging between 25 and 63 MPa, 7.6 and 16 GPa, and 1.2 and 1.3, respectively. The σ values of the poly(4HCA-co-DHCA)s with $C = 25–50$ mol% were comparable to those of PC [41] and to all the conventional environmentally degradable polymers reported so far [40], whereas E was much higher than they were. The photoreaction increased the σ and E values to maximum values of 104 MPa and 19 GPa, respectively, presumably due to the cross-linkage formation (Table 12.2). The changes in ε values were negligible. In contrast, the thermotropic temperatures were maintained regardless of the photoreaction.

Furthermore, after burying copolymer rectangles in soil for 10 months, deformation (Figure 12.17a) and an 8–13% weight loss were detected. In contrast, PC control samples that were buried alongside the copolymers showed neither deformation nor weight loss. Moreover, SEM photos of the deformed poly(4HCA-co-DHCA) compacts demonstrate that the presence of many holes with a diameter of about 10 μm contrasts with the smooth surface of a hydrolytically degraded polymer, as PC buried alongside showed a flat surface without holes (Figure 12.17b). These findings could indicate an environmental degradation of the copolymer samples by natural microorganisms rather than exposure to rainwater.

Poly(4HCA-co-DHCA)s are hyperbranched polymers that have many rigid branches, just as in lignin. Thus we tried to hybridize the copolymers with celluloses

12.6 LC Copolymers for High Heat-Resistant Polymers

Table 12.2 Synthetic conditions of poly(4HCA-co-DHCA)s, and their performance.

C in feed[a] (mol%)	M_w (×10⁴)	M_n/M_w	C (mol%)	Yield (wt%)	Degree of crystallization[a] (%)	σ[b] (MPa)	E[b] (GPa)	ε[b]	T_g[b] (°C)	T_m[b] (°C)	T_{10} (°C)	Bio-degradability[c]
0	–	0.8	0	84	91	X	X	X	N	220	300	Slow
25	4.4	2.4	21	80	38	38(64)	11(17)	1.3	169	225	290	Slow (fast)
40	4.8	1.6	38	79	10	63(88)	11(15)	1.3	157	250	305	Slow (fast)
50	9.1	2.8	45	80	9	50(104)	16(19)	1.2	124	260	310	Medium (fast)
75	8.0	2.6	77	79	0	25(29)	7.6(9.2)	1.2	115	220	315	Fast
100	7.0	2.4	100	69	0	X	X	X	114	N	320	Fast
PHB	–	–	–	–	–	40	4.0	–	–12	177	–	Fast
PCL	–	–	–	–	–	24	0.27	–	–60	57–60	–	Fast
PLA	–	–	–	–	–	68	2.1	–	55	149	–	Fast
PC	–	–	–	–	–	55	2.0	–	157	225	–	None

[a] Numerals refer to the molar percentage of monomer composition, C, of the DHCA units in poly(4HCA-co-DHCA)s.
[b] The abbreviations σ, E, and ε refer to mechanical strength, Young's modulus, and maximal strain, respectively.
[c] The data in parentheses were obtained after UV irradiation for 4 h (wavelength >250 nm, strength 130 mW cm).

(a)

(b)

Figure 12.17 The compact of P(4HCA-co-DHCA) copolymer of C = 21% taken out from soil after burial for 10 months. (a) Macroscopic digital image; (b) FE (field emission) SEM image of the sample surface, showing the microbes (arrows) on the compact.

to increase their mechanical properties and then measure their thermodynamic properties [42].

12.6.2
Biohybrids

Kenaf fibers are very stiff and too long and thick to mix with the copolymers on the nanometer scale. Millimeter-scaled broken pieces of kenaf fibers were dispersed over the resins after polymerization–simultaneous hybridization. These dispersions introduced an inhomogeneity, functioning as mechanically weak points, which are bad for the performance of the hybrid; the mechanical strength of resins was in fact decreased after hybridization using this type of kenaf fibers.

Next we milled the kenaf hybrids before use. However, as the fibers werevery stiff, it was difficult to reduce their size using a mechanical mill with numerous blades, even if liquid nitrogen had been used to freeze the fibers. One of the best methods for grinding down the kenaf fibers into nanometer-scaled fibers was using a ceramic pestle. An SEM image illustrated the existence of nanofibers 1–5 nm in thickness and 20–200 nm in length. In this study, the nanofibers were used as fillers to reinforce the copolymer resins.

In addition, we tried cinnamoylation of cellulose fibers, aiming to increase the miscibility of the fibers with the copolymers. The ground fibers were dispersed into dimethylacetamide (DMAc), and then cinnamoyl chloride was added. The reaction mixture was stirred overnight to give dark-brown powdery nanofibers. The cinnamoylation of the nanofiber surface was confirmed by IR analyses, abbreviated as *c-kenaf*.

As direct mixing may be difficult for *hybridization with kenaf fibers*, we carried out a polymerization–simultaneous hybridization, that is, 4HCA and DHCA were copolymerized in the presence of celluloses. Here the feed composition of 4HCA/DHCA was fixed at 60 : 40 mol/mol. The reaction solution increased its viscosity gradually and became dark brown. The color was darker than the resins prepared without celluloses, presumably due to the impurities, such as lignins, contaminating the celluloses. However IR analyses demonstrated that ester linkages between 4HCA and/or DHCA were successfully formed. On the other hand, the grafting of the monomers onto the cellulose were not confirmed by IR studies.

The effects of hybridization of the copolymers with the *nanofibers* were investigated in terms of mechanical properties and melting behaviors. The hybrids were successfully processed into rectangular compacts in a brass template by hot pressing. The mechanical strength of the resins was measured by three-points bending tests. Without fillers, the mechanical strengths of the non-oriented copolymer resins were 31 MPa, which increased to 41 MPa on hybridization using the kenaf nanofiber (10 wt%). The kenaf nanofibers have a reinforcing effect on the hyperbranching copolymers. We tried to prepare the hybrids by mixing the melting copolymer with just the nanofibers. However, the mechanical strengths were not increased through the hybridization methods. These results indicated that it was difficult for the hyperbranching chains to interact with the celluloses. On the other hand, the hyperbranching chains might become entangled with the cellulose chains when the hyperbranching chains were propagated.

Dynamic mechanical properties were measured in order to determine the softening temperature of the resins and also the viscoelasticity. The copolymer resins and each hybrid showed two peaks for tan δ around -50 and $150\,°C$ (Figure 12.18). The peak at the lower temperature can be assigned to local relaxation and the other at the higher temperature can be assigned to the glass transition. While the relaxation at around $-50\,°C$ did not change regardless of hybridization, softening temperatures increased from 139 to $154\,°C$ on increasing the composition of the kenaf nanofibers up to 10 wt%. This result indicated that the cellulose directly interacts with the hyperbranch copolymers on the molecular level, for example, cross-linking physically. The interaction may be based on hydrogen and/or π-hydrogen interaction of phenols with the hydroxyls of the sugar units. However, kenaf nanofibers aggregated through their own strong interchain interactions to form apparent nanofiber networks, which reduced the miscibility of the copolymer chains with the nanofibers, giving the left shoulders of the tan δ peaks. The softening temperature of the hybrid with the c-kenaf nanofiber was slightly higher than that of the corresponding hybrid with normal nanofibers. The cinnamoylation can increase the interaction of the nanofibers with hyperbranch polymers, as expected, presumably due to the high compatibility between the

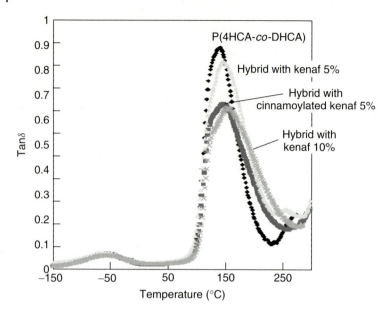

Figure 12.18 Temperature dependence of loss tangent, tan δ, of rectangular compacts of poly(4HCA-co-DHCA) (C = 40) hybrids with kenaf cellulose in dynamic mechanical analyses.

cinnamoyl groups of the nanofibers and the copolymers. Moduli in torsion and complex viscosities at room temperature increased with hybridization, presumably due to the reinforcement by kenaf networks. From these results, we can summarize the effects of cinnamoylation of kenaf nanofibers; the cinnamoylation enhances the interaction of the copolymers with the kenaf nanofibers to increase the softening temperature.

12.7
Conclusion

We successfully prepared high-performance and functional polymers from renewable starting materials, such as coumarates, which are rigid, photoreactive phenols. In particular, poly(p-coumaric acid-co-caffeic acid) showed extremely high heat resistance and high mechanical toughness and fantastic functions such as photo-tunable hydrolytic properties. The copolymers could lead to the development of novel, environmentally friendly engineering plastics for use in automobiles, aircraft, electronic devices, and other materials. We finally propose the positive application of AB_x-type multifunctional phytomonomers as renewable starting materials. These types of monomers exist widely as amino acids, glycolipids, and other metabolites, and will take on the new aspects of highly functional and highly efficient bio-base polyesters and/or polyamides.

Acknowledgments

This research was mainly supported by a Grant-in-Aid for J.S.T. (Practical application research, Kaneko Project). I appreciate the valuable discussions with Prof. Mitsuru Akashi and Prof. Michiya Matsusaki on investigating cell compatibility, and thank Dr Dong Jian Shi, Ms Hang Thi Tran, and Mr Takanori Ishikura for helping with the experiments.

References

1. Stevens, E.S. (2002) *Green Plastics: An Introduction to the New Science of Biodegradable Plastics*, Princeton University Press, Princeton.
2. Vert, M. (2005) *Biomacromolecules*, **6**, 538.
3. Taniguchi, I. and Kimura, Y. (2001) *Biopolymers*, **3b**, 431.
4. Saulnier, B., Ponsart, S., Coudane, J., Garreau, H., and Vert, M. (2004) *Macromol. Biosci.*, **4**, 232.
5. Yamamoto, M., Witt, U., Skupin, G., Beimborn, D., and Mueller, R.-J. (2002) *Biopolymers*, **4**, 299.
6. Nagarajan, V., Singh, M., Kane, H., Khalili, M., and Bramucci, M.J. (2006) *Polym. Environ.*, **14**, 281.
7. Okano, K., Kondo, A., and Noda, H. (2006) *Eco Ind.*, **11**, 43.
8. Imai, Y. and Yokota, R. (2002) *Saishin Polyimides*, NTS Press, Tokyo.
9. Yano, H., Sugiyama, J., Nakagaito, A.N., Nogi, M., Matsuura, T., Hikita, M., and Handa, K. (2005) *Adv. Mater.*, **17**, 153.
10. Thielemans, W. and Wool, R. (2005) *Biomacromolecules*, **6**, 1895.
11. Imai, Y. (1995) *High Perform. Polym.*, **7**, 337.
12. Madhavamoorthi, P. (2004) *Synth. Fibers*, **33**, 16.
13. Nagata, M. and Hizakae, S. (2003) *Macromol. Biosci.*, **3**, 412.
14. Kimura, K., Inoue, H., Kohama, S.-I., Yamashita, Y., and Sakaguchi, Y. (2003) *Macromolecules*, **36**, 7721.
15. Jin, X., Carfagna, C., Nicolais, L., and Lanzetta, R. (1995) *Macromolecules*, **28**, 4785.
16. Kricheldrorf, H.R. and Stukenbrock, T. (1998) *J. Polym. Sci. A Polym. Chem.*, **36**, 2347.
17. Reina, A., Gerken, A., Zemann, U., and Kricheldrorf, H.R. (1999) *Macromol. Chem. Phys.*, **200**, 1784.
18. Krawinkel, T. and Kricheldrorf, H.R. (1998) *Macromolecules*, **31**, 1016.
19. Muzafarov, E.N. and Zolotareva, E.K. (1989) *Biochem. Physiol. Pflanz.*, **184**, 363.
20. Stampe, J. and Ziegler, A. (1995) *Macromolecules*, **28**, 5306.
21. Pillai, C.K.S., Sherrington, D.C., and Sneddon, A. (1992) *Polymer*, **33**, 3968.
22. Kaneko, T., Tran, H.T., Shi, D.J., and Akashi, M. (2006) *Nat. Mater.*, **5**, 996.
23. Ricarda, N., Anthony, J.M., and Cathie, M. (2004) *Nat. Biotechnol.*, **22**, 746.
24. Yang, C.-M., Lee, C.-N., and Chou, C.-H. (2002) *Bot. Bull. Acad. Sinica (Taiwan)*, **43**, 299.
25. (a) Kyndt, J.A., Meyer, T.E., Cusanovich, M.A., and Van Beeumen, J.J. (2002) *FEBS Lett.*, **512**, 240; (b) Kyndt, J.A., Hurley, J.K., Devreese, B., Meyer, T.E., Cusanovich, M.A., Tollin, G., and Van Beeumen, J.J. (2004) *Biochemistry*, **43**, 1809.
26. Kort, R., Phillips-Jones, M.K., van Aalten, D.M.F., Haker, A., Hoffer, S.M., Hellingwerf, K.J., and Crielaard, W. (1998) *Biochim. Biophys. Acta*, **1385**, 1.
27. Hellingwerf, K.J., Hendriks, J., and Gensch, T.J. (2002) *Biol. Phys.*, **28**, 395.
28. Boerjan, W., Ralph, J., and Baucher, M. (2003) *Annu. Rev. Plant Biol.*, **54**, 519.
29. Cain, R.B., Bilton, R.F., and Darrah, J.A. (1968) *Biochem. J.*, **108**, 797.
30. Nagel, W. and Hiller, W. (1933) *Fett. Chem. Umsch.*, **40**, 49.

31. (a) Kaneko, T., Matsusaki, M., Tran, H.T., and Akashi, M. (2004) *Macromol. Rapid Commun.*, **25**, 673; (b) Matsusaki, M., Kishida, A., Stainton, N., Ansell, C.G.W., and Akashi, M. (2001) *J. Appl. Polym. Sci.*, **82**, 2357.
32. Tran, H.T., Matsusaki, M., Shi, D.J., Kaneko, T., and Akashi, M. (2008) *J. Biomater. Sci. Polym. Ed.*, **19**, 75.
33. (a) Demus, D., Goodby, J., Gray, G.W., Spiess, H.-W., and Vill, V. (eds) (1998) *Handbook of Liquid Crystals*, Wiley-VCH Verlag GmbH, Weinheim; (b) Chung, T.-S. (1986) *Polym. Eng. Sci.*, **26**, 901; (c) Fu, K., Nematsu, T., Sone, M., Itoh, T., Hayakawa, T., Ueda, M., Tokita, M., and Watanabe, J. (2000) *Macromolecules*, **33**, 8367; (d) Griffin, B.P. and Cox, M.K. (1980) *Br. Polym. J.*, **12**, 147.
34. Silverstein, R.M. and Webster, F.X. (1998) *Spectrometric Identification of Organic Compounds*, 6th edn, John Wiley & Sons, Inc., New York.
35. Tanaka, Y., Tanabe, T., Shimura, Y., Okada, A., Kurihara, Y., and Sakakibara, Y. (1975) *Polym. Lett. Ed.*, **13**, 235.
36. Flory, P.J. (1952) *J. Am. Chem. Soc.*, **74**, 2718.
37. Matsusaki, M., Tran, H.T., Kaneko, T., and Akashi, M. (2005) *Biomaterials*, **26**, 6263.
38. Kaneko, T., Tran, H.T., Matsusaki, M., and Akashi, M. (2006) *Chem. Mater.*, **18**, 6220.
39. Weng, W., Markel, E.J., and Dekmezian, A.H. (2001) *Macromol. Rapid Commun.*, **22**, 1488.
40. Gross, R.A. and Kalra, B. (2002) *Science*, **297**, 803.
41. Pochan, J.M. and Pochan, D.F. (1980) *Macromolecules*, **13**, 1577.
42. Kaneko, T. and Ishikura, T. (2008) *Trans. MRS-J.*, **33**, 501.

13
Enzymatic Polymer Synthesis in Green Chemistry
Andreas Heise and Inge van der Meulen

13.1
Introduction

Biotechnology is playing a major role in the field of green chemistry. The application of nature's toolset is a fast growing area in several industries such as food, fine chemicals, and polymers. Recent advances in generic techniques have made *biotechnology* much more broadly applicable for the synthesis and modification of chemicals. In particular *in vitro* enzymatic catalysis has seen a steady increase in (industrial) applications, and enzyme catalysis has established itself as an indispensable tool in the synthesis of small molecules.

Examples can be found in the production of pharmaceutical intermediates where biotechnology is generating significant turnover and reducing the environmental impact [1]. The success of *enzyme catalysis* in pharmaceutical synthesis is based on the selectivity and efficiency of enzymes; promoting reactions that are not easily accessible by conventional techniques. This can lead to increased purity and the reduction of waste and lower energy requirements through simplified synthetic routes. Examples are the replacement of tedious protection/deprotection chemistry (chemo- and regioselectivity) and asymmetric synthesis of chiral compounds (enantioselectivity). A specific example is the biotechnological route to the antibiotic Cephalexin, which is practiced on an industrial scale with high environmental and cost benefits compared with the traditional chemical synthesis (material savings 65%, energy savings 65%; cost reduction 50%) [2].

In recent years enzyme catalysis has also been successfully applied in *polymer synthesis*. Notably, three of the six enzyme groups have been reported in enzymatic polymerization *in vitro*, that is, oxidoreductases, transferases, and hydrolases. Some oxidoreductases such as peroxidase, laccase, and bilirubin oxidase have been used as catalysts for the oxidative polymerizations of phenol and aniline derivatives to produce novel polyaromatics. Transferases are enzymes transferring a group from one compound (donor) to another compound (acceptor). Several transferases such as phosphorylases and synthases have been found to be effective for catalyzing *in vitro* synthesis of polysaccharides and polyesters. Hydrolases including glycosidases, lipases, and priteases are enzymes catalyzing a bond-cleavage reaction by hydrolysis. Under synthetic conditions they have been employed as catalysts for the reverse reaction, that is, the bond-forming reaction, for example for polyesters. Hydrolases, and in particular lipases, are the most successful class of enzymes in polymer forming reactions.

The motivation for using *enzymes in polymer synthesis* was initially mainly scientific curiosity, but when this technology started to produce results comparable to conventional polymerizations, the potential was recognized. Specifically, the possibility to make polymers that are not available through conventional methods and their natural character make enzymes a promising catalytic system. As for the latter, it has to be noted that employing a natural catalyst does not necessarily qualify a process as green or sustainable. The process has to be viewed in its entirety, including the solvents used, raw materials, and the energy balance.

In this chapter, selected examples illustrating the application of enzymes in polymer synthesis will be discussed. The goal is not to provide a complete overview of enzyme catalysis in polymer synthesis but to emphasize selected examples, where enzyme catalysis could potentially contribute to a greener reaction process. This involves the combination with green solvents and renewable raw materials. The interested reader is also referred to recent review articles on enzymatic polymer synthesis [3–8].

13.2
Polymers

13.2.1
Polycondensates

Among *polycondensates*, the synthesis of polyesters is the main focus of enzymatic polymerization. The two main routes to polyesters are the ring-opening polymerization (ROP) of lactones and the polycondensation of dicarboxylic acids and diols or of hydroxyl-acids (or their derivatives). Typically, these reactions are acid or metal catalyzed and for some polyesters are conducted on a large industrial scale. Lipases have been used very successfully in *enzymatic synthesis of polyesters*. Lipases belong to the class of hydrolases that catalyze the hydrolysis of various bonds, for example esters, in their natural aqueous environment. Owing to the reversible nature of these reactions, the equilibrium can be controlled by the reaction conditions and thus hydrolases can also be used to catalyze the bond forming reaction. This is typically achieved under non-aqueous conditions, namely in organic media. The stability of the lipases in a variety of organic media and over a wide temperature range is therefore a critical condition for conducting enzymatic polyester synthesis. In particular *Candida Antarctica* lipase B (CALB) in its free form or immobilized on macroporous resin (Novozym® 435) allowed breakthroughs in the enzymatic polyester synthesis due to its exceptionally high activity and robustness under ester forming conditions in organic solvents [9, 10].

13.2.1.1 Polyesters by Ring-Opening Polymerization

Since the discovery of lipase catalyzed ring-opening polymerizations of lactones in 1993 by two independent groups, that is, Uyama, Kobayashi, and coworkers [11] and the group of Knani [12], this is by far the most studied enzymatic polymerization. Reaction parameters investigated are, for example, the enzyme origin [13–17], concentration [18], temperature [19, 20], organic solvent [13, 18, 21, 22], and water content [23, 24]. Of note is the fact that the reaction kinetics and the achievable molecular weights of the enzymatic ROP increase with the lactone ring size, while it is vice versa in metal catalyzed ROP due to the absence of ring strain in larger lactones [25, 26]. This opens up opportunities for the polymerization of macrolactones, which can be derived from natural sources.

A prominent example is pentadecalactone (PDL), which belongs to the class of naturally occurring macrocyclic musks and is used in the fragrance industry. Its eco-friendliness led to an increased demand and the development of improved synthetic routes, which make PDL commercially available in larger amounts [27]. Poly(pentadecalactone) (PPDL), with a significantly high molecular weight, can only be obtained by lipase catalyzed ROP. Molecular weights (M_n) of up to 150 kg mol^{-1} were reported [28, 29] employing Novozym 435 and up to 200 kg mol^{-1} in mini-emulsion with *Pseudomonas Cepacia* (PC) lipase [30]. PPDL is of potential technical interest as it resembles the properties of polyethylene (PE); PPDL is a semi-crystalline polymer with a melting-point around 100 °C and a glass transition temperature of −27 °C [31–33]. The crystallization behavior, the crystal structure of PPDL, and the mechanical properties reveal large similarities with PE [33, 34]. As with PE, PPDL is neither enzymatically nor hydrolytically degradable in buffer solutions [35], but can be expected to degrade under harsher hydrolytic conditions, as are often found in the environment (e.g., soil) and in recycling processes. The first application related PPDL reports were on end-functionalized low-molecular weight PPDL for coating applications [36] and high molecular weight PPDL spun into fibers revealing a tensile strength of up to 0.74 GPa [29] (Scheme 13.1).

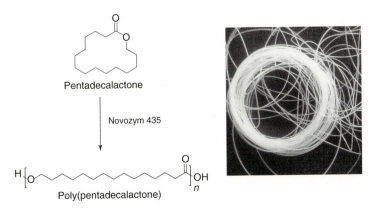

Scheme 13.1 Enzymatic ROP of pentadecalactone (left) and fibers spun from poly(pentadacalactone) (right) [29]. (Reproduced with permission from [29].)

Unsaturated natural macrocycles ambrettolide and globalide were also polymerized by Novozym 435 and investigated as potential biomaterials. The *crystalline polymers* (melting-point 46–55 °C) were non-toxic and could be cross-linked in the melt yielding fully amorphous transparent materials with a gel content of 97% [35].

With respect to the polymerization of medium-size *lactones* such as caprolactone, the *enzymatic polymerization* does not provide any structural advantage over the chemical process. In contrast to metal-mediated ROP, only limited molecular weight and endgroup control can be exercised in enzymatic ROP due to the mechanistic nature of the polymerization (monomer activation) [37, 38]. Polymer structures such as block copolymers are thus difficult to realize in *enzymatic ROP* [39]. In order to overcome these limitations, enzymatic ROP has been combined with various chemical polymerizations, which allows for the synthesis of more complex structures, for example block copolymers. Several examples of the enzymatic initiation from hydroxy-terminated macroinitiators, for example, polybutadiene [40] and PEG [41–44] are known, but in the majority of the reports the enzymatic ROP is conducted first.

Recently it has been mainly controlled polymerization techniques that have been applied for the subsequent macroinitiation, which generally results in well-defined block copolymers. Atom transfer radical polymerization (ATRP) represents the methods most widely applied in this synthetic strategy. In the initial report, the combination of enzymatic ROP and ATRP using the dual initiator yielding a block copolymer of styrene and caprolactone (CL) was described [45]. This concept was further investigated with respect to the reaction conditions, the initiator structure, and variations of the monomers [24, 46–56]. However, all concepts involving ATRP rely on a metal catalyst. Full metal-free and thus greener approaches to *block copolymers* were realized by the combination of enzymatic ROP with nitroxide mediated living free radical polymerization (Scheme 13.2) [57]. With this system it was also possible to successfully conduct a one-pot chemoenzymatic cascade polymerization from a mixture containing a dual initiator, CL, and styrene. Moreover, it was shown that this approach is compatible with the stereoselective polymerization of 4-methylcaprolactone for the synthesis of chiral block copolymers (Scheme 13.2). A metal free synthesis of block copolymers using a radical chain transfer agent as a dual initiator in enzymatic ROP to yield poly(CL-*b*-styrene) was also reported recently [58].

Very recently the first example of the metal-free combination of enzymatic and chemical carbene catalyzed ROP for the formation of block copolymers (Scheme 13.3) was described [59]. The approach took advantage of the fact that CALB has a high catalytic activity for lactones but not for lactides (LLAs), while it is vice versa for carbenes. In the applied synthetic strategy, the enzymatic polymerization of CL was conducted first. Addition of LLA and carbene to the reaction mixture led to the *macroinitiation* directly from the hydroxyl endgroup of PCL allowing the synthesis of PCL/PLLA block copolymers.

The modification of functional polymer backbones was also investigated as a green route to *graft copolymers*. A significant contribution was made by Keul and Moeller using well-defined polyglycidols as promising biomedical materials

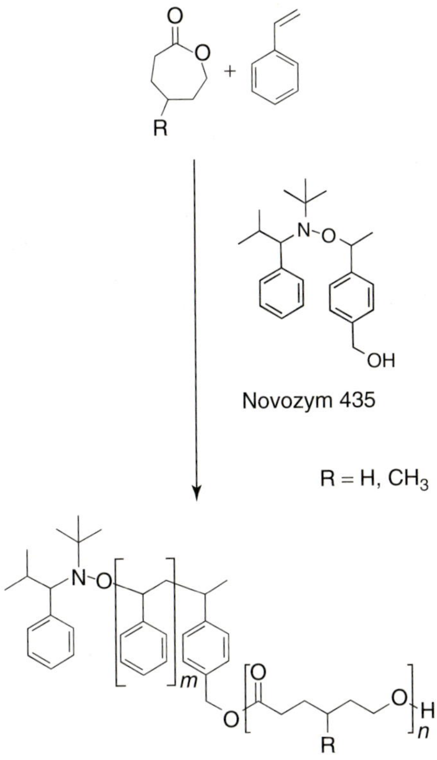

Scheme 13.2 One-pot enzymatic ring opening and living free radical cascade polymerization [57].

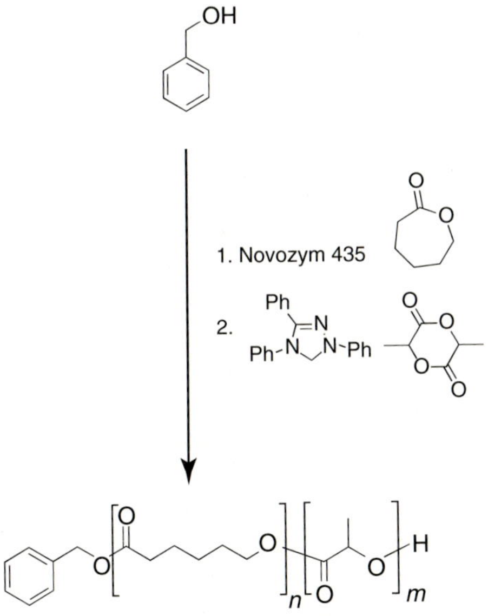

Scheme 13.3 Block copolymers by combination of enzymatic ROP and carbene catalyzed ROP [59].

[60]. In an initial paper these authors compared the chemical and enzymatic ROP of CL from linear and star-shaped polyglycidol obtained from anionic ROP [61]. It was found that the zinc-catalyzed ROP of CL from the multifunctional polyglycidol resulted in a quantitative initiation efficiency, while with enzymatic ROP the PCL (polycaprolactone) was only grafted to 15–20% of the hydroxyl groups. The enzymatically obtained partly grafted polymer, on the other hand, has a hydrophilic polyether head with a hydrophobic PCL tail. The difference in the initiation efficiency is due to the different polymerization mechanisms. While the chemical ROP is endgroup activated and controlled, the enzymatic ROP is monomer activated. In the latter, the steric constraints in the nucleophilic attack of the hydroxyl groups on the EAM (enzyme-activated monomer) prevents the reactions of all hydroxyl groups. This was also confirmed by a study of Heise and coworkers on the enzymatic ROP from polystyrene containing 10% of 4-vinylbenzyl alcohol [62]. While quantitative enzymatic modification of the hydroxyl groups with the small vinyl acetate was observed, the grafting of PCL produced only a 50–60% grafting efficiency. Moreover, the results suggested that the grafting action is a combination of grafting from and grafting onto by transesterification of PCL.

The incomplete *enzymatic grafting* from polyalcohols opens up opportunities to synthesize unique structures, as was shown in several examples by Moeller and coworkers. For example, heterografted molecular bottle brushes were synthesized starting from PCL grafted polyglycidol [63]. The grafting yield in this step was about 50%. Selective acetylation of the hydroxyl groups at the PCL graft ends was achieved via enzymatic reaction with vinyl acetate without acetylation of the remaining hydroxyl groups at the polyglycidol backbone. The remaining hydroxyl groups on the polyglycidol backbone were used to initiate chemically catalyzed ROP of lactide to produce a heterografted polymer comprising PCL and PLA grafts. When the endgroups of the PCL were capped with acrylates, the resulting materials could be formulated into UV cross-linked microspheres (Scheme 13.4) [64].

13.2.1.2 Polyesters by Condensation Polymerization

Lipases such as CALB are very efficient catalysts in *polycondensations* and as such have the potential to replace conventional metal catalysts and reduce the reaction temperature. Several examples have been reported in the literature using typical aliphatic diols and diacids or hydroxyacids [12, 65–68]. Similar to conventional polycondensations, the removal of the condensation by-product, such as water, is crucial to obtaining high monomer conversion and high molecular weights. As this technology is common practice in the polymer industry, *enzymatic polycondensation* could replace chemical polycondensation without further capital investment. It is thus not surprising that the feasibility of large-scale industrial enzymatic polycondensation was investigated early on. Baxenden Chemical developed a process that uses CALB to catalyze the polycondensation reaction of aliphatic diols and diacids at much lower temperature (60 °C) [69]. According to the Baxenden information, the lipase-catalyzed process, when compared with the conventional process, could eliminate the use of organic solvents and inorganic acids, and yields energy savings of about 2000 MW annually at full industrial-scale operation [2]. However,

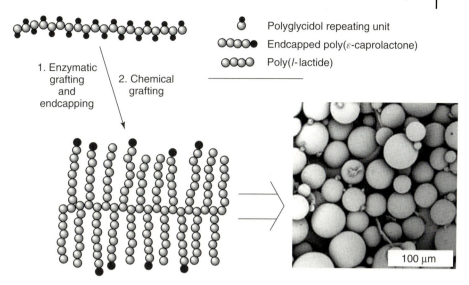

Scheme 13.4 Chemoenzymatic synthesis of heterografted polymers comprising PCL and PLA grafts on a polyglycidol backbone. After acrylation these polymers can be formulated into cross-linked microspheres [64]. (Reproduced with permission from [64].)

this process has not been implemented on a technical scale yet. In can only be speculated that the contribution from the cost of the enzyme to the overall process costs for commodity products, such as aliphatic polyesters, is still too high.

Another strategy in lipase catalyzed polycondensation is the use of *green monomers* as starting materials, such as glycerol. One example is the synthesis of epoxide-containing polymers from natural unsaturated aliphatic fatty acids from plant oils [70]. In this approach the lipase catalyzes not only the polycondensation but also the epoxidation of unsaturated groups in the presence of catalytic amounts of carboxylic acids under mild conditions [71–73]. In the reported procedure, the synthesis of epoxidated fatty acid side chains on a divinyl sebacate/glycerol backbone was performed via two routes (Scheme 13.5). In the first route, the polyester was formed first and epoxidation was subsequently carried out on the polymer. In the second route, epoxidation was followed by polymerization. Upon curing of these polymers via the epoxy groups, biodegradable films from renewable plant oils could be made. Another way to make epoxide-containing polymers is by ROP, starting from unsaturated macrolactones. Converting the double bond into epoxides and sequential polymerization results in polyesters with epoxide functionalities in the backbone [74].

Gross and coworkers have synthesized *terpolymers* by incorporating the green monomers sorbitol and glycerol into polyesters [75]. Incorporation of these polyols into polyesters is a feasible strategy to obtain highly functional polymers (Scheme 13.6). When producing these polymers in a chemical manner, elaborate protection–deprotection steps are necessary to avoid cross-linking between the

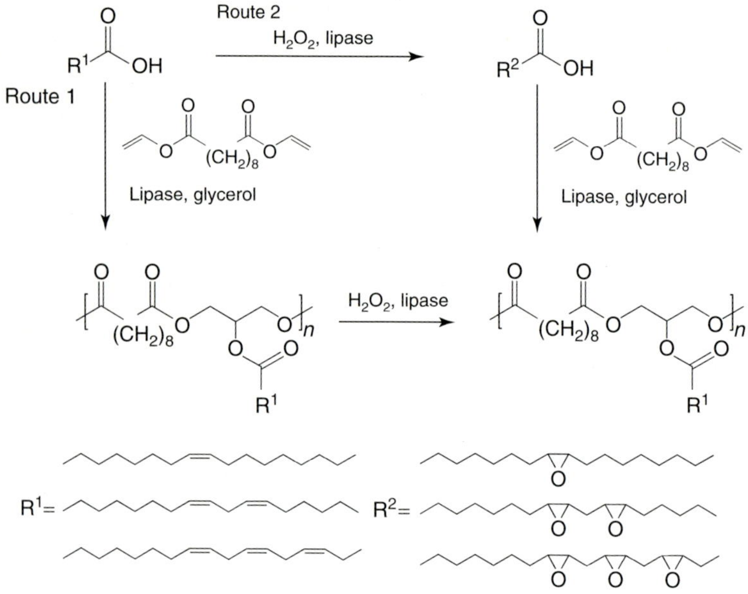

Scheme 13.5 Enzymatic synthesis of epoxide-containing polyesters via two routes [70].

polyol units. Using regioselective lipases or proteases such as Novozym 435 as a catalyst, cross-linking can be prevented. One synthetic challenge is the low solubility of the monomers adipic acid, glycerol, and sorbitol in non-polar organic solvents and the significant decrease of enzyme activity in polar organic solvents. The use of binary or ternary solvent mixtures circumvented this problem and the branching of glycerol polyesters could be controlled by adjusting the feed ratio of the monomers [76].

High molecular weight polyricinoleate ($M_w = 100\,600\,\text{g mol}^{-1}$) was synthesized using *condensation polymerization* with lipase *Pseudomonas cepacia* (Scheme 13.7) [77]. It was shown that the main chain double bonds could be cross-linked using dicumyl peroxide at elevated temperatures. This allows for the synthesis of soft materials. This curing concept was also applied to non-branched aliphatic polyesters derived from the ROP of unsaturated macrolactones [35].

13.2.2
Polyphenols

Polyphenols are formed by the *oxidative polymerization* of the corresponding phenol monomer. Owing to the oxidative character of the polymerization reactions, the choice of enzyme is from the class of oxidoreductase (Class I) [67]. Most of these enzymes contain a low valent metal as the catalytic center. Oxidoreductases that are commonly used are laccase [78, 79], bilirubin oxidase [80–82], and horseradish peroxidase (HRP), which is the most widely used enzyme for these reactions. The mechanism of HRP catalyzed polymerization has been extensively studied

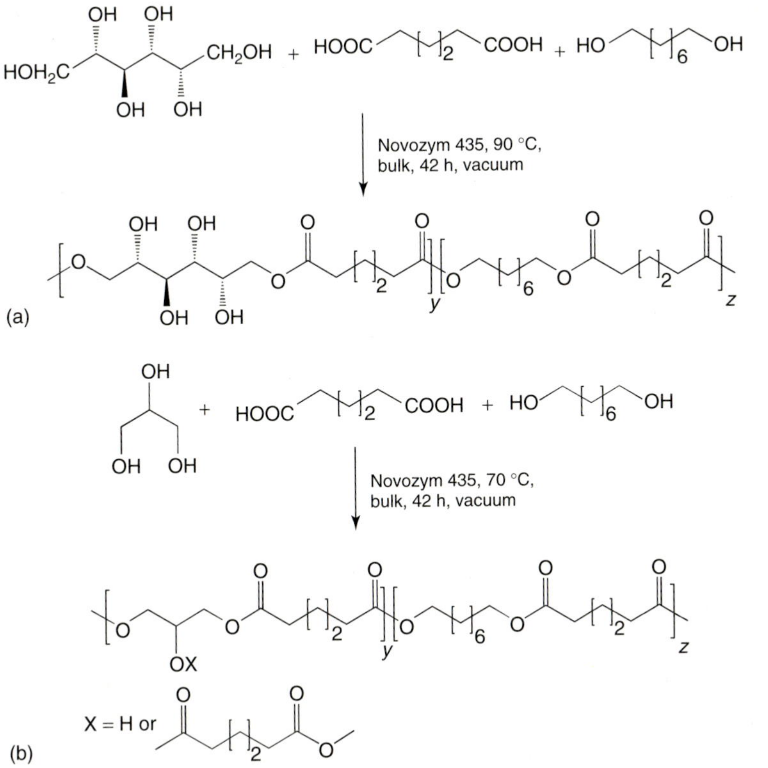

Scheme 13.6 (a) Polyester formation with sorbitol and (b) polyester synthesis using glycerol [77].

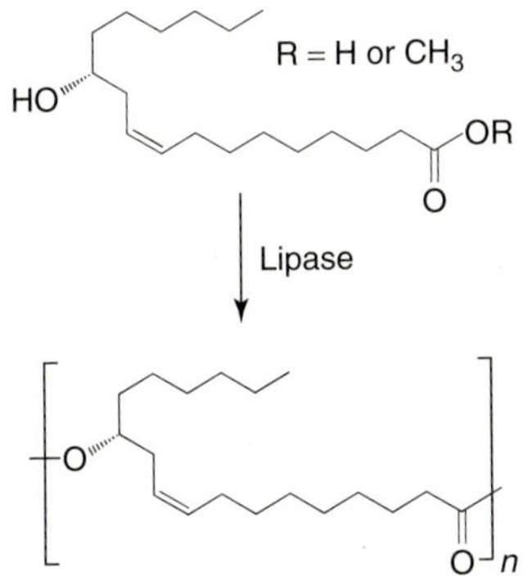

Scheme 13.7 Synthesis of polyricinoleate.

in the past [83, 84]. It catalyzes the decomposition of hydrogen peroxide at the expense of aromatic proton donors, which allows for polymerization. Using a polar organic solvent mixed with a buffer, polyphenols were produced efficiently by this process. The enzymatic approach proved to be a good alternative for conventional high temperature formaldehyde based synthetic routes. However, synthesis in a

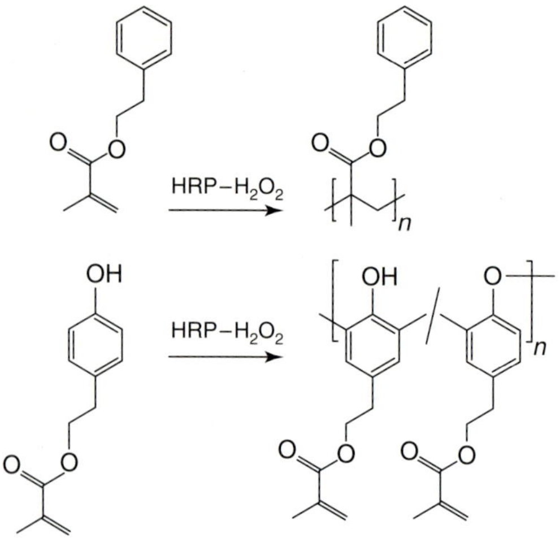

Scheme 13.8 Chemoselective polymerization of the phenol moiety by HRP in the polymerization of 2-(4-hydroxyphenyl)ethyl methacrylate [87].

completely aqueous media only produced polymers in low yields. Better results were obtained in the presence of templating molecules. For example, it was found that the use of cyclodextrins enhances polymer production [85]. Kobayashi and coworkers used poly(propylene glycol)-poly(ethylene glycol)-poly(propylene glycol) (PPG-PEG-PPG; Pluronics) triblock copolymers as a template for polyphenol polymerization. HRP-catalyzed synthesis of phenolic polymers in the presence of Pluronic F68 (EG_{76}-PG_{29}-EG_{76}) resulted in ultra-high molecular weight products ($M_w > 10^6$ g mol^{-1}) [86]. The so-called enzymatic template polymerization thus offers routes towards efficient phenol polymerization in green media.

HRP is not only active in the polymerization of polyphenols, but it can also catalyze the formation of polyvinyl polymers. The chemoselectivity between the phenol and vinyl polymerization was studied by the polymerization of 2-(4-hydroxyphenyl)ethyl methacrylate (Scheme 13.8) [87]. This monomer contains both functional groups that can react in peroxidase polymerization in one molecule. It was found that the phenol moiety is chemoselectively polymerized by HRP, while in the absence of the phenol moiety the acrylate was polymerized.

13.2.3
Vinyl Polymers

The group of Derango et al. was the first to report enzymatic mediated synthesis of *vinyl polymers* [88]. They reported the synthesis of polyacrylamide in the presence of bisacrylamide using xanthine oxidase, chloroperoxidase or alcohol oxidase as a catalyst. Another oxidoreductase used to mediate the polymerization of vinyl compounds was glucose oxidase in the presence of Fe^{2+} and dissolved oxygen [89]. The system HRP–β-diketones–H_2O_2 was used to polymerize styrene

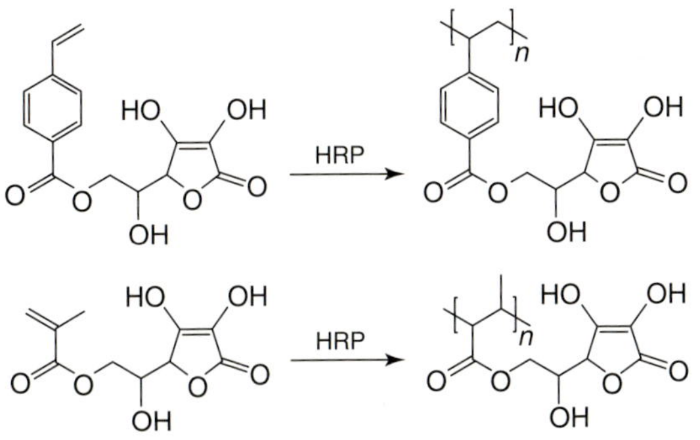

Scheme 13.9 Enzymatic synthesis of vitamin C functionalized vinyl polymers [94].

[90] and methyl methacrylate [91]. Polymerization of acrylamide can also be catalyzed by manganese peroxidase in the presence of acetylacetone [92]. Enzymatic polymerization of vinyl polymers was shown to be possible with many different vinyl monomers and styrene derivatives and monomers derived from acrylates were investigated thoroughly. An interesting polymer based on vitamin C has been synthesized by a *double enzymatic process* (Scheme 13.9) [93, 94]. The vinyl monomer was produced from ascorbic acid via transesterification with CALB and then polymerized by HRP. Vitamin C can act as a free radical scavenger and the polymeric form was shown to be almost as active as the free form. By incorporating vitamin C into polymers, the shelf-life of labile components in food and pharmaceuticals can be improved.

For greener processes, enzymatic vinyl polymerization has been attempted in aqueous emulsions. The *emulsion polymerization* of styrene could be performed using HRP–H_2O_2–Acac (acetylacetonate) as an initiation system. Stable polymer colloids and nanospheres of about 30–50 nm were obtained at room temperature [95]. However, compared with traditional chemically catalyzed emulsion polymerization of styrene, the yields and molecular weight obtained in the enzyme catalyzed reaction were lower. The HRP catalyzed oxidation requires enzyme binding to the substrate. Hence, the oxidation activity of HRP strongly depends on the substrate [96].

13.2.4
Polyanilines

Polyaniline (PANI) is one of the most important *conductive polymers*. It is widely used in conducting and optical applications (organic LEDs, rechargeable batteries, antistatic coatings). Common methods of PANI synthesis are by means of chemical or electrochemical oxidation. Both methods require harsh reaction conditions, high temperatures, highly toxic solvents, extreme pH, and strong oxidants. Solubility and processability of the products are low; therefore treatment with fuming sulfuric acid is needed [97, 98]. However, when enzymes are used in this polymerization process, harsh conditions can be avoided. Polymerization takes place at room

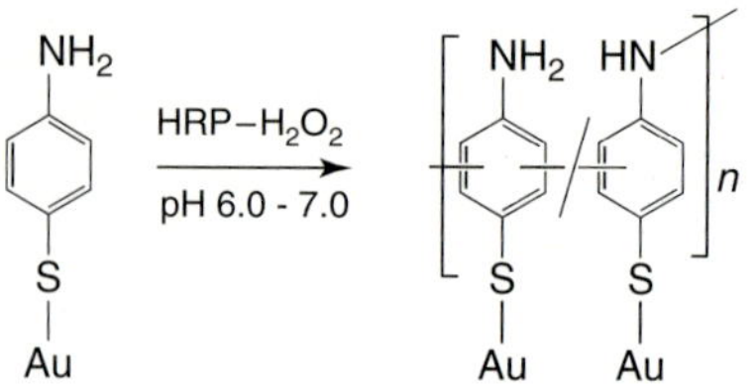

Scheme 13.10 Enzymatic polymerization of 4-aminothiophenol on a gold surface [110].

temperature, in aqueous organic medium at neutral pH. This alternative method provides soluble and processable conducting polymers [99]. The processability, linearity of polymers and the molecular weight are improved by using modified enzymatic polymerization techniques in aqueous media. Various aspects, including solvent mixture [100, 101], water-soluble monomers [102, 103], the use of micelles [104] or reverse micelles [105, 106], reactions at the air–water interface [107], and the use of polyanions [108, 109] have been studied.

Among others, Xu and Kaplan have investigated the use of enzymatically obtained PANIs in conducting applications [110]. Compared with metal, carbon, and semi-conducting materials, PANI and related polymers may provide alternative, more flexible *nanodevices*. In this study, enzymatic polymerization and dip-pen nanolithography (DPN) patterned reactive monomers were combined to explore ambient surface reactions for the formation of conducting polymers (Scheme 13.10). The polymerization was carried out on the DPN patterned monomers by dipping the substrate into the H_2O_2–enzyme stock solution for about 30 s followed by rinsing with distilled water. Polymers with a mass up to 1000 Da were collected.

Template-assisted polymerization is another way of improving the solubility of PANIs [111–113]. For example, the *enzymatic template synthesis* of self-doped PANI in the presence of cationic templates using a carboxylated aniline monomer and a polycationic template (Scheme 13.11) could be performed at room temperature. The complex depicted in Scheme 13.11 showed a conductivity of 0.3 S cm^{-1} with additional doping with hydrochloric acid. Biological materials such as DNA [114], lignosulfonate, and ionic liquids (ILs) [115] were also successfully investigated as

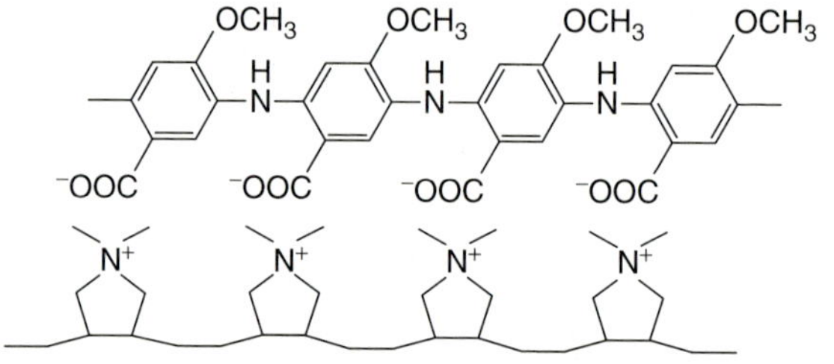

Scheme 13.11 Complex formation between ionized poly(3-amino-4-methoxybenzoic acid) and a poly(diallyldimethyl ammonium chloride) template at high pH [113].

templates. However, conductivities reported are not as high as the one reported for the template shown in Scheme 13.11.

13.3
Green Media for Enzymatic Polymerization

13.3.1
Ionic Liquids

The interest in ILs as *green reaction media* is due to their properties: near-zero vapor pressure [116], thermal stability [117], tunable polarity, tunable hydrophobicity, and solvent miscibility. Using ILs as reaction media for enzymatic reactions was first reported in 2000 [118–120]. ILs are an interesting alternative to replace organic solvents in the enzymatic conversion of small molecules [121, 122]. Lipases, for example, show good activity and, in some instances, improved enantioselectivity when employed in pure IL for the kinetic resolution of enantiomers by transesterification [123–127]. Besides the enzymatic synthesis of small molecules, also enzymatic polymerization was attempted in ILs. Nara *et al.* investigated the lipase (*Pseudomonas cepacia*) catalyzed polycondensation of diethylene octane-1,8-dicarboxylate and 1,4-butanediol in 1-butyl-3-methylimidazolium hexafluorophosphate ([BMIM][PF$_6$]) [128]. However, only low molecular weight oligomers could be obtained even at elevated temperatures.

In a second report Uyama *et al.* investigated both the CALB catalyzed ring-opening polymerization of ε-caprolactone and the polycondensation of diethyl adipate and diethyl sebacate, respectively, with 1,4-butanediol [129]. ROP in [BMIM][PF$_6$] resulted in low molecular weight oligomers ($M_n < 850$ g mol^{-1}) when the reaction was conducted for less than 3 d, while a higher molecular weight ($M_n = 4200$ g mol^{-1}) could only be obtained after a reaction time of 7 d. The molecular weights of the polycondensates did not exceed 1500 g mol^{-1} in either [BMIM][PF$_6$] or 1-butyl-3-methylimidazolium tetrafluoroborate [BMIM][BF$_4$]. Heise and coworkers reported higher molecular weights for enzymatic ROP in ILs ($M_n = 12\,700$ g mol^{-1} in [BMIM][BF$_4$]) obtained within a shorter reaction time (24 h) (Scheme 13.12) [130]. Polycondensation of dimethyl adipate and dimethyl sebacate in an open system at a temperature close to the boiling-point of the condensation product resulted in an M_n of up to 5900 g mol^{-1}.

Besides CALB and PC, other enzymes were screened for activity in ILs. *Yarrowia lipolytica* lipase (YLL), *Candida rugosa* lipase (CRL), and porcine pancreatic lipase (PPL) were employed successfully as catalysts in the enzymatic ROP of CL in 1-ethyl-3-methylimidazolium tetrafluoroborate ([EMIM][BF$_4$]), [BMIM][BF$_4$], 1-butylpyridinium tetrafluoroborate ([BuPy][BF$_4$]), 1-butylpyridinium trifluoroacetate ([BuPy][CF$_3$COO]), and 1-ethyl-3-methylimidazolium nitrate ([EMIM][NO$_3$]). PCL with molecular weights in the range of 300–9000 g mol^{-1} were reported [131].

Recently Dordick and coworkers have presented the polymerization of polyphenols using soybean peroxidase (SBP) as a catalyst in ILs [132]. Phenolic polymers

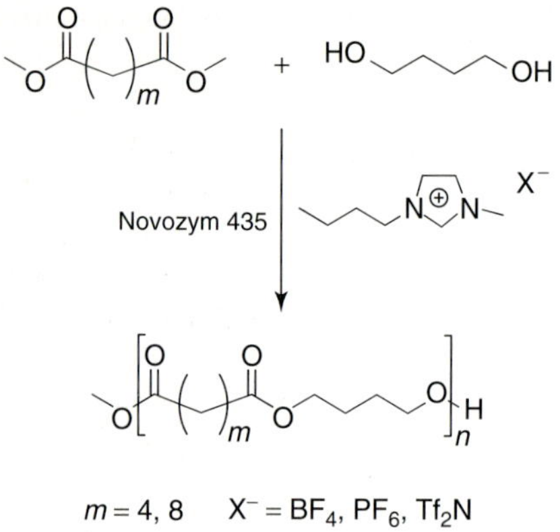

Scheme 13.12 Enzymatic polycondensation in ionic liquids [130].

$m = 4, 8 \quad X^- = BF_4, PF_6, Tf_2N$

with number average molecular weights ranging from 1200 to 4100 g mol^{-1} were obtained depending on the composition of the reaction medium and the nature of the phenol. Specifically, SBP was highly active in methylimidazolium-containing ILs, including [BMIM][BF$_4$], and 1-butyl-3-methylpyridinium tetrafluoroborate [BMPy][BF$_4$] with the IL content as high as 90% (v/v); the balance being aqueous buffer.

One major drawback of using ILs is that a second phase (normally a VOC, volatile organic compound) is needed to obtain the products. Blancard and coworkers demonstrated that scCO$_2$ can be used as this second phase in polymerization in ILs [133].

13.3.2
Supercritical Carbon Dioxide

Supercritical carbon dioxide (scCO$_2$) has received much attention as a green reaction medium due to its cheap, environmentally friendly, non-toxic, and non-flammable character. Moreover, the critical parameters needed are relatively easy to obtain ($T_c = 31\,°C$, $P_c = 73.8$ bar; 1 bar $= 10^5$ Pa). The use of enzymes in scCO$_2$ was pioneered by the groups of Randolph [134] and Hammond [135] in 1985. Since then, enzymes have been utilized extensively in scCO$_2$ for processes ranging from acidolysis to chiral synthesis of esters, and there is an extensive literature describing recent progress [136–138]. In some instances deactivation of native enzymes in this medium was reported and therefore different methods have been investigated to stabilize the enzymes [139–142]. Solutions were found through the use of microemulsions or reverse micelles [143, 144] to stabilize the enzyme in the water phase, while the high diffusivity of scCO$_2$ was maintained. Other methods used are lipid-coated enzymes [145], sol gels [146], immobilized enzymes [147], cross-linked enzyme crystals (CLECs) [148–150], and cross-linked enzyme aggregates (CLEAs) [151–153].

In the last decade scCO$_2$ also became an established medium for polymerizations [154], including enzymatic polymerizations. Most of the enzymatic polymerizations performed in scCO$_2$ are enzymatic ROP of lactones, often in combination with ATRP or RAFT (reversible addition–fragmentation chain transfer) [155]. Most intensively studied has been the Novozym 435 catalyzed ROP of ε-caprolactone [156, 157]. CL is soluble in the supercritical fluid containing the dispersed enzyme beads, and PCL precipitates during the polymerization reaction. Despite the insolubility of the product in the medium, it is very effectively plasticized by scCO$_2$, a fact that significantly improves the mass transport of monomer and the progress of polymerization. By balancing the solubilizing power of the solvent and the best temperature for enzyme activity in scCO$_2$, optimum conditions were found; a relatively low scCO$_2$ density of 0.50 g cm^{-3}, a pressure of 1500 psi (103 bar), a temperature of 35 °C, a catalyst loading of 10 wt%, and a reaction time of 24 h [156]. A big advantage of the scCO$_2$ approach is that the whole process can be carried out using only scCO$_2$. This means that both the synthesis of the polymer and any subsequent recycling or cleaning steps can be achieved *in-situ*. The polymer can be purified at the completion of the reaction by supercritical extraction of residual monomer, low molecular weight oligomers, and cyclic products by simply flushing with scCO$_2$. Through this strategy the use of organic solvents to purify the polymer is eliminated. Reusing the enzyme in different polymerization cycles led to reproducible high molecular weight polymer and retention of high yields.

Similar to conventional solvents, the *synthesis of block copolymers* by combination of enzymatic ROP with other techniques was investigated. Howdle and coworkers reported that a one-pot, simultaneous synthesis of block copolymers by enzymatic ROP and ATRP employing a bifunctional initiator, CL, and MMA is possible in supercritical CO$_2$ [158]. They could show that CL acts as an scCO$_2$ co-solvent, which was crucial to allow the radical polymerization to remain homogeneous and controlled. The unique ability of scCO$_2$ to solubilize highly fluorinated species was utilized by extending this methodology to the synthesis of novel copolymers consisting of a semifluorinated block of poly(1H,1H,2H,2H-perfluorooctyl methacrylate) (FOMA) and PCL (Scheme 13.13) [159]. Block copolymers were successfully synthesized by a two-step process based on the sequential monomer addition. Parallel experiments in conventional solvents did not yield any block copolymers due to the limited solubility of FOMA in these solvents. These, and other enzymatic polymerization procedures in scCO$_2$, have recently been reviewed [160].

Similarly successful was the combination of enzymatic ROP with RAFT in scCO$_2$ for the synthesis of PCL-PS (polystyrene) block copolymers employing a

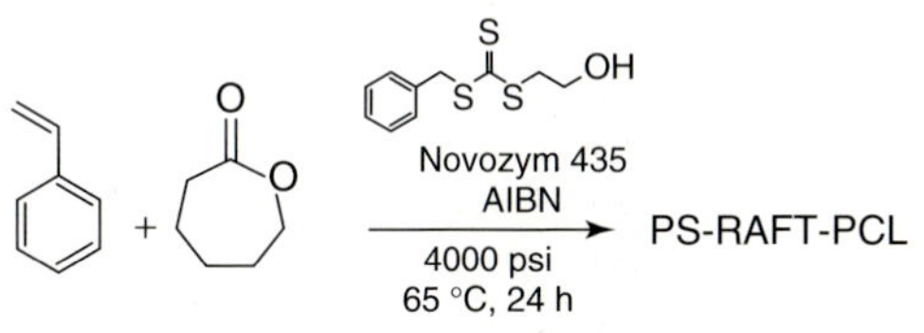

Scheme 13.13 Chemoenzymatic polymerization of PS-*b*-PCL block copolymer in supercritical carbon dioxide [161].

bifunctional RAFT initiator [161]. As RAFT does not require a metal catalyst this process, in combination with the green solvent CO_2, it is environmentally desirable.

The first synthesis of *graft copolymers* in $scCO_2$ was reported from a poly-HEMA [poly(hydroxyethyl methacrylate)]–PMMA [poly(methyl methacrylate)] random copolymer by the combination of ATRP and enzymatic ROP of CL [155]. These workers first synthesized the random copolymer containing 13 and 30% HEMA by ATRP in $scCO_2$. While PCL grafting was confirmed, not all hydroxyl groups participated in the grafting reaction (33%). Similar to the situation in conventional solvents, steric effects were thought to reduce the grafting efficiency. The results were similar when both the ATRP reaction and enzymatic grafting were carried out simultaneously in one pot, that is, all components for the radical and the enzymatic reaction were present at the same time. This approach resulted in about 40% functionalization of the hydroxyl groups. In another study, they confirmed the limitation to the grafting efficiency to be due to two reasons. (i) The poly(MMA-*co*-HEMA) probably had a blocky structure because of the different reactivity ratios of the monomers in the radical polymerization. (ii) The hydroxy groups are too close to the polymer backbone. Consequently, the grafting efficiency was significantly improved when highly randomized poly(MMA-*co*-HEMA) from starved-feed polymerization was used (80%) and when a PEG spacer was introduced between the polymer backbone and the hydroxyl group (100%) [162].

Other polymerization systems in $scCO_2$ have been reported of late [163]. One example is the HRP-mediated free radical polymerization of water-soluble acrylamide in an *inverse emulsion*. The polymerization takes place in water droplets formed in $scCO_2$, which are either stabilized as reversed micelles using perfluoropolyether (PFPE) ammonium carboxylate (PFPE-COO$^-$NH$_4^+$) or, in the absence of stabilizer, using very high shear. HRP, 2,4-pentadione, and hydrogen peroxide were used to polymerize acrylamide. The results obtained with polymerization using $scCO_2$ were comparable to results obtained with conventional free radical polymerization. Moreover, in the applied system, only $scCO_2$ and water are used to obtain the desired product.

13.4
Conclusions and Outlook

The enzymatic synthesis of polymers is a serious alternative to the chemical synthesis. Significant progress has been achieved in recent years on this relatively young polymerization technique. As with all polymerization techniques, enzymatic polymerizations have their advantages and drawbacks and it remains to be seen whether it will have an impact on industrial polymer synthesis. It is clear that many challenges still have to be overcome, even for polymer systems in which the enzymatic processes has significantly advanced, as is the case for polyester synthesis. While lipases are perfectly capable of producing aliphatic polyesters of industrial specification, the process is economically not yet feasible. Moreover, aromatic polyesters such as PET (polyethylene terephthalate), which occupy the largest market segment among polyesters, can not so far be synthesized enzymatically.

Nevertheless, enzymes hold significant promises with respect to green polymer chemistry. However, the advantages of enzymes are not directly transferable from small molecule synthesis to polymers. The success of biotechnology in small molecule synthesis is mainly based on the reduction in the reaction steps due to the enzymatic selectivity. This translates directly into a reduced environmental impact. Polymers are usually obtained in a one-step reaction without protection/deprotection steps, which reduces the potential for significant savings (cost and environmental impact) through the simple replacement of a chemical reaction catalyst by an enzyme. To make a process greener and more environmentally benign, the enzymatic polymerization has to be integrated with other green technologies, such as green solvents ($scCO_2$, water, etc.) or renewable raw materials. While a great deal of fundamental research has been done in the past and will still be needed in the future on this topic, we are approaching a transition towards the development of integrated green processes in polymer science and enzymatic polymerization is one element of it.

References

1. Patel, N.R. (2007) *Biocatalysis in the Pharmaceutical and Biotechnology Industries*, CRC Press, Boca Raton.
2. OECD (2001) *The Application of Biotechnology to Industrial Sustainability*, OECD, Paris.
3. Matsumura, S. (2006) *Adv. Polym. Sci.*, **194**, 95.
4. Kobayasi, S. and Uyama, H. (2006) *Adv. Polym. Sci.*, **194**, 133.
5. Varma, I.K., Albertsson, A.-C., Rajkhowa, R., and Srivastava, R.K. (2005) *Prog. Polym. Sci.*, **30**, 949.
6. Kobayashi, S. (2009) *Macromol. Rapid Commun.*, **30**, 237.
7. Kobayashi, S. and Makino, A. (2009) *Chem. Rev.*, **109**, 5288.
8. Srivastava, R.K. and Albertsson, A.-C. (2008) *Adv. Drug Deliv. Rev.*, **60**, 1077.
9. Mei, Y., Miller, L., Gao, W., and Gross, R.A. (2003) *Biomacromolecules*, **4**, 70.
10. Kirk, O. and Christensen, M.W. (2002) *Org. Process Res. Dev.*, **6**, 446.
11. Uyama, H., and Kobayashi, S. (1993) *Chem. Lett.*, 1149.
12. Knani, D., Gutman, A.L., and Kohn, D.H. (1993) *J. Polym. Sci. A Polym. Chem.*, **31**, 1221.
13. Namekawa, S., Suda, S., Uyama, H., and Kobayashi, S. (1999) *Int. J. Biol. Macromol.*, **25**, 145.
14. Kobayashi, S., Takeya, K., Suda, S., and Uyama, H. (1998) *Macromol. Chem. Phys.*, **199**, 1729.
15. Noda, S., Kamiya, N., Goto, M., and Nakashio, F. (1997) *Biotech. Lett.*, **19**, 307.
16. Nakaoki, T., Mei, Y., Miller, L.M., Kumar, A., Kalra, B., Miller, M.E., Kirk, O., Christensen, M., and Gross, R.A. (2005) *Ind. Biotechnol.*, **1**, 126.
17. Hunsen, M., Azim, A., Mang, H., Wallner, S.R., Ronkvist, A., Xie, W., and Gross, R.A. (2007) *Macromolecules*, **40**, 148.
18. Cordova, A., Iversen, T., Hult, K., and Martinelle, M. (1998) *Polymer*, **39**, 6519.
19. Mei, Y., Kumar, A., and Gross, R.A. (2002) *Macromolecules*, **35**, 5444.
20. Kumar, A. and Gross, R.A. (2000) *Biomacromolecules*, **1**, 133.
21. MacDonald, R.T., Pulapura, S.K., Svirkin, Y.Y., Gross, R.A., Kaplan, D.L., Akkara, J., Swift, G., and Wolk, S. (1995) *Macromolecules*, **28**, 73.

22. Dong, H., Cao, S.-G., Li, Z.-Q., Han, S.-P., You, D.-L., and Shen, J.-C. (1999) *J. Polym. Sci. A Polym. Chem.*, **37**, 1265.
23. Matsumoto, M., Odachi, D., and Kondo, K. (1999) *Biochem. Eng. J.*, **4**, 73.
24. de Geus, M., Peeters, J., Wolffs, M., Hermans, T., Palmans, A.R.A., Koning, C.E., and Heise, A. (2005) *Macromolecules*, **38**, 4220.
25. Duda, A., Kowalski, A., Penczek, S., Uyama, H., and Kobayashi, S. (2002) *Macromolecules*, **35**, 4266.
26. van der Mee, L., Helmich, F., de Bruijn, R., Vekemans, J.A.J.M., Palmans, A.R.A., and Meijer, E.W. (2006) *Macromolecules*, **39**, 5021.
27. Panten, J., Surburg, H., and Hölscher, B. (2008) *Chem. Biodivers.*, **5**, 1011.
28. Kumar, A., Kalra, B., Dekhterman, A., and Gross, R.A. (2000) *Macromolecules*, **33**, 6303.
29. de Geus, M., van der Meulen, I., Goderis, B., Vanhecke, K., Dorschu, M., van der Werff, H., Koning, C.E., and Heise, A. (2010) *Polym. Chem.*, **1**, 525.
30. Taden, A., Antonietti, M., and Landfester, K. (2003) *Macromol. Rapid Commun.*, **24**, 512.
31. Lebedev, B. and Yevstropov, A. (1984) *Makromol. Chem.*, **185**, 1235.
32. Focarete, M.L., Scandola, M., Kumar, A., and Gross, R.A. (2001) *J. Polym. Sci. B Polym. Phys.*, **39**, 1721.
33. Skoglund, P. and Fransson, A. (1998) *Polymer*, **39**, 3143.
34. Gazzano, M., Malta, V., Focarete, M.L., Scandola, M., and Gross, R.A. (2003) *J. Polym. Sci. B Polym. Phys.*, **41**, 1009.
35. van der Meulen, I., de Geus, M., Antheunis, H., Deumens, R., Joosten, B.E.A.J., Koning, C.E., and Heise, A. (2008) *Biomacromolecules*, **9**, 3409.
36. Simpson, N., Takwa, M., Hult, K., Johansson, M., Martinelle, M., and Malmstrom, E. (2008) *Macromolecules*, **10**, 3613.
37. Takwa, M., Xiao, Y., Simpson, N., Malmstrom, E., Hult, K., Koning, C.E., Heise, A., and Martinelle, M. (2008) *Biomacromolecules*, **9**, 704.
38. Xiao, Y., Takwa, M., Hult, K., Koning, C.E., Heise, A., and Martinelle, M. (2009) *Macromol. Biosci.*, **9**, 713.
39. Srivastava, R.K. and Albertsson, A.C. (2007) *Macromolecules*, **40**, 4464.
40. Kumar, A., Gross, R.A., Wang, Y.B., and Hillmyer, M.A. (2002) *Macromolecules*, **35**, 7606.
41. Panova, A.A. and Kaplan, D.L. (2003) *Biotechnol. Bioeng.*, **84**, 103.
42. He, F., Li, S.M., Vert, M., and Zhuo, R.X. (2003) *Polymer*, **44**, 5145.
43. Srivastava, R.K. and Albertsson, A.-C. (2006) *Macromolecules*, **39**, 46.
44. Kaihara, S., Fisher, J.P., and Matsumura, S. (2009) *Macromol. Biosci.*, **9**, 613.
45. Meyer, U., Palmans, A.R.A., Loontjens, T., and Heise, A. (2002) *Macromolecules*, **35**, 2873.
46. de Geus, M., Peters, R., Koning, C.E., and Heise, A. (2008) *Biomacromolecules*, **9**, 752.
47. de Geus, M., Schormans, L., Palmans, A.R.A., Koning, C.E., and Heise, A. (2006) *J. Polym. Sci. A Polym. Chem.*, **44**, 4290.
48. Peeters, J., Palmans, A.R.A., Veld, M., Scheijen, F., Heise, A., and Meijer, E.W. (2004) *Biomacromolecules*, **5**, 1862.
49. Sha, K., Li, D.S., Wang, S.W., Qin, L., and Wang, J.Y. (2005) *Polym. Bull.*, **55**, 349.
50. Sha, K., Li, D., Li, Y., Ai, P., Wang, W., Xu, Y., Liu, X., Wu, M., Wang, S., Zhang, B., and Wang, J. (2006) *Polymer*, **47**, 4292.
51. Sha, K., Qin, L., Li, D.S., Liu, X.T., and Wang, J.Y. (2005) *Polym. Bull.*, **54**, 1.
52. Sha, K., Li, D., Li, Y., Zhang, B., and Wang, J. (2008) *Macromolecules*, **41**, 361.
53. Sha, K., Li, D., Li, Y., Liu, X., Wang, S., and Wang, J. (2008) *Polym. Int.*, **57**, 211.
54. Sha, K., Li, D., Li, Y., Liu, X., Wang, S., Guan, J., and Wang, J. (2007) *J. Polym. Sci. A Polym. Chem.*, **45**, 5037.
55. Zhang, B., Li, Y., Wang, W., Chen, L., Wang, S., and Wang, J. (2009) *Polym. Bull.*, **62**, 643.

56. Zhang, B., Li, Y., Sun, J., Wang, S., Zhao, Y., and Wu, Z. (2009) *Polym. Int.*, **58**, 752.
57. van As, B.A.C., Thomassen, P., Kalra, B., Gross, R.A., Meijer, E.W., Palmans, A.R.A., and Heise, A. (2004) *Macromolecules*, **37**, 8973.
58. Kerep, P. and Ritter, H. (2007) *Macromol. Rapid Commun.*, **28**, 759.
59. Xiao, Y., Coulembier, O., Koning, C.E., Heise, A., and Dubois, P. (2009) *Chem. Commun.*, 2472.
60. Keul, H. and Moeller, M. (2009) *J. Polym. Sci. A Polym. Chem.*, **47**, 3209.
61. Hans, M., Gasteier, P., Keul, H., and Moeller, M. (2006) *Macromolecules*, **39**, 3184.
62. Duxbury, C.J., Cummins, D., and Heise, A. (2007) *Macromol. Rapid Commun.*, **28**, 235.
63. Hans, M., Keul, H., Heise, A., and Moeller, M. (2007) *Macromolecules*, **40**, 8872.
64. Hans, M., Xiao, Y., Keul, H., Heise, A., and Moeller, M. (2009) *Macromol. Chem. Phys.*, **210**, 736.
65. Dai, S., Xue, L., Zinn, M., and Li, Z. (2009) *Biomacromolecules*, **10**, 3176.
66. Okumara, S., Iwai, M., and Tominaga, Y. (1984) *Agric. Biol. Chem.*, **48**, 2805.
67. Rodney, R.L., Allinson, B.T., Beckman, E.J., and Russell, A.J. (1999) *Biotechnol. Bioeng.*, **65**, 485.
68. Uyama, H., Inada, K., and Kobayashi, S. (2000) *Polym. J.*, **32**, 440.
69. Taylor, A. and Binns, F. (1998) Enzymatic synthesis of polyesters using lipase. WO/1998/055642.
70. Uyama, H., Kuwabara, M., Tsujimoto, T., and Kobayashi, S. (2003) *Macromolecules*, **4**, 211.
71. Björkling, F., Godtfredsen, S.E., and Kirk, O. (1990) *J. Chem. Soc. Chem. Commun.*, 1301.
72. Björkling, F., Frykman, H., Godtfredsen, S.E., and Kirk, O. (1992) *Tetrahedron*, **48**, 4587.
73. Hilker, I., Bothe, D., Prüss, J., and Warnecke, H.-J. (2008) *Chem. Eng. Sci.*, **56**, 427.
74. Veld, M.A.J., Palmans, A.R.A., and Meijer, E.W. (2007) *J. Polym. Sci. A Polym. Chem.*, **45**, 5968.
75. Kumar, A., Kulshrestha, A.S., Gao, W., and Gross, R.A. (2003) *Macromolecules*, **36**, 8219.
76. Kulshrestha, A.S., Gao, W., and Gross, R.A. (2005) *Macromolecules*, **38**, 3193.
77. Ebata, H., Toshima, K., and Matsumura, S. (2007) *Macromol. Biosci.*, **7**, 798.
78. Akkara, J.A., Kaplan, D.L., John, V.T., and Tripathy, S.K. (1996), In: *Polymeric Materials Encyclopedia* (ed. J.C. Salamone), CRC Press, New York.
79. Kobayashi, S., Shoda, S., and Uyama, H. (1996) Enzymatic polymerisation, In: *Polymeric Materials Encyclopedia*, vol. 3 (ed. J.C. Salamone), CRC Press, New York, p. 2102.
80. Dordick, J., Marletta, M.A., and Klibanov, A.M. (1987) *Biotechnol. Bioeng.*, **15**, 31.
81. Akkara, J.A., Sencal, K.J., and Kaplan, D.L. (1991) *J. Polym. Sci. A Polym. Chem.*, **29**, 1561.
82. Dordick, J. (1989) *Enzyme Microb. Technol.*, **11**, 194.
83. Hewson, W.D. and Dunford, B. (1976) *J. Biol. Chem.*, **251**, 6036.
84. Hewson, W.D. and Dunford, B. (1976) *J. Biol. Chem.*, **251**, 6043.
85. Mita, N., Tawaki, S., Uyama, H., and Kobayashi, S. (2002) *Macromol. Biosci.*, **2**, 127.
86. Kim, Y.-J., Koichiro, S., Hiroshi, U., and Kobayashi, S. (2008) *Polymer*, **49**, 4791.
87. Uyama, H., Lohavisavapanich, C., Ikeda, R., and Kobayashi, S. (1998) *Macromolecules*, **31**, 554.
88. Derango, R.A., Chiang, L., Dowbenko, R., and Lasch, J.G. (1992) *Biotechnol. Tech.*, **6**, 523.
89. Iwata, H., Hata, Y., Matsuda, T., and Ikada, Y. (1991) *J. Polym. Chem. Polym. Chem. Ed.*, **29**, 1217.
90. Singh, A., Ma, D., and Kaplan, D.L. (2000) *Biomacromolecules*, **1**, 592.
91. Kalra, B. and Gross, R.A. (2000) *Biomacromolecules*, **1**, 501.
92. Iwahara, K., Hirata, M., Honda, Y., Watanabe, T., and Kuwahara, M. (2000) *Biotechnol. Lett.*, **22**, 1355.
93. Singh, A. and Kaplan, D.L. (2003) *Adv. Mater.*, **15**, 1291.

94. Singh, A. and Kaplan, D.L. (2004) *J. Macromol. Sci. A Pure Appl. Chem.*, **41**, 1377.
95. Shan, J., Kitamura, Y., and Yoshizawa, H. (2005) *Colloid Polym. Sci.*, **284**, 108.
96. Kwon, H.-S., Chung, E., Lee, D.-I., Lee, C.-H., Ahn, I.-S., and Kim, J.-Y. (2009) *J. Appl. Polym. Sci.*, **112**, 2935.
97. Chen, S.A. and Hwang, G.W. (1994) *J. Am. Chem. Soc.*, **116**, 7939.
98. Chen, S.A. and Hwang, G.W. (1996) *Macromolecules*, **29**, 3950.
99. Xu, P., Singh, A., and Kaplan, D.L. (2006) *Adv. Polym. Sci.*, **194**, 69.
100. Ikeda, R., Uyama, H., and Kobayashi, S. (1996) *Macromolecules*, **29**, 3053.
101. Wang, P. and Dordick, J.S. (1998) *Macromolecules*, **31**, 941.
102. Alva, K.S., Marx, K.A., Kumar, J., and Tripathy, S.K. (1996) *Macromol. Rapid Commun.*, **17**, 859.
103. Alva, K.S., Kumar, J., Marx, K.A., and Tripathy, S.K. (1997) *Macromolecules*, **30**, 4024.
104. Liu, W., Kumar, J., Tripathy, S., and Samuelson, L.A. (2002) *Langmuir*, **18**, 9696.
105. Premachandran, R., Banerjee, S., Wu, X.-K., John, V.T., McPherson, G.L., Akkara, J.A., and Kaplan, D.L. (1996) *Macromolecules*, **29**, 6452.
106. Premachandran, R., Banerjee, S., John, V.T., McPherson, G.L., Akkara, J.A., and Kaplan, D.L. (1997) *Chem. Mater.*, **9**, 1342.
107. Bruno, F.F., Akkara, J.A., Samuelson, L.A., Kaplan, D.L., Marx, K.A., Kumar, J., and Tripathy, S.K. (1995) *Langmuir*, **11**, 889.
108. Liu, W., Kumar, J., Tripathy, S., Senecal, K.J., and Samuelson, L.A. (1999) *J. Am. Chem. Soc.*, **121**, 71.
109. Liu, W., Cholli, A.L., Nagarajan, R., Kumar, J., Tripathy, S., Bruno, F.F., and Samuelson, L.A. (1999) *J. Am. Chem. Soc.*, **121**, 11345.
110. Xu, P. and Kaplan, D.L. (2004) *Adv. Mater.*, **16**, 628.
111. Kim, S.-C., Kumar, J., Bruno, F.F., and Samuelson, L.A. (2006) *J. Macromol. Sci. A Pure Appl. Chem.*, **43**, 2007.
112. Rumbau, V., Pomposo, J.A., Eleta, A., Rodriguez, J., Grande, H., Mecerreyes, D., and Ochoteco, E. (2007) *Biomacromolecules*, **8**, 315.
113. Rumbau, V., Pomposo, J.A., Alduncin, J.A., Grande, H., Mecerreyes, D., and Ochoteco, E. (2007) *Enzyme Microb. Technol.*, **40**, 1412.
114. Nagarajan, R., Roy, S., Kumar, J., Tripathy, S.K., Dolukhanyan, T., Sung, C., Bruno, F.F., and Samuelson, L.A. (2001) *J. Macromol. Sci. A Pure Appl. Chem.*, **38**, 1519.
115. Rumbau, V., Marcilla, R., Ochoteco, E., Pomposo, J.A., and Mecerreyes, D. (2006) *Macromolecules*, **39**, 8547.
116. Earle, M.J., Esperancüa, J.M.S.S., Gilea, M.A., Canongia Lopes, J.N., Rebelo, L.P.N., Magee, J.W., Seddon, K.R., and Widegren, J.A. (2006) *Nature*, **439**, 831.
117. Kosmulski, M., Gustafsson, J., and Rosenholm, J.B. (2004) *Thermochim. Acta*, **412**, 47.
118. Cull, S.G., Holbrey, J.D., Vargas-Mora, V., Seddon, K.R., and Lye, G.J. (2000) *Biotechnol. Bioeng.*, **69**, 227.
119. Erbeldinger, M., Mesiano, A.J., and Russell, A.J. (2000) *Biotechnol. Prog.*, **16**, 1129.
120. Madeira Lau, R., Van Rantwijk, F., Seddon, K.R., and Sheldon, R.A. (2000) *Org. Lett.*, **2**, 4189.
121. Van Rantwijk, F., Madeira Lau, R., and Sheldon, R.A. (2002) *Curr. Opin. Biotechnol.*, **13**, 565.
122. Park, S. and Kazlauskas, R.J. (2003) *Curr. Opin. Biotechnol.*, **14**, 432.
123. Park, S. and Kazlauskas, R.J. (2001) *J. Org. Chem.*, **66**, 8395.
124. Schöfer, S.H., Kaftzik, N., Wasserscheid, P., and Kragl, U. (2001) *Chem. Commun.*, 425.
125. Kim, K.W., Song, B., Choi, M.Y., and Kim, M.J. (2001) *Org. Lett.*, **3**, 1507.
126. Oh, C.R., Choo, D.J., Shim, W.H., Lee, D.H., Roh, E.J., Lee, S., and Song, C.E. (2003) *Chem. Commun.*, 1100.
127. Zhao, H., Luo, R.G., and Malhotra, S.V. (2003) *Biotechnol. Prog.*, **19**, 1016.
128. Nara, S.J., Harjani, J.R., Salunkhe, M.M., Mane, A.T., and Wadgaonkar, P.P. (2003) *Tetrahedron Lett.*, **44**, 1371.
129. Uyama, H., Takamoto, T., and Kobayashi, S. (2002) *Polym. J.*, **34**, 94.

130. Marcilla, R., de Geus, M., Mecerreyes, D., Duxburry, C.J., Koning, C.E., and Heise, A. (2006) *Eur. Polym. J.*, **42**, 1215.
131. Barrera-Rivera, K.A., Marcos-Fernandez, A., Vera-graziano, R., and Martinez-Richa, A. (2009) *J. Polym. Sci. A Polym. Chem.*, **47**, 5792.
132. Eker, B., Zagorevski, D., Zhu, G., Linhardt, R.J., and Dordick, J.S. (2009) *J. Mol. Catal. B, Enzym.*, **59**, 177.
133. Blancard, L.A., Hancu, D., Beckman, E.J., and Brennecke, J.F. (1999) *Nature*, **399**, 28.
134. Randolph, T.W., Blanch, H.W., Prausnitz, J.M., and Wilke, C.R. (1985) *Biotechnol. Lett.*, **7**, 325.
135. Hammond, D.A., Karel, M., Klibanov, A.M., and Krukonis, V.J. (1985) *Appl. Biochem. Biotechnol.*, **11**, 393.
136. Jessop, P.G. and Leitner, W. (eds) (1999) *Chemical Synthesis Using Supercritical Fluids*, Wiley-VCH Verlag GmbH, Wienheim.
137. Mesiano, A.J., Beckman, E.J., and Russell, A.J. (1999) *Chem. Rev.*, **99**, 623.
138. Hobbs, H.R. and Thomas, N.R. (2007) *Chem. Rev.*, **107**, 2786.
139. Kamat, S., Barrera, J., Beckman, E.J., and Russell, A.J. (1992) *Biotechnol. Bioeng.*, **40**, 158.
140. Kamat, S., Critchley, G., Beckman, E.J., and Russell, A.J. (1995) *Biotechnol. Bioeng.*, **46**, 610.
141. Carvalho, J.B., de Sampaio, T.C., and Barreiros, S. (1996) *Biotechnol. Bioeng.*, **49**, 399.
142. Mase, N., Sako, T., Horiwaka, Y., and Takabe, K. (2003) *Tetrahedron Lett.*, **44**, 5175.
143. Ikushima, Y. (1997) *Adv. Colloid Interface Sci.*, **71–72**, 259.
144. Kane, M.A., Baker, G.A., Pandey, S., and Bright, F.V. (2000) *Langmuir*, **16**, 4901.
145. Mori, T., Li, M., Kobayashi, A., and Okahata, Y. (2002) *J. Am. Chem. Soc.*, **124**, 1188.
146. Novak, Z., Habulin, M., Krmelj, V., and Knez, Z. (2003) *J. Supercrit. Fluids*, **27**, 169.
147. Matsuda, T., Watanabe, K., Harada, T., Nakamura, K., Arita, Y., Misumi, Y., Ichikawa, S., and Ikariya, T. (2004) *Chem. Commun.*, 2286.
148. Matsuda, T., Tsuji, K., Kamitanaka, T., Harada, T., Nakamura, K., and Ikariya, T. (2005) *Chem. Lett.*, **34**, 1102.
149. Fontes, N., Almeida, M.C., Garcia, S., Peres, C., Pertridge, J., Halling, P.J., and Barreiros, S. (2001) *Biotechnol. Prog.*, **18**, 355.
150. Harper, N. and Barreiros, S. (2002) *Biotechnol. Prog.*, **18**, 1451.
151. Sheldon, R.A., Schoevaart, R., and van Langen, L.M. (2005) *Biocatal. Biotransformation*, **23**, 141.
152. Hobbs, H.R., Kondor, B., Stephenson, P., Sheldon, R.A., Thomas, N.R., and Poliakoff, M. (2006) *Green Chem.*, **8**, 816.
153. Dijkstra, Z.J., Weyten, H., Willems, L., and Keurentjes, J.T.F. (2006) *J. Mol. Catal. B, Enzym.*, **39**, 112.
154. Beckman, E.J. (2004) *J. Supercrit. Fluids*, **28**, 121.
155. Villarroya, S., Zhou, J., Thurecht, K.J., and Howdle, S.M. (2006) *Macromolecules*, **39**, 9080.
156. Loeker, F.C., Duxbury, C.J., Kumar, R., Gao, W., Gross, R.A., and Howdle, S.M. (2004) *Macromolecules*, **37**, 2450.
157. Thurecht, K.J., Heise, A., de Geus, M., Villarroya, S., Zhou, J., Wyatt, M.F., and Howdle, S.M. (2006) *Macromolecules*, **39**, 7967.
158. Duxbury, C.J., Wang, W.X., de Geus, M., Heise, A., and Howdle, S.M. (2005) *J. Am. Chem. Soc.*, **127**, 2384.
159. Zhou, J., Villarroya, S., Wang, W., Wyatt, M.F., Duxbury, C.J., Thurecht, K.J., and Howdle, S.M. (2006) *Macromolecules*, **39**, 5352.
160. Villarroya, S., Thurecht, K.J., Heise, A., and Howdle, S.M. (2007) *Chem. Commun.*, 3805.
161. Thurecht, K.J., Gregory, A.M., Villarroya, S., Zhou, J., Heise, A., and Howdle, S.M. (2006) *Chem. Commun.*, 4383.
162. Villarroya, S., Dudek, K., Zhou, J., Irvine, D.J., and Howdle, S.M. (2008) *J. Mater. Chem.*, **18**, 989.
163. Villarroya, S., Thurecht, K.J., and Howdle, S.M. (2008) *Chem. Commun.*, 863.

14
Green Cationic Polymerizations and Polymer Functionalization for Biotechnology

Judit E. Puskas, Chengching K. Chiang, and Mustafa Y. Sen

14.1
Introduction

Cationic polymerization is defined as a chain growth reaction with positively charged active centers. The total volume of polymers produced commercially by cationic polymerization is estimated to be around 3–3.5 million tons per annum, about 3% of the total synthetic polymer market. Based on statistics alone, cationic polymerization is a minor player in the field [1]. However, this book chapter will highlight the significance of cationic polymerization in terms of the history of polymer science and *green chemistry*. Our interpretation of *green polymerizations* applies to those performed under environmentally friendly reaction conditions or using monomers or other ingredients, such as catalysts, from renewable resources. In this chapter we will discuss *green* cationic polymerizations. Natural rubber (NR) biosynthesis, the ultimate green cationic polymerization, will be discussed in more detail because we have been investigating this system over the last 10 years. Our research in this area led us to enzyme-catalyzed polymer functionalization, which also proceeds with the involvement of ionic species. As enzymes play an important role, enzyme catalysis in general will be discussed first.

14.2
Enzyme Catalysis

It was recognized over 30 years ago that enzymes were not restricted to their natural aqueous reaction media [2, 3] and that they could efficiently act as *in vitro* catalysts for the transformation of a wide range of substrates in organic solvents [4–6]. So far about 3000 enzymes have been identified and classified, by the International Union of Biochemistry and Molecular Biology, into six categories according to the type of reactions they can catalyze (Table 14.1) [7]. Among these, oxidoreductases and hydrolyzes are the most widely used *catalysts in biotransformations*. For example, in the 1987–2003 time period, about 85% of enzyme research was performed with oxidoreductases (25%) and hydrolases (60%) [4].

Green Polymerization Methods: Renewable Starting Materials, Catalysis and Waste Reduction
Edited by Robert T. Mathers and Michael A. R. Meier
Copyright © 2011 WILEY-VCH Verlag GmbH & Co. KGaA, Weinheim
ISBN: 978-3-527-32625-9

Table 14.1 Classification of enzymes.

Enzyme class	Reaction type	Typical reaction
Oxidoreductases	Redox reactions in which H and O atoms or electrons are transferred between molecules. This class includes dehydrogenases, oxygenases, oxidases, and peroxidases	$RH + O_2 + 2H^+ + 2e^- \rightarrow ROH + H_2O$
Transferases	Transfer of groups such as aldehydic, ketonic, acyl, sugar, phosphoryl, or methyl	$AB + C \rightarrow A + BC$
Hydrolases	Hydrolysis and formation of esters, amides, lactones, lactams, epoxides, nitriles, anhydrides, glycosides, and organohalides. This class includes lipases, esterases, proteases, and glycosidases.	$AB + H_2O \rightarrow AOH + BH$ $R^1C(O)OR^2 + R^3OH \rightarrow R^1C(O)OR^3 + R^2OH$
Lyases	Non-hydrolytic addition and elimination of small molecules on C=C, C=N, and C=O bonds.	$RC(O)COOH \rightarrow RC(O)H + CO_2$
Isomerases	Isomerization such as epimerization, racemization, and rearrangement.	$AB \rightarrow BA$
Ligases	Formation and cleavage of C–O, C–S, C–N, C–C bonds with simultaneous cleavage of ATP.	$X + Y + ATP \rightarrow XY + ADP + Pi$

Although the word "*biotransformation*" refers to any type of chemical conversion that is mediated by either whole cells (microbial, plant, or animal) or enzymes derived from them, *whole cell biocatalysis* has limitations, such as requiring expensive equipment, tedious recovery, and multiple side reactions, so it is practiced less frequently. In fact, whole cell systems are advantageous over isolated enzymes only when the transformation requires multiple enzymes and the regeneration of cofactors is necessary [8]. For example, *in vitro* natural rubber (NR) biosynthesis requires the use of active rubber particles that contain the membrane-bound rubber transferase enzyme; this will be discussed in detail. Today both whole cells and isolated enzymes are used in many different industrial applications [9, 10].

The use of enzymes in organic synthesis has several advantages over conventional chemical catalysis, including the ability to operate under mild conditions, high enantio-, regio-, and chemoselectivity, and catalyst recyclability and biocompatibility, which render them environmentally friendly alternatives to conventional chemical catalysts. These characteristics, especially the high selectivity, have resulted in many applications in the food and pharmaceutical industries. The synthesis of the low-calorie sweetener, aspartame, which is produced by DSM/Tosoh on a kiloton scale, can be given as an example of highly selective biotransformations [10].

The use of enzymes in polymer chemistry is relatively new [11]. Lipases effectively catalyze ring-opening polymerization (ROP) [12–14] so they will be discussed in more detail.

14.2.1
Lipases

Lipases belong to the group of hydrolases, and catalyze the hydrolysis of triglycerides to fatty acids and glycerol at the lipid–water interface *in vivo* (Figure 14.1).

They display almost no activity provided the substrate is in a dissolved monomeric state. However, when the substrate concentration is beyond its solubility limit, that is, critical micelle concentration, a lipophilic phase forms, and a sharp increase in lipase activity takes place. This phenomenon is called *"interfacial activation"* [15, 16]. Interfacial activation is attributed to a rearrangement process within the enzyme [17]. The X-ray structures of lipases usually show a "closed" conformation where a lid blocks the active site. This lid opens when the enzyme is exposed to an interface of a biphasic water–oil system and thus the catalytic activity of the lipase increases [18]. The closed and open conformations of the human pancreatic lipase–procolipase (a cofactor) complex are given in Figure 14.2 as an example [11]. Although the interfacial activation and the presence of a lid covering the active site have been generally used to classify an enzyme as a true lipase, some lipases such as *Candida antarctica* lipase B [19], and lipase from *Pseudomonas aeruginosa* [20] do not show interfacial activation even though they contain a small lid.

Lipases differ significantly in their amino acid sequences. However, X-ray structural investigations have shown that all lipases have similar 3D-structures [21–25] (see Figure 14.2) and that they are the members of the so-called α/β-hydrolase fold [26]. The α/β-*hydrolase fold* consists of a core of eight mostly parallel β sheets, which are surrounded on both sides by α-helices. Based on their active site, lipases are classified as serine hydrolases. The catalytic machinery consists of a nucleophile serine (Ser), an aspartate (Asp), or glutamate (Glu) that is hydrogen bonded to a histidine (His), and several oxyanion stabilizing residues [27].

Lipase-catalyzed hydrolysis involves ionic species. The mechanism of the hydrolysis of a butyric acid ester catalyzed by the lipase from *Candida rugosa* is given in

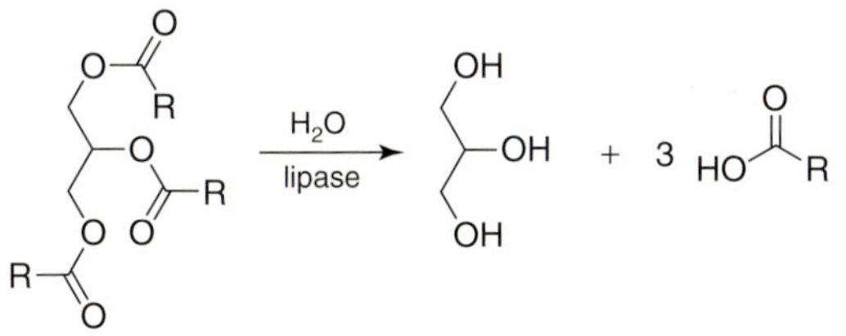

Figure 14.1 Lipase-catalyzed hydrolysis of triglycerides.

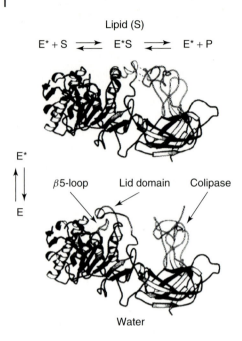

Figure 14.2 The structure of the human pancreatic lipase–procolipase in the closed (E) and open (E*S) conformation. S = substrate, P = product). (Reprinted with permission from *Trends Biotechnol.* 1997, **15**, 32–38. Copyright © 1997 Elsevier [16]).

Figure 14.3 [28]. Firstly, the nucleophilic serine hydroxyl group attacks the carbonyl group of the substrate forming the first tetrahedral intermediate. The collapse of this tetrahedral intermediate releases the first product, an alcohol, and leaves an acyl–enzyme complex. In the next step, water attacks the acyl–enzyme complex to form a second tetrahedral intermediate. The collapse of this tetrahedral intermediate yields the second product, butyric acid. The oxyanions of the intermediates are stabilized by two or three hydrogen bonds, the so-called *oxyanion hole*. In a transesterification reaction, instead of water, an alcohol nucleophile attacks the acyl–enzyme complex thereby yielding a new ester; this will be discussed in more detail for *Candida antarctica* lipase B-catalyzed transesterifications.

Similarly, amines and hydrogen peroxide can act as the nucleophiles leading to amides and peracids, respectively [4]. In most instances, the formation of acyl-enzyme is fast, therefore deacylation is the rate-determining step [29]. The ability to act efficiently on a broad range of natural and "unnatural" esters while retaining high enantio- or regioselectivity makes lipases ideal catalysts not only for food and oil processing, but also for the preparation of enantiopure pharmaceuticals and synthetic intermediates [30, 31]. In fact, lipase-catalyzed biotransformations constitute about 40% of all biotransformations reported to date [4]. The most commonly employed lipases in organic synthesis are: porcine pancreatic lipase, lipase from *Pseudomonas cepacia*, lipase from *Candida rugosa* (formerly notated as *Candida cylindracea*), lipase from *Rhizomucor meihei* (also known as *Mucor meihei*), and lipase B from *Candida antarctica* (CALB) [29].

Enzyme-catalyzed transesterification reactions are especially useful for the preparation of *optically active compounds* by asymmetrization or resolution of racemic

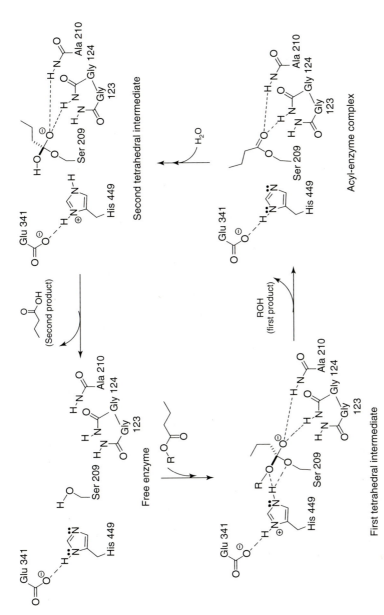

Figure 14.3 Hydrolysis of a butyric acid ester catalyzed by the lipase from *Candida rugosa*. (Adapted from Bornscheuer and Kazlauskas [29]).

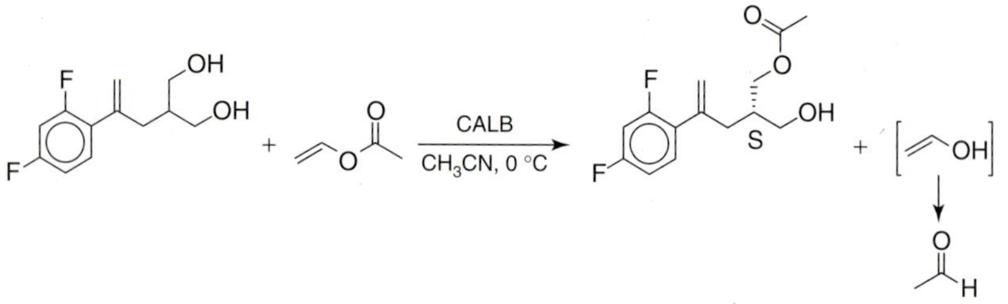

Figure 14.4 Asymmetrization of a prochiral diol by CALB-catalyzed enantioselective transesterification.

or prochiral substrates. Figure 14.4 shows an example of *enantioselective transesterification* starting from a racemic mixture [32]. Although both lipases [30, 33] and proteases [34], the enzymes that hydrolyze peptide bonds *in vivo*, have been used as transesterification catalysts, the latter found limited application due to high substrate selectivity.

Yadav and Trivedi [35] compared the catalytic activity of various commercially available lipases in the transesterification of vinyl acetate with *n*-octanol (Figure 14.5 and Table 14.2). This reaction is irreversible because the vinyl alcohol product immediately tautomerizes into acetaldehyde.

It was observed that CALB was the most efficient lipase. It gave 82% conversion, whereas *Mucor meihei* lipase and *Pseudomonas* lipase gave only 18 and 8% conversions, respectively, within 90 min. The inactivity of *Candida rugosa* lipase and porcine pancreatic lipase was attributed to denaturation by the acetaldehyde by-product [36].

The *structure of the alcohol* is an important parameter affecting the initial rate and overall conversions in enzymatic transesterification. It was observed that straight-chain alcohols gave better conversion compared with aromatic and

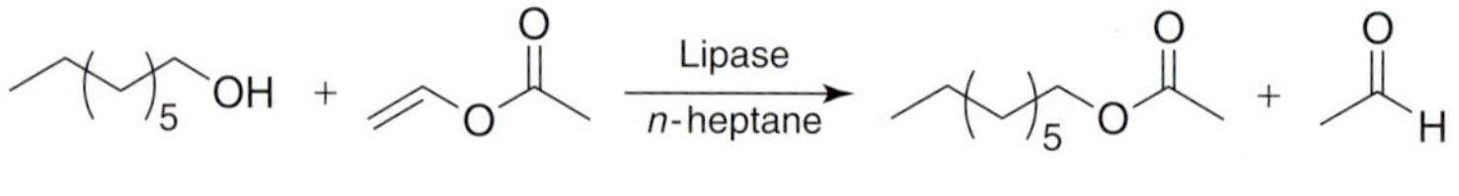

Figure 14.5 Lipase-catalyzed transesterification of vinyl acetate with *n*-octanol.

Table 14.2 Comparison of the activity of commercially available lipases in the transesterification of vinyl acetate with *n*-octanol at 30 °C.

Enzyme	Activity (μmol min^{-1} mg^{-1})
Candida antarctica lipase B	11.3
Mucor meihei lipase	1.30
Pseudomonas lipase	0.93
Candida rugosa lipase	0.00
Porcine pancreatic lipase	0.00

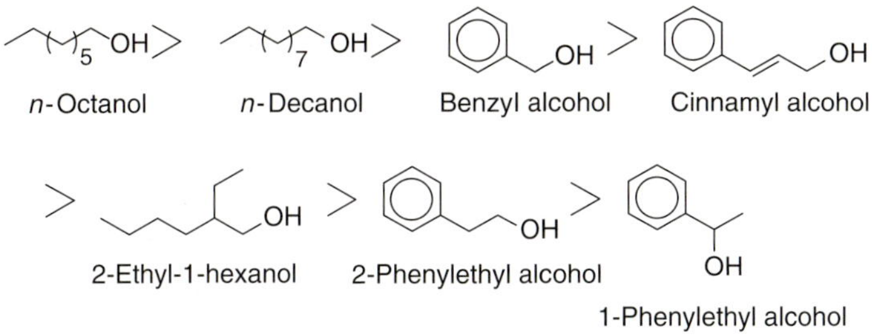

Figure 14.6 The reactivity of various alcohols in CALB-catalyzed transesterification of vinyl acetate.

branched-chain alcohols in the CALB-catalyzed transesterification of vinyl acetate due to less steric hindrance around the hydroxyl group (Figure 14.6) [35]. Furthermore, aromatic alcohols with saturated shorter side chains (e.g., benzyl alcohol) were more reactive than those with longer but unsaturated side chains (e.g., cinnamyl alcohol); however, no explanation for this finding was given.

Although an increase in ester concentration was shown to increase the rate and conversion of transesterification [35, 37], an increase in alcohol concentration might cause reduced rates and conversions, due to competitive inhibition by the alcohol that can bind reversibly to the enzyme active site and prevent the binding of the ester substrate [35, 38]. The driving force for alcohol binding might be the high polarity of the region around the active serine site of the enzyme [39].

It is known that hydrophobic solvents are generally favored over hydrophilic solvents because the latter can strip away the tightly bound water layer from the enzyme surface, which is required to maintain the catalytically active conformation of the enzyme [4]. Although many parameters describing the *hydrophobicity of a solvent* such as the Hildebrandt solubility parameter (δ), the dielectric constant (ε), and the dipole moment (μ), have been proposed [40], the most reliable results were obtained by the logarithm of the partition coefficient (log P) of a given solvent [41]. The *partition coefficient* is defined as the ratio of the equilibrium concentrations of a dissolved substance in a two-phase system consisting of two largely immiscible solvents, usually n-octanol and water. A solvent with a (log P) value greater than 2 ensures high retention of the activity of the enzyme [41]. The work of Riva and coworkers [42] can be given as an example of the direct correlation between the enzymatic activity and solvent hydrophobicity. CALB-catalyzed transesterification of vinyl acetate with cyclohexanol (Figure 14.7) was quantitative in all solvents except for dimethylformamide, which had a log P of -1.0; also, the enzyme showed the highest reactivity in hexane, the solvent with the highest log P (Table 14.3).

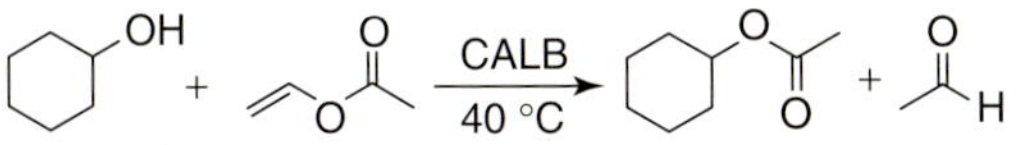

Figure 14.7 CALB-catalyzed transesterification of vinyl acetate with cyclohexanol.

Table 14.3 Effect of solvent hydrophilicity on enzyme activity.

Solvent[a]	Log P	Initial rate (μmol min^{-1})	Conversion (%) 5 h	Conversion (%) 24 h
Hexane	3.5	2.14	100	–
THF	0.49	1.17	95	100
Acetonitrile	−0.33	0.76	85	100
DMF	−1.0	0	–	–

[a] THF, tetrahydrofuran; DMF, dimethylformamide.

The ionization state of the enzyme, which is a function of pH, determines its conformation and thus its activity. On the other hand, the concept of pH value in organic solvents is challenged [43]. It has been found that enzymes in organic solvents have a so-called "*pH memory*" which means that the catalytic activity of the enzyme reflects the pH of the last aqueous solution from which the enzyme was recovered [44, 45]. This behavior stems from the fact that the ionization state of the charged groups of the enzyme does not change on both dehydration and subsequent placement in organic solvents. Therefore, it is important to employ enzymes that have been recovered by lyophilization or precipitation from aqueous solutions at their pH optimum [44, 45]. Alternatively, the optimization of the ionization state can be achieved by adding buffers consisting of acids and their conjugate bases to the reaction media [46].

The simplest way of preparing enzymes for use in organic solvents is *lyophilization* or precipitation from aqueous solutions. However, diffusional limitations due to protein–protein stacking in organic media and lyophilization-induced enzyme denaturation cause diminished enzymatic activities [47]. When lyophilization is carried out in the presence of lyoprotectants [48] [sugars, poly(ethylene glycol)], substrate-resembling ligands [49], inorganic salts [50], and crown ethers [51], large enhancements of enzyme activity has been observed. *Enzyme immobilization* is another possible way of increasing the catalytic activity of enzymes. Enzyme aggregation in organic solvents can be prevented by immobilization and thus the surface area of the enzyme can be increased, which in turn can improve the catalytic activity [52]. In fact, immobilization can also enhance thermal stability by stabilizing the tertiary structure of the protein, ease separation from the reaction mixture and enable repeated or continuous use of the enzyme [4, 53]. These attributes make enzyme immobilization a preferred method for the improvement of enzyme properties. The most widely used immobilization techniques can be categorized as follows [3]:

- **Adsorption**: Enzymes are non-covalently attached to an insoluble matrix such as synthetic resins, activated charcoal, Celite, ion-exchange resins, controlled-pore glass, alumina, and silica [54–57]. Adsorption forces are relatively weak and include Van der Waals forces, ionic forces, and hydrogen bonding.

- **Covalent attachment**: Enzymes are attached covalently onto a macroscopic carrier through amino-, carboxy-, sulfhydryl, hydroxyl-, or phenolic groups of the protein. Porous glass, polysaccharides, ceramic, and poly(vinyl acetate) based synthetic copolymers are often used as carriers for covalent immobilization [58–60].
- **Cross-linking**: Enzymes are attached onto each other by covalent bonds using bifunctional reagents such as α,ω-glutardialdehyde, dimethyl adipimidate, and hexamethylene diisocyanate [61, 62].
- **Entrapment**: Enzymes are entrapped in a macroscopic matrix such as agar and alginate gels, carrageenan, silica gel, and polyacrylamide [63–66].
- **Membrane confinement**: Enzymes are enclosed in restricted compartments that are bordered by semi-permeable membranes. This system is a close imitation of biological immobilization within a living cell where many enzymes are membrane-bound in order to provide a safe environment for them. The semi-permeable membranes in these systems are formed by surfactants, polyamides, or polyethersulfones [67–69].

In enzyme-catalyzed ROP and polymer functionalization, CALB is used most often, so this enzyme will be discussed in more detail.

14.2.2
Candida antarctica Lipase B

The basidiomycetous yeast *Candida antarctica* produces two different lipases, A and B [70]. The two lipases are very different. Lipase A is highly thermostable whereas lipase B is less thermostable. Lipase A is more active toward large triglycerides and lipase B is much more active toward a broad range of esters, amides, and thiols [53]. In contrast to lipase A, lipase B (CALB) is a very well characterized catalyst and its use is highly diversified [71]. CALB is composed of 317 amino acid residues and has a molecular weight of 33 kDa (Figure 14.8).

The *amino acid sequence* and the active site structure of the enzyme were resolved in 1994 [24]. The active site of the enzyme consists of serine (Ser105), histidine (His224), and aspartate (Asp187) as the catalytic triad. Its oxyanion hole is stabilized by three hydrogen bonds: one from glutamine (Gln106) and two from threonine (Thr40) units. The structures of these amino acid residues are given in Figure 14.9.

The catalytic serine residue is located at the bottom of a deep and narrow pocket, which is approximately 10×4 Å wide and 12 Å deep [19]. The region around the serine residue is considerably polar, but the inner walls of the pocket are very hydrophobic. The pocket is composed of two channels: one hosting the acyl- and the other hosting the alcohol-moiety of the ester substrate (Figure 14.10), with the first channel being the more spacious one [19]. Therefore, the enzyme is expected to have a much higher degree of selectivity toward alcohol substrates [72]. For instance, although primary alcohols are generally excellent substrates for CALB, tertiary alcohols such as *t*-butanol are so inert toward CALB that they can be used effectively as solvents in CALB-catalyzed reactions [73].

In almost all *biotransformations catalyzed by CALB*, the enzyme is immobilized on different macroporous carrier materials, as a result of the advantages of enzyme

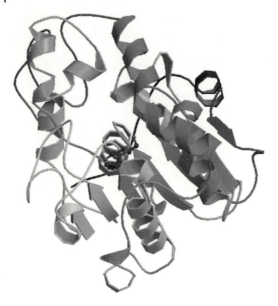

Figure 14.8 3D-structure of CALB. (Reprinted with permission from Research Collaboratory for Structural Bioinformatics-Protein Data Bank (RSCB PDB) PDB ID: 1TCB, Uppenberg, J., Hansen, M.T., Patkar, S., Jones, T.A. *Structure*, 1994, **2**, 293–308 [24]).

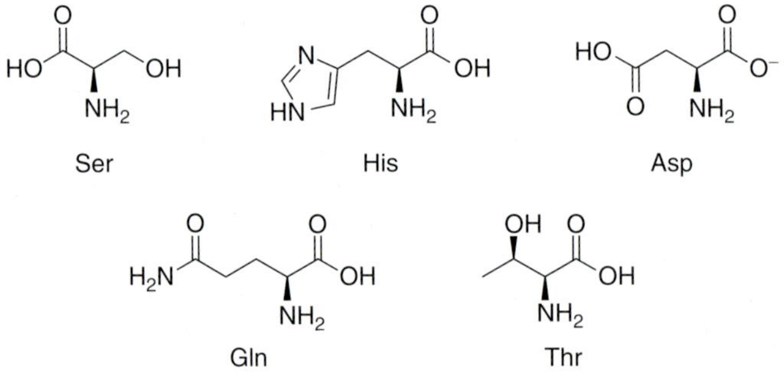

Figure 14.9 Structures of the amino acid residues forming the catalytic triad and the oxyanion hole of CALB.

immobilization outlined in the previous section. Novozym® 435, which is produced by Novozymes A/S, Denmark, is the most widely employed form of immobilized CALB. The production involves the transfer of the gene coding of lipase from *Candida antarctica* to the host organism *Aspergillus oryzae*. The enzyme, which is produced by the host organism, is then immobilized on a macroporous acrylic resin by physical adsorption. The product consists of bead-shaped particles of 0.3–0.9 mm diameter with a water content of 1.3% (w/w) and a protein content of 20% (w/w) [74, 75].

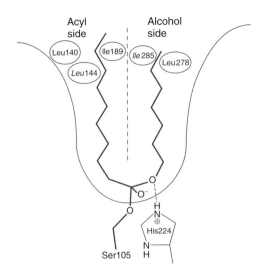

Figure 14.10 The active site pocket of CALB partitioned into two sides: an acyl side and an alcohol side. (Reprinted with permission from *Biochemistry*, 1995, **34**, 16838–16851. Copyright © 1995 American Chemical Society [19]).

14.2.3
CALB-Catalyzed Transesterification Reactions

Transesterification reactions are generally reversible. In order to change the reversible nature of the reaction into an irreversible type, the nucleophilicity of the leaving group of the acyl donor should be depleted by the introduction of electron-withdrawing groups, such as trifluoroethyl- or trichloroethyl-, into the ester [76]. Alternatively, the use of oxime esters [77], thioesters [78], and anhydrides [57] as activated acyl donors have been proposed. The use of enol esters [79], such as vinyl or isopropenyl esters, appears to be the most useful as they liberate unstable enols as by-products, which rapidly tautomerize to give the corresponding aldehydes or ketones (Figure 14.4). Therefore, the reaction becomes essentially irreversible. It was shown that acyl transfer reactions using enol esters were 100–1000 times faster than the reactions using non-activated esters such as ethyl acetate [79]. Vinyl esters are favored over isopropenyl esters because of less steric hindrance and thus higher reaction rates [80]. Acetaldehyde, which forms during the reactions with vinyl esters, is known to inactivate the lipases from *Candida rugosa* and *Geotrichum candidum* by forming a Schiff's base with the lysine residues of the protein; however most lipases, including CALB, tolerate the liberated acetaldehyde [36]. The catalytic cycle of the CALB-catalyzed transesterification of vinyl acetate with *n*-octanol is visualized as an example in Figure 14.11 [81], based on the mechanism shown in Figure 14.3. The different shadings represent the carbonyl and hydroxyl pockets of the enzyme.

Firstly, the nucleophilic serine (Ser105) residue attacks the carbonyl group of the vinyl acetate, forming a tetrahedral intermediate that is stabilized by the oxyanion hole of the enzyme via three hydrogen bonds: one from glutamine (Gln106) and two from threonine (Thr40) units. In the second step, the ester bond is cleaved to form the first product, vinyl alcohol, which will tautomerize to acetaldehyde, and

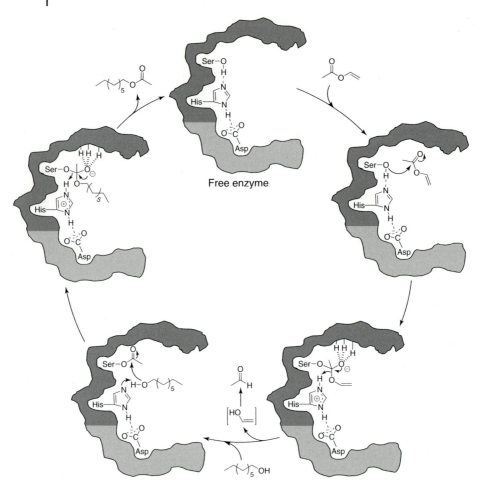

Figure 14.11 Illustration of the mechanism of CALB-catalyzed transesterification of vinyl acetate with n-octanol [81]. The different shading represents the two enzyme pockets (see Figure 14.10).

the acyl–enzyme complex. In the third step, the reactant alcohol, *n*-octanol, attacks the acyl–enzyme complex to form a second tetrahedral intermediate, which is again stabilized by the oxyanion hole. In the last step, the tetrahedral intermediate collapses and the enzyme is deacylated to form the desired product, octyl acetate. The nucleophilic attack by the Ser105 is mediated by the His224–Asp187 pair. Similarly to many other lipases, the last step, that is, the deacylation, was shown to be the rate limiting step in CALB-catalyzed transesterification. It was demonstrated by Garcia-Alles and Gotor [82] that the reaction rates for ester substrates differing only in the leaving group were the same. Similarly, Martinelle and Hult [39] observed the same reaction rates using vinyl- and ethyl octanoate as the acyl donors and (*R*)-2-octanol as the acyl acceptor in the presence of CALB.

We will advance a similar mechanism for the lipase-catalyzed ROP of lactones, which will be discussed in the next section.

14.3
"Green" Cationic Polymerizations and Polymer Functionalization Using Lipases

14.3.1
Ring-Opening Polymerization

The enzymatic ROP of cyclic monomers was first reported in 1993 [83, 84]. ε-Caprolactone (CL) was polymerized in the presence of the lipase from *Pseudomonas fluorescens* in bulk for 10 d (Figure 14.12) and a polymer with a molecular weight of $M_n = 7000$ g mol^{-1} was obtained [83]. The polymerization of the same monomer (CL) in *n*-hexane using porcine pancreatic lipase as the catalyst and methanol as the initiator afforded a polymer with a degree of polymerization up to 35 [84]. Since then a wide variety of cyclic monomers, that is, lactones, lactides and macrolides (large ring lactones), cyclic carbonates, cyclic phosphates, cyclic depsipeptides, and oxiranes, have been enzymatically polymerized, but the lipase-catalyzed ROP of lactones has been the most extensively investigated type of enzymatic ROP polymerization [85–88].

By analogy with other lipase-catalyzed reactions, it was suggested that the mechanism of the ROP of lactones involved the formation of a lipase–lactone complex, followed by the formation of an acyl–enzyme intermediate. Subsequently, this intermediate would be deacylated by water or alcohol in the initiation step and by the terminal hydroxyl group of a growing polymer chain in the propagation step (Figure 14.13) [89].

Gross *et al.* [90] recently reported that CALB, which was inhibited irreversibly at the serine (Ser105) residue with diethyl *p*-nitrophenyl phosphate, failed to initiate the ROP of CL. Therefore, they concluded that the *lipase-catalyzed polymerization proceeds by the catalysis at the active serine residue of the lipase and not by other chemical or non-specific protein mediated processes*. Figure 14.14 is our proposal for the representation of the catalytic cycle, based on the involvement of ionic intermediates as shown in Figure 14.11. Thus the nucleophilic serine (Ser105) residue (aided by the His224–Asp187 pair) would attack the carbonyl group of the lactone, forming an intermediate "*activated monomer*." Ring-opening of this activated monomer would yield the acyl–enzyme complex. Subsequently the –OH group of the initiator (R_i–OH) would attack the acyl–enzyme complex and the

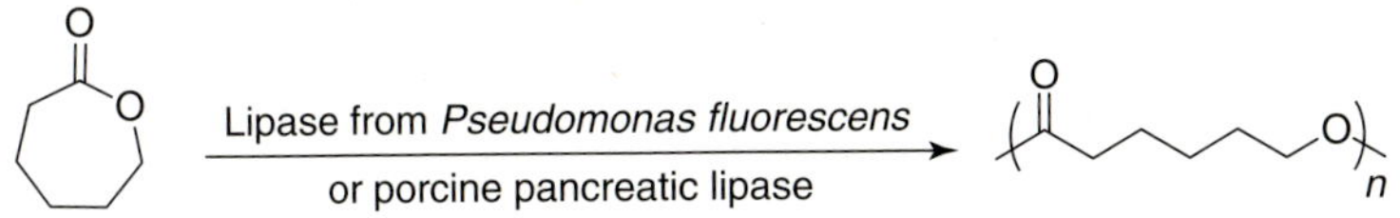

Figure 14.12 Lipase-catalyzed, ring-opening polymerization of ε-caprolactone.

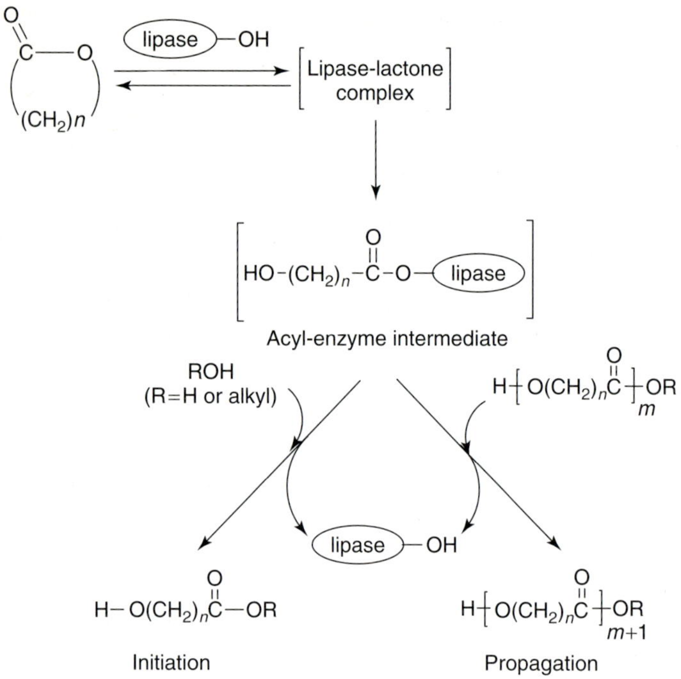

Figure 14.13 Mechanism suggested for the lipase-catalyzed ring-opening polymerization of lactones [89].

enzyme would be deacylated to a hydroxyacid ester. Water can also serve as an initiator, yielding a hydroxyacid in the first step.

Propagation would proceed through repeating the cycle with deacylation by an HO-capped polymer (P_n–OH) of the acyl–enzyme complex that forms from the activated monomer.

This proposal is based on the analogy with Figure 14.1 and will need to be substantiated, but the involvement of ionic species seems very likely.

One interesting feature of the enzyme-catalyzed ROP of lactones is that the rate of polymerization as well as the degree of polymerization is independent of the ring strain. On the other hand, the *polymerization by chemical catalysts* is governed by the ring strain, that is, the smaller the lactone ring, the faster it can be polymerized [91]. In fact, it was observed that the rate of polymerization increased with the increase in lactone ring size in *Pseudomonas fluorescens* lipase- and CALB-catalyzed lactone polymerizations [91–94]. This could be due to the stronger recognition of the larger ring lactones by the enzyme due to their higher hydrophobicity, as lipases inherently catalyze the hydrolysis of hydrophobic fatty acid esters *in vivo* [95].

Cordova *et al.* [96] utilized both initiator and terminator methods to prepare functionalized poly(ε-caprolactone)s using unsaturated alcohols, unsaturated fatty acids and esters, vinyl esters, and phenolic compounds in the presence of CALB. Methyl 6-O-poly(ε-caprolactone)-β-D-glucopyranoside was synthesized by

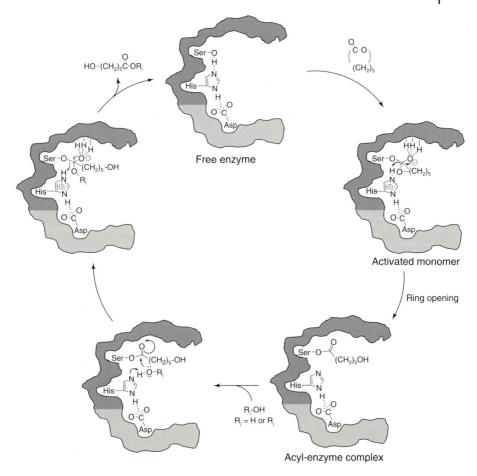

Figure 14.14 Suggested mechanism for the lipase-catalyzed ring-opening polymerization of lactones.

CALB-catalyzed regioselective acylation at the primary hydroxyl group of methyl β-D-glucopyranoside and ROP of CL [97]. Similarly, a hexahydroxyl-functional dendrimer was used as the initiator for the preparation of poly(ε-caprolactone) monosubstituted dendrimer via CALB-catalyzed regiospecific acylation [98].

The chemoselectivity of CALB toward alcohols over thiols in *transacylation* was exploited in the synthesis of thiol-functionalized poly(ε-caprolactone)s using either 2-mercaptoethanol as the initiator or γ-thiobutyrolactone and 3-mercaptopropionic acid as terminators [99]. Polyesters with 70–90% thiol functionality were obtained. The use of glycidol as the initiator for the ROP of CL in the presence of CALB afforded epoxy-functionalized poly(ε-caprolactone) (Figure 14.15). Based on MALDI-TOF MS (matrix-assisted laser desorption/ionization time-of-flight mass spectrometry) analysis, it was concluded that CALB selectively catalyzed the ring opening of CL, while the epoxy ring of the glycidol remained intact.

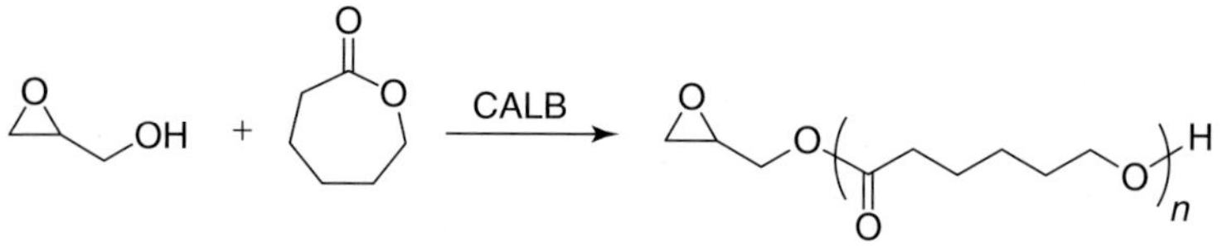

Figure 14.15 Enzymatic synthesis of epoxy-functionalized poly(ε-caprolactone).

14.3.2
Enzyme-Catalyzed Polymer Functionalization

Enzymatic catalysis in the *functionalization of preformed polymers* has not been utilized extensively despite the advantages enzymes offer. We established the conditions under which quantitative functionalization of preformed polymers was achieved [100–102]. Here we will illustrate this novel technology on polyisobutylenes (PIBs) synthesized by carbocationic polymerization. Low molecular weight liquid PIBs were functionalized under solventless conditions. As PIB is biocompatible, the functionalized polymers are prime candidates for biomedical applications.

The ability of enzymes to function in bulk reaction media is well-known and has been exploited in various transformations, including esterification and transesterification of small molecules [103–106], and the synthesis of polymers by condensation and ROPs [12]. Apart from being an environmentally benign method, the use of enzymes in solvent-free systems increases the commercial feasibility of the process by avoiding costly solvent evaporation and recycling.

Two PIBs fitted with a 3-hydroxyl-2-methylpropyl-terminus [PIB–CH_2–C(CH_3)–CH_2–OH] were reacted with vinyl methacrylate in bulk in the presence of CALB (10 wt% relative to the total weight of reactants) (Figure 14.16). The first polymer, Glissopal–OH, was obtained by hydroboration/oxidation of Glissopal® 2300 (from BASF).

The ^{1}H- and ^{13}C-NMR spectra of the methacrylation product of neat Glissopal–OH are shown in Figure 14.17. The resonances of the methylene protons adjacent to the methacrylate endgroup are observed at $\delta = 3.84$–4.02 ppm (j_1 and j_2) and the vinyl [$\delta = 6.13$ ppm (h_1) and 5.56 ppm (h_2)] and methyl [$\delta = 1.97$ ppm (i)] protons of the newly formed methacrylate-end appear at the expected positions. The ^{13}C-NMR also confirmed the structure of the methacrylated polymer where the carbonyl carbon is observed at $\delta = 167.73$ ppm (j), and resonances of the vinyl [CH_2=C(CH_3)-], methyl [CH_2=C($\underline{C}H_3$)-], and the alpha

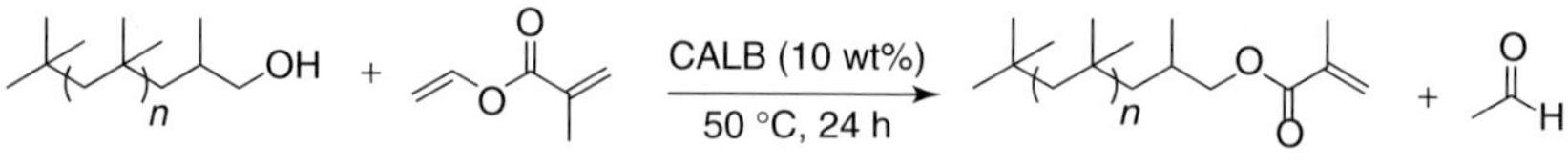

Figure 14.16 CALB-catalyzed methacrylation of hydroxyl-terminated polyisobutylenes [PIB–CH_2–C(CH_3)–CH_2–OH] in bulk.

14.3 "Green" Cationic Polymerizations and Polymer Functionalization Using Lipases

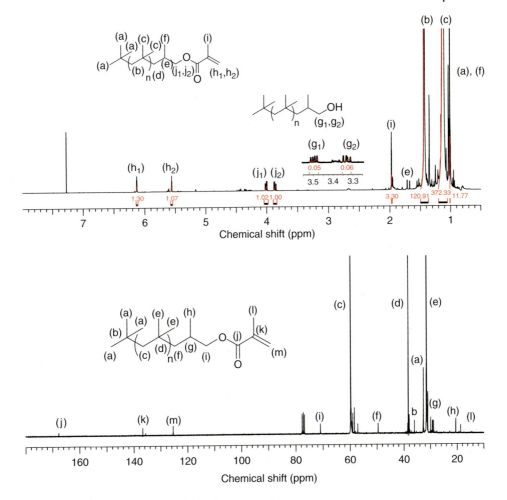

Figure 14.17 ^{1}H-NMR spectrum of the CALB-catalyzed methacrylation product of Glissopal–OH in bulk (top) and its corresponding ^{13}C-NMR spectrum (bottom) (solvent, CDCl$_3$).

carbon [CH$_2$=$\underline{\text{C}}$(CH$_3$)-] of the methacrylate group are observed at δ = 125.38 (m), 18.60 (l), and 136.71 ppm (k), respectively.

The other PIB that was reacted with vinyl methacrylate in bulk was derived from α-(tert-butyl)-ω-(2-chloro-2-methylpropyl)polyisobutylene through dehydrochlorination [107] followed by hydroboration/oxidation [108]. The α-(tert-butyl)-ω-(2-chloro-2-methylpropyl)polyisobutylene was synthesized via TMPCl–BCl$_3$ initiated carbocationic polymerization of isobutylene [109, 110]. Similarly to Glissopal–OH, this polymer also reacted efficiently with CALB in bulk. These examples illustrate that low molecular weight liquid polymers can be effectively methacrylated under solvent-free conditions by CALB-catalyzed transesterification of vinyl methacrylate.

The same process was used to transesterify vinyl methacrylate with poly(dimethyl siloxanes) [102].

While the chemistry of lipase catalysis in polymerization and polymer functionalization is not fully understood, we have a worse case with regard to the chemistry of NR biosynthesis. *Natural rubber biosynthesis* remains one of the most mysterious natural processes as the membrane-bound rubber transferase enzyme has never been isolated and its structure is unknown. The next section will discuss NR biosynthesis, the *"ultimate green cationic polymerization."*

14.4
Natural Rubber Biosynthesis – the Ultimate Green Cationic Polymerization

Natural rubber is a critical and strategic industrial raw material for manufacturing a wide variety of products, ranging from medical devices and personal protective equipment to aircraft tires. The most important source of NR has been *Hevea brasiliensis* (i.e., the Brazilian rubber tree). In 1998, 10 million tons of NR were produced worldwide for commercial use, from which about 15% was consumed in the U.S.A. [111]. While the U.S.A. is self-sufficient in synthetic rubber production, with substantial export activities, no NR is produced domestically. The development of a U.S.A.-based supply of NR was recognized in the "Critical Agricultural Materials Act of 1984" (Laws 95–592 and 98–284). The Act recognizes that NR latex is a commodity of vital importance to the economy and the defense of the nation. It is important to emphasize that synthetic polyisoprene (PIP) does not match the performance of imported *Hevea* NR in several applications, so NR remains irreplaceable.

Figure 14.18 Examples of rubber producing plants. (Reprinted with permission from Maureen Whalen, USDA-ARS © 2008).

14.4 Natural Rubber Biosynthesis – the Ultimate Green Cationic Polymerization

Our research group set out to study the *biosynthesis of natural rubber* from the polymer chemistry point of view. NR is obtained from latex, an aqueous emulsion present in the laticiferous vessels (ducts) or parenchymal (single) cells of rubber-producing plants. Although more than 2500 plant species are known to produce NR, there is only one important commercial source, *Hevea brasiliensis* (the Brazilian rubber tree). The rubber latex from *H. brasiliensis* is harvested by "*tapping*" the rubber tree. An incision is made on the trunk and latex oozes out of the incision. This white liquid is collected and then coagulated to yield high molecular weight (>1 million g mol^{-1}) polymer. The rubber from guayule, *Parthenium argentatum*, grown in the Southwestern U.S.A., is being developed as a non-allergenic NR mainly for rubber gloves [112, 113]. The guayule rubber is produced by a green aqueous-based extraction process on a commercial scale from the stems of the shrub [114]. Russian dandelion, *Taraxacum kok-saghyz*, is also being developed as a rubber-producing crop in the northern U.S.A. [115, 116]. Figure 14.18 shows a selection of rubber producing plants that have been of research and commercial interest.

14.4.1
Anatomy of the NR Latex, and Structure of Natural Rubber

The productivity of *Hevea* trees is usually as high as 50–100 g latex per day in a mature tree [117]. Depending on the seasonal effects and the state of the soil, the average composition of *Hevea brasiliensis* NR latex is: 25–35 wt% *cis*-1,4-polyisoprene; 1–1.8 wt% protein; 1–2 wt% carbohydrates; 0.4–1.1 wt% lipids 0.5–0.8 wt% amino acids, and 50–70 wt% water [118]. The latex particles with a diameter of 0.1–10 μm [119] (Figure 14.19a) are stabilized by a membrane of phospholipid monolayer (Figure 14.19b) [120]. The *cis*-prenyltransferase enzyme is a membrane-bound amphiphilic enzyme, yet to be isolated and fully characterized [119, 121].

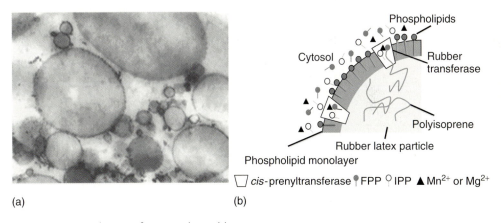

Figure 14.19 Visualization of NR particles and latex structure.

14.4.1.1 Structure of Natural Rubber

Infrared (IR), NMR spectroscopy, and X-ray studies have shown that NR is polyisoprene in the *cis*-1,4 configuration [119, 122, 123]. It has been known since the 1950s that the chain elongation of rubber molecules proceeds by the addition of the isopentenyl pyrophosphate (IPP) to polyisoprenoid-based initiators [124, 125]. Based on the mechanism of low molecular weight terpenoid biosynthesis, the initiator was assumed to be dimethylallyl pyrophosphate (DMAPP) [126], but Tanaka did not find dimethyl allyl head groups in *Hevea* rubber. The general agreement from the literature is that, depending on the plant source, NR contains 1–3 units of *trans* isoprene (IP) from the initiator. In the case of *Hevea* rubber, the initiator is farnesyl pyrophosphate ($n = 3$), followed by >5000 *cis*-only IP units and a variety of endgroups: hydroxyls, aldehydes, amines, and so on. The structure is visualized in Figure 14.20, although Tanaka stated that the structure of the head groups and endgroups is ambiguous [119].

The current understanding of NR structure is based upon naturally occurring model compounds [119, 127]. Figure 14.21 shows the ^{13}C-NMR spectrum of a naturally occurring low molecular weight (2.4×10^4 g mol^{-1}) rubber, where the signals of the trans-initiator units, the dimethyl allyl head group and ester endgroups can be seen.

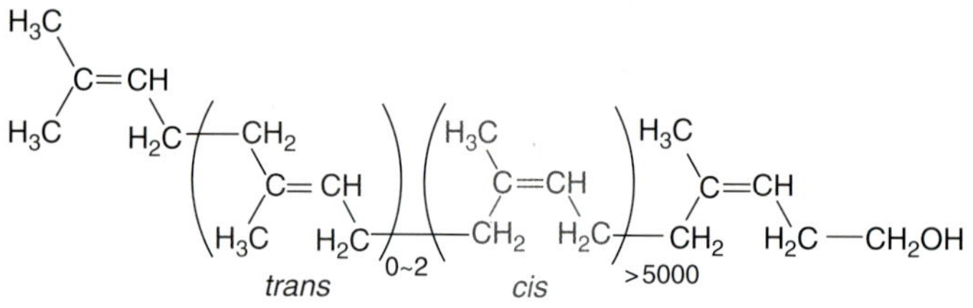

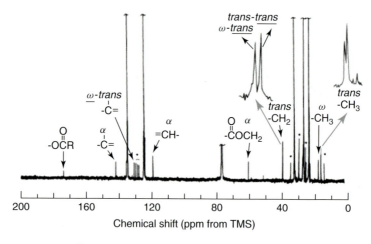

Figure 14.20 Chemical structure of NR.

Figure 14.21 ^{13}C-NMR of low molecular weight NR from *L. volemus* [119].

However, the addition of radioactive ^{14}C DMAPP into fresh *Hevea* latex did not form new rubber chains containing the radioactive dimethyl allyl head group [125]. ^{13}C-NMR spectroscopy of high molecular weight NR from *Hevea* and *P. argentatum* could not detect dimethyl allyl head groups [128]. In accordance with this, Tanaka concluded that the currently accepted initiation mechanism remains unproven [119–121].

14.4.2
Biochemical Pathway of NR Biosynthesis

The *biosynthesis of NR* is catalyzed by a rubber transferase enzyme, *cis*-prenyltransferase, which is integrated into the phospholipid monolayer that surrounds the latex particles [120, 124, 127, 129]. The phospholipid monolayer stabilizes the particles to prevent aggregation in the aqueous medium. The hydrophobic polymer chains reside within the latex particles, and polymerization proceeds at the active sites of the enzyme [120, 130]. The *rubber transferase enzyme* is proposed to be amphiphilic, with hydrophilic regions facing the cytosol to allow the access of hydrophilic building blocks and the hydrophobic regions extending into the interior of the rubber particle.

14.4.2.1 Monomer
The building block (*monomer*) of NR is IPP. Figure 14.22 shows the structure of IPP, which can be considered as an adduct of pyrophosphoric acid ($H_4P_2O_7$) and IP.

At the physiological pH of 7.4, IPP is a stable dianion [131]. The synthesis of IPP *in vivo* can proceed by two different pathways: the mevalonate (MVA) pathway and the non-MVA (deoxy-xylulose) pathway. These two distinct pathways have evolved in different organisms. In eukaryotes, IPP is derived from acetyl-CoA (coenzyme A), while in prokaryotes and plant chloroplasts, IPP is derived from 1-deoxy-D-xylulose-5-phosphate [132]. In higher plants, the MVA pathway operates mainly in the cytoplasm and mitochondria, whereas the non-MVA pathway operates in the plastids [133]. It is highly likely that the IPP polymerized by the rubber transferase derived from the MVA pathway that occurs in the cytosol [134]. However, IPP produced by the 1-deoxy-D-xylulose-5-phosphate pathway may also diffuse from the plastids into the cytosol [135].

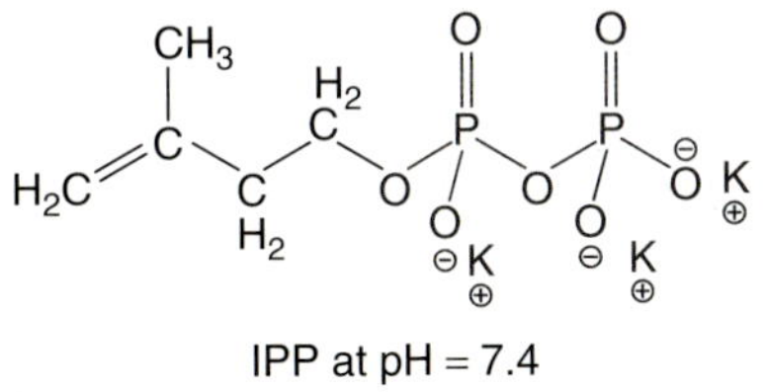

IPP at pH = 7.4

Figure 14.22 Structure of isopentenyl pyrophosphate IPP at pH = 7.4.

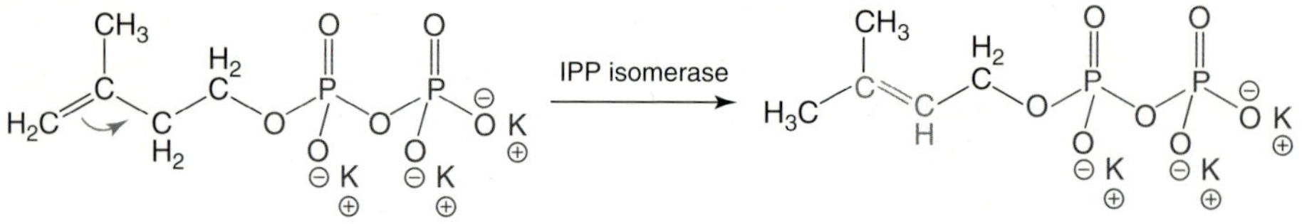

Figure 14.23 IPP isomerization to yield DMAPP in NR biosynthesis.

14.4.2.2 Initiators

Allylic diphosphates are the *initiators* of chain growth. IPP is isomerized to 1,1-dimethylallyl pyrophosphate (DMAPP) by IPP isomerase, presented in Figure 14.23.

DMAPP can then add 1–3 IPP units, catalyzed by *trans*-prenyltransferases [136], to form oligomeric allylic pyrophosphates (APPs). These APPs are termed geranyl pyrophosphates (GPPs) farnesyl pyrophosphates (FPPs), and geranylgeranyl pyrophosphates (GGPPs), respectively, and they all can serve as initiators. Their chemical structures are shown in Figure 14.24.

IPP and APP are termed "*substrate*" and "*cosubstrate*", respectively, in the biochemical literature, which would be termed *monomer* and *initiator*, respectively, in polymer chemistry. Enzymatic activity requires the presence of divalent cations, such as Mg^{2+} or Mn^{2+}, known as "*activity cofactors*" [127, 129, 136, 137]. While the exact role of the cofactors Mg^{2+} and Mn^{2+} is still unclear, Scott *et al.* recently demonstrated that only cofactor-activated IPP monomer will interact with the enzyme, while the FPP initiator may bind, even in the absence of cofactors

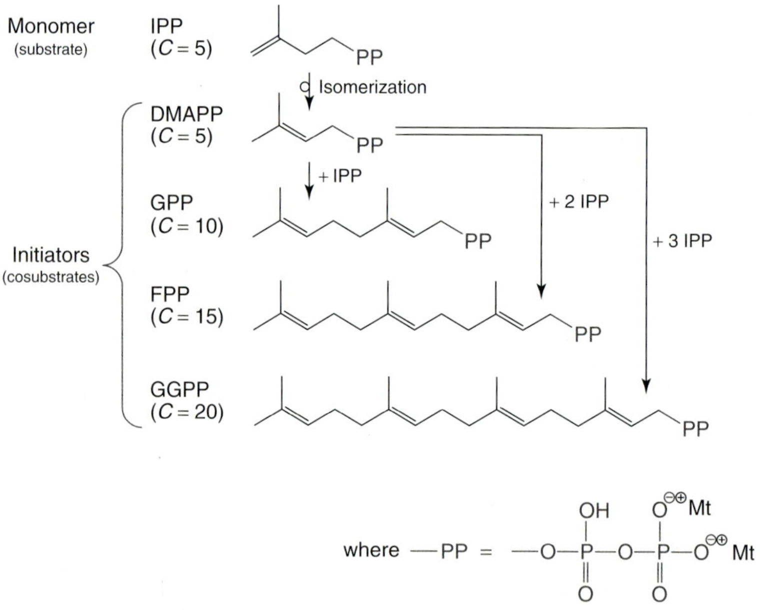

Figure 14.24 Structure of the allylic oligoisoprene pyrophosphates (APPs).

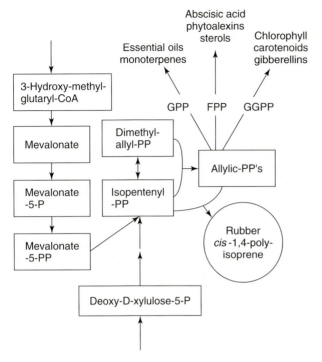

Figure 14.25 Natural rubber biosynthesis pathway, showing the IPP synthesis via the mevalonate or the xylulose pathway [130].

[137]. They also showed that the metal ion concentration *in vitro* can affect the rubber biosynthesis of *Ficus elastica*, *Hevea brasiliensis*, and *Parthenium argentatum*. Although a metal ion is required for rubber biosynthesis, an excess of metal ions interacts with the rubber transferase, inhibiting its activity [137]. da Costa *et al.* later suggested that *H. brasiliensis* could use the cytosolic magnesium concentration as a regulatory mechanism for rubber biosynthesis and molecular weight *in vivo* [138]. Figure 14.25 shows a flow diagram of NR biosynthesis reproduced from the biochemical literature [130]. The resulting molecular weight and molecular weight distributions are species-dependent; however, the biochemical pathway does not explain the control of the molecular weight parameters *in vivo* [139]. The broad molecular weight distribution is the result of continuous initiation, branching, and/or cross-linking via acid-catalyzed cyclization, or by "abnormal" functional groups (aldehydes, epoxides, and amines) [140, 141].

14.4.2.3 Catalyst: Rubber Transferase

Prenyltransferases are enzymes that catalyze the synthesis of linear prenyl diphosphates involved in the biosynthesis of various isoprenoid compounds: sterols, terpenes, NR, and others. Based on the configuration of IP units in the products, prenyltansferases are classified into two groups: *trans*- and *cis*-prenyltransferases.

In both prokaryotes and eukaryotes, *trans*-prenyltransferases catalyze the formation of geranyl diphosphate (GPP, C_{10}), farnesyl diphosphate (FPP, C_{15}), and geranylgeranyl diphosphate (GGPP, C_{20}), which serve as initiating molecules to produce many other longer chain length isoprenoids compounds necessary for cellular growth and survival. The *structural genes* for FPP synthase [142–145] and GGPP synthase [146–148] have been cloned from various organisms and have been characterized. Mutational analyses and X-ray crystallographic investigations of the structure of *trans*-prenyltransferases revealed crucial amino acid residues in the conserved domains for the mechanism of chain length determination [149–152].

Conversely, only three *cis*-prenyltransferase genes have recently been successfully cloned. Two genes are the structural genes for undecaprenyl diphosphate synthase that catalyzes the formation of undecaprenyl diphosphate (C_{55}), which serves as a glycosyl carrier lipid during the biosynthesis of cell wall polysaccharide components in *Escherichia coli* and *Micrococcus luteus* [153, 154]. The other is the *cis*-prenyltransferase responsible for the biosynthesis of dolichols used for the glycosylation of proteins in yeast [155]. Although the *cis*- and *trans*-prenyltransferases catalyze similar reactions of the sequential condensation between isopentenyl diphosphate (IPP) and allylic diphosphates in the presence of Mg^{2+} ions, the sequences were found to have no similarity between *cis*- and *trans*-prenyltransferases. This makes identification of other *cis*-prenyltransferases gene sequences extremely difficult. The *cis*-prenyltransferases cloned from *E. coli*, *M. luteus*, and yeast share a low level of sequence homology (~30% identity) among them.

Genetic sequences of rubber transferase, a *cis*-prenyltransferase, remain unidentified because it is a membrane-associated enzyme in low abundance [120]. The fact that the activity of rubber particles is rapidly lost upon disruption of their structural integrity poses one of the great challenges in sequencing the enzyme [129, 156]. Generally, rubber transferase is accepted as an amphiphilic enzyme residing at the interface of the latex (Figure 14.19b). Propagation occurs when the initiator or the NR molecule is bound at the specific catalytic site within the funnel-like crevice of the enzyme and is activated by cofactors, and IPP enters from the aqueous phase.

This understanding arises from the analysis of the 3D-structure of various prenyltransferase enzymes and other biochemical studies concerning the mechanisms of isoprenoid syntheses [157]. For example, the binding sites for the APP and IPP in an avian *trans*-prenyltransferase were located within the hydrophilic regions of the enzyme, whereas chain growth was proposed to take place within a hydrophobic pocket positioned toward the bottom end of the conical enzyme (Figure 14.26).

Poulter has investigated the mechanism of polyisoprenoid biosynthesis and established that *"elongation"* proceeds by a carbocationic mechanism, as shown in Scheme 14.1.

This mechanism is accepted in the biochemical community. However, the existence of carbocations is viewed with skepticism in the polymer community. Although carbocationic polymerizations were reported in aqueous media [158], they generally yield low molecular weights.

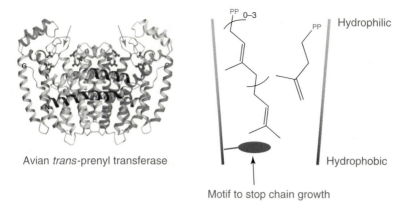

Figure 14.26 A scheme of the active sites in an avian *trans*-prenyltransferase [157]. The oval represents a large amino acid motif proposed to stop chain growth.

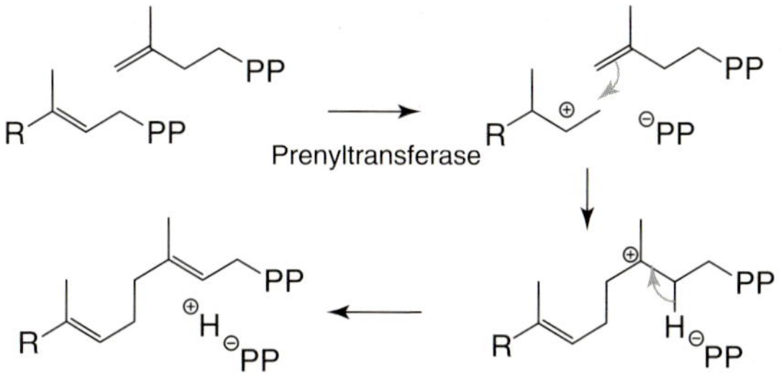

Scheme 14.1 Poulter's mechanism of prenylation through a tertiary carbocation intermediate.

14.4.3
Chemical Mechanism of Natural Rubber Biosynthesis

Strikingly, the exact chemical macro- and microstructures of NR remain unknown; the current understanding of NR structure is based upon the described naturally occurring low molecular weight polyisoprenoids [127]. While the concepts of NR biosynthesis are understood from a biochemical point of view, understanding of this process in terms of polymer chemistry principles is lacking. Figure 14.27 shows the biosynthesis of NR in terms of the initiation and propagation steps proposed by Tanaka in polymer chemical symbolism [140].

Our group recently proposed that NR biosynthesis proceeds by a *carbocationic mechanism* (Figure 14.28) [159]. According to this proposal, initiation starts by an enzyme (and divalent cofactor) assisted ionization of the carbon-oxygen bond of the initiator and yields an allylic carbocation plus a pyrophosphate counteranion; the enzyme plus cofactor(s) coordinate with the pyrophosphate "protecting" group and mediate the formation of the initiating carbocation. According to polymer chemical convention, the enzyme plus cofactors constitute the *coinitiating system*. Ionization

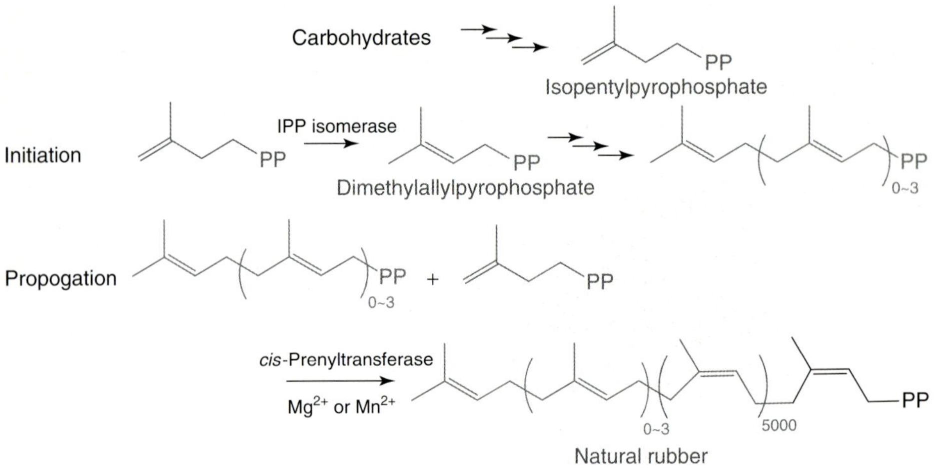

Figure 14.27 NR biosynthesis chemical scheme.

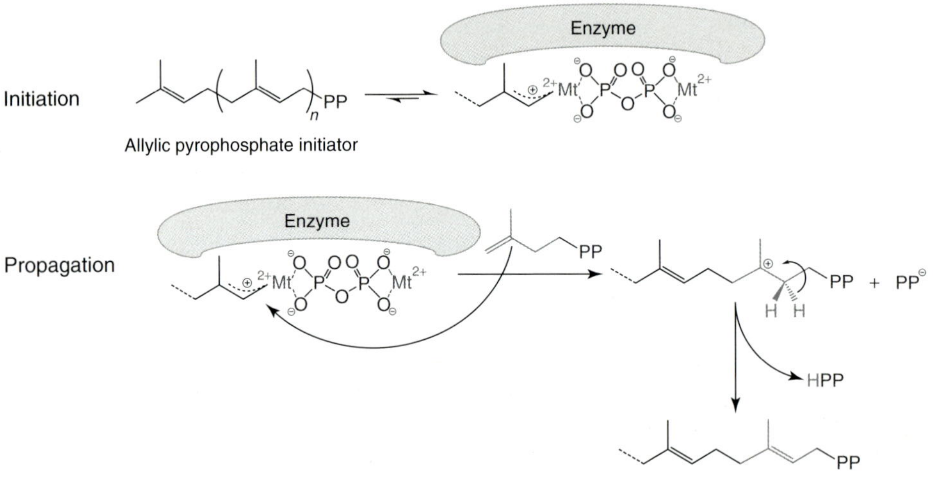

Figure 14.28 Proposed mechanism of NR biosynthesis.

at the chain end is favored by resonance stabilization of the allylic carbocation and increasing entropy of the system. Subsequently, the IPP adds to the allylic carbocation, yielding a tertiary carbocation, which, via proton elimination, regenerates the trisubstituted allylic pyrophosphate [159]. This mechanism applies to the formation of trans-1,1-dimethylallylic initiators (natural initiators invariably are trans), catalyzed by trans-prenyltransferase, and to the incorporation of the cis-units, that is, propagation (Figure 14.29). With regards to trans- or cis-stereoregulation, we proposed that the specific enzyme functions as the template. The incorporation of each IPP unit is always accompanied by the loss of pyrophosphoric acid (HPP) (or its salts).

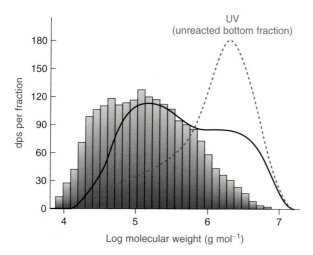

Figure 14.29 SEC trace of BF (dashed line) and *in vitro* synthesized NR (solid line). (Reproduced from Tangpakdee et al. [162]).

14.4.4
In vitro NR Biosynthesis

The first *in vitro polymerization* system using WRPs (washed rubber particles) from *H. brasiliensis* was described by Archer et al. in the 1980s. They incubated ^{14}C-IPP in the presence of unlabeled neryl pyrophosphate (NPP) or geranyl pyrophosphate (GPP) initiators in a suspension of WRP isolated from living *H. brasiliensis* latex and showed new rubber formation [124].

To date, genetic sequences of the rubber transferase remain unidentified because it is a membrane-bound enzyme in low abundance [120]. Conventional continuous assays to determine the enzymatic activities are a challenge due to the fact that the activity of the rubber particles is rapidly lost upon disruption of their structural integrity [129, 156]. The current method for determining the activity of rubber transferases is by radiometric assay, where the activity is calculated based on the incorporation rate of ^{14}C IPP monomer into higher molecular weight rubber produced *in vitro* [129, 160, 161].

Benedict et al. produced *in vitro* guayule rubber using WRP, synthetic IPP as monomer, and DMAPP as initiator in a reconstituted latex [160]. Their WRP was prepared from stems of *P. argentatum*; after removal of the bark and impurities, the rubber latex was centrifuged and the top rubber particulate layer was collected and purified by repeated washing with buffer and centrifugation as the WRP. These workers used SEC (size-exclusion chromatography) coupled with a scintillation spectrometer to demonstrate that radioactive IPP became incorporated into NR [160]. They observed that rubber was formed with a peak molecular weight of $\sim 10^5$ g mol^{-1} within 15 min, and that the rubber was able to grow to $\sim 10^6$ g mol^{-1} in 3 h [160].

Tanaka's group established a novel method for *in vitro* rubber biosynthesis using the fresh bottom fraction (BF) of the *latex* [162, 163]. They employed a muslin cloth to filter out some of the coagulants and then centrifuged the liquid latex. The latex partitioned into three major fractions: the top particulate layer, a middle clear phase, which was called C-serum (CS) and a BF marked BF [163]. The BF was used for *in vitro* NR biosynthesis and it was observed that more than ~10 wt% new rubber formed with the addition of very small amounts of IPP or FPP to fresh BF [163]. It was also found that new rubber formed by the incubation of BF without the addition of IPP or FPP, concluding that the BF contains all the enzymes and precursors necessary to produce rubber [164]. The formation of new rubber was confirmed by the incorporation of ^{14}C-labeled IPP into the resulting rubber [162]. Figure 14.29 compares the UV traces of the endogenous NR from BF and the *in vitro* NR rubber. The BF has a high molecular weight fraction around $\sim 10^6$ g mol^{-1}, with a lower molecular weight tail. The newly formed rubber produced a peak at about $\sim 10^5$ g mol^{-1}. The radioactive traces also revealed that while the ^{14}C-IPP was incorporated into new chains, it also added to pre-existing chains in the lower molecular weight tail fraction of the BF [162].

Wititsuwannakul's group in Thailand also developed their unique washed bottom fraction particles (WBPs) for *in vitro* NR biosynthesis [165]. This system also incorporated radioactive IPP into NR. More recently, this group cloned two suspected gene sequences of *Hevea* rubber transferase and expressed them in *E. Coli* [121]. It was found that a combination of one of the clones with WBP resulted in enhanced NR growth.

Cornish *et al.* developed the method to produce WRP that is utilized in our studies. This method is an improvement over that of Benedict's, in that the rubber particles from the top fraction are collected and purified by repeated washes with buffer, centrifugation, and re-suspension of the top fraction in a buffer [129]. Her group determined the molecular weight of *in vitro* NR by means of dual-labeled liquid scintillation spectrometry (SS) [27]. By introducing both radioactive IPP monomer and FPP initiator into *in vitro* NR biosynthesis and calculating the incorporation rate of the radioactive materials, an average molecular weight of the newly formed rubber was calculated from the ratio of ^{14}C-labeled monomer to ^{3}H-labeled initiator [139, 166]. It is important to note that the proposed molecular weight calculation assumes that chain growth starts only from the synthetic FPP initiator and the IPP monomer does not add to pre-existing rubber [139, 166]. The reported molecular weights were in the 10^4–10^5 g mol^{-1} range [139, 166], somewhat lower than those reported by Tanaka's and Wititsuwannakul's group.

The *in vitro* systems described above produced NR in milligram amounts because the IPP monomer needs to be synthesized and is fairly unstable. Based on our study of NR biosynthesis over the last decade we have recently discovered that feeding synthetic IP into latex produces NR! More details about this discovery will be reported elsewhere.

In the final part of this chapter we will discuss synthetic cationic polymerizations involving IP, a monomer available from renewable resources.

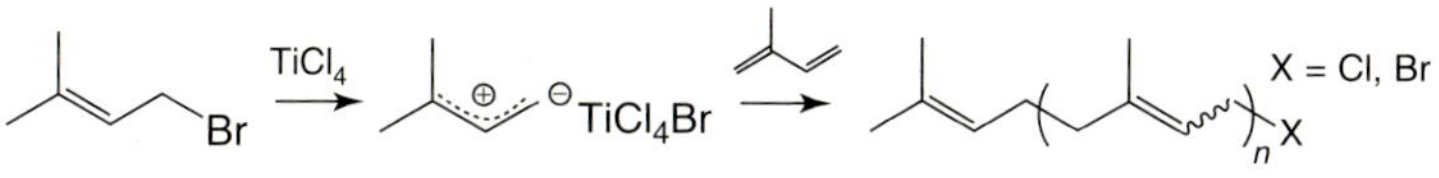

Scheme 14.2 Nature-inspired cationic IP polymerization.

14.5
Green Synthetic Cationic Polymerization and Copolymerization of Isoprene

The cationic polymerization of IP yields low molecular weight products, but interestingly only 1,4-enchainment (mostly trans) is formed [167]. Based on our NR biosynthesis studies, our international research team developed bio-inspired synthetic routes toward the synthesis of polyisoprenes based on *carbocationic polymerization*. We reported IP polymerization initiated by dimethyl allyl bromide (DMABr)–TiCl$_4$ and showed that using this strategy, 1,4-oligoisoprene carrying a dimethyl allyl head group is produced in both cis- and trans-configurations, together with cyclized sequences (Scheme 14.2). However, the copolymerization of IP with isobutylene yields an important industrial product, butyl rubber [1].

Butyl rubber is a random copolymer of isobutylene and a small amount (0.7–2.2 mol%) of IP. The IP incorporates exclusively in 1,4-enchainment, mostly trans- with traces of cis-configuration. This copolymer was the first example of low unsaturation elastomers. The starting point of the development of butyl rubber was a collaborative research effort carried out by Otto (I.G. Farbenindustrie) and Thomas (Standard Oil Development Co.) in 1933–1935. Thomas suggested the use of the AlCl$_3$–ethyl chloride polymerization system, instead of BF$_3$–ethylene. Later ethyl chloride was replaced by methyl chloride as diluent due to its higher stability [168, 169]. Currently the AlCl$_3$-coinitiated copolymerization of isobutylene with IP in methyl chloride is the basic technology for the production of butyl rubber. Although numerous incremental process improvements have been reported since World War II, production of butyl rubber is still based on the original technology: a continuous slurry process at low temperatures (−90 to −100 °C) using methyl chloride as diluent and a solution of AlCl$_3$ dissolved in methyl chloride as the initiating system. Small amounts of water or HCl as cationogen sources are used to promote initiation. The reactor resembles a heat exchanger, with the slurry circulating inside the pipes, which are cooled by boiling ethylene in the jacket. The product is separated from the diluent and unreacted monomers by coagulating it in hot water. The diluent is recycled into the process after purification and drying. The IB and IP monomers and the MeCl diluent currently come from synthetic sources, but they all are available from renewable resources!

IB is currently obtained from the C$_4$ refinery stream.[1] The stream is reacted with water with acid catalysis, the *tert*-butanol forming from IB is separated by distillation, followed by cracking at high temperature into very pure IB. According to some sources *tert*-butanol is available from natural fermentation of waste sludge

---

1) Product of crude oil distillation, containing a mixture of C$_4$ hydrocarbons such as isobutylene, *cis*- and *trans*-2-butenes, 1-butene, and butane.

containing ethers [MBTE (methyl tertiary butyl ether) or EBTE (ethyl tertiary butyl ether)] used as oil additives [170, 171], although may not be biodegradable in contrast with its other isomers.

IP is currently produced from the C_5 feed stream.[2)] It is important to note that the IP is often more expensive than NR, which makes synthetic PIP less attractive. However, IP is also available from renewable resources. Plants release more than 600 kilotons of IP into the atmosphere every year. Genencor and Goodyear have announced the industrial and cost-effective production of bio-IP [172] from 2013.

Thus the monomers could be available from renewable resources, which makes the butyl process somewhat "greener." Another "green" industrial process is the production of polyterpene resins by carbocationic polymerization using IP-based monomers isolated from turpentine oil (α- or β-pinene) or etheral oils of lemon, orange, mandarin, or caraway (*d,l*-limonene or dipentene and *d*-limonene) [173].

In summary, "green" chemistry, including polymer chemistry, is a critical area of research and technology. We all must contribute to the efforts of leaving a "greener" environment to the future generations.

Acknowledgments

This material is based upon work supported by the National Science Foundation under the grants **CHE-0616834 (GOALI), DMR-0509687**, and **#0804878**. We wish to thank the Ohio Board of Regents and the National Science Foundation (**CHE-0341701** and **DMR-0414599**) for funds used to purchase the NMR instrument used in this work. We also wish to thank the National Science Foundation (DMR #0509687) for funds used to upgrade our SEC instrument. The fresh *Hevea* latex was the generous gift of Dr Rogério Manoel Biagi Moreno, Embrapa Agricultural Instrumentation, Saõ Carlos-Saõ Paulo, Brazil. The authors are grateful for the contribution of Dr Lucas Dos Santos.

References

1. Puskas, J.E. and Kaszas, G. (2003) in *Encyclopedia of Polymer Science and Technology*, 3rd edn, vol. 5 (ed. J.I. Kroschwitz), John Wiley & Sons, Inc., Hoboken, pp. 382–418.
2. Klibanov, A.M., Samokhin, G.P., Martinek, K., and Berezin, I.V. (1977) *Biotechnol. Bioeng.*, **19**, 1351–1361.
3. Butler, L.G. (1979) *Enzyme Microb. Technol.*, **1**, 253–259.
4. Faber, K. (2004) *Biotransformations in Organic Chemistry*, 5th edn, Springer-Verlag, New York.
5. Wong, C.H. and Whitesides, G.M. (1994) *Enzymes in Synthetic Organic Chemistry*, Pergamon, Oxford.
6. Koeller, K.M. and Wong, C.H. (2001) *Nature*, **409**, 232–240.
7. International Union of Biochemistry and Molecular Biology (1992) *Enzyme Nomenclature*, Academic Press, San Diego.

2) A mixture of isoprene, cyclopentadiene, piperylene (1,2-pentadiene) along with various mono-olefins.

8. Pannuri, S., DiSanto, R., and Kamat, S. (2004), In: *Kirk-Othmer Encyclopedia of Chemical Technology*, 5th edn, vol. 3 (ed. J.I. Kroschwitz), John Wiley & Sons, Inc., Hoboken, pp. 668–683.
9. Kirk, O., Dahmus, T., Borchert, T.V., Fuglsang, C.C., Olsen, H.S., Hansen, T.T., Lund, H., Schiff, H.E., and Nielsen, N.K. (2004), In: *Kirk-Othmer Encyclopedia of Chemical Technology*, 5th edn, vol. 10 (ed. J.I. Kroschwitz), John Wiley & Sons, Inc., Hoboken, pp. 248–317.
10. Schmid, A., Dordick, J.S., Hauer, B., Kiener, A., Wubbolts, M., and Witholt, B. (2001) *Nature*, **409**, 258–268.
11. Matsumura, S. and Takahashi, J. (1986) *Makromol. Chem. Rapid Commun.*, **7**, 369–373.
12. Gross, R.A., Kumar, A., and Kalra, B. (2001) *Chem. Rev.*, **101**, 2097–2124.
13. Kobayashi, S., Uyama, H., and Kimura, S. (2001) *Chem. Rev.*, **101**, 3793–3818.
14. Kobayashi, S. and Makino, A. (2009) *Chem. Rev.*, **109**, 5288–5353.
15. Sarda, L. and Desnuelle, P. (1958) *Biochim. Biophys. Acta*, **30**, 513–521.
16. Verger, R. (1997) *Trends Biotechnol.*, **15**, 32–38.
17. Rubin, B. (1994) *Nat. Struct. Biol.*, **1**, 568–572.
18. Ransac, S., Carriere, F., Rogalska, E., Verger, R., Marguet, F., Buono, G., Melo, E.P., Cebral, J.M.S., Egloff, M.P.E., van Tilbeurgh, H., and Cambillau, C. (1996), In: *Molecular Dynamics of Biomembranes* (ed. A.F. Op den Kamp), Springer, Heidelberg, pp. 265–304.
19. Uppenberg, J., Oehrner, N., Norin, M., Hult, K., Kleywegt, G.J., Patkar, S., Waagen, V., Anthonsen, T., and Jones, T.A. (1995) *Biochemistry*, **34**, 16838–16851.
20. Jaeger, K.E., Ransac, S., Koch, H.B., Ferrato, F., and Dijkstra, B.W. (1993) *FEBS Lett.*, **332**, 143–149.
21. Blow, D. (1990) *Nature*, **343**, 694–695.
22. Brady, L., Brzozowski, A.M., Derewenda, Z.S., Dodson, E., Dodson, G., Tolley, S., Turkenburg, J.P., Christiansen, L., Huge-Jensen, B., and Norskov, L. (1990) *Nature*, **343**, 767–770.
23. Schrag, J.D., Li, Y., Wu, S., and Cygler, M. (1991) *Nature*, **351**, 761–764.
24. Uppenberg, J., Hansen, M.T., Patkar, S., and Jones, T.A. (1994) *Structure*, **2**, 293–308.
25. Winkler, F.K., D'Arcy, A., and Hunziker, W. (1990) *Nature*, **343**, 771–774.
26. Ollis, D.L., Cheah, E., Cygler, M., Dijkstra, B., Frolow, F., Franken, S.M., Harel, M., Remington, S.J., and Silman, I. (1992) *Protein Eng.*, **5**, 197–211.
27. Dodson, G.G., Lawson, D.M., and Winkler, F.K. (1992) *Faraday Discuss.*, **93**, 95–105.
28. Cygler, M., Grochulski, P., Kazlauskas, R.J., Schrag, J.D., Bouthillier, F., Rubin, B., Serreqi, A.N., and Gupta, A.K. (1994) *J. Am. Chem. Soc.*, **116**, 3180–3186.
29. Bornscheuer, U.T. and Kazlauskas, R.J. (2006) *Hydrolases in Organic Synthesis: Regio- and Stereoselective Biotransformations*, 2nd edn, Wiley-VCH Verlag GmbH, Weinheim.
30. Gotor, V. (2000) *Biocatal. Biotransformation*, **18**, 87–103.
31. Krishna, S.H. and Karanth, N.G. (2002) *Catal. Rev., Sci. Eng.*, **44**, 499–591.
32. Morgan, B., Dodds, D.R., Zaks, A., Andrews, D.R., and Klesse, R. (1997) *J. Org. Chem.*, **62**, 7736–7743.
33. Santaniello, E., Ferraboschi, P., and Grisenti, P. (1993) *Enzyme Microb. Technol.*, **15**, 367–382.
34. Bordusa, F. (2002) *Chem. Rev.*, **102**, 4817–4867.
35. Yadav, G.D. and Trivedi, A.H. (2003) *Enzyme Microb. Technol.*, **32**, 783–789.
36. Weber, H.K., Stecher, H., and Faber, K. (1995) *Biotechnol. Lett.*, **17**, 803–808.
37. Yadav, G.D. and Lathi, P.S. (2005) *J. Mol. Catal. B, Enzym.*, **32**, 107–113.
38. Rizzi, M., Stylos, P., Riek, A., and Reuss, M. (1992) *Enzyme Microb. Technol.*, **14**, 709–714.
39. Martinelle, M. and Hult, K. (1995) *Biochim. Biophys. Acta*, **1251**, 191–197.
40. Laane, C., Boeren, S., Hilhorst, R., and Veeger, C. (1987), In: *Biocatalysis in Organic Media* (eds C., Laane, J. Tramper, and M.D. Lilly), Elsevier, Amsterdam, pp. 65–84.

41. Laane, C., Boeren, S., Vos, K., and Veeger, C. (1987) *Biotechnol. Bioeng.*, **30**, 81–87.
42. Danieli, B., Luisetti, M., Sampognaro, G., Carrea, G., and Riva, S. (1997) *J. Mol. Catal. B, Enzym.*, **3**, 193–201.
43. Valivety, R.H., Brown, L., Halling, P.J., Johnston, G.A., and Suckling, C.J. (1990), In: *Opportunities in Biotransformations* (eds L.G. Copping, R.E. Martin, J.A. Pickett, C. Bucke, and A.W. Bunch), Elsevier, London, pp. 81–87.
44. Zaks, A. and Klibanov, A.M. (1985) *Proc. Natl. Acad. Sci. U.S.A.*, **82**, 3192–3196.
45. Zaks, A. and Klibanov, A.M. (1988) *J. Biol. Chem.*, **263**, 3194–3201.
46. Xu, K. and Klibanov, A.M. (1996) *J. Am. Chem. Soc.*, **118**, 9815–9819.
47. Griebenow, K. and Klibanov, A.M. (1995) *Proc. Natl. Acad. Sci. U.S.A.*, **92**, 10969–10976.
48. Debulis, K. and Klibanov, A.M. (1993) *Biotechnol. Bioeng.*, **41**, 566–571.
49. Russell, A.J. and Klibanov, A.M. (1988) *J. Biol. Chem.*, **263**, 11624–11626.
50. Khmelnitsky, Y.L., Welch, S.H., Clark, D.S., and Dordick, J.S. (1994) *J. Am. Chem. Soc.*, **116**, 2647–2648.
51. Broos, J., Kakodinskaya, I.K., Engbersen, J.F.J., Verboom, W., and Reinhoudt, D.N. (1995) *J. Chem. Soc. Chem. Commun.*, 255–256.
52. Mateo, C., Palomo, J.M., Fernandez-Lorente, G., Guisan, J.M., and Fernandez-Lafuente, R. (2007) *Enzyme Microb. Technol.*, **40**, 1451–1463.
53. Kirk, O. and Christensen, M.W. (2002) *Org. Process Res. Dev.*, **6**, 446–451.
54. Kato, T. and Horikoshi, K. (1984) *Biotechnol. Bioeng.*, **26**, 595–598.
55. Wiegel, J. and Dykstra, M. (1984) *Appl. Microbiol. Biotechnol.*, **20**, 59–65.
56. Miyawaki, O. and Wingard, L.B. (1984) *Biotechnol. Bioeng.*, **26**, 1364–1371.
57. Bianchi, D., Cesti, P., and Battistel, E. (1988) *J. Org. Chem.*, **53**, 5531–5534.
58. Marek, M., Valentova, O., and Kas, J. (1984) *Biotechnol. Bioeng.*, **26**, 1223–1226.
59. Cannon, J.J., Chen, L.F., Flickinger, M.C., and Tsao, G.T. (1984) *Biotechnol. Bioeng.*, **26**, 167–173.
60. Weetall, H.H. and Mason, R.D. (1973) *Biotechnol. Bioeng.*, **15**, 455–466.
61. Khan, S.S. and Siddiqi, A.M. (1985) *Biotechnol. Bioeng.*, **27**, 415–419.
62. Kaul, R., D'Souza, S.F., and Nadkarni, G.B. (1984) *Biotechnol. Bioeng.*, **26**, 901–904.
63. Karube, I., Kawarai, M., Matsuoka, H., and Suzuki, S. (1985) *Appl. Microbiol. Biotechnol.*, **21**, 270–272.
64. Qureshi, N. and Tamhane, D.V. (1985) *Appl. Microbiol. Biotechnol.*, **21**, 280–281.
65. Umemura, I., Takamatsu, S., Sato, T., Tosa, T., and Chibata, I. (1984) *Appl. Microbiol. Biotechnol.*, **20**, 291–295.
66. Fukui, S. and Tanaka, A. (1984) *Adv. Biochem. Eng. Biotechnol.*, **29**, 1–33.
67. Bednarski, M.D., Chenault, H.K., Simon, E.S., and Whitesides, G.M. (1987) *J. Am. Chem. Soc.*, **109**, 1283–1285.
68. Kragl, U., Vasic-Racki, D., and Wandrey, C. (1993) *Indian J. Chem. B*, **32B**, 103–117.
69. Luisi, P.L. (1985) *Angew. Chem. Int. Ed. Engl.*, **24**, 439–450.
70. Hoegh, I., Patkar, S., Halkier, T., and Hansen, M.T. (1995) *Can. J. Bot.*, **73** (Suppl. 1), S869–S875.
71. Anderson, E.M., Larsson, K.M., and Kirk, O. (1998) *Biocatal. Biotransfor.*, **16**, 181–204.
72. Arroyo, M. and Sinisterra, J.V. (1994) *J. Org. Chem.*, **59**, 4410–4417.
73. Woudenberg-van Oosterom, M., Van Rantwijk, F., and Sheldon, R.A. (1996) *Biotechnol. Bioeng.*, **49**, 328–333.
74. Nakaoki, T., Mei, Y., Miller, L.M., Kumar, A., Kalra, B., Miller, M.E., Kirk, O., Christensen, M., and Gross, R.A. (2005) *Ind. Biotechnol.*, **1**, 126–134.
75. Novozyme® 435 product information from Novozymes A/S. http://www.novozymes.com/en (Accessed Apr 9, 2009).
76. Kirchner, G., Scollar, M.P., and Klibanov, A.M. (1985) *J. Am. Chem. Soc.*, **107**, 7072–7076.
77. Ghogare, A. and Kumar, G.S. (1989) *J. Chem. Soc. Chem. Commun.*, 1533–1535.

78. Oehrner, N., Martinelle, M., Mattson, A., Norin, T., and Hult, K. (1994) *Biocatal.*, **9**, 105–114.
79. Wang, Y.F., Lalonde, J.J., Momongan, M., Bergbreiter, D.E., and Wong, C.H. (1988) *J. Am. Chem. Soc.*, **110**, 7200–7205.
80. Faber, K. and Riva, S. (1992) *Synthesis*, 895–910.
81. Sen, M.Y. (2009) Green Polymer Chemistry: Functionalization of Polymers Using Enzymatic Catalysis PhD dissertation. The University of Akron.
82. Garcia-Alles, L.F. and Gotor, V. (1998) *Biotechnol. Bioeng.*, **59**, 684–694.
83. Uyama, H. and Kobayashi, S. (1993) *Chem. Lett.*, 1149–1150.
84. Knani, D., Gutman, A.L., and Kohn, D.H. (1993) *J. Polym. Sci. A Polym. Chem.*, **31**, 1221–1232.
85. Albertsson, A. and Srivastava, R.K. (2008) *Adv. Drug Deliv. Rev.*, **60**, 1077–1093.
86. Varma, I.K., Albertsson, A.C., Rajkhowa, R., and Srivastava, R.K. (2005) *Prog. Polym. Sci.*, **30**, 949–981.
87. Matsumura, S. (2006) *Adv. Polym. Sci.*, **194**, 95–132.
88. Kobayashi, S. (2009) *Macromol. Rapid Commun.*, **30**, 237–266.
89. Uyama, H., Takeya, K., and Kobayashi, S. (1995) *Bull. Chem. Soc. Jpn.*, **68**, 56–61.
90. Mei, Y., Kumar, A., and Gross, R. (2003) *Macromolecules*, **36**, 5530–5536.
91. Duda, A., Kowalski, A., Penczek, S., Uyama, H., and Kobayashi, S. (2002) *Macromolecules*, **35**, 4266–4270.
92. Namekawa, S., Suda, S., Uyama, H., and Kobayashi, S. (1999) *Int. J. Biol. Macromol.*, **25**, 145–151.
93. Kumar, A., Kalra, B., Dekhterman, A., and Gross, R.A. (2000) *Macromolecules*, **33**, 6303–6309.
94. Van der Mee, L., Helmich, F., De Bruijn, R., Vekemans, J.A.J.M., Palmans, A.R.A., and Meijer, E.W. (2006) *Macromolecules*, **39**, 5021–5027.
95. Kobayashi, S. (2006) *Macromol. Symp.*, **240**, 178–185.
96. Cordova, A., Iversen, T., and Hult, K. (1999) *Polymer*, **40**, 6709–6721.
97. Cordova, A., Iversen, T., and Hult, K. (1998) *Macromolecules*, **31**, 1040–1045.
98. Cordova, A., Hult, A., Hult, K., Ihre, H., Iversen, T., and Malmstroem, E. (1998) *J. Am. Chem. Soc.*, **120**, 13521–13522.
99. Hedfors, C., Oestmark, E., Malmstroem, E., Hult, K., and Martinelle, M. (2005) *Macromolecules*, **38**, 647–649.
100. Sen, M.Y., Puskas, J.E., Ummadisetty, S., and Kennedy, J.P. (2008) *Macromol. Rapid Commun.*, **29**, 1598–1602.
101. Puskas, J.E., Sen, M.Y., and Kasper, J.R. (2008) *J. Polym. Sci. A Polym. Chem.*, **46**, 3024–3028.
102. Puskas, J.E., Sen, M.Y., and Seo, K.S. (2009) *J. Polym. Sci. A Polym. Chem.*, **47**, 2959–2976.
103. Vosmann, K., Wiege, B., Weitkamp, P., and Weber, N. (2008) *Appl. Microbiol. Biotechnol.*, **80**, 29–36.
104. Weitkamp, P., Weber, N., and Vosmann, K. (2008) *J. Agric. Food Chem.*, **56**, 5083–5090.
105. Weitkamp, P., Vosmann, K., and Weber, N. (2006) *J. Agric. Food Chem.*, **54**, 7062–7068.
106. Chang, C.S. and Wu, P.L. (2007) *J. Biotechnol.*, **127**, 694–702.
107. Kennedy, J.P., Chang, V.S.C., Smith, R.A., and Ivan, B. (1979) *Polym. Bull.*, **1**, 575–580.
108. Ivan, B., Kennedy, J.P., and Chang, V.S.C. (1980) *J. Polym. Sci. Polym. Chem. Ed.*, **18**, 3177–3191.
109. Hayat-Soytas, S. (2009) Living Carbocationic Polymerization of Isobutylene by Expoxide/Lewis Acid Systems: The Mechanism of Initiation PhD dissertation. The University of Akron.
110. Kaszas, G., Gyor, M., Kennedy, J.P., and Tudos, F. (1983) *J. Macromol. Sci. Chem.*, **A18**, 1367–1382.
111. U.S. Department of Commerce. Office of Trade and Industry Information (1998) Trade Stat Express, Washington, DC.
112. Cornish, K. (1996) US Patent 5, 580,942.
113. Siler, D.J., Cornish, K., and Hamilton, R.G. (1996) *J. Allergy Clin. Immunol.*, **98**, 895–902.
114. Cornish, K., Wood, D.F., and Windle, J.J. (1999) *Planta*, **210**, 85–96.

115. Buchanan, R.A., Cull, I.M., Otey, F.H., and Russell, C.R. (1978) *Hydrocarbon and Rubber Producing Crops Evaluation of 100 US Plant Species*, USDA, ARS, Washington, DC.
116. Mooibroek, H. and Cornish, K. (2000) *Appl. Microbiol. Biotechnol.*, **53**, 355–365.
117. Rao, P.S., Saraswathyamma, C.K., and Sethuraj, M.R. (1998) *Agric. For. Meteorol.*, **90**, 235–245.
118. Subramaniam, A. (1995) *Immunol. Allergy Clin. North Am.*, **15**, 1–20.
119. Tanaka, Y. (2001) *Rubber Chem. Technol.*, **74**, 355–375.
120. Cornish, K. (2001) *Nat. Prod. Rep.*, **18**, 182–189.
121. Asawatreratanakul, K., Zhang, Y.W., Wititsuwannakul, D., Wititsuwannakul, R., Takahashi, S., Rattanapittayaporn, A., and Koyama, T. (2003) *Eur. J. Biochem.*, **270**, 4671–4680.
122. Murakami, S., Senoo, K., Toki, S., and Kohjiya, S. (2002) *Polymer*, **43**, 2117–2120.
123. Makani, S., Brigodiot, M., Marechal, E., Dawans, F., and Durand, J.P. (1984) *J. Appl. Polym. Sci.*, **29**, 4081–4089.
124. Archer, B.L. and Audley, B.G. (1967) *Adv. Enzymol. Relat. Areas Mol. Biol.*, **29**, 221–257.
125. Lynen, F. and Henning, U. (1960) *Angew. Chem.*, **72**, 820–829.
126. Poulter, C.D., Satterwhite, D.M., and Rilling, H.C. (1976) *J. Am. Chem. Soc.*, **98**, 3376–3377.
127. Tanaka, Y., Gik-Hwee, E., Ohya, N., Nishiyama, N., Tangpakdee, J., Kawahara, S., and Wititsuwannakul, R. (1996) *Phytochemistry*, **41**, 1501–1505.
128. Benedict, C.R., Greer, P.J., and Foster, M.A. (2008) *Ind. Crops Prod.*, **27**, 225–235.
129. Cornish, K. and Backhaus, R.A. (1990) *Phytochemistry*, **29**, 3809–3813.
130. Cornish, K. (2001) *Phytochemistry*, **57**, 1123–1134.
131. Poulter, C.D. and Rilling, H.C. (1978) *Acc. Chem. Res.*, **11**, 307–313.
132. Kellogg, B.A. and Poulter, C.D. (1997) *Curr. Opin. Chem. Biol.*, **1**, 570–578.
133. Dubey, V.S., Bhalla, R., and Luthra, R. (2003) *J. Biosci.*, **28**, 637–646.
134. Chappell, J. (1995) *Annu. Rev. Plant Biol.*, **46**, 521–547.
135. Lichtenthaler, H.K. (1999) *Annu. Rev. Plant Biol.*, **50**, 47–65.
136. Cornish, K. (1993) *Eur. J. Biochem.*, **218**, 267–271.
137. Scott, D.J., da Costa, B.M.T., Espy, S.C., Keasling, J.D., and Cornish, K. (2003) *Phytochemistry*, **64**, 123–134.
138. da Costa, B.M.T., Keasling, J.D., and Cornish, K. (2005) *Biomacromolecules*, **6**, 279–289.
139. Cornish, K., Castillon, J., and Scott, D.J. (2000) *Biomacromolecules*, **1**, 632–641.
140. Tanaka, Y. (1989) *Prog. Polym. Sci.*, **14**, 339–371.
141. Bhowmick, A.K. and Stephens, H.L. (2000) *Handbook of Elastomers*, CRC Press, Boca Raton.
142. Pan, Z., Herickhoff, L., and Backhaus, R.A. (1996) *Arch. Biochem. Biophys.*, **332**, 196–204.
143. Wilkin, D.J., Kutsunai, E.P.A., and Edwards, P.A. (1990) *J. Biol. Chem.*, **265**, 4607–4614.
144. Koyama, T., Obata, S., Osabe, M., Takeshita, A., Yokoyama, K., and Uchida, M. (1993) *J. Biochem.*, **113**, 355–363.
145. Adiwilaga, K. and Kush, A. (1996) *Plant Mol. Biol.*, **30**, 935–946.
146. Zhu, X.F., Suzuki, K., Saito, T., Okada, K., Tanaka, K., Nakagawa, T., Matsuda, H., and Kawamukai, M. (1997) *Plant Mol. Biol.*, **35**, 331–341.
147. Math, S.K., Hearst, J.E., and Poulter, C.D. (1992) *Proc. Natl. Acad. Sci. U.S.A.*, **89**, 6761–6764.
148. Ohnuma, S., Suzuki, M., and Nishino, T. (1994) *J. Biol. Chem.*, **269**, 14792–14797.
149. Ohnuma, S., Narita, K., Nakazawa, T., Ishida, C., Takeuchi, Y., Ohto, C., and Nishino, T. (1996) *J. Biol. Chem.*, **271**, 30748–30754.
150. Tarshis, L.C., Yan, M., Poulter, C.D., and Sacchettini, J.C. (1994) *Biochemistry*, **33**, 10871–10877.
151. Tarshis, L.C., Proteau, P.J., Kellogg, B.A., Sacchettini, J.C., and Poulter, C.D. (1996) *Proc. Natl. Acad. Sci. U.S.A.*, **93**, 15018–15023.

152. Ohnuma, S., Nakazawa, T., Hemmi, H., Hallberg, A.M., Koyama, T., Ogura, K., and Nishino, T. (1996) *J. Biol. Chem.*, **271**, 10087–10095.
153. Shimizu, N., Koyama, T., and Ogura, K. (1998) *J. Biol. Chem.*, **273**, 19476–19481.
154. Apfel, C.M., Takacs, B., Fountoulakis, M., Stieger, M., and Keck, W. (1999) *J. Bacteriol.*, **181**, 483–492.
155. Sato, M., Sato, K., Nishikawa, S., Hirata, A., Kato, J., and Nakano, A. (1999) *Mol. Cell. Biol.*, **19**, 471–483.
156. Xie, W., McMahan, C.M., DeGraw, A.J., Distefano, M.D., Cornish, K., Whalen, M.C., and Shintani, D.K. (2008) *Phytochemistry*, **69**, 2539–2545.
157. Liang, P.H., Ko, T.P., and Wang, A.H.J. (2002) *Eur. J. Biochem.*, **269**, 3339–3354.
158. Satoh, K., Kamigaito, M., and Sawamoto, M. (2000) *Macromolecules*, **33** (13), 4660–4666.
159. Puskas, J.E., Gautriaud, E., Deffieux, A., and Kennedy, J.P. (2006) *Prog. Polym. Sci.*, **31**, 533–548.
160. Benedict, C.R., Madhavan, S., Greenblatt, G.A., Venkatachalam, K.V., and Foster, M.A. (1990) *Plant Physiol.*, **92**, 816–821.
161. Mau, C.J.D., Scott, D.J., and Cornish, K. (2000) *Phytochem. Anal.*, **11**, 356–361.
162. Tangpakdee, J., Tanaka, Y., Ogura, K., Koyama, T., Wititsuwannakul, R., and Wititsuwannakul, D. (1997) *Phytochemistry*, **45**, 269–274.
163. Tangpakdee, J., Tanaka, Y., Ogura, K., Koyama, T., Wititsuwannakul, R., Wititsuwannakul, D., and Asawatreratanakul, K. (1997) *Phytochemistry*, **45**, 261–267.
164. Tangpakdee, J., Tanaka, Y., Ogura, K., Koyama, T., Wititsuwannakul, R., and Chareonthiphakorn, N. (1997) *Phytochemistry*, **45**, 275–281.
165. Wititsuwannakul, D., Rattanapittayaporn, A., and Wititsuwannakul, R. (2003) *J. Appl. Polym. Sci.*, **87**, 90–96.
166. Castillon, J. and Cornish, K. (1999) *Phytochemistry*, **51**, 43–51.
167. Gaylord, N.G. (1970) *Pure Appl. Chem.*, **23**, 305–326.
168. Kennedy, J.P. and Ivan, B. (1992) *Designed Polymers by Carbocationic Macromolecular Engineering: Theory and Practice*, Hanser, Munich.
169. Puskas, J.E. and Kaszas, G. (1996) *Rubber Chem. Technol.*, **69**, 462–475.
170. Mormile, M.R., Liu, S., and Suflita, J.M. (1994) *Environ. Sci. Technol.*, **28**, 1727–1732.
171. Finneran, K.T. and Lovley, D.R. (2001) *Environ. Sci. Technol.*, **35**, 1785–1790.
172. Fall, R.R., Kuzma, J., and Nemecek-Marshall, M. (1998) US Patent 5, 849,970.
173. Puskas, J.E. and Kaszas, G. (2000) *Prog. Polym. Sci.*, **25**, 403–452.

Further Reading

Burger, K. and Smit, H.P. (1997) *The Natural Rubber Market: Review, Analysis, Policies and Outlook*, Woodhead Publishing, Cambridge.

Elias, H.G. (1997) *An Introduction to Polymer Science*, Wiley-VCH Verlag GmbH, Weinheim.

Fisher, J.C. and Pry, R.H. (1971) *Technol. Forecast. Soc. Change*, **3**, 75–88.

Hosler, D., Burkett, S.L., and Tarkanian, M.J. (1999) *Science*, **284**, 1988–1991.

Morton, M. (1981) *J. Macromol. Sci. A Pure Appl. Chem.*, **15**, 1289–1302.

Slack, C. (2003) *Noble Obsession: Charles Goodyear, Thomas Hancock, and the Race to Unlock the Greatest Industrial Secret of the Nineteenth Century*, Hyperion Books, New York.

Index

a

ABA triblock copolymer 19
acid-catalyzed condensation 33
acid-catalyzed polycondensation
– furfuryl alcohol (FA) 37
acid-catalyzed polymerization
– furfuryl alcohol (FA) 36
acrolein 62
– hydration 76
acrylated methyl oleate (AMO) 23
acrylic acid 62
acrylic resins 14
acrylonitrile (AN) 71, 243
activity cofactor 334
acyclic diene metathesis (ADMET) polymerization 15ff., 130f.
acyclic triene metathesis (ATMET) polymerization 22
additive 107
adhesives
– hot-melt 113
– terpene based 113
adsorption 320
alcohols 175
aliphatic polycarbonate
– biodegradable 172
alkene
– cyclic 117
alkene monomer
– ring-opening metathesis polymerization (ROMP) 118
alkyd resins 12
alkylaluminoxane coinitiator 116
allylic pyrophosphate (APP) 334
aluminoxanes 111
aluminum
– $AlCl_3$–$SbCl_3$ system 111
– $AlCl_3$–$SbCl_3$–H_2O initiating system 108
– based catalyst 63ff.
– bimetallic aluminum salen derivative 170
– $Cu/ZnO/Al_2O_3$ catalyst 75
– dimeric aluminum salen based complex 183
– monomeric aluminum salen based complex 183
– (salen)Al(III)Cl system 184
– (salen/salan)aluminum 207
– tetraphenylporphyrin–aluminum compound 183
– Zn–Al mixed oxide 60
amino-alkoxy(bisphenolate)yttrium 207
4-aminothiophenol 302
ammoxidation 71
amphiphilic monomer 78
aniline 177
anionic initiator 176
anionic ring-opening polymerization 176
antimony halide 107
– $AlCl_3$–$SbCl_3$ system 111
– $AlCl_3$–$SbCl_3$–H_2O initiating system 108
aqueous media
– polymer materials 140
aqueous polymerization 103
aramides 42
arm-first method 244f.
aromatic component 265
atom transfer radical polymerization (ATRP) 129ff., 238, 249
azide–alkyne ("click") cycloaddition
– copper-catalyzed 138
5-azido-1,4:3,6-dianhydro-L-iditol 230
5-azido-2-O-benzyl-1,4:3,6-dianhydro-L-iditol 230

b

BEMP 210
5-benzyloxytrimethylene carbonate 179
5,6-benzo-2-methylene-1,3-dioxepane 243
beta (β)-silicon effect 110
5,5-(bicyclo[2.2.1]hept-2-en-5,5-ylidene)-1,3-dioxan-2-one (NBC) 177
bilirubin oxidase 298
bio-synthetic hybrid material 129ff.
– controlled and living polymerization in water 129ff.
– methods and applications 129ff.
biocatalyst 167
biocompatible catalytic process 166
biocompatible materials 222
biocompatible metal
– catalyst 182
bioconjugate
– synthetic–natural polymer bioconjugate 247
biodegradable polymer
– CRP 238
biodegradable/renewable materials 214
biodegradability 237
biodegradation process
– thermoplastic elastomer 194
biodiesel 144
– composition 147
– effect of feedstock 155
– polymer production 145
– polymerization solvent 143ff.
– sustainable solution polymerization 143ff.
biohybrids 286
biomass 221
– glycidol 65
– lifecycle of buildup and degradation 222
biomaterial 221
– liquid crystalline (LC) copolymer 274
biomonomer 276
– high-performance polymer from phenolic biomonomer 265ff.
bioplastics 282
biotechnology
– polymer functionalization 313
biotransformation 15, 313f.
– catalyst 313
[N,N'-bis(3,5-di-*tert*-butylsalicylidene)-1,2-ethylenediimine]Ca(II) 192
1,3-bis(2,6-diisopropylphenyl)-imidazol-2-ylidene 194
exo,exo-2,3-bis(methoxymethyl)-7-oxanorbornene 132
5,5′-bis(oxepane-2-one) 213
bis(phenolato)scandium complex 207

(1R, 2R)-N,N'-bis(salicylidene)-1,2-diaminocyclohexane 183
N,N'-bis(salicylidene)-1,2-phenylenediimine-aluminum(III) chloride 186
bisfuranyl monomer 40
blending partner 213
block copolymer
– synthesis 305
bromoacrylated methyl oleate (BAMO) 23
2-(2′-bromoisobutyryloxy)ethyl 2″-methacryloyloxyethyl disulfide 250
2-(2′-bromopropionyloxy)ethyl acrylate 250
bulk erosion 196
bulkier counteranion 108
butadiene 81f.
butyl acrylate (BA) 146
butyl rubber 341
α-(*tert*-butyl)-ω-(2-chloro-2-methylpropyl)polyisobutylene 329
N'-*tert*-butyl-N,N,N',N',N'',N'-hexamethylphosphorimidic triamide (P$_1$-t-Bu) 210
1-butyl-3-methylimidazolium hexafluorophosphate [BMIM][PF6] 303
1-butyl-3-methylimidazolium tetrafluoroborate [BMIM][BF$_4$] 303
2-*tert*-butylimino-2-diethylamino-1,3-dimethylperhydro-1,3,2-diazaphosphorine (BEMP) 210
butyric acid ester 317

c

c-kenaf 287
caffeic acid (3,4-dihydroxycinnamic acid, DHCA) 267
calcium
– Ca(II) salen complex 192
– tris(pyrazolyl)calcium 207
camphenic carbenium ion 99f.
Candida antarctica 16, 179ff.
– lipase B (CALB) 315ff.
– transesterification reaction 323f.
Candida rugosa 315f.
Candida tropicalis 16
caprolactone (CL) 294, 303
capsule
– nano-sized 243
carbanion initiator 129
carbocationic mechanism 337
carbocationic polymerization 111, 341
carbohydrate
– fermentable 168
carbon dioxide
– supercritical (scCO$_2$) 304

carbonatation
– direct   61
carbonates
– cyclic organic   59
– renewable six-membered cyclic   165ff.
– six-membered cyclic   61, 165ff.
carboxylation   170
carboxylic acid
– functional   65
catalysis   4, 165ff.
– enzymatic   178, 313
catalyst
– biocompatible metal   182
– biotransformation   313
– cationic-based   174
– chemical stability   133
– Cu/ZnO/Al$_2$O$_3$ 75
– CuO/SiO$_2$ 74
– group 1 and 2 based   191
– group 4–12 based   186
– group 13 based catalyst   182
– group 14 based catalyst   182
– homogeneous   167
– lanthanide-based   190
– lipophilic   133
– metallo-organic   207
– nanosized gold   67
– non-toxic   191
– polymerization by chemical catalyst   326
– Pt-Au/C   68
– Pt/NaY zeolite   74
– ROP of six-membered cyclic carbonates   182
– rubber transferase   335
– second generation   209
– single-component   190
– VSb-based   71
– water soluble ROMP catalyst   133
catalytic process
– biocompatible   166
– green catalyst method   172
– importance   4
– selective conversion of glycerol into functional monomers   57ff.
catalytic system
– biocompatible   173
cationic olefin polymerization   93ff.
– concept   93
– initiation   93ff.
– propagation   95
cationic polymerization   33, 96ff., 313ff., 341
– dipentene   104
– green   313ff., 330, 341
– termination   98
cationic ring-opening polymerization   173
ceiling temperature   171
cell compatibility   273
cell-adhesion property
– homopolymer   273
chain transfer   96ff.
– facile   206
chain transfer agent   116
– renewable   124
chain transfer constant   122f.
chain transfer to solvent constant   149
chain transfer to solvent rate constant   150
chain-length distribution (CLD) method   149
chitosan   40
ω-chloro fatty acid esters   16
cholic acid (CA)   276
– co-monomer   276
– P(4HCA-co-CA)   277
chromium
– (salen)Cr(III)Cl complex   186
Cleland's reagent   240
click cycloaddition   135ff.
– copper-catalyzed azide–alkyne cycloaddition   138
clicked pyridine ligand   135
cloud point   113
co-monomer
– lithocholic acid   274
coating
– hot-melt   113
coinitiator   95
coinitiating system   337
comb-shaped copolymer   50
commercially viable   115
composite materials   14
condensation
– acid-catalyzed   33
condensation polymerization   298
conductive polymer   301
conjugated polymers   39
controlled/living radical polymerization (CRP)   235ff.
– biodegradable polymer   238
coordination–insertion polymerization   205
coordination–insertion ring-opening polymerization   181
copolymer
– graft   306
– hydrolytic behavior   281
copolymerization   31, 274
– isoprene   341

Index

copper
- azide–alkyne ("click") cycloaddition 138
- Cu/ZnO/Al$_2$O$_3$ catalyst 75
- CuO/SiO$_2$ catalyst 74

core first method 244
cosubstrate 334
coumarate
- homopolymer 267
- phytomonomer 266

m-coumaric acid (3-hydroxycinnamic acid, 3HCA) 267
p-coumaric acid (4-hydroxycinnamic acid, 4HCA) 266
counteranion
- bulkier 108
- nucleophilic 110
coupling
- reversible 52
covalent attachment 321
Cp*TiMe$_3$–B(C$_6$F$_5$)$_3$ 101
cross-linked degradable polymer 252
cross-linked enzyme aggregate (CLEA) 304
cross-linked enzyme crystal (CLEC) 304
cross-linked materials 12
cross-linking 52, 321
cross-metathesis 16
- fatty alcohol with methyl acrylate 17
- sequence 17
cyclic alkene 117
cyclic carbonate
- renewable six-membered 165ff.
- six-membered 61, 165ff.
cyclic olefin polymerization 130f.
cyclic organic carbonate
- transcarbonatation 59
[2+2] cycloaddition 131
β-cyclodextrin 82
cyclooctadiene (COD) 131
cyclooctene (COE) 131

d

decarboxylation
- side reaction product 174
degenerative transfer polymerization 238
degradability 237
degradable functional group 239
degradable polymer
- CRP 238
- hyperbranched 250
- star polymer 244
degradable polymeric segment 242
degree of polymerization (DP) 41, 149
Degussa–Dupont process 76, 166

dehydration
- glycerol 62
dendrimer 327
- hexahydroxyl-functional 327
1-deoxy-D-xylulose-5-phosphate 333
depolymerization 46f.
deprotection 49
2,3-dialkyl-1,4-anhydro-D-erythritol 225
diamino di- and monoanhydroalditols 227f.
1,4:3,6-dianhydrosorbitol 224
1,4-diazabicyclo[2.2.2]octane (Dabco) 177
1,8-diazabicyclo[5.4.0]undec-7-ene (DBU) 177, 194, 209
dicumyl peroxide (DCP) 213
dicyclopentadiene (DCPD) 119f.
Diels–Alder polycondensation
- non-linear 51
- reversible 48ff.
Diels–Alder polymerization
- non-linear 49
Diels–Alder reaction 45ff.
- furan polymer 45
diethyl adipate 303
diethyl sebacate 303
diethylcarbonate (DEC) 171
difuran diacid monomer 42
difuran monomer 34
3,4-dihydroxycinnamic acid (DHCA) 267
diisocyanates 213
1,3-diisopropyl-4,5-dimethyl-imidazol-2-ylidene 194
(β-diketiminatato)zinc 207
1,3-dimethyl-2-imidazolidinone (DMI) 77
5,5-dimethyl-1,3-dioxan-2-one (DTC) 173
1,1-dimethylallyl pyrophosphate (DMAPP) 332ff.
trans-1,1-dimethylallylic initiator 338
2-(N,N-dimethylamino)ethyl methacrylate (DMAEMA) 243
4-dimethylaminopyridine (DMAP) 177, 208
N,N-dimethylaniline 177
dimethylcarbonate (DMC) 169
- preparation from renewable resources 169
2,2-dimethyltrimethylene carbonate 183
2-(2,4-dinitrophenylthio)ethyl 2-bromoisobutyrate 241
1,3-dioxan-2-one 177
dip-pen nanolithography (DPN) patterned reactive monomer 302
dipentene 92, 104f., 342
- cationic polymerization 104
disiloxane 109
disulfide 239

disulfide cross-linked gel   252
– polyMMA-based   253
drug carrier   235
– polymeric   236
drug-delivery vehicle   53
drying alkyd resin   13
drying oils   12

e

economical green production   115
E-factor   5
elastomer   113
– biodegradation process of thermoplastic elastomer   194
– non-toxic thermoplastic   172
electrophilic substitution   33
elongation   336
emulsion
– inverse   306
emulsion polymerization   301
– styrene   301
enchainment
– probability of heterotactic enchainment   208
– probability of isotactic enchainment   208
ene–yne cross-metathesis   17
entrapment   321
environmentally benign processes   3
enzymatic catalysis   178, 313
enzymatic grafting   296
enzymatic polycondensation   296
enzymatic polymer synthesis   291ff.
– green chemistry   291ff.
enzymatic polymerization   294, 302f.
– green media   303
enzymatic ring-opening polymerization   178
enzymatic template synthesis   302
enzyme
– activity   320
– classification   314
– cross-linked enzyme aggregate (CLEA)   304
– cross-linked enzyme crystal (CLEC)   304
– immobilization   320
– polymer functionalization   328
enzyme-activated monomer (EAM)   179
epoxidized plant oil derivative   13
epoxy resins   13
erosion
– bulk   194
– surface   194
esterification
– acid-catalyzed   79

ethers
– THF   223
– dianhydrosorbitol   224
ethyl *tertiary* butyl ether (EBTE)   342
ethylene
– polymerization   122
ethylene oxide
– hydroformylation   76
exo-olefinic group   105

f

FA, *see* furfuryl alcohol
FA–lignin thermoset   38
farnesyl pyrophosphate (farnesyl diphosphate, FPP)   334ff.
– synthase   336
fatty acid   11
– hydroxylated   78ff.
– synthesis of ω-functionalized fatty acid   16
fatty acid alkyl ester (FAAE)   145ff.
– composition   147
fatty acid methyl ester (FAME)   144ff.
– effect of FAME solvent on polymerization kinetics   148
– solvent   158f.
fatty alcohol
– cross-metathesis with methyl acrylate   17
fermentable carbohydrate   168
ferulic acid (3-methoxy-4-hydroxycinnamic acid, MHCA)   267
fixed-bed reactor   75
floor temperature   171
Flory–Stockmayer equation   50
2-formylfuran (F)   30
free radical polymerization   148
– solvent effect   149
Friedel–Crafts alkylation   96
fully furanic polyamide   43
fully furanic polyurethane   44
functional group
– degradable   239
– polymer   239
functionalization
– preformed polymer   328
– protein   136
furan   29ff.
– progenitors of polymer siblings   29ff.
furan compound
– first-generation   29f.
furan dicarboxylic acid (FDCA)   35ff.
furan macromolecular synthesis   46
furan monomer   29f.
– synthesis   31
furan polyamide   42

furan polyester   41
furan polymer   45
– Diels–Alder reaction   45
furan–aliphatic polyurethane   44
furan–aromatic polyurethane   44
furfural (F)   30
– derivative   30
– monomer   31
furfuryl alcohol (FA)   30
– acid-catalyzed polycondensation   37
– acid-catalyzed polymerization   36
– FA–lignin thermoset   38
– polymerization   37
furyl oxirane (FO)   45

g

gas phase reaction
– glycerol   75
geranyl pyrophosphate (geranyl diphosphate, GPP)   334ff.
geranylgeranyl pyrophosphate (geranylgeranyl diphosphate, GGPP)   334ff.
– synthase   336
giant amphiphiles   140
glass transition temperature   41
glyceric acid   65f.
glycerol   57ff., 167
– acid-catalyzed esterification with juniperic acid   79
– catalytic oxidation to glyceric acid   65f.
– conversion into acrolein/acrylic acid   62
– conversion into acrylonitrile   71
– conversion into glycidol   63f.
– conversion into 1,3-propanediol   76f.
– conversion into propylene glycol   72ff.
– dehydration   62
– direct carbonatation   61
– oxidation to functional carboxylic acid   65
– oxidation/anionic polymerization   69
– oxidative-assisted polymerization   68
– oxycarbonylation   61
– selective esterification with hydroxylated fatty acids   80
– selective conversion into functional monomers   57ff.
– selective coupling with functional monomers   78
– selective oxidation   68
– telomerization with butadiene   81f.
glycerol carbonate   58f.
glycerolysis
– urea   60
glycidol   63f.
– biomass   65

glycosilicone   230
glycosilicon-polyamide type structure   232
gold catalyst
– nanosized   67
graft copolymer   294, 306
– (bio)degradable backbone   246
graft polymer (polymer brush)
– degradable   245
graft-shaped copolymer   50
grafting
– enzymatic   296
grafting-from technique   136ff., 245
grafting-through-methodology   245
grafting-to methodology   136ff., 245
green catalyst method
– catalytic process   172
green cationic polymerization   325ff.
– lipase   325
– natural rubber   330
green chemistry   3, 57, 106, 313
– enzymatic polymer synthesis   291ff.
green industrial process   342
green media   303
– enzymatic polymerization   303
green monomer   297
green polymer chemistry   4
green polymerization   3, 313ff.
green polymerization method   4
– status and outlook   3
green polymerization solvent   160
green polymerization system   91, 117
green production
– economical   115
– process   104
green solvent   144
group 1 and 2 based catalyst   191
group 4–12 based catalyst   186
group 8 transition metals   132
group 13 based catalyst   182
group 14 congeners   109
group 14 based catalyst   182
group transfer polymerization (GTP)   129
gum turpentine   92

h

3HCA (3-hydroxycinnamic acid)   267
4HCA (4-hydroxycinnamic acid)   266
HDPE (high-density polyethylene)   15
heat-set polymer   212
Henry relationship   123
heterocycle   31
heterofermentative process   202
heterograft polymer   247
heterotactic enchainment   208

Hevea brasiliensis 331ff.
high heat-resistant polymer
– LC copolymer 282
high molecular weight PLA 203
high softening point resin 105
high temperature polymerization 153
high-performance polymer from phenolic biomonomer 265ff.
Hildebrand solubility parameter 123
homofermentative process 202
homogeneous catalyst 167
homopolymer
– cell-adhesion property 273
– LC property 267
– melting behavior 270
– ordered structure 270
homopolymerization 153
horseradish peroxidase (HRP) 298
hot-melt adhesives and coatings 113
Hoveyda–Grubbs catalyst 134
– generation 3 catalyst 135
– second-generation catalyst 16
hybrid construct 37
hydration
– acrolein 76
hydroformylation
– ethylene oxide 76
α/β-hydrolase fold 315
hydrolysis
– lipase-catalyzed 315
hydrolytic behavior
– copolymer 281
hydrolytic engineering polymer 281
hydrophilicity
– enzyme activity 320
– solvent 320
hydrophobicity
– solvent 319
3-hydroxycinnamic acid (3HCA) 267
4-hydroxycinnamic acid (4HCA) 266
N-(2-hydroxyethyl) methacrylamide 236
2-hydroxyethyl methacrylate (HEMA) 240
hydroxymethylfurfural (2-hydroxymethyl-5-formylfuran, HMF) 34f.
– industrial production 35ff.
N-hydroxysuccinimidyl (NHS)
– aldehyde 136
– ester 136
hyperbranched architecture 277ff.
hyperbranched degradable polymer 250
hyperbranched LC polyarylates 279

i

ideal green
– processes 105
– monomer 106
immortal polymerization 189
in situ polymerization 38
inhibitor 33, 146
inimer 250ff.
initiating system 108
initiation 93ff.
– physical method 95
initiator 95, 334
– anionic 176
– difunctional 243
initiator system 105f.
initiator-fragment incorporation radical copolymerization 250
injection moulding 212
interfacial activation 316
interfacial polymerization 43
inverse emulsion 306
iodine value 157
iodine-transfer polymerization 238
ionic liquid (IL) 144, 303f.
isobutylene 96, 116
isomerization 106
isomerization–polymerization 99
isopentenyl pyrophosphate (isopentenyl diphosphate, IPP) 332ff.
isoprene (IP) 332
– copolymerization 341
– green synthetic cationic polymerization 341
N-isopropylacrylamide (NIPAM, NIPAAm) 136, 243
isosorbide 224
isotactic enchainment
– probability 208

j

juniperic acid 79

k

kenaf fibers 286
– hybridization 287
Kevlar 35

l

laccase 298
lactide
– cycle 202
– degradation 204
– *rac*-lactide 208
– ROP 203f.

lactone 293, 295
lanthanide guanidinate complex
– homoleptic 191
lanthanide-based catalyst 190
Lewis acid coinitiator 95ff.
limiting oxygen index (LOI) 21
d,l-limonene 342
d-limonene 91f., 117ff., 342
l-limonene 91f.
linear (bio)degradable polymer 239
lipase 315ff.
– catalyzed hydrolysis 315
– green cationic polymerization 325
– immobilized 179
lipase B from *Candida antarctica* (CALB) 316ff.
liquid crystalline (LC) copolymer
– biomaterials 274
– high heat-resistant polymer 282
– photofunctional polymer 279
liquid crystalline (LC) polyarylates
– hyperbranched 279
liquid crystalline (LC) state 265ff.
– homopolymer 267
liquid phase
– glycerol 73
lithocholic acid
– co-monomer 274
– P(4HCA-co-LCA) 274ff.
living free radical method 136
living free radical polymerization 138
living polymerization 129ff.
– application 129ff.
– in water 129
– method 129ff., 238
– β-pinene 102
living-like radical polymerization 238
luminescence 40
lumped parameter 152
lumped rate parameter 153
lyophilization
– enzyme 320

m

macroinitiation 294
macroinitiator 242
– difunctional 243
macromolecular architecture 50
macromonomer 246
Mark–Houwink–Sakarada (MHS) equation 119
material chemistry 58
Mayo method 149
melt polymerization 174

melting behavior
– homopolymer 270
melting temperature 41
membrane confinement 321
8-p-menthene 104
m-menthene repeat unit 99
metal salen complex 184ff.
metallo-organic catalyst 207
metallocene catalyst 122
metallocene polymerization 122
– monoterpene 121
metallocene–MAO catalyst system 121
metallocyclobutane 131
– transition metal-based 131
metathesis chemistry 130
methacrylate
– sugar-based 243
methacrylated fatty acid (MFA) 24
2-methacryloyloxyethyl phosphorylcholine (MPC) 241
N-methacryloyltyrosinamide 236
1,6-methano[10]annulene moiety 40
3-methoxy-4-hydroxycinnamic acid (MHCA) 267
methyl acrylate 17
5-methyl furfural (MF) 30
– polycondensation 39
methyl methacrylate (MMA) 146, 245
methyl oleate (MOA) 16
methyl 6-O-poly(ε-caprolactone)-β-D-glucopyranoside 326
methyl ricinoleate 18
methyl tert-butyl ether (MTBE) 169, 342 auf s. 342 findet sich eigentlich mbte, aber ich denke das ist identisch und nur ein Tippfehler
methyl undecenoate (MUA) 23
methyl 10-undecenoate 20
5-methyl-5-benzyloxycarbonyl-1,3-dioxan-2-one 179
7-methyl-1,5,7-triazabicyclo-[4.4.0]dec-5-ene (MTBD) 193, 210
methylaluminoxane (MAO) 101, 116
exo-methylene cyclohexane 99
methylene moiety 37
5-methylene-2-phenyl-1,3-dioxolan-4-one 243
methylene-4-vinylcyclohexane 104f.
mevalonate (MVA) pathway 333
MF, *see* 5-methyl furfural
model prediction 154f.
molecular parameter 211
molecular photodimerization 40

molecular weight   119
– control   118
– distribution (MWD)   173, 237
molybdenum-containing complex   131
monomer   146, 333f.
– activated monomer   325
– dip-pen nanolithography (DPN) patterned reactive monomer   302
– phytochemical   266
– solubility   268
monomer stoichiometry   49
monomer synthesis   15, 165ff.
monoterpene (MT)   91ff., 116
– acyclic   92
– aromatic   92
– bicyclic   92
– chain transfer agent   116
– metallocene polymerization   121
– monocyclic   92
– monomer in polymer chemistry   91ff.
– physical properties   116
– polymerization solvents   91ff.
– solvent   116ff.
MPC (2-methacryloyloxyethyl phosphorylcholine)   241
multiple healing   52

n

NAD (nicotinamide adenine dinucleotide)   168
nano-sized capsule   243
nanocages   140
nanodevices   302
nanofibers   287
nanoparticles
– monodispersed   282
– photoreaction   282
– photosensitive degradable   282
natural rubber (NR)   330ff.
– anatomy of the NR latex   331
– C-serum (clear phase) (CS)   340
– fresh bottom fraction (BF) of the latex   340
– structure   332
natural rubber biosynthesis   314, 330ff.
– biochemical pathway   333
– chemical mechanism   337
– *in vitro*   339
nematic phase   269ff.
– poly(*p*-coumaric acid) (P4HCA)   269
NHC (*N*-heterocyclic carbene)   194, 209
NIPAM, *see* *N*-isopropylacrylamide
NIPAAm, *see* *N*-isopropylacrylamide
ω-nitrile fatty acid ester   16

nitroxide-mediated polymerization (NMP)   129, 238
non-anionic mechanism   129
non-cross-linked polymer   15
non-MVA (deoxy-xylulose) pathway   333
norbornene   131ff.
– Zeonor (hydrogenated norbornene-based copolymer)   130
Norsorex (polynorbornene)   130
novel materials   37
Novozym® 435   322

o

P-octa-2,7-dienyl-*P,P,P*-[tri(3-sulfonato-phenyl)-phosphonium hydrogencarbonate] trisodium salt   82
oil additive   342
olefin metathesis   15, 130
– reaction   131
olefin polymerization
– cationic   93
exo-olefinic group   105
oleochemistry   57
oligo(furylene vinylene)   40
oligomer   39
– linear   36
ordered structure   270
organic–inorganic hybrid   37
organocatalytic ring-opening polymerization   193, 208
organogermanium system   109
organosilicon halide   109
organotin halide   107
oxetane
– coupling   188
oxidation
– catalytic   65f.
– glycerol   65ff.
– selective   68
oxidative polymerization   298
oxidoreductase   298
oxyanion hole   316
oxycarbonylation
– glycerol   61
ozonolysis   15

p

P3HCA, *see* poly(*m*-coumaric acid)
P4HCA, *see* poly(*p*-coumaric acid)
palladium   81
– based catalyst   82
– Pd(acac)$_2$ 81
– Pd/TOMPP catalyst   82f.
– Pd/TPPS complex   82

PANI, see polyaniline
Parthenium argentatum 331ff.
partition coefficient 319
PEG, see polyethylene glycol
PEGylated pyridine 135
pellethen® 221
pentadecalactone (PDL) 293
perfluoropolyether (PFPE) ammonium carboxylate 306
Perlon 226
permissible exposure limit (PEL) 145
petroleum solvent
– replacing 116
pH memory 320
phenolic biomonomer
– high-performance polymer 265ff.
N-phenylmaleimide–FA adduct 52
phosphazene base 210
phosphine ligand 82
photochemical behavior
– PFV dimer 40
photodimerization
– molecular 40
photofunctional polymer 279
– LC copolymer 279
photoluminescence
– oligo(furylene vinylene) 40
photoperoxidation 15
photoreaction
– nanoparticles 282
photoreaction conversion 279
photoresist
– negative 42
photosensitive degradable nanoparticles 282
phototunable hydrolyzes 279
phytomonomer
– coumarate 266
pinene 117
α-pinene 91ff., 123, 342
– cationic polymerization 106
– copolymer 112
– polymerization 110f.
– resin 112
β-pinene 91ff., 342
– cationic polymerization 98
– co- and terpolymers 103
– living polymerization 102
– polymerization 101
PLA, see poly(lactic acid)
plant oils 11ff.
– renewable feedstock for polymer science 11ff.
platinum
– Pt-Au/C catalyst 68

– Pt/NaY 74
Pluronics 300
polyA-S_2-polyB 241
polyA-S_2-2-Py 241
poly(acryloyl glucosamine)-b-polyNIPAAm 254
polyamide 42f., 226
– chiral 228
– fully furanic 43
polyamides X,20 19
poly(3-amino-4-methoxybenzoic acid) 302
polyaniline (PANI) 301f.
polyB-S_2-2-Py 241
poly(butylene adipate-co-terephthalate)(PBAT) 214
poly(butylene succinate) (PBS) 214
poly(β-butyrolactone) (PBL) 214
poly(ε-caprolactone) (PCL) 213, 327
– polyCL-b-polyAA 243
– poly(CL-b-styrene) 294, 305
polycarbonate
– biodegradable aliphatic 172
– degradation 196
– recycling system 179
– renewable 179
polycarboxylate 69
polycondensates 292
polycondensation 33ff., 296
– acid-catalyzed 33ff.
– enzymatic 296
– MF 39
poly(m-coumaric acid) (P3HCA) 268ff.
poly(p-coumaric acid) (P4HCA) 267ff.
– nematic phase 269
– P(DHCA-co-4HCA) nanoparticles 282
– P(4HCA-co-CA) 277, 288
– P(4HCA-co-DHCA) 279ff.
– P(4HCA-co-LCA) 274ff.
– synthetic conditions of P(4HCA-co-DHCA) 285
– thermotropic property 269
poly(DCPD) (polydicyclopentadiene) polymer
– physical properties 121
poly(3,4-dihydroxycinnamic acid) (PDHCA) 268ff.
– P(DHCA-co-4HCA) nanoparticles 282
– P(4HCA-co-DHCA) 279ff.
– synthetic conditions of P(4HCA-co-DHCA) 285
polydimethylsiloxane (PDMS) 231
– carbohydrate-modified 232
– carbohydrate-segmented 233
– glycostructure-segmented 231
poly($para$-dioxanone) (PPD) 214

poly(1,5-dioxepan-2-one) (PDXO)   214
polydispersity   173
polydispersity index (PDI)   173
poly(dodecyloate)   18
poly(DTC)   176
polyester   41
– bio-based   265
– biodegradable   213
– condensation polymerization   296
– enzymatic synthesis   292
– ring-opening polymerization   293
polyether   223
polyethylene (PE)   166
poly(ethylene adipate)   50
poly(2,5-ethylene furandicarboxylate) (PEF)   41
polyethylene glycol (PEG)   134
– PEG-200   83
poly(ethylene terephthalate) (PET)   35ff., 168
poly-FA   37f.
polyfuran   39
poly(2,5-furan dicarboxylate)   42
poly(2,5-furylene vinylene) (PFV)   39
– dimer   40
polyglycerol   64
polyglycidols   294
polyHEMA   139, 240, 250
– P(DHCA-co-4HCA) nanoparticles   282
– poly-HEMA [poly(hydroxyethyl methacrylate)]–PMMA [poly(methyl methacrylate)] random copolymer   306
– poly(MMA-co-HEMA)   306
– pullulan–polyHEMA graft copolymer conjugates   250
poly(hydroxyalkanoates)   214
poly{4-hydroxybenzoic acid (HBA)-co-5-hydroxynaphthoic acid (HNA)}   270
poly[(R)-3-hydroxybutyrate] (PHB)   214
poly(hydroxyethyl methacrylate) (polyHEMA)   139, 240, 250
poly N-isopropylacrylamide (polyNIPAAm, PNIPAM)   240f.
– poly(acryloyl glucosamine)-b-polyNIPAAm   254
– polyNIPAM   138f.
– polyNIPAAm-b-polyMPC-S$_2$-poly-MPC-b-polyNIPAAm (MPC=2-methacryloyloxyethyl phosphorylcholine)   241
poly(lactic acid)/poly(lactide) (PLA)   201ff., 265
– blending   213
– degradation   204

– isotactic   208
– lifecycle   202
– molecular parameter   211
– PLA-b-PM-b-PLA thermoplastic elastomers   215
– processing effect   211
– property   210
– robust renewable material   201ff.
– ROP   203f.
– thermal performance   211
poly(D-lactic acid) (PDLA)   215
– PDLA-b-PM-b-PDLA block copolymer   215
poly(D,L-lactic acid) (PDLLA)   211
poly(L-lactide) (PLLA)   204
– PLLA-b-poly(1,5-dioxepan-2-one)-b-PLLA triblock copolymer   214
– PLLA-b-PM-b-PLLA   215
poly(lactide)/poly(ε-caprolactone) blend   213
poly(menthide) (PM)   215
polyMePEGMA   252
– polyMePEGMA-b-poly(5′-O-methacryloyl uridine)   252
polymer   292
– average molecular weight   157
– biodegradable   235ff.
– blends   213
– branched   244
– characterization   148
– composition   158
– conductive   301
– conjugated   39
– crystalline   294
– degradable   235ff.
– degradable functional group   239
– degradable polymeric segment   242
– environmentally friendly   265
– functionalization of preformed polymer   328
– furan-based   35
– furfuryl alcohol   35
– graft polymer (polymer brush)   245ff.
– high-performance polymer from phenolic biomonomer   265ff.
– hydrolytic engineering   281
– hyperbranched   119, 250
– microstructure   158
– miktoarm star copolymer   244
– multiple cleavable group   243
– photofunctional   279
– polymeric segment   243
– profluorescent 3-azidocoumarin-terminated   140
– saccharide-derived functional   221ff.
– star-shaped polymeric molecule   244

polymer (contd.)
- reversible cross-linking 52
- tacticity 204
polymer bioconjugate
- synthetic–natural 247
polymer blend 213
polymer brush 245ff.
polymer chemistry
- monomer 91ff.
polymer functionalization
- biotechnology 313ff.
- enzyme-catalyzed 328
- lipase 325
polymer materials
- aqueous media 140
polymer molecular weight 97
polymer precursor 166
polymer process 144
polymer production
- biodiesel 145
- technology 144
polymer science 3
- plant oils as renewable feedstock 11ff.
polymer synthesis 18
- enzymatic 291ff.
polymer–protein biohybrid 140
polymer–protein conjugate 137ff.
polymerization 31, 46f., 101ff., 147
- acid-catalyzed 36
- anionic 69
- aqueous 103
- cationic 33, 68, 96ff., 313ff.
- chemical catalyst 326
- controlled 235ff.
- degree of polymerization 41, 149
- elevated temperature 145
- enzymatic 294, 302f.
- FA 37
- green 313ff.
- green cationic 325
- green synthetic cationic 341
- immortal 189
- in situ 38
- interfacial 43
- isobutylene 116
- kinetics 157
- linear 46
- living 102, 129ff., 235ff.
- non-linear 49
- oxidative 298
- oxidative-assisted 68
- qualitative aspect 46
- radical 235ff.

- rate of polymerization 151ff.
- solution 113, 143ff.
- solvent-free 189
- template-assisted 302
polymerization chemistry
- metal complex 205
polymerization solvent 143ff.
- monoterpene 91ff.
polymerization system
- dipentene 105
poly(MMA-co-HEMA) 306
poly(methacryloyloxyethyl phosphorylcholine) (polyMPC) 139
poly(3-methoxy-4-hydroxycinnamic acid) (PMHCA) 268ff.
polyMMA-b-polySty 252
polyNIPAM, see poly N-isopropylacrylamide
polyNIPAAm-b-polyMPC-S$_2$-polyMPC-b-poly-NIPAAm (MPC=2-methacryloyloxyethyl phosphorylcholine) 241
polynorbornene (Norsorex) 15, 130
polyolefin
- unsaturated 130
polyoligo(ethylene glycol) acrylate (OEGA) 241
polyols 14
polyoxybutylene 223
- functionalized 225
- saccharide-doted 226
poly(para-dioxanone) (PPD) 214
poly(pentadecalactone) (PPDL) 293
poly(1H,1H,2H,2H-perfluorooctyl methacrylate) (FOMA) 305
polyphenol 298
poly(β-pinene) 99ff.
poly[poly(ethylene glycol) methacrylate] [poly(PEGMA)] 138
poly(propylene carbonate) (PPC) 214
poly(propylene glycol)-poly(ethylene glycol)-poly(propylene glycol) (PPG-PEG-PPG; Pluronics) 300
polysaccharide 29
polysaccharide (pullulan)–synthetic polymer bioconjugate 249
polystyrene
- poly(CL-b-styrene) 294, 305
- polySty-b-polyOEGA 241
polytransesterification 41
poly(tetramethylene adipate-co-terephthalate) (PTAT) 214
poly(trimethylene carbonate) (poly(TMC), PTMC) 171ff., 214
poly(trimethylene terephthalate) (PTT) 76, 168

polyurea 229
– chiral 229
polyurethane 43, 229
– chiral 229
– furan–aliphatic 44
– furan–aromatic 44
– fully furanic 44
– polyoxybutylene-based 222
– sugar-derived AB-type 229
– synthesis 14
poly(vinyl alcohol) (PVA) 40
poly(vinyl chloride) (PVC) 166
prenyltransferase 335
– *cis*-prenyltransferase 335
– *trans*-prenyltransferase 335ff.
pressure-sensitive tape 113
processing condition 212
processing effect 211
profluorescent 3-azidocoumarin-terminated polymer 140
propagation 95
1,3-propanediol 76f., 166
– biological pathway 169
– preparation from renewable resources 166
propylene glycol 72f.
protein functionalization 136
protein–polymer biohybrid 140
protein–polymer conjugate 137ff.
β-proton elimination 96f.
Pseudomonas aeruginosa 315
Pseudomonas fluorescens lipase 325f.
pullulan 249
pullulan–polyHEMA graft copolymer conjugates 250
pyridine 177
– PEGylated 135
– ruthenium-coordinating 135
pyridine ligand
– clicked 135
– labile 135
2-pyridyl 2′-methacryloyloxyethyl disulfide 241
4-pyrolidinopyridine (PPY) 208
pyrophosphoric acid (HPP) 338

q
quinuclidine 177

r
γ-radiation 103
radical copolymerization 14
radical method
– living free 136

radical polymerization 237f.
– controlled/living (CRP) 235ff.
radical ring-opening polymerization (RROP) 237ff.
radical–solvent complex formation 153
RAFT, *see* reversible addition–fragmentation chain transfer
rate constant 151
rate of polymerization 151ff.
recombinant strain 168
renewable feedstock
– plant oils for polymer science 11ff.
renewable polycarbonate production and recycling system 179
renewable resources 5, 119, 166ff.
– dimethylcarbonate 169
– 1,3-propanediol 166
retardation 153
reversible addition–fragmentation chain transfer (RAFT) polymerization 130ff., 238ff.
reversible polymer cross-linking 52
ring-closing metathesis (RCM) 130f.
ring-opening metathesis polymerization (ROMP) 15, 117f., 130ff.
– alkene monomer 118
– living 130
ring-opening polymerization (ROP) 171, 325
– anionic 176
– catalyst 182
– cationic 173
– coordination–insertion pathway 181
– enzymatic 178, 294
– lactide 203f.
– monoterpene 117
– organocatalytic 193, 208
– reaction 165
– renewable six-membered cyclic carbonate 165ff.
– six-membered cyclic carbonate 172ff.
– thermodynamic properties 171
room-temperature ionic liquid (RTIL), *see also* ionic liquid 144
RROP, *see* radical ring-opening polymerization
rubber transferase 333ff.
ruthenium
– metathesis catalyst 120
– Rh(CO)$_2$(acac) 167
ruthenium benzylidene catalyst 132ff.
ruthenium complex
– cationic 167
ruthenium-based structure 132

s

saccharide-derived functional polymer 221ff.
salen
– bimetallic aluminum salen derivative 170
– dimeric aluminum salen based complex 183
– calcium(II) salen complex 192
– monomeric aluminum salen based complex 183
– (salen)Al(III)Cl system 184
– (salen)Cr(III)Cl complex 186
– (salen/salan)aluminum 207
– (salen)Sn(IV)(X)(Y) complex 185
scandium
– bis(phenolato)scandium complex 207
Schiff-base nickel complex 101f.
self-condensing vinyl polymerization (SCVP) 250
self-metathesis 16
Shell process 76, 166
shell–cross-linked micelle 243
β-silicon effect 110
silicon halide 111
silicone
– hydrophilically modified 230
silicone surfactant 230
single-component catalyst 190
small interfering ribonucleic acid (siRNA) 241
smart particle 138
solid catalyst
– acid 62
– activity 60
solubility
– monomer 268
solution polymerization 113, 143ff.
– biodiesel as a polymerization solvent 143ff.
solvent 116
– hydrophilicity 320
– hydrophobicity 319
– monoterpene 116
solvent effect 149
solvent recovery technique 160
solvent-free polymerization
– TMC 189
Sorona® [poly(trimethyleneterephthalate)] 76, 168
(−)-sparteine 209
spiro-bis-dimethylencarbonate 213
stable free radical polymerization (SFRP) 238
star polymer
– degradable 244
stereocomplex formation 212
sterically hindered pyridine (SHP) 97
strippable imaging material 52
styrene (Sty) 146
– emulsion polymerization 301
styrene–butadiene rubber (SBR) 103
substitution
– electrophilic 33
substrate 334
sugar 29
sugar-based methacrylates 243
sulfate turpentine 92
supercritical carbon dioxide (scCO$_2$) 304
supercritical water 144
supermacroinitiator 252ff.
surface erosion 196
surgical materials
– hydrophilic 221
sustainable development 3
sustainable reaction conditions 5
sustainable solution polymerization 143ff.
synthetic–natural polymer bioconjugate 247

t

tackifiers 103
tackifying agents 113
tape
– pressure-sensitive 113
Taraxacum kok-saghyz 331
tecoflex® 221
telomerization
– glycerol 81
template-assisted polymerization 302
termination
– cationic polymerization 98
terpene 91ff.
– terpene based adhesives 113
terpenic resins 112
– application 113
– characteristics 112
– commercial production and markets 113ff.
– demand 115
– environmental aspects of production 115
– overview 92
– production data 114
– softening point 112
terpolymers 297
α-(*tert*-butyl)-ω-(2-chloro-2-methylpropyl)poly-isobutylene 329
N'-*tert*-butyl-N,N,N',N',N'',N''-hexamethyl-phosphorimidic triamide (P$_1$-*t*-Bu) 210
2-*tert*-butylimino-2-diethylamino-1,3-dimethyl-perhydro-1,3,2-diazaphosphorine (BEMP) 210

tetrafuran–bismaleimide system 52
tetraphenylporphyrin–aluminum compound 183
thermal degradation 46
thermal performance
– PLA 211
thermal reversibility 45
thermoplastic elastomer (TPE) 214
– biodegradation process 194
– non-toxic 172
– PLA-based 214
thermoplastic polymer 265
thermoreversible network 52
thermosetting resin 269
thermotropic property
– poly(p-coumaric acid) (P4HCA) 269
thiourea 209
tin
– (salen)Sn(IV)(X)(Y) complex 185
– tin(II) bis(2-ethyl-hexanoate) 182, 205
– tin(II) octanoate 205
– tin(II) octoate 182
titanium 122
– Cp*TiMe$_3$–B(C$_6$F$_5$)$_3$ 101
TMC, see trimethylene carbonate 180
p-toluenesulfonic acid (PTSA) 79
TOMPP (tris-(o-methoxyphenyl)phosphine) 82
TPPTS [3,3′,3″-phosphinidynetris(benzenesulfonic acid)trisodium salt] ligand 81
transacylation 327
transcarbonatation
– cyclic organic carbonate 59
transesterification 165
– CALB-catalyzed 323f.
– enantioselective 318
transfer to solvent effectiveness 151
transition metal
– group 8 132
transition metal-based metallocyclobutanes 131
1,5,7-triazabicyclo[4.4.0]dec-1-ene (TBD) 19, 193, 210
triethylamine 177
trimethylene carbonate (TMC) 165, 180, 195
– α,ω-dihydroxy dimer (DTMC) 179
– α,ω-dihydroxy trimer (TTMC) 179
– solvent-free polymerization 189
– synthesis 171

tris(pyrazolyl)calcium 207
tungsten-containing complex 131
turnover frequency (TOF) 173
turnover number (TON) 5
turpentine oil 342

u
undec-10-enol 18
undecylenyl undecenoate 18
urea
– glycerolysis 60

v
vanadium
– VSb-based catalyst 71f.
Vestenamer 130
vinyl acetate (VAc) 146
vinyl polymer 300
viscosity
– intrinsic 119
volatile organic solvents 6

w
washed bottom fraction particle (WBP) 340
washed rubber particle (WRP) 339
water soluble ROMP catalyst 133
water trapping reagent 170
WATPOLY polymerization simulator 152
– database 152
whole cell biocatalysis 314
wide angle X-ray diffraction (WAXD) 270f.
wood turpentine 92

y
Young's modulus 212
yttrium
– amino-alkoxy(bisphenolate)yttrium 207

z
zeolite 74
Zeonor (hydrogenated norbornene-based copolymer) 130
Ziegler–Natta catalyst 101ff.
zinc
– Cu/ZnO/Al$_2$O$_3$ catalyst 75
– (β-diketiminatato)zinc 207
– Zn–Al mixed oxide 60
– zinc alkoxide 190
zirconium 122, 207
– zirconium phosphate (γ-ZrP) catalyst 60